Lecture Notes in Electrical Engineering

Volume 1150

The book series *Lecture Notes in Electrical Engineering* (LNEE) publishes the latest developments in Electrical Engineering—quickly, informally and in high quality. While original research reported in proceedings and monographs has traditionally formed the core of LNEE, we also encourage authors to submit books devoted to supporting student education and professional training in the various fields and applications areas of electrical engineering. The series cover classical and emerging topics concerning:

- Communication Engineering, Information Theory and Networks
- Electronics Engineering and Microelectronics
- Signal, Image and Speech Processing
- Wireless and Mobile Communication
- Circuits and Systems
- Energy Systems, Power Electronics and Electrical Machines
- Electro-optical Engineering
- Instrumentation Engineering
- Avionics Engineering
- Control Systems
- Internet-of-Things and Cybersecurity
- Biomedical Devices, MEMS and NEMS

For general information about this book series, comments or suggestions, please contact leontina.dicecco@springer.com.

To submit a proposal or request further information, please contact the Publishing Editor in your country:

China

Jasmine Dou, Editor (jasmine.dou@springer.com)

India, Japan, Rest of Asia

Swati Meherishi, Editorial Director (Swati.Meherishi@springer.com)

Southeast Asia, Australia, New Zealand

Ramesh Nath Premnath, Editor (ramesh.premnath@springernature.com)

USA, Canada

Michael Luby, Senior Editor (michael.luby@springer.com)

All other Countries

Leontina Di Cecco, Senior Editor (leontina.dicecco@springer.com)

**** This series is indexed by EI Compendex and Scopus databases. ****

Ferran Martín · Enrique Bronchalo
Editors

Coupled Structures for Microwave Sensing

Springer

Editors
Ferran Martín
Departament d'Enginyeria Electrònica
Universitat Autònoma de Barcelona
Bellaterra, Spain

Enrique Bronchalo
Universidad Miguel Hernandez de Elche
Elche, Alicante, Spain

ISSN 1876-1100 ISSN 1876-1119 (electronic)
Lecture Notes in Electrical Engineering
ISBN 978-3-031-53860-5 ISBN 978-3-031-53861-2 (eBook)
https://doi.org/10.1007/978-3-031-53861-2

This Springer imprint is published by the registered company Springer Nature Switzerland AG
The registered company address is: Gewerbestrasse 11, 6330 Cham, Switzerland

Preface

Within the paradigms of the Internet of Things (IoT) and Industry 4.0, there has been an increasing research activity for the development of high-performance low-cost sensors and sensor networks in recent years. Among the different technologies useful for sensing (e.g., electrochemical, optics, acoustics, or microwaves), microwave technology has attracted a significant interest within the Sensors Community. Thus, microwave sensors (especially planar sensors) have been the subject of an intensive and progressively growing research in the last decade, with an exponentially increasing number of journal and conference papers devoted to this topic, and a proliferation of projects and contracts with companies. Various reasons explain the growing interest for planar microwave sensors, including their low cost and profile, their small size, or the possibility of sensor implementation on flexible substrates, including polymeric and organic substrates, as well as fabric. Their compatibility with other technologies (e.g., microfluidics, textiles, micromachining, 3D printing, etc.), and with both additive (e.g., inkjet- or screen-printing) and subtractive (e.g., photoetching) fabrication processes are also key factors that justify such growing interest. Additionally, microwaves are very sensitive to the electromagnetic properties of the materials to which they interact. Thus, microwave sensors are very useful for the dielectric characterization of materials (solids or liquids), and for the measurement of many variables related to material permittivity. Planar microwave sensors can operate by contact or contactless with the material under test (MUT) or analyte, and can be wirelessly connected to the reader (of special interest in many IoT applications), in schemes based on the so-called sensing tags (which act as a "smart skin", able to provide information of the material or sample under study). Let us also mention that the necessary associated electronics for signal generation, post-processing, and, eventually, communication purposes of planar sensors can be integrated within the sensor's substrate. This represents a simplicity in system design and reduces the overall sensor costs. Finally, microwaves, adopting the broadest sense of the term (i.e., including ultra-high frequency—UHF—waves, millimeter waves, and THz waves), are very versatile, as far as many different types of materials and analytes, as well as a wide variety of physical, chemical, and biological variables, can be measured, monitored, or characterized by means of microwave sensors (e.g., material properties and composition, vital constants, bacterial growth, medical variables, physical variables, such as temperature and humidity, displacements and velocities, etc.).

Several indicators determine the performance of planar microwave sensors (linearity, dynamic range, sensitivity, resolution, accuracy, repeatability, selectivity, etc.), but probably the most representative performance indicator is the sensitivity, intimately related to resolution, and critical in applications where tiny variations of the input variable (measurand) should be detected. Thus, sensitivity optimization has been the "battle horse" for researchers involved in microwave sensing for many years, and it is still a hot topic of research. Many strategies for sensitivity enhancement in planar microwave sensors have been reported in the recent literature, including active feedback in frequency-variation sensors, step-impedance configurations terminated with resonant elements in phase-variation sensors, or schemes based on resonance–anti-resonance in coupling-modulation sensors, to cite some of them. It has been recently demonstrated that coupled structures, including coupled lines and coupled resonators, are very interesting in order to implement highly sensitive microwave sensors. Thus, various international groups have dedicated a significant effort to optimize the sensitivity (as well as other performance indicators) in planar microwave sensors using schemes based on coupled lines and coupled resonators. This book is intended to provide an overview of such sensing schemes and strategies through the contribution of the most relevant and reputed international Groups in these topics.

The book contains two introductory chapters, one devoted to the topic of planar microwave sensors (Chapter "Planar Microwave Sensors"), and the other one (Chapter "Coupled Planar Microwave Resonators and Transmission-Line Structures") focused on coupled microwave structures. The main aim of these chapters is to familiarize the reader with the main concepts used in the subsequent chapters. Moreover, Chapter "Planar Microwave Sensors" provides an up-to-date state of the art in planar microwave sensors useful to appreciate, and understand, the potential of coupled structures for sensing. The book then covers in the following chapters the main proposed strategies for sensitivity optimization by using coupled structures, including coupled lines and directional couplers, as well as coupled distributed and semi-lumped resonators. Such strategies encompass coupling schemes used to enhance the variation of the frequency, the phase, the magnitude, or the quality factor (the typical output variables of the considered sensors) with the dielectric properties of the materials under study. A diversity of schemes, such as weakly coupled structures, electromagnetic induced transparency (EIT)-like structures, coupled line directional couplers, electrically/magnetically coupled resonators, etc., will be reviewed throughout the different book chapters. Essentially, the considered sensors in the different chapters are permittivity sensors. Nevertheless, a final book chapter will also be dedicated to sensors based on magneto-inductive (MI) wave transmission lines for contactless resonant imaging of conductive environments.

Chapter "Permittivity Sensors Based on Coupled Line Sections" focuses on coupled-line section sensors excited by a differential signal and is devoted to the dielectric characterization of solid and liquid samples. It is shown in the chapter that by locating the samples under study in the region between the coupled lines, where there is a high density of electric field lines, the sensitivity can be enhanced. Various types of configurations are considered, including edge-coupled and broadside-coupled structures, operating in reflection as well as in transmission, and an exhaustive analysis is carried out for all of them. It is shown that, from such analysis, using a systematic procedure, the complex permittivity of the materials under study can be obtained.

In Chapter "Coupled-Line Directional Coupler Permittivity Sensors", the authors utilize a microstrip coupled-line directional coupler for dielectric constant sensing. The material under test is placed on top of the coupled lines, and the dielectric properties are retrieved from the coupler's response, particularly from the coupling and isolation levels. It is shown in the chapter that high sensitivity can be achieved, especially when the isolation level is considered.

Reflective-mode phase-variation permittivity sensors based on coupled resonators (either distributed or semi-lumped), and operating at a single frequency, is the topic of Chapter "Reflective-Mode Permittivity Sensors Based on Distributed and Semi-Lumped Coupled Resonators". Unprecedented sensitivities can be achieved by weakly coupling the resonant elements, a consequence of the proximity between the split resonances when the coupling is weak. It is shown in the chapter that by tuning the operating frequency in the region between such resonances, the sensitivity increases as the inverse square of the coupling coefficient. The effects of losses are also analyzed, and it is concluded that losses might benefit sensitivity, provided the sensor is adequately designed.

In Chapter "Transmission-Mode Permittivity Sensors Based on Coupled Resonators", transmission-mode sensors consisting of a transmission line loaded with a pair of coupled resonators are presented. The frequency response of these sensors exhibits two zeros and a transmission peak in between. Such sensors can be considered to be based on the microwave analog of electromagnetic induced transparency (EIT), a well-known phenomenon in the field of optics, where a transmission zero in the response of a single resonator reverses to a peak (transparency) when an additional resonator is coupled to the existing one. It is demonstrated in the chapter that such transmission peak is very sensitive to variations in the dielectric properties of the samples under study, placed on top of the coupled resonators. The chapter also includes sensing structures with multiple transparent peaks based on spoof localized surface plasmon (spoof-LSP) resonators.

Analytical methods to retrieve the permittivity of dielectric samples by means of symmetrically coupled resonators are covered in Chapter "Analytical Methods to Retrieve the Permittivity by Means of Coupled Resonators". By using a circuit model of the coupled resonators and the responses of reference materials with known dielectric properties, the permittivity and the loss tangent of the sample under test can be expressed in terms of the coupling coefficient between the resonators. Moreover, the different geometries of the electric field in the even and odd resonances are used to measure the dielectric properties of samples with uniaxial anisotropy.

Chapter "Planar Microwave Sensors Based on Coupled Ring Resonators and Applications", focused on sensors based on coupled ring resonators, shows how the coupling between the ring resonators can enhance the sensitivity, dynamic range, selectivity, and detection range in microwave sensors. The chapter illustrates how utilizing coupled ring configurations can optimize the detection of physical quantities, such as UV radiation, mass of material samples, etc. The chapter also includes wireless sensors, based on antenna configurations used to excite the resonant elements. Practical applications are also reported in the chapter.

Chapter "Solute Concentration Sensing in Aqueous Solutions with Coupled Microstrip Resonators" focuses on the application of sensors based on coupled microstrip resonators to the determination of the solute concentration in aqueous solutions. After presenting the dielectric and conductive properties of these solutions, based on broadband experimental measurements, the chapter presents an exhaustive analysis of the proposed sensors, where the influence of the mutual and self-capacitances of the resonators on the sensitivity is discussed. The chapter also reports two practical examples of sensors applied, respectively, to binary (water–glucose) and ternary (water–sucrose–sodium chloride) solutions.

Finally, Chapter "Sensing Using Magnetoinductive Waves" is devoted to microwave sensing based on magneto-inductive (MI) waves, a type of wave supported by an array of magnetically coupled resonators. Recently, MI waves have been used in contactless resonant imaging of conductive environments, both in the frequency and in the time domain. The underlying physics is that a perturbation due to the presence of a conductive object in the vicinity of a meta-atom (a resonant element of the array) causes reflections of MI waves within the array. The focus in the chapter is time-domain reflectometry, where accurate single and multiple defect localization and characterization is demonstrated. Potential applications for real-time contactless monitoring of inhomogeneous conductive environments ranging from in situ 3D printing quality monitoring to medical imaging are discussed.

To end this preface, let us mention that the contributors of the different chapters included in the book have been given full freedom to conceive and structure their respective assigned chapters at their convenience. For this main reason, certain overlapping between some chapters, as well as some differences in the considered terminology and nomenclature from chapter to chapter, have been accepted by the book Editors. It is the Editors' hope that the present book results of interest to Academia and Industry, especially to researchers, postgraduate students, and professionals involved in microwave and sensor technologies.

Barcelona, Spain — Ferran Martín
Elche, Alicante, Spain — Enrique Bronchalo
December 2023

Acknowledgments

The Editors would like to express their most sincere gratitude to the contributors of the present book. It has been really a pleasure to work with them on this project, with the result of a complete and up-to-date manuscript on the topic of microwave sensors based on coupled structures. We would like also to thank Anna Cedenilla for her effort in handling the permissions of reproduced figures, as well as for her assistance to the book contributors on this subject. The contents presented in this book have been the result of a research activity carried out by authors of the different chapters in the last years. The Editors would like to thank, on behalf of all the book contributors, the funding Agencies and Institutions that have supported our activity through many research projects and contracts. A list of the most representative is given below:

- Grants PID2019-103904RB-I00, PID2022-139181OB-I00, and PID2022-139783OB-I00, Agencia Estatal de Investigación, Ministerio de Ciencia e Innovación, Spain (MCIN/AEI 10.13039/501100011033).
- Grant 2019/34/E/ST7/00342, National Science Centre, and Subvention of the Institute of Electronics, AGH University of Science and Technology, Krakow, Poland.
- Grant 62301104, National Natural Science Foundation of China, Grant 2022RQ034, Dalian Youth Science and Technology Star Project, Grant K202202, State Key Laboratory of Millimeter Waves, and Grant JYTMS20230170, Basic Scientific Research Project of Education Department of Liaoning Province.
- SYMETA grant EP/N010493/1, Engineering and Physical Sciences Research Council (EPSRC), United Kingdom.
- Grant CIAPOS/2021/267, APOSTD Program, Conselleria d'Innovació, Universitats, Ciència i Societat Digital, Generalitat Valenciana, and European Social Fund.
- Grants RYC2022-036257-I and RYC2022-035819-I (Ramón y Cajal Program), grant IJC2019-040786-I (Juan de la Cierva Program), and FPU Grant FPU20/05700 (Ayudas para la formación de profesorado universitario), Ministerio de Ciencia e Innovación, and Ministerio de Universidades, Spain, and European Union.

The authors of some chapters would like also to express their gratitude to several persons for their support to the research activities reported in this book:

- The authors of Chapters 1 and 5 are grateful to Gerard Sisó and Javier Hellin for fabricating the prototypes, and to the Administrative Staff of the Departament d'Enginyeria Electrònica (Marta Mora, Mari Carmen Cosano, Mònica Mateu, Laura Romera, and Ton Vendrell), Universitat Autònoma de Barcelona, for their support concerning management and administrative issues.
- The authors of Chapter 6 are grateful to Mengqi Li and Suiqing Xu for their support.
- The authors of Chapter 9 would like to thank Anyela Aquino Velasquez, Thierry Aubry, Ernesto Ávila-Navarro, Odette Sandrine Bakam Nguenouho, Jessica Benedicto, Jean Luc Carré, Rubén Martínez, Juan M. Morcillo de Miguel, Christian Person, Patrick Queffelec, Philippe Roquefort, José M. Sabater-Navarro, Bernard Simon, Mahdi Srour, and UBO Open Factory team.
- The authors of Chapter 10 are grateful to Grahame Faulkner for assistance with the setup and to the colleagues from the OxiMeta network for fruitful discussions.

The Editors are also indebted to their respective families for their patience and understanding during the preparation of this manuscript and for their continuous and unconditional support. Thank you very much!

Ferran Martín
Enrique Bronchalo

Contents

Planar Microwave Sensors

Ferran Martín

Abstract In this introductory chapter to planar microwave sensors, the main sensing approaches and working principles are presented, and some representative illustrative prototype examples are reported. Some strategies devoted to sensor performance optimization (manly the sensitivity) are discussed, excluding those strategies based on coupled structures, which are the subject of the subsequent chapters in the book. The chapter also highlights how machine learning can be applied to improve sensor robustness and selectivity. Finally, a section is devoted to outline those aspects of coupled structures that make them useful for sensing.

Microwave sensors have been the subject of an increasing research activity in recent years. Such growing interest for the development of novel sensing elements based on microwave technology has been motivated by a series of causes, including the advent of the fourth industrial revolution [1–5], also known as Industry 4.0, the progressive deployment of the so-called Internet of Things (IoT) [6–10], and the proliferation of Smart Systems [11] in a variety of scenarios (e.g., smart cities, smart health, smart agriculture, or structural health monitoring—SHM, to cite some of them). It is obvious that in an intelligent and interconnected world, aligned with the digital transformation associated with the previous paradigmatic (and revolutionary) concepts (Industry 4.0, IoT, and Smart Systems), sensors and sensor networks are key components. Thus, data collection from the environment or from a certain system is fundamental in order to retrieve information from that system and take actions (typically autonomously) when necessary. For example, structural damage in civil and urban infrastructures caused, e.g., by corrosion, or by extreme ambient conditions, might represent a critical risk for citizens (unfortunately, there are various

F. Martín (✉)
Departament d'Enginyeria Electrònica, CIMITEC, Universitat Autònoma de Barcelona, 08193 Bellaterra, Spain
e-mail: Ferran.Martin@uab.cat

F. Martín and E. Bronchalo (eds.), *Coupled Structures for Microwave Sensing*,
Lecture Notes in Electrical Engineering 1150,
https://doi.org/10.1007/978-3-031-53861-2_1

well-known examples of bridges or buildings that have collapsed). Thus, the continuous monitoring of structural health in civil and urban infrastructures (for automatically detecting any potential damage without human intervention) using sensors and sensor networks is of the highest interest, and the research activity in this field has experienced an exponential growth in recent years [12–30].

Even though, besides microwaves, there are many other technologies useful for sensing, including optics, ultrasounds, mechanical, magnetic, electrical, electrochemical, etc., microwave sensors (adopting the broadest sense of the term, from ultra-high frequency—UHF—sensors up to THz sensors) exhibit a relevant advantage, i.e., their inherent wireless connectivity. This is an important aspect within the framework of the IoT and digital world, where sensor networks wirelessly connected to a central unit are necessary. The so-called radiofrequency identification (RFID) sensors, also designated as sensing tags, are considered an enabling technology for IoT, and an increasing research activity, especially for the implementation of chipless and battery-free RFID-sensors, is going on (see [31–43] and [44], Chap. 7). Thus, although optical sensors probably dominate the sensors' market, and there are many applications where such sensors exhibit superior performance (as compared with other sensing technologies),[1] the (unique) wireless connectivity of microwave sensors has been a key aspect towards the expansion of such sensing technology (both at research level and commercially) and penetration in the market. Nevertheless, there are many other advantages of microwave sensors over their optical counterparts. For example, microwaves are very sensitive to the dielectric properties of the materials to which they interact. Thus, microwave sensors are canonical permittivity sensors, able to provide the complex dielectric constant of the so-called material under test (MUT) [45–49], as well as many other variables related to it, such as material composition, or certain physical variables related to material's permittivity (e.g., temperature and humidity)[2] [50–52]. Moreover, the presence of defects in samples correlates with the average, or effective, dielectric constant "seen" by the sensing element (typically a transmission line, a waveguide, or a resonator). Thus, microwave sensors are very useful for defect detection (a canonical application might be structural health monitoring). Another advantage of microwaves over optical radiation concerns the fact that microwaves penetrate certain substances (dielectrics), and therefore can be useful to retrieve information non-intrusively and non-invasively. Thus, microwave sensors can potentially be useful in medical applications, for example for monitoring vital signals [53, 54], or for the detection of malignant tissues or tumors, among many others[3] [55, 56]. Another benefit that derives from the capability of microwaves to

---

[1] For example, optical sensors exhibit very high sensitivity and selectivity for the detection of many types of biological analytes, such as toxins, drugs, antibodies, proteins, viruses, etc.

[2] In order to enhance the sensitivity of the sensor to temperature or humidity, functional materials exhibiting a strong dependence of the permittivity with those variables can be used. Examples of such materials are polyvinyl alcohol (PVA), with a dielectric constant very sensitive to humidity, or polyamide, a material that exhibits a temperature dependent dielectric constant. This aspect will be further considered in Sect. 2.1.4.

[3] Obviously, there are several commercial techniques for the diagnosis of cancer, e.g., magnetic resonance imaging (MRI), computerized tomography (CT), positron emission tomography (PET),

penetrate dielectric media (not transparent in general for light) is the possibility to detect buried objects. This is the case of the so-called ground penetrating RADAR (GPR), a type of remote microwave sensor commercially available and useful in dozens of applications (e.g., in archaeology, civil engineering, etc.) [57–59].

Further advantages of microwaves for sensing result by implementing the sensors in planar technology [44]. Thus, planar microwave sensors exhibit low cost and profile, as compared to non-planar sensors (such as cavity sensors [45–49], waveguide sensors [60, 61], or coaxial probes [62–66]). Moreover, planar sensors can be implemented in a variety of substrates, including rigid (e.g., low-loss commercial microwave substrates or general-purpose substrates, such as FR4) and flexible substrates (e.g., polymeric substrates, organic substrates, such as paper, and textile substrates, such as fabric), and by means of both subtractive (e.g., photoetching) and additive (e.g., screen- or inkjet-printing) fabrication processes. Therefore, planar microwave sensors can be useful for the implementation of conformal elements that adapt to the surface of the sample under study (e.g., bracelet sensors [67]), or can be of interest for the implementation of wearables (integrated in garments and clothes) [68–71], or green sensors (e.g., on paper substrates combined with organic conductive inks [72]). Additional advantages of planar technology for sensing include the possibility of integration of the associated electronics (needed for signal generation, post processing, and, eventually, for communication purposes) in the same substrate (with the consequent reduction in cost and overall dimensions), as well as the compatibility of planar sensing structures with microfluidics, micromachining, and 3D-printed elements (e.g., mechanic holders). Thus, liquid monitoring and characterization with planar sensors is possible by combining the sensitive elements (e.g., semi-lumped resonators) with micro-channels [73–75], micro-pipes [76], or 3D-printed holders (pools) [77, 78] containing the liquid under test. Applications of such sensors include the identification of impurities in liquid samples, the determination of volume fraction of components in liquid mixtures, or the detection of deterioration in certain agrifood products by usage (e.g., oils subjected to frying processes), among others.

The present book focuses on the design and implementation of planar microwave sensors based on coupled structures (coupled lines, directional couplers, and coupled resonators) as sensing elements, as well as on the different reported strategies for performance optimization (mainly to boost up the sensitivity, or derivative of the output variable with the input variable). Unprecedented sensitivities with such sensors, proposed in recent years, have been demonstrated (to be reported later in this book). Nevertheless, there are many other approaches for the implementation of planar microwave sensors. Such approaches, obeying to different working principles, have their own advantages and drawbacks, which should be discussed in order to better understanding the high potential of planar microwave sensors based on coupled structures, the subject of this monograph. Thus, the present introductory chapter briefly analyzes (in Sect. 1) the different types of planar microwave sensors

---

or ultrasounds, among others, but these techniques are, in general, very expensive, and, in some cases, annoying for the patients.

grouped by working principle and provides some illustrative examples (excluding those sensors based on coupled structures, exhaustively studied in the subsequent chapters of this book).[4] The chapter also dedicates a unit (Sect. 2) to present some strategies for performance optimization (with especial emphasis on sensitivity enhancement). Section 3 concisely outlines the potential of machine learning (ML) for selectivity and robustness improvement in microwave sensors (a topic that has attracted the interest of the Microwave Sensors Community). Finally, Sect. 4 briefly highlights the potential of coupled structures for microwave sensing, as a link between the planar microwave sensors overviewed in this chapter and those to be studied in detail in the subsequent chapters. A general overview of other sensor technologies, including optical, acoustic, mechanical, magnetic, electrical, electrochemical, as well as microwave (non-planar and remote) sensors, is out of the scope of this book, but the interested reader can find detailed information in other sources, e.g., in [44]. A detailed list (and description) of the main sensor performance indicators (sensitivity, resolution, dynamic range, linearity, error, etc.) is also given in [44].

1 Classification of Planar Microwave Sensors

Microwave sensors can be divided in two main categories: remote and non-remote. Remote sensors, mainly RADARs (radio detection and ranging) and radiometers, gather information of an object (e.g., the velocity of a moving target) or physical phenomenon/system (e.g., rainfall detection in a certain region) without physical contact with it [57–59, 79–82]. Moreover, the target is typically located at significant distances from the sensing element. By contrast, in non-remote microwave sensors, the measurand (e.g., the dielectric properties of a certain MUT, or any physical, chemical, or biological variable of interest) is retrieved by proximity (not necessarily by contact), through the in situ interaction of microwaves with matter. Non-remote microwave sensors can be further divided between non-planar and planar. The latter, those of interest in the present book, as mentioned, are implemented by either additive or subtractive fabrication processes on a planar substrate, and their advantages over the non-planar counterparts (mainly cavity, waveguide, and coaxial sensors) have been indicated before. There are many different classification criteria for planar microwave sensors, including the application type, the frequency of operation, or the working principle. In this section, the adopted classification scheme for planar microwave sensors follows their working principle, since such categorization provides physical insight on the mechanisms governing sensor's functionality and performance. Nevertheless, there is a set of binary classification schemes (applicable in most cases to other sensor types) that should be briefly presented, since such schemes are exhaustively used throughout this book.

[4] Nevertheless, other binary classification schemes of planar microwave sensors are also included in Sect. 1.

1.1 Binary Classification Schemes

Planar microwave sensors can be binary classified as follows: contact/contactless, invasive/non-invasive, intrusive/non-intrusive, wired/wireless, single-ended/differential, resonant/non-resonant, and reflective-mode/transmission-mode. Let us next give some more details of each binary classification scheme.

1.1.1 Contact and Contactless Sensors

In contact sensors, the MUT is in direct contact with the sensitive element, e.g., a planar resonator or a transmission line section, whereas in contactless sensors, such contact is inexistent. Nevertheless, proximity between the sample and the sensitive element is required in contactless sensors, so that the MUT modifies the electromagnetic properties of the sensitive element (e.g., the resonance frequency or quality factor in resonant-type sensors, or the phase of the reflection or transmission coefficient in non-resonant sensors) through the interaction with the electromagnetic fields generated by such sensitive element.

In contact sensors, there are two main issues that should be briefly discussed: (i) the air gap between the sensitive element and the MUT (for solid MUTs), and (ii) substrate absorption, when the MUT is a liquid. The air gap is the cause of inaccuracies in the determination of the permittivity of the MUT. To circumvent the presence of the air gap, a usual procedure is to pressure the MUT against the sensitive element (e.g., by means of Teflon screws, or by means of clamps). Substrate absorption might also generate false measurement results since it alters the permittivity of the substrate. The usual approach to avoid substrate absorption is to coat the sensitive region with a narrow dry film able to repel liquids (for instance, in [74], a 0.12-mm thick glass film, with dielectric constant $\varepsilon_r = 5.5$, was used). Note that with the presence of such film, the MUT is prevented to be in direct contact with the sensitive part of the sensor. Nevertheless, such dry film can be considered part of the sensing element, and therefore such sensors are contact devices, provided the MUT (a liquid) is in contact with such film (the typical situation in microfluidic sensors, in liquid sensors based on mechanical holders, or in submersible sensors).

In some contactless sensors, the sensitive element is not in direct contact with the MUT, but with an element that contains the MUT, for example, a pipe or a channel, where a certain liquid flows. There are examples of sensors where such elements (pipes or channels) are positioned on top of the sensor substrate [83], in contact with (or close to) the sensitive region, whereas in other cases, small micro-channels or micro-pipes are embedded within the sensor substrate [84]. There are also cases where a pipe containing the liquid under study crosses transversally the sensor substrate (i.e., with the axis of the pipe orthogonally oriented with regard to the substrate plane) [76, 85], but typically such sensors exhibit a limited sensitivity, as compared to those sensors where the pipe or channel axis is parallel to the substrate plane. Figure 1 illustrates all these possibilities.

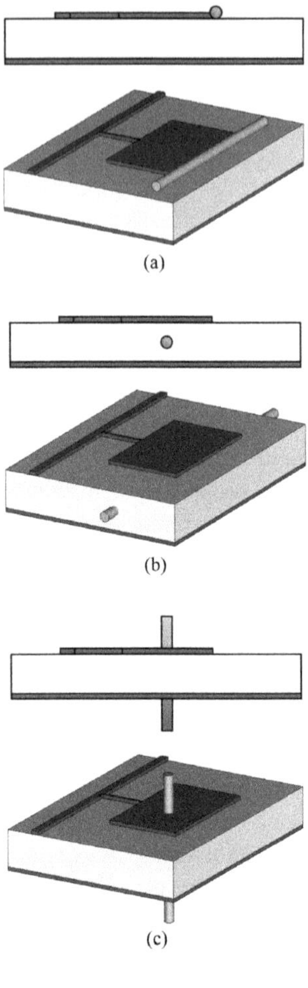

Fig. 1 Cross-sectional and perspective views (schematic) of contactless liquid sensors based on circularly shaped pipes (the element driving the liquid under study) placed on top of the substrate (**a**), embedded in the substrate (**b**), or transversally oriented with regard to the substrate, crossing it (**c**)

1.1.2 Invasive and Non-invasive Sensors

Such terminology mainly refers to liquid sensors and the difference lies in whether the sensitive element of the sensor "invades" the liquid under study or not. Invasive sensors are necessarily contact sensors, whereas in non-invasive sensors, the sensitive element is placed outside of the pipe or channel containing the fluid. Figure 2 illustrates the difference between invasive and non-invasive sensors.

1.1.3 Intrusive and Non-intrusive Sensors

Non-intrusive sensors, either invasive or non-invasive, do not interfere with the flow profile. By contrast, intrusive sensors protrude into the flow and distort the flow profile, as it can be appreciated in Fig. 2. In general, non-intrusive and non-invasive

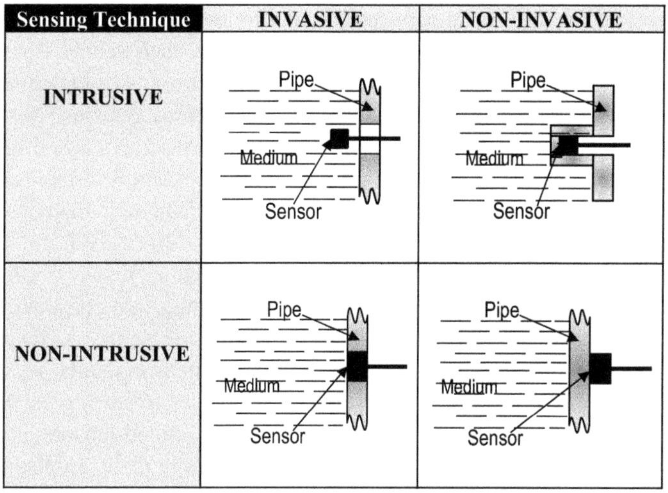

Fig. 2 Distinction between invasive/non-invasive and intrusive/non-intrusive sensors. Reprinted with permission from [44], copyright 2023 Wiley

sensors are preferred in order to prevent from any potential alteration of the liquid under study caused by direct contact with the sensing element, and to avoid perturbing the element driving the liquid (a tube, a pipe, or a channel, typically).

1.1.4 Wired and Wireless Sensors

Wired/wireless sensors should not be confused with contact/contactless sensors. The latter terminology, as it has been explained in Sect. 1.1.1, is used to differentiate whether the sensitive element is in direct contact with the MUT, or not. By contrast, the "wireless" or "wired" characteristic refers to the type of communication (data transfer) between the sensing element and the central unit. Thus, both wired and wireless sensors can be either contact or contactless, depending on the specific transduction mechanism (contact or contactless) to retrieve the sensing data from the sample under study or stimulus. Wireless sensor networks (WSNs) and RFID sensors are examples of wireless sensors.

1.1.5 Single-Ended and Differential Sensors

Single-ended sensors, also called absolute sensors, consist of a single sensing element that provides the (absolute) output variable in response to a certain stimulus or measurand. Differential-mode, or relative, sensors are based on two independent (typically identical) sensing elements (the sensing pair), each one responding to its own stimulus or measurand. However, in differential sensors, the considered input variable is the difference between the input variable present in either sensor, whereas the difference between the output variable in each sensor is the considered output (differential) signal. Differential sensors are costly and oversized, as compared to their single-ended counterparts, for obvious reasons. Nevertheless, differential-mode sensors are robust against cross-sensitivities caused by ambient factors (e.g., temperature or humidity), since such environmental variables are "seen" as common-mode stimuli by such sensors. This is the main advantageous aspect of differential-mode sensors over single-ended sensors. To reduce costs and sometimes to alleviate the effects of sensor imbalance (causing inaccurate results),[5] single-ended sensors can operate "differentially" by simply performing two consecutive single-ended measurements (two-step process differential measurements). As far as ambient factors do not change in the time lapse between the two measurements, sensor robustness against such cross-sensitivities is preserved. However, retrieving the differential sensor data by means of a two-step process (using single-ended sensors) is not a real-time measurement, and such approach cannot be applied in many applications (e.g., in measurements where the input variable changes very fast or should be monitored continuously). Differential sensors are of special interest in applications where differences between a sample (sample under study) and a reference (REF) sample should be detected. In such comparator functionality, tiny differences between the MUT sample and the REF sample can be discriminated provided the sensor exhibits high sensitivity (intimately related to resolution) and sensor imbalances are negligible. The authors recommend the review paper [86], focused on balanced transmission lines, circuits, and sensors, to those readers interested in differential-mode sensors.

1.1.6 Resonant and Non-resonant Sensors

Planar microwave resonant sensors are typically implemented by means of a transmission line coupled to a planar resonator (the sensitive element) [77] or loaded with it [87]. Nevertheless, there are other possible configurations of planar resonant sensors, e.g., a planar antenna loaded with an electrically small resonant element [88], or simply a resonator, or a set of resonators, etched on a dielectric substrate, and excited by a certain interrogation signal (this is the approach used in backscattering sensing RFID tags) [18, 52]. Figure 3 depicts all these possible configurations. There are many types of planar resonant elements useful for sensing, including distributed resonators (e.g., open ended or short-circuited quarter- and half-wavelength transmission lines) and semi-lumped resonators, either metallic (e.g., the split ring resonator, SRR, or the step-impedance resonator, SIR, among others) or slotted (e.g., the complementary split ring resonator, CSRR, or the dumbbell-shaped defect ground structure, DB-DGS, resonator, among others) [44].

---

[5] By sensor imbalance, we mean potential differences between the sensors constituting the differential sensor pair, typically caused by fabrication related tolerances. Such imbalances might generate a non-negligible output (differential) signal, despite the fact that the input differential signal is null. Thus, the accuracy in the manufacturing process is critically important in differential-mode sensors.

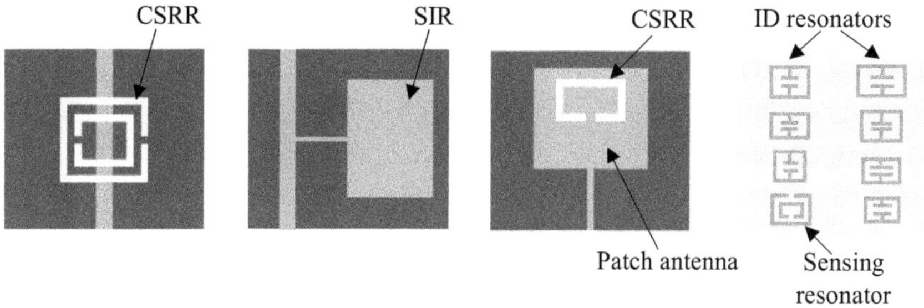

Fig. 3 Topologies of some possible configurations of resonant sensors. **a** Transmission line coupled to a semi-lumped resonant element (CSRR); **b** transmission line loaded with a semi-lumped resonator (SIR); **c** patch antenna loaded with a resonant element (CSRR); **d** back-scattered chipless-RFID sensor based on a set of resonators for identification, and one resonator for sensing. The upper metal level is depicted in soft grey, whereas the lower metal level is depicted in dark grey

By contrast, planar non-resonant sensors typically consist of a transmission line section, which acts as sensitive element, and, eventually, of other planar elements. To reduce the dimensions of the sensing region, a common procedure is to meander the sensing line [89]. Nevertheless, there are other useful approaches to decrease the size of the sensor, for example, by implementing the sensitive part with artificial transmission lines, including slow-wave transmission lines [90], composite right/left handed (CRLH) transmission lines [91], or electroinductive-wave (EIW) transmission lines [92]. Such artificial lines exhibit a strong dispersion, with the result of a small phase or group velocity, and hence dimensions. Nevertheless, it should be clarified that despite the fact that sensors implemented by means of EIW transmission lines, and based on the same principles as those of ordinary lines, have been reported [92], such sensors cannot actually be considered non-resonant, since EIW transmission lines consist of a chain of electrically coupled (slotted) resonators (CSRRs in [92]). These planar sensor topologies, aimed to reduce the size of the sensing region, are depicted in Fig. 4.

In general, it is considered that planar resonant sensors are more sensitive, accurate, smaller, and cheaper than non-resonant sensors. Nevertheless, resonant sensors cannot be applied to the dielectric characterization of materials over broad frequency bands (as required in dielectric spectroscopy [93, 94]), as consequence of their narrowband nature.[6] The typical (canonical) output variable in resonant sensors is the resonance frequency, influenced by the dielectric properties of the MUT. Nevertheless, sometimes, the magnitude of the resonant notch, or peak, as well as the quality factor, or the phase of the reflection or transmission coefficient, can be used for sensing. In non-resonant sensors, the typical output variable is the magnitude and/

[6] For broadband characterization of materials, non-resonant sensors should be used. However, it should be clarified that artificial lines, such as slow-wave, CRLH, or EIW transmission lines, inherently exhibit a limited transmission band. Therefore, such artificial lines are not useful to retrieve the dielectric characteristics of materials over broad bands.

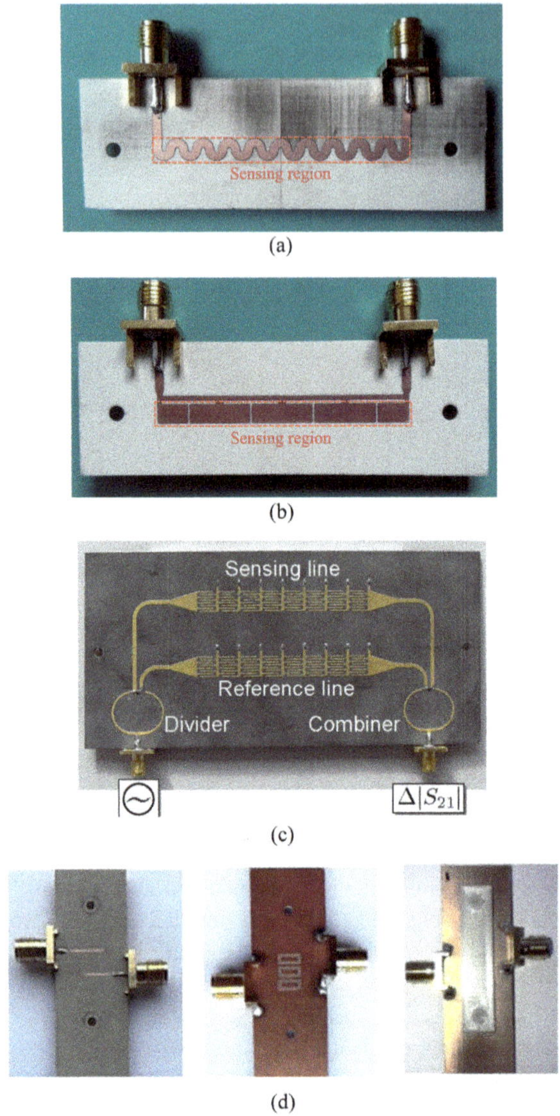

Fig. 4 Examples of topologies (photographs) of planar non-resonant sensors based on non-ordinary lines. **a** Meander line-based sensor; **b** slow-wave transmission line based sensor; **c** CRLH line based differential sensor; **d** EIW transmission line based sensor. In (**d**), the top (left figure) and bottom (center figure) views of the EIW sensor, as well as the sensing region (chain of three rectangular CSRRs) loaded with a dielectric sample (right figure), are depicted. Reprinted with permission from [90] (**a**) and (**b**), [91] (**c**), and [92] (**d**); copyright 2021, 2009 and 2020 IEEE

or phase of the reflection or transmission coefficient of the transmission line used as sensitive element, also influenced by the MUT.[7]

[7] The presence of the MUT in contact or in proximity to the sensing line modifies the characteristic impedance and the complex propagation constant of the line, which in turn affects the reflection and transmission coefficient of such line.

1.1.7 Reflective-Mode and Transmission-Mode Sensors

The last binary categorization of planar microwave sensors distinguishes between reflective- and transmission-mode sensors. The former are one-port structures, either resonant or non-resonant, and the output variable is a certain characteristic of the reflection coefficient (e.g., the resonance frequency, resonance magnitude and/or quality factor, or the phase and/or magnitude of the reflection coefficient measured at a certain frequency/ies). Transmission-mode sensors are two-port structures where the sensing data is naturally retrieved through the information contained in the transmission coefficient (nevertheless, the sensing data can also be collected from the reflection coefficient, as complementary, or redundant, information).

At prototype (laboratory) level, vector network analyzers (VNAs) are commonly used equipment to infer the reflection and/or the transmission coefficient of the sensor. However, the use of such expensive measurement instruments is prohibitive (in general) in real operational environment. Magnitude (envelope) and/or phase detectors constitute the solution for retrieving the sensing data without the use of VNAs [95]. Nevertheless, in reflective-mode sensors, the same (and unique) port is used for sensor excitation and for collecting the sensing data simultaneously. Therefore, solutions based on circulators or directional couplers are needed in order to physically separating the input and output signals, a necessary condition to retrieve the sensing data through the use of envelope and/or phase detectors [96, 97].

1.2 Classification by Working Principle

The purpose of this section is to briefly present the different working principles of planar microwave sensors, as well as some illustrative examples (for a detailed analysis/study of each sensor type, the authors recommend the book [44]). The main working principles, or sensor types (classified according to such principles and presented in dedicated subsections), are frequency-variation sensors, frequency-splitting sensors, coupling modulation sensors, and phase-variation sensors. Rather than a working principle, differential sensing is a mode of operation. Nevertheless, differential-mode sensors are also the subject of a specific subsection, since there are specificities in such sensors that are worthy to be discussed. Figure 5 illustrates a representative topology (sketch) and the working principle of the different sensor types, to be discussed in the next sections.

1.2.1 Frequency-Variation Sensors

Frequency variation is probably the most extended working principle in planar microwave sensors [55, 56, 75–77, 98–136]. The main reason is that frequency-variation sensors exhibit a good combination of size, cost, accuracy, and sensitivity.

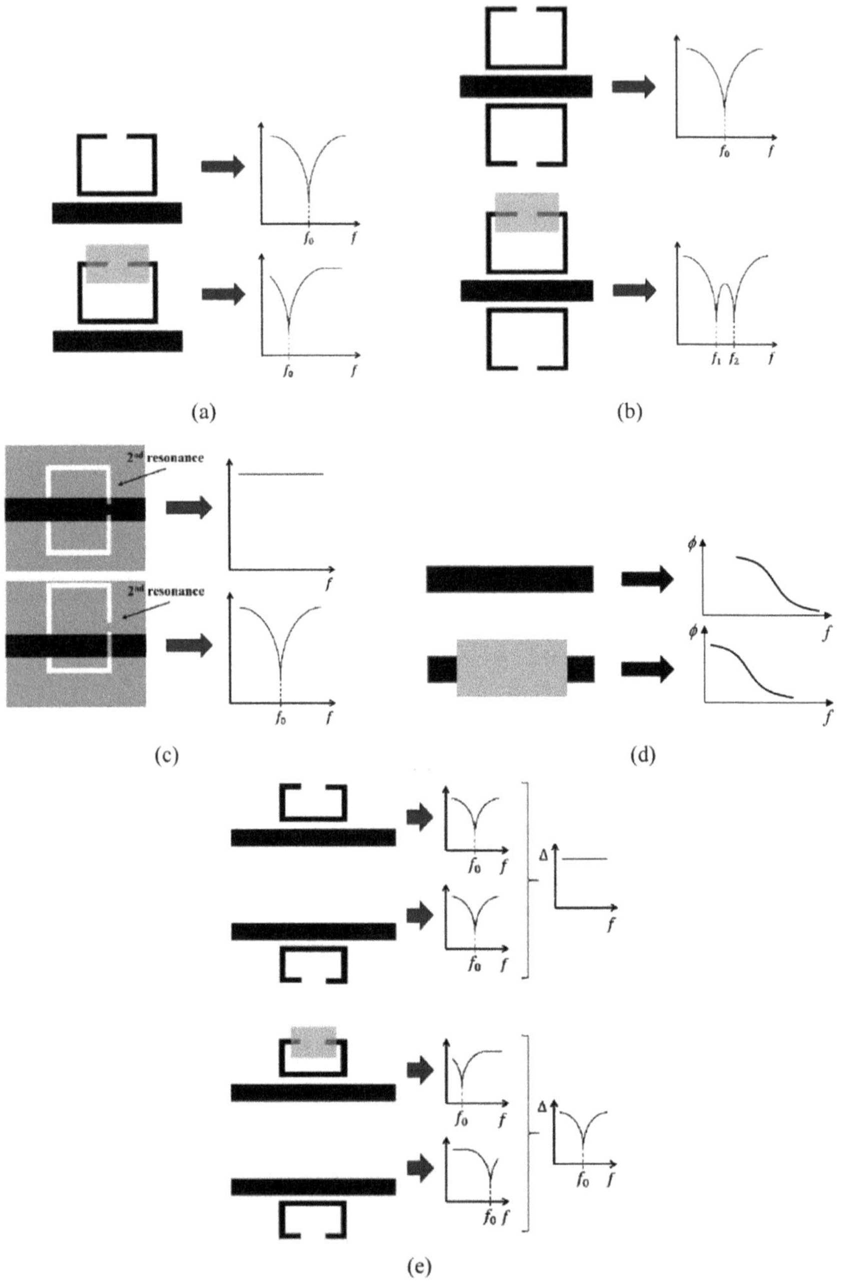

Fig. 5 Illustration of the working principle of the different types of planar microwave sensors and representative topologies. **a** Frequency variation; **b** frequency splitting; **c** coupling modulation; **d** phase variation; **e** differential sensing

Frequency-variation sensors are resonant-type devices typically consisting of a transmission line coupled or loaded with a planar resonator (either metallic or slotted) that can operate either in reflection or in transmission (or in reflection and transmission simultaneously). The canonical output variable in these sensors is the resonance frequency, intimately related with the dielectric constant of the MUT, ε_{MUT}, or with any other physical variable affecting the dielectric constant of a certain material (functional material, or film) in contact or in proximity with the sensing resonator. The variation in the resonance frequency is due to the changes in the capacitance of the resonant element caused by perturbations in the value of ε_{MUT}. An additional output variable in frequency-variation sensors is the magnitude, or quality factor, of the resonance notch (notched-type sensors) or peak (bandpass-type sensors). Such output variable correlates with the loss factor (loss tangent or imaginary part of the complex permittivity) of the MUT, but it typically exhibits cross sensitivity with the dielectric constant.

Frequency-variation sensors are robust against electromagnetic interference (EMI) and noise, as far as the main output variable in such sensors is a frequency (the resonance frequency), quite immune to the effects of EMI and noise. However, frequency-variation sensors require a sweeping interrogation signal able to cover the output dynamic range (or frequency span generated by the variation of the resonance frequency caused by the changes in the ε_{MUT} within the considered input dynamic range).[8] Such interrogation signal should be generated by means of a voltage-controlled oscillator (VCO) in operational environment.[9] Thus, a wideband VCO, or an array of VCOs (e.g., managed by a microcontroller), might be needed in order to cover the whole (potentially wide) output dynamic range in highly sensitive frequency-variation sensors in which a significant variation of the dielectric constant of the MUT (or a certain related variable) is expected. The use of such wideband VCOs, or VCO arrays, represents a penalty in terms of cost, at least as compared with other planar microwave sensors operating at a single frequency (single-frequency sensors), such as coupling-modulation or phase-variation sensors.

As mentioned in Sect. 1.1.6, metallic and slotted resonators are useful for the implementation of planar resonant sensors, and examples of frequency-variation sensors based on both types of resonators can be found in the literature (see the references indicated at the beginning of this subsection). Nevertheless, in sensors based on microstrip lines loaded with slotted resonators (e.g., CSRRs or DB-DGS resonators), etched in the ground plane beneath the excitation line, any potential interference between the MUT and the excitation line is prevented, due to the backside isolation caused by the ground plane. Such isolation also extends to potential related accessories, such as mechanic holders, fluidic channels, or pipes, necessary for liquid

---

[8] Note that the output dynamic range is determined by the input dynamic range and by the sensor sensitivity, or variation of the resonance frequency of the sensing resonator with the dielectric constant of the MUT.

[9] At laboratory level, a signal generator or a VNA is typically used for the generation of the interrogation signal of the sensor.

sensing. Thus, slotted resonators in combination with microstrip technology are very attractive for microwave sensing.

As an example of frequency-variation sensor, let us report a transmission-mode device consisting of a microstrip line loaded with a DB-DGS resonator, transversally etched in the ground plane, as Fig. 6 illustrates [122]. This structure, with substrate parameters and dimensions indicated in the caption of Fig. 6, exhibits good sensitivity of the resonance frequency (the canonical output variable) with the dielectric constant of the MUT. The reason is that this structure behaves as a parallel resonant tank series connected to the line, so that a variation in the capacitance of the DB-DGS resonator caused by a change in the dielectric constant of the MUT directly affects the resonance frequency. By contrast, in CSRR-based sensors, the coupling capacitance between the line and the resonator, not affected by the MUT, degrades the sensitivity [44, 137] (this aspect will be further discussed in Sect. 2.1.1). Figure 7 depicts the simulated and measured responses corresponding to different MUTs (indicated in the figure), all semi-infinite,[10] placed on top of the DB-DGS sensing resonator.[11] From an analytical method presented in [122], where the second output variable is the notch depth at resonance, the dielectric constant and the loss tangent of the MUT can be retrieved. The results reported in [122] are in reasonably good agreement with the nominal values of the different samples. Nevertheless, cavity sensors, in general, provide better accuracy in the determination of the loss tangent, especially for low-loss materials.

1.2.2 Frequency-Splitting Sensors

In frequency-splitting sensors, the canonical output variable is also a frequency, but unlike frequency-variation sensors, frequency-splitting sensors use two identical sensing resonators, one for the MUT (or measurand of interest), and the other one for the so-called reference (REF) sample (or measurand) [71, 74, 87, 138–147]. The simplest configuration of a frequency-splitting sensor is a transmission line symmetrically loaded with (or coupled to) the pair of (identical but not necessarily symmetric) resonators (see Fig. 8) [138–140]. In such configuration, the necessary proximity between both resonators (needed for resonator's excitation by the signal line) generates inter-resonator coupling, which tends to degrade sensor sensitivity at small perturbations [138–140]. Thus, unless certain stratagems for coupling cancellation are considered [144, 145], the cascaded configuration [87], or the divider/combiner configuration [74, 141], are preferred (see Fig. 9).[12] The working principle

---

[10] By semi-infinite we mean a MUT of sufficient thickness and transverse dimensions to guarantee that the electromagnetic field generated by the DB-DGS does not reach the boundaries of the MUT. The semi-infinite MUT approximation is not a general requirement for sensing, but a requisite of the analytical method reported in [122].

[11] For DI water, a mechanical holder (acting as a pool or container) was fabricated by means of a 3D printer and attached to the ground plane of the sensing structure (see further details in [122]).

[12] The divider/combiner configuration, similar to the cascaded configuration, prevents from inter-resonator coupling, since the resonant elements are significantly separated. However, in the divider/

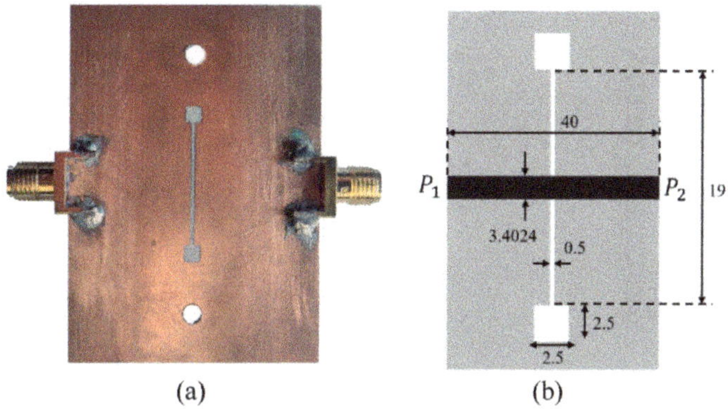

Fig. 6 Photograph from the backside (**a**) and layout (**b**) of a planar frequency-variation sensor based on a DB-DGS (slotted) resonator and devoted to permittivity sensing. The dimensions are indicated in mm. The sensor was implemented in the *Rogers 4003C* substrate with dielectric constant $\varepsilon_r = 3.55$, thickness $h = 1.524$ mm, and loss factor tan $\delta = 0.0022$. The ground plane in (**b**) is depicted in grey. Reprinted with permission from [122]; copyright 2022 IEEE

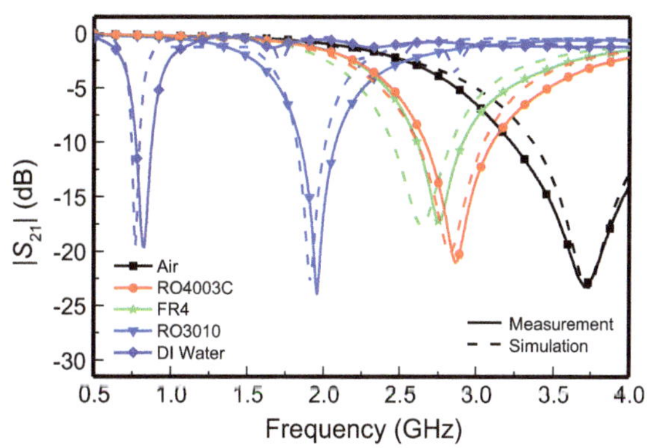

Fig. 7 Measured and simulated transmission coefficient of the bare DB-DGS-based sensor of Fig. 6, and sensor loaded with stacks of *FR4*, *Rogers RO3010*, and *Rogers 4003C*, as well as DI water. Reprinted with permission from [122]; copyright 2022 IEEE

---

combiner frequency-splitting sensor, the notches appear, in general, as consequence of an interfering phenomenon. Namely, at the resonance frequency of the resonator of one of the branches, signal propagation through that branch is precluded. However, signal propagation is not necessarily reflected back to the source, since it can be transmitted, or partially transmitted, through the other (parallel) branch. Thus, in general, the notches are not given by the resonance frequencies of the resonators (except for the symmetric case), but by those frequencies where the signals propagating at both branches destructively interfere at the output T-junction. This mode of operation (signal interference) also degrades sensor sensitivity at small perturbations [141]. However, if the line section between the T-junctions and the plane of the resonators is conveniently chosen, a short at the resonance frequency of the resonators when they are loaded with the REF sample is generated at the input and output T-junctions, and under these circumstances, sensitivity degradation is circumvented, as it is demonstrated in [141]. If the sensing resonant elements are CSRRs or SIRs, the line section between the T-junctions and the plane of the resonators should be a half-wavelength [or $\theta_1 = \pi$, see Fig. 9b]. By contrast, the electrical length of such line sections should be $\theta_1 = \pi/2$

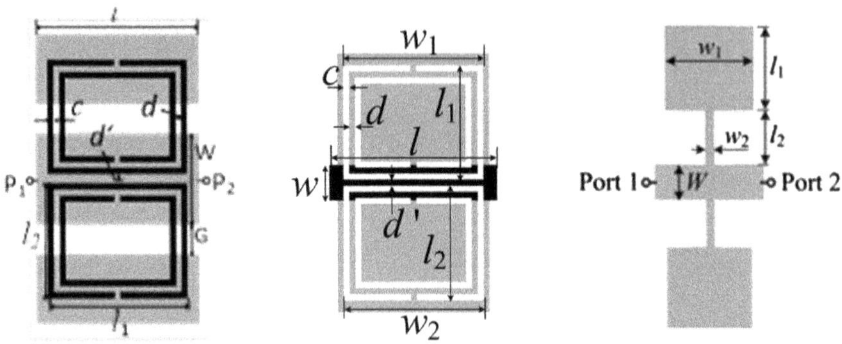

Fig. 8 Different configuration of frequency-splitting sensors consisting of a transmission line loaded with (or coupled to) a pair of identical symmetrically placed resonators. **a** Coplanar waveguide (CPW) transmission line (in grey color) magnetically coupled to a pair of SRR (depicted in black), etched in the backside of the substrate; **b** microstrip line (depicted in black) electrically coupled to a pair of CSRRs, etched in the ground plane (depicted in grey color); **c** microstrip line loaded with a pair of SIRs

of frequency-splitting sensors is very simple. If the dielectric loads on both sensing resonators are identical, a single resonance in the form of a notch (according to the configurations of Figs. 8 and 9) appears. However, when symmetry is truncated, two notches, with frequency separation related to the level of asymmetry, arise.[13] Thus, the canonical output variable in frequency-splitting sensors is the frequency difference (or split) between the two resonances. The difference in the notch magnitude can also be used as an additional variable for sensing (this is necessary in applications devoted to the determination of the complex permittivity of the sample under study [74]).

Similar to differential-mode sensors, frequency-splitting sensors are robust against cross sensitivities to common-mode stimuli (e.g., ambient factors), as far as the output and input variables in frequency-splitting sensors are of differential nature. Nevertheless, frequency-splitting sensors are not true differential-mode sensors, since they are not implemented by means of two independent (single-ended) sensors, but through a pair of sensing resonators and a common microwave structure.

As a prototype example of frequency-splitting sensor, let us briefly report the microfluidic device presented in [74], devoted to the characterization of liquid samples, in particular to the determination of the complex permittivity of mixtures of ethanol and DI water. The topology of the sensor is based on the splitter/combiner configuration, and the sensing resonators are SRRs (see Fig. 10, where a photograph of the whole sensor, including the fluidic channels, is also depicted). By injecting pure

(i.e., a quarter-wavelength) for SRRs. The difference is explained by the fact that CSRRs and SIRs generate a short at resonance, whereas SRRs open the line when they resonate.

[13] Nevertheless, frequency-splitting sensors based on bandpass structures are also possible [146]. In this case, the resonances manifest as peaks in the frequency response of the sensor.

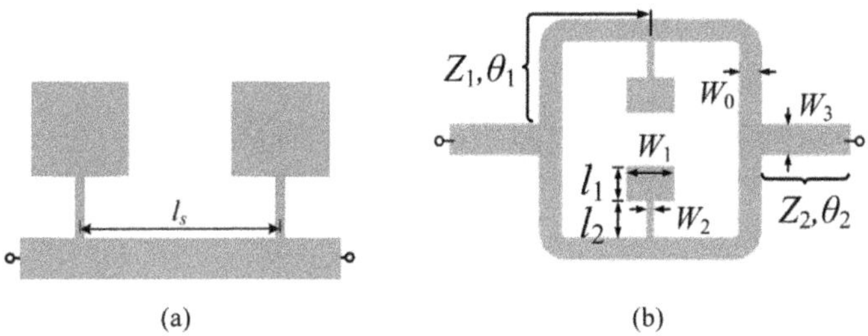

Fig. 9 Configurations of a frequency-splitting sensor based on a pair of SIRs. **a** Cascade configuration of a microstrip line loaded with a pair of SIRs; **b** splitter/combiner configuration of a microstrip line loaded with a pair of SIRs in either branch

DI water in the REF channel and mixtures of DI water and ethanol (with different volume fractions) in the MUT channel, the frequency responses of the device change, see Fig. 11a. In all the cases, two notches appear, since the load of the SRRs is asymmetric. Note that the notch magnitudes are very limited due to the effect of losses of the considered liquid samples. Figure 11b depicts the differential (splitting) frequency, Δf_z, as well as the differential notch depth, ΔS_{21}, the considered output variables, as a function of the volume fraction of ethanol. Note that the sensitivity in the limit of small perturbations (small volume fractions of ethanol) is somehow degraded. The reason is that such sensor was not optimized, since it was conceived as a general-purpose liquid sensor (i.e., the electrical length of the line sections between the T-junctions and the SRRs central position was not set to $\theta_1 = \pi/2$ at the resonance frequency of the SRR loaded with DI water). Nevertheless, it was demonstrated in [74] that the sensor is useful to provide the complex permittivity (real and imaginary parts, i.e., the input variables) of mixtures of DI water and ethanol. For that purpose, a mathematical model was developed, where a linear dependence between the output and input variables was assumed. Calibration of the model was carried out from the knowledge of the complex permittivity of pure ethanol (100% volume fraction), and by considering an additional volume fraction of ethanol (e.g., 10% ethanol), with the complex permittivity of that mixture inferred from the complex permittivity of the constituent components using, e.g., the Weiner model [148] (see further details in [74]). Using that model and the output variables of Fig. 11b, the complex permittivity as a function of the volume fraction of ethanol was obtained. The results, depicted in Fig. 12, reveal that the real and the imaginary parts of the permittivity are circumscribed between the static lower (WL) and upper (WU) limits of the Weiner model, thereby validating the functionality of the sensor for the determination of the complex permittivity of liquid samples.

1.2.3 Coupling-Modulation Sensors

The working principle of coupling-modulation sensors is the variation of the coupling level between, at least, a planar resonant element and a transmission line (excitation

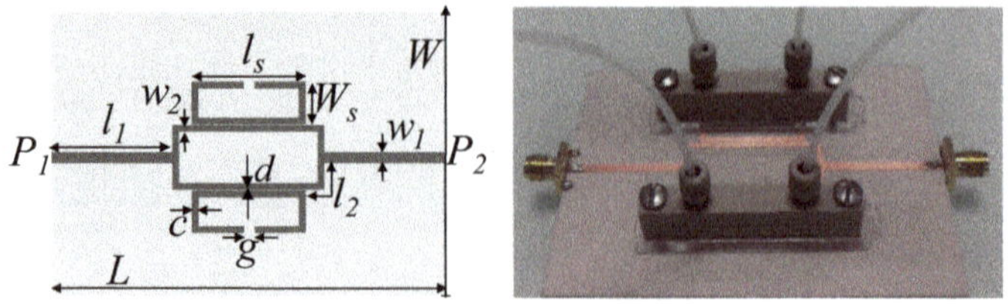

(a) (b)

Fig. 10 Layout (**a**) and photograph (**b**) of a microfluidic frequency-splitting sensor based on the splitter/combiner configuration and two SRRs. Dimensions are: $L = 86$ mm, $W = 62$ mm; inverter dimensions are $l_1 = 27$ mm and $w_1 = 2.22$ mm; SRR dimensions are $l_s = 25$ mm, $W_s = 9$ mm, $c = 1.4$ mm, $g = 2.4$ mm; the slot separation between the lines and the SRRs is $d = 0.2$ mm; the dimensions of the transmission lines sections between the T-junctions and the SRRs are $l_2 = 9.21$ mm, $w_2 = 1.34$ mm. The sensor was implemented in the *Rogers RO3010* substrate with dielectric constant $\varepsilon_r = 10.2$, thickness $h = 1.27$ mm and loss tangent $\tan\delta = 0.0023$. The fluidic channels, located on top of the SRR gap region (the more sensitive part), were made of polydimethylsiloxane (PDMS). Further details on channel dimensions and other mechanical accessories can be found in [74]. Reprinted with permission from [74]; copyright 2017 IEEE

Fig. 11 Measured transmission coefficient (magnitude) for different mixtures of DI water and ethanol (MUT liquid), considering DI water as REF liquid (**a**), and representation of Δf_z and $\Delta|S_{21}|$ as a function of the ethanol concentration (**b**). Reprinted with permission from [74]; copyright 2017 IEEE

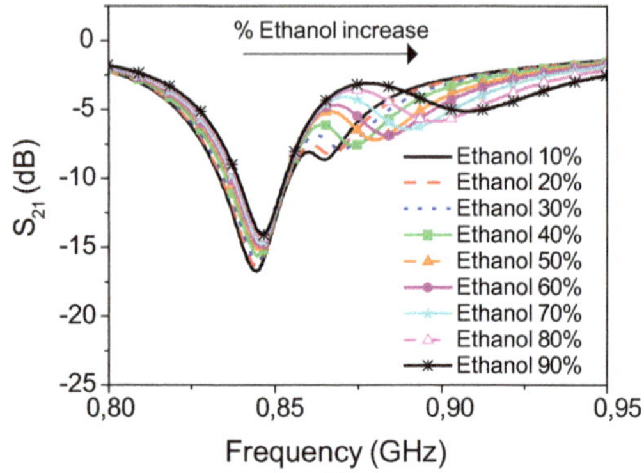

(a)

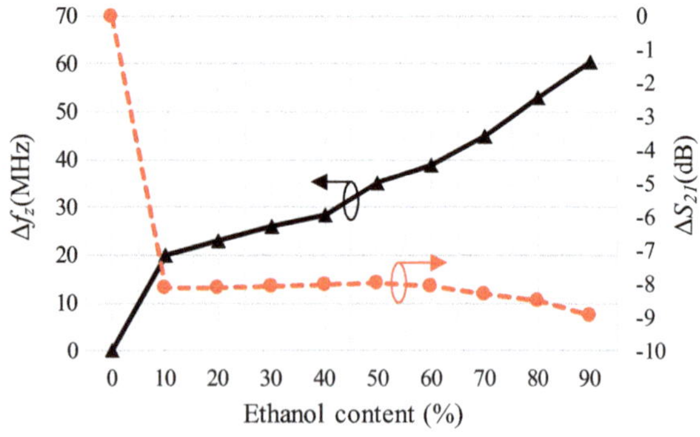

(b)

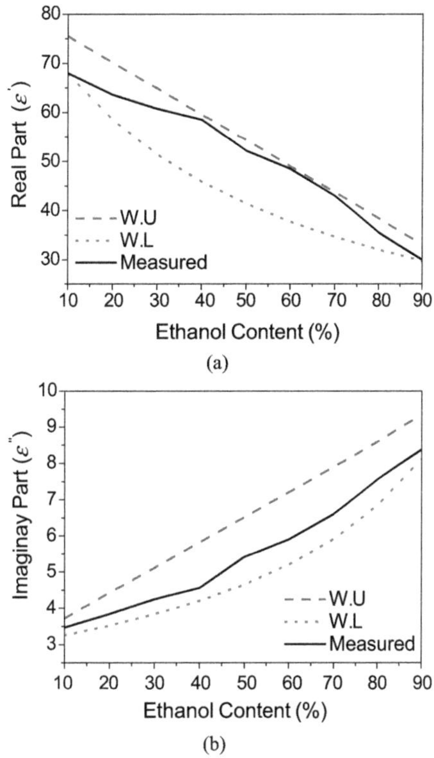

Fig. 12 Extracted value for the real (**a**) and imaginary (**b**) parts of the complex dielectric constant in mixtures of DI water/ethanol. The static Weiner model (upper—WU- and lower—WL-limits) is also included for comparison purposes. Reprinted with permission from [74]; copyright 2017 IEEE

line, or host line) caused by the measurand. The effect of such modulation in the electromagnetic coupling between the resonant/s element/s and the line is a modification of the frequency response, and particularly the magnitude of either the transmission (transmission-mode sensors) or reflection (reflective-mode sensors) coefficient at a certain (operating) frequency. Thus, coupling-modulation sensors are single-frequency devices, where the harmonic signal needed for sensing can be simply generated by means of a microwave oscillator tuned to the desired frequency, or through a narrowband VCO. The canonical output variable in these sensors is the magnitude of either the transmission or reflection coefficient, depending on the operation mode of the sensor (transmission or reflection), at the operating frequency. For sensitivity optimization, it is convenient to set this frequency to the value where the variation of the input variable generates a maximum excursion of the magnitude of the transmission or reflection coefficient. Since the coupling between the host line and the sensing resonator/s depends on their relative position and/or orientation, coupling modulation sensors are especially suitable for the measurement of displacements and velocities (either linear or angular) [149–176]. For such application (displacement/velocity sensing), the resonance frequency of the sensing resonator/s does not vary significantly. Moreover, the maximum excursion of the magnitude of the transmission, or reflection, coefficient occurs at that frequency, or at a frequency close to it. Consequently, in coupling-modulation sensors devoted to displacement/velocity sensing, the input harmonic signal is usually tuned to the resonance frequency of the sensing resonator.

Certain coupling-modulation sensors exploit the electromagnetic symmetry properties of transmission lines loaded with a single symmetric resonator [44, 149–158]. The key idea behind this sensing strategy is to use a planar resonator exhibiting a symmetry plane of electromagnetic nature (either an electric or a magnetic wall) different to that of the axial symmetry plane of the line. Under these conditions, the line is transparent provided the symmetry plane of the resonant element is in perfect alignment with the symmetry plane of the line (resulting in a perfectly symmetric structure). The reason is that electromagnetic coupling between the line and the resonator is inexistent when the symmetry planes are of different electromagnetic sort (despite the fact that the resonator and the line are closely spaced) [44, 149, 158]. However, if symmetry is disrupted by a relative displacement (linear, angular, or both) between the host line and the resonant element, the coupling is activated, and a notch in the frequency response (at the resonance frequency of the sensing resonator) arises (see Fig. 13). The level of asymmetry determines the magnitude of the notch. Hence, the notch magnitude is the natural output variable in coupling-modulation displacement sensors based on electromagnetic symmetry properties. Obviously, for displacement sensing, it is necessary that the host line and the sensing resonator are etched in independent substrates (in relative motion). Although most coupling-modulation sensors based on symmetry disruption are devoted to the measurement of displacements and velocities, symmetry perturbation can also be achieved by asymmetrically loading the resonant element (e.g., by a pair of unequal dielectric samples). Nevertheless, this approach is rarely used for the dielectric characterization of materials, since, for that purpose, there are methods that are more accurate (e.g., frequency variation or phase variation).

In other coupling-modulation sensors, also devoted to motion sensing, the level of proximity between the line and the resonator (not necessarily symmetric) modulates their mutual coupling. Such coupling is maximum when the resonator is on top of the line,[14] and null if the resonator is sufficiently distant from the line. An interesting type of displacement and velocity sensors, designated as electromagnetic encoders [159–176], are implemented by etching a chain (linear or circular) of resonant elements in a movable dielectric substrate. If such chain is transversally displaced over a transmission line (the stator) at short distance, the frequency response of the line is modulated by chain motion. More specifically, if the line is fed by a harmonic signal tuned to the resonance frequency of the resonant elements, an amplitude modulated (AM) signal with an envelope function with as many dips, and peaks, per time unit as resonant elements crossing the line axis, will be generated in the output port of the line. The time lapse between adjacent dips (and peaks), easily identifiable by means of an envelope (AM) detector, determines the chain velocity, whereas the chain position can be inferred from the cumulative number of dips (or peaks) from

[14] It is assumed that these sensors do not exploit electromagnetic symmetry properties, contrarily to those of the preceding paragraph.

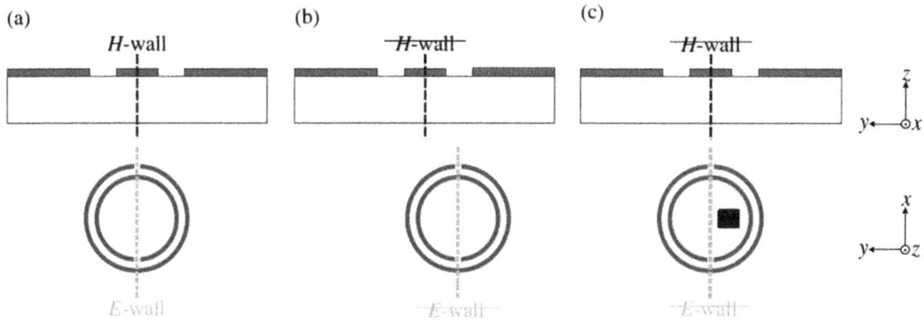

Fig. 13 Sketch showing the effects of symmetry disruption in a CPW transmission line loaded with a SRR etched in the backside of the substrate. **a** Perfect symmetry, preventing line-to-resonator coupling; **b** symmetry cancelation by means of a lateral displacement of the SRR, thereby activating coupling; **c** symmetry perturbation by means of an asymmetric dielectric loading, also activating coupling. The CPW and the SRR (etched in the back substrate side) are depicted in cross-sectional and bottom views, respectively, for better understanding

a reference (REF) position. These displacement/velocity sensors are designated as incremental encoders (examples can be found, e.g., in [163, 164]). Nevertheless, quasi-absolute and absolute encoders have also been reported (e.g., in [161, 171, 172, 175]). In such encoders, at least two chains of resonant elements are used. One chain is generally used for synchronization purposes, whereas the other chain (or chains) is (are) encoded through the presence or absence of resonant elements at predefined positions of the chain, and such identification (ID) code determines the absolute position of the chain (without the need of knowing the initial chain position, or REF position).

It should be mentioned that the distinction between quasi-absolute and absolute electromagnetic encoders obeys to the number of bits per encoder position. Some encoders exhibit a unique bit per position (determined by the absence or presence of resonator at the predefined positions of the position chain). In such cases (or if the number of bits per encoder position is very small), it is obvious that the position cannot be determined "absolutely", because the number of encoded positions that can be differentiated by the ID code is 2^n, where n is the number of bits per encoder row. Nevertheless, it is possible to assign a higher number of bits to each position (or row) by also including a subset of certain number of bits corresponding to the previous $n-1$ positions. Such $n-1$ previous bits, plus the bit given by the corresponding position, provides a total number of n bits. Note that n should satisfy

$$n \geq \log_2\left(\frac{L}{p}\right) \tag{1}$$

where L is the total encoder length (assuming that we consider a linear encoder) and p is the encoder period (so that L/p is the total number of different positions).[15] Of course, any sequence of n adjacent bits of the whole encoder should not repeat, and this is possible by encoding the position chain according to the so-called De Bruijn sequence [177]. However, it is obvious that in order to determine the absolute position of the encoder after a system reset, the encoder must displace n periods. For that reason, such electromagnetic encoders are designated as quasi-absolute. In true absolute encoders, the number of bits per position (excluding the bits assigned to preceding positions) should be high enough to satisfy expression (1), so that the encoder position can be "absolutely" determined (especially after a system reset) without the need of reading the bits of the preceding positions.

In the quasi-absolute encoders reported in [161, 171], an encoded chain (following the De Bruijn sequence, and called position chain) determines the encoder position. An additional (parallel) chain, called velocity chain, with all the resonators present and situated at the same axial positions than those of the position chain, provides the instantaneous encoder velocity (given by the time lapse between adjacent pulses), as well as the necessary synchronism to read the position chain. To generate the two envelope functions (one periodic, and associated to the velocity chain, and the other one containing the ID code of the encoder), two independent harmonic signals are needed. An additional signal is necessary if discerning the encoder direction of motion is a requirement. In such case, rather than an additional (third) encoder chain, the solution is simply to add an additional sensing element in the stator. Such element should be able to detect the resonators of the velocity chain, but with certain led or lag (depending on the motion direction of the encoder) with regard to the signal (envelope function) generated by the other sensing element devoted to determining the velocity and synchronism. The two or three harmonic signals can be injected to the stator by means of a combiner, and then separated at the output port by means of a multiplexer (diplexer or triplexer). Finally, the (two or three) envelope functions can be retrieved by cascading an envelope (AM) detector to the output ports of the multiplexer [a sketch of this system is depicted in Fig. 14a]. Alternatively, a scheme based on injecting the required (two or three) harmonic signal at different time intervals, by means of a VCO managed by a microcontroller, is also possible [see Fig. 14b]. In that case, a single AM detector suffices, and it is considered the best solution (applied in the quasi-absolute electromagnetic encoder system reported in [171]).

Figure 15 depicts the layouts and photographs of the reader and encoder designed and fabricated in [171]. The reader (stator) is a microstrip line loaded with three CSRRs tuned to different (resonance) frequencies (stator dimensions and substrate details are given in the caption of Fig. 15). Note that two CSRRs (the CSRR devoted to detect the velocity, and the CSRR devoted to determine the motion direction) are located at the same transverse plane of the line (hence the lag or lead between the corresponding signals provides the motion direction). The encoder, with the two chains previously indicated, was implemented by means of screen-printing on a

[15] The encoder period, p, determines the spatial resolution of the encoder.

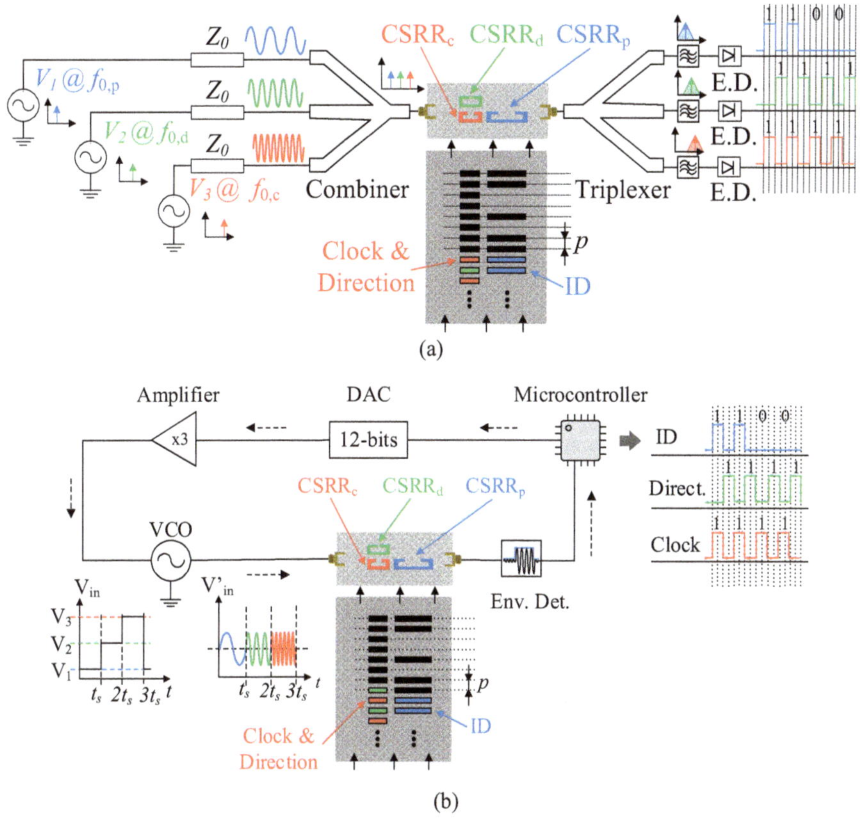

Fig. 14 Sketch of a quasi-absolute encoder system with synchronous reading based on a combiner/multiplexer (**a**) and based on a VCO managed by a microcontroller (**b**)

rubber elevator belt. Figure 16 shows the photograph of the whole system, whereas Fig. 17 depicts the signal present at the output port of the AM detector (which contains the three AM demodulated signals), as well as the three (separated) envelope functions (achieved by the microcontroller). The encoder was displaced at a velocity of 40 mm/s with regard to the stator. Note that the peaks/valleys in the envelope function of the position chain perfectly correlate with the bits of the different encoder positions.

Several efforts to increase the number of bits per encoder position, and thus implement true absolute encoders, have also been made. For that purpose, it is necessary to combine time encoding (the bits of the position chain are read sequentially, row by row, in a time-division multiplexing scheme) with another domain, e.g., frequency encoding. Frequency encoding is achieved by considering resonators of different sizes, a solution reported, e.g., in [173–176]. For example, 2 bits per row were achieved in [173] by considering resonators of four different sizes. This system does not require synchronism, as far as a certain resonator is always present at the different

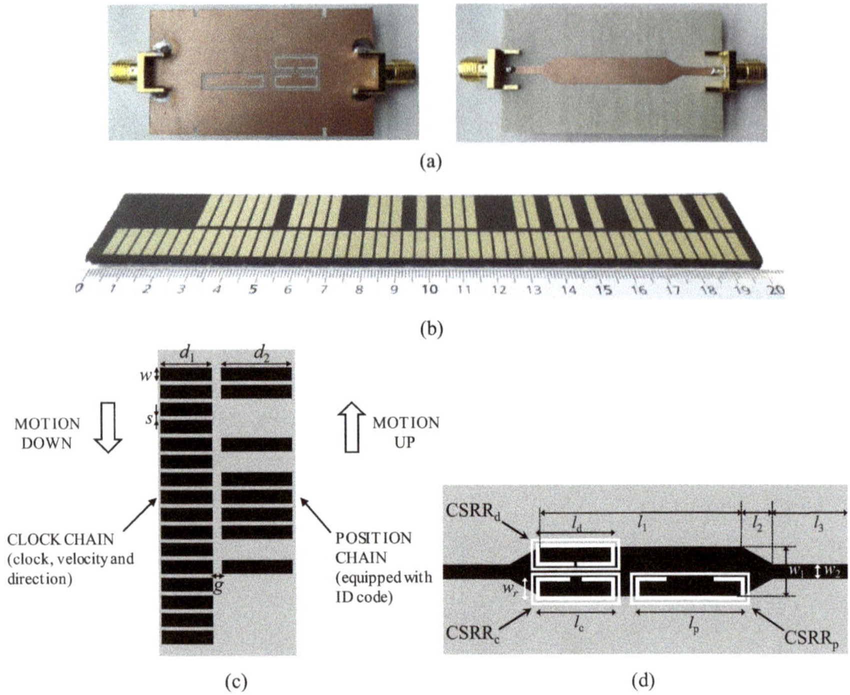

Fig. 15 Photograph of the fabricated reader, including the bottom (left) and top (right) views (**a**), and encoder (**b**), and layout of the encoder (**c**) and reader (**d**) that include the relevant dimensions. Encoder dimensions are: $d_1 = 11.5$ mm; $d_2 = 15.9$ mm, $w = 3$ mm, $s = 1$ mm and $g = 1.9$ mm. Reader dimensions are: $l_1 = 26.6$ mm; $l_2 = 3.8$ mm, $l_3 = 10$ mm, $w_1 = 6.4$ mm, $w_2 = 1.9$ mm, $l_c = l_d = 10.5$ mm; $l_p = 14.5$ mm, $w_r = 2.9$ mm. CSRR slots width is $c = 0.5$ mm, and ring splits are $s_d = 0.4$ mm, $s_c = 1.6$ mm and $s_p = 6.2$ mm. The reader was implemented in the *Rogers 4003C* substrate with thickness $h = 0.2$ mm, dielectric constant $\varepsilon_r = 3.38$, and loss factor tan $\delta = 0.0022$. Reprinted with permission from [171]; copyright 2022 MDPI

encoder positions. For encoder reading, as many harmonic signals as resonator sizes are needed. In a system like the one reported in [173], consisting of a microstrip line with a gap (the reader), and encoders implemented by means of linear resonators transversally oriented with regard to the chain axis, and parallel to the reader line axis, it is not possible to significantly increase the number of bits. Note that this would require an excessive number of independent harmonic signals, and resonator sizes (e.g., eight harmonics signals for a 3-bit, per position, encoder), which does not seem reasonable. However, it is possible to consider multiple encoder position chains (of resonators with different sizes, and hence frequencies), and add certain complexity to the reader, in order to substantially enhance the number of bits per encoder position. This solution was reported in [174, 175], by considering readers based on a power splitter configuration. In [175], the achieved number of bits per row is 8.78, which means that 440 different positions can be encoded, a reasonable number for many applications. In that system, four encoder chains and a 1:4 power

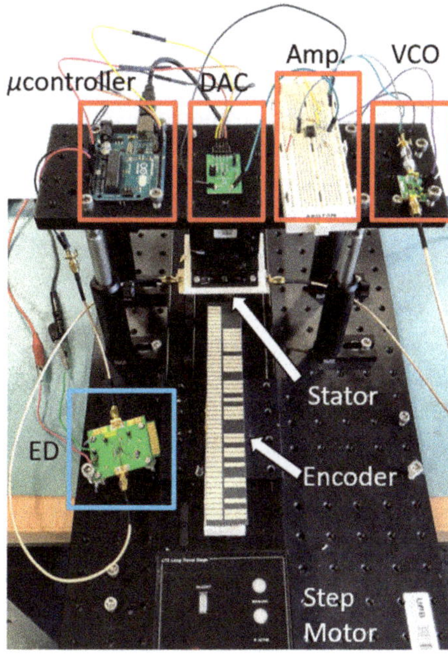

Fig. 16 Photograph of the quasi-absolute encoder system with synchronous reading based on an advanced architecture with signal generation and processing by means of a microcontroller. Reprinted with permission from [171]; copyright 2022 MDPI

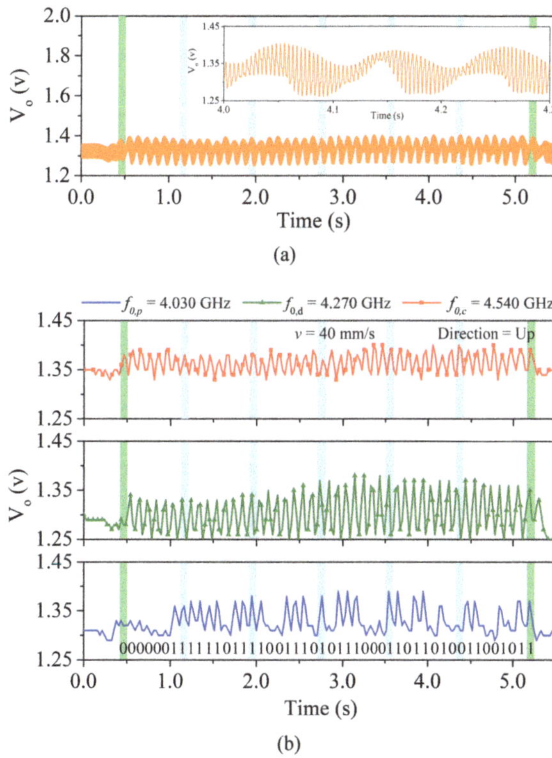

Fig. 17 **a** Measured signal at the output port of the envelope (AM) detector; **b** measured envelope functions for the velocity (and clock for synchronism), direction, and position signals. The frequencies of thc harmonic signals for the clock, direction and position are indicated in (**b**) as $f_{0,c}$, $f_{0,d}$, and $f_{0,p}$, respectively. Reprinted with permission from [171]; copyright 2022 MDPI

splitter (i.e., with four output ports) was considered. However, this system is relatively complex to explain in an introductory chapter like the present one. Therefore, let us report in the next paragraphs the encoder system presented in [174], based on encoders with two position chains and a power splitter reader with two output ports (1:2 power splitter).

Figure 18 depicts the reader and one specific encoder corresponding to the system reported in [174]. The perspective view is useful to understand how the encoder must be displaced over the reader at short distance. One of the chains provides 5 different states per position, corresponding to the four different strip lengths, plus an additional state associated to the absence of strip. The same applies to the second chain. Consequently, this gives $5 \times 5 = 25$ different states per encoder position. However, the simultaneous absence of inclusion in both chains for a given position should be avoided, since at least one strip is required for the (necessary) synchronism of the encoder. Thus, the total number of states is $25 - 1 = 24$, corresponding to 4.58 bits per position. For the sequential detection of the specific state (or ID code) associated to each position (encoder reading), the encoder is displaced as depicted in Fig. 18a. The input signal is prevented to be transmitted to the output port (either 2 or 3) when a strip lies on top of the access line of that port, provided the frequency of such signal is tuned to the value associated to such specific strip resonator (which depends on its length). Namely, when a strip is on top of a line, a notch in the frequency response, controlled by the strip length, is generated. This means that for encoder reading, four harmonic signals (corresponding to the four strip sizes) are needed, and such signals should be injected to the input port of the splitter. For that purpose, either a combiner or a VCO managed by a microcontroller (in order to inject the harmonic feeding signals at different time intervals) can be considered.

The detection of each specific strip is revealed by the reduction in the amplitude of the corresponding injected (carrier) signal at the corresponding output port (port 2 for the left chain, and port 3 for the right chain). In other words, each injected carrier signal is amplitude modulated (AM) by encoder motion, i.e., by strip overlapping with the access lines of the output ports, and the envelope functions contain the ID code. Obviously, the absence of strip in any of the chains (but not in both chains simultaneously, as justified due to synchronization) does not reduce the amplitude of the feeding signals at the corresponding port, and therefore can also be detected. Note that for retrieving the ID code for each encoder position (or row), the envelope functions of each carrier signal should be inferred at both output ports of the splitter (the architecture of this system is depicted in Fig. 19). Note that encoder reading proceeds sequentially, row by row, as the encoder is displaced over the reader. However, the states (or ID codes) corresponding to each row are generated by frequency encoding (through the length of the strips). For that main reason, it is considered that the proposed encoder system exploits both the frequency and the time domains simultaneously (thus, this type of systems can be designated as hybrid electromagnetic encoders).

When the strips are on top of the output arms of the reader, such strips generate notches at frequencies of $f_{0,1} = 4.550$ GHz, $f_{0,2} = 5.160$ GHz, $f_{0,3} = 5.885$ GHz, and $f_{0,4} = 6.540$ GHz for strips designated as 1, 2, 3, and 4, respectively. Therefore, these

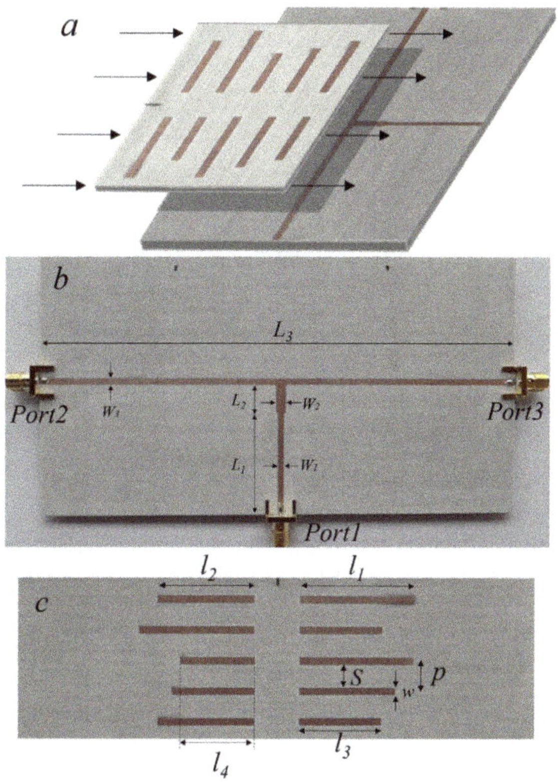

Fig. 18 Perspective view of the proposed encoder system (**a**), and photograph of the power splitter microstrip reader (**b**) and a specific encoder with two chains of inclusions (**c**). The encoder is uncoated in the back substrate side, whereas the reader has a ground plane metallization in the back substrate size. Dimensions (in mm) are: $L_1 = 31.20$, $L_2 = 8.80$, $L_3 = 150.00$, $W_1 = 1.80$, $W_2 = 2.98$, $W_3 = 1.80$, $l_1 = 24.00$, $l_2 = 21.00$, $l_3 = 18.00$, $l_4 = 16.00$, $S = 5.00$, $p = 6.50$, $w = 1.50$. Note that l_i with $i = 1, 2, 3, 4$ indicates the strip length. The reader and the encoder were implemented on the *Rogers 4003C* substrate with dielectric constant $\varepsilon_r = 3.55$, thickness $h = 0.81$ mm, and loss factor tan $\delta = 0.0022$. Reprinted with permission from [174]; copyright 2022 EuMA

are the frequencies of the required four harmonic signals. For validation purposes, the encoder of Fig. 18 was displaced over the reader in [174] with a velocity of 10 mm/s. The eight envelope functions (four at each output port of the reader) recorded for each frequency, when the tag was displaced over the reader, are depicted in Fig. 20.[16] The figure reveals that a perfect correlation between the dips in the envelope functions and the crossings of the corresponding strip resonators over the output access lines exist, thereby validating the approach (see further details in [174]).

To summarize this subsection devoted to coupling-modulation sensors, let us mention that, in general, displacement sensors based on electromagnetic encoders exhibit a larger dynamic range as compared to those dynamic ranges of displacement sensors based on electromagnetic symmetry properties (of the order of resonator dimensions, and hence small). As compared to frequency-variation or frequency-splitting sensors, coupling-modulation sensors are less immune to the effects of EMI and noise, since these sensors are based on magnitude measurements, less tolerant

[16] It should be mentioned that in [174], rather than implementing the complete system sketched in Fig. 19, the envelope functions at each output arm of the splitter for each frequency ($f_{0,1} = 4.550$ GHz, $f_{0,2} = 5.160$ GHz, $f_{0,3} = 5.885$ GHz, and $f_{0,4} = 6.540$ GHz) were inferred independently, as a first proof-of-concept.

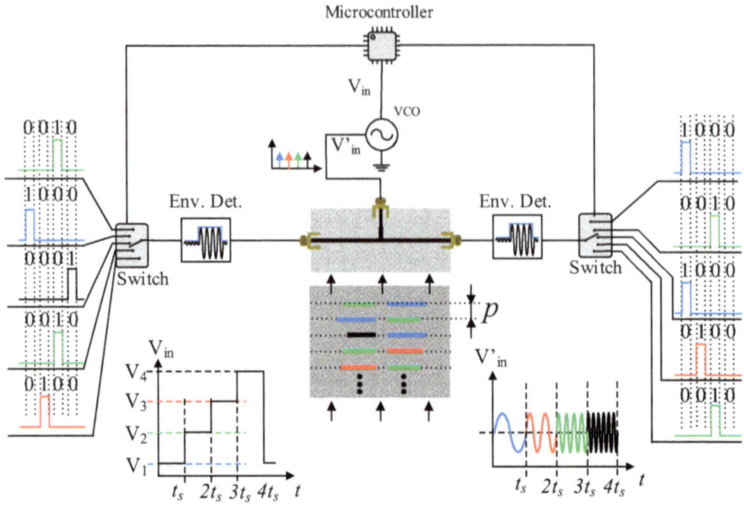

Fig. 19 Sketch of the hybrid time/frequency domain encoder system based on a two-chain encoder and power splitter reader with two output ports. The VCO generates four harmonic signals, managed by a microcontroller, in order to detect the different sizes of the chain inclusions (frequency encoding is achieved through the size of the inclusions, as indicated in the text)

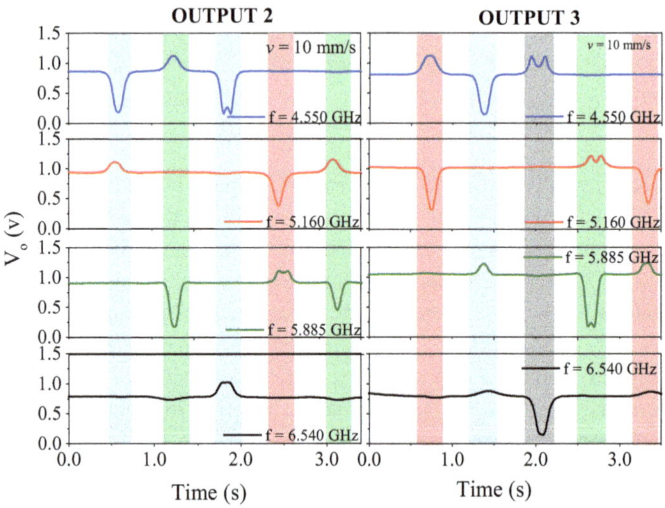

Fig. 20 Envelope functions for the four frequencies and for the two output ports, for the encoder depicted in Fig. 18. Reprinted with permission from [174]; copyright 2022 EuMA

to such effects. However, as mentioned, coupling-modulation sensors operate at a single frequency, and this reduces the cost of the associated electronics needed for signal generation.[17] Note that the envelope detectors required for retrieving the output variables (displacement and velocity) in electromagnetic encoders, as well as other potentially required electronic devices, such as VCOs or microcontrollers, are also low-cost components.

1.2.4 Phase-Variation Sensors

Phase-variation sensors constitute a very interesting approach for microwave sensing since they combine the advantages of the previously discussed sensors, at least in many cases. Namely, similar to frequency measurements, phase measurements are robust against EMI and noise, and phase variation sensors can be designed to operate at a single frequency. The canonical output variable in these sensors is the phase of either the transmission or the reflection coefficient of the considered sensing structure at the operating frequency.

There are many types of phase-variation sensors. Transmission-mode phase-variation sensors are typically based on meandered lines [89, 178] (i.e., non-resonant sensing structures[18]). The reason is that the sensitivity in these sensors increases with the length of the line and operational frequency. Thus, through line meandering, long lines (and hence high sensitivity) can be achieved with a reasonable shape factor. Other approaches to enhance sensor sensitivity in transmission-mode phase-variation sensors are based on the high dispersion that artificial lines (e.g., slow-wave, EIW, or CRLH lines) exhibit [90–92], see Fig. 4.

There are also resonant-type phase-variation sensors, in particular, reflective-mode devices consisting of a transmission line terminated with a resonator (either distributed or semi-lumped), acting as the sensitive element [179–184]. By setting the operational frequency to the resonance frequency of the resonant element, the slope of the phase of the reflection coefficient at that frequency can be made very high. The result is a significant excursion of the phase of the reflection coefficient by varying the dielectric constant of the MUT, i.e., a high sensitivity. Moreover, as it will be later discussed in this chapter (in Sect. 2.1.2), the sensitivity can be significantly enhanced by cascading high/low impedance inverters (in practice implemented by means of quarter-wavelength transmission lines) to the sensing element. Sensitivities as high as 528.7° have been reported in [179], by considering an open-ended high-impedance quarter-wavelength resonator (equivalent to a grounded series resonator). Other reflective-mode resonant-type phase-variation sensors based on open complementary split ring resonators (OCSRRs) [182], step impedance resonators (SIRs)

[17] Exceptions are certain electromagnetic encoders (quasi-absolute or absolute), which require several harmonic signals for encoder reading, as explained. Nevertheless, a single VCO, managed by a microcontroller, suffices, at least in most cases, for the generation of such signals. This does not represent a significant increase of the cost of the associated electronics.

[18] However, most phase-variation sensors do not exploit the broadband potentiality of non-resonant sensors.

[183], and slot resonators [78, 184], have been reported. The advantage of such sensors is the small size of the sensing (semi-lumped) resonators, which contributes to reduce significantly the sensing region. Transmission mode phase-variation sensors based on step-impedance lines are reported in [185], but the achievable sensitivity in such sensors is limited as compared to those sensitivities achievable with the reflective-mode counterparts.

The very high sensitivities of phase-variation sensors, and particularly of those sensors based on a cascade of resonator-terminated high/low impedance inverters, are very useful to detect small changes in the measurand (typically the dielectric constant, but also other related variables) in the vicinity of the reference value. Thus, phase-variation sensors are interesting in diverse applications, such as detection of small levels of impurities in liquids, determination of small volume fractions of solute in diluted solutions, defect detection, monitoring of sample degradation, etc. For example, it was demonstrated in [78] that the tiny changes experienced by edible oils by usage after various frying processes can be detected by means of a reflective-mode phase-variation sensor based on a slot resonator. This prototype device will be reported next as a good (and representative) example of a sensor able to detect small changes experienced in samples (intimately related to its high sensitivity). Nevertheless, let us first mention that despite the fact that most phase-variation sensors are devoted to dielectric characterization, there are also examples of phase-variation sensors applied to the measurement of displacements, either linear or angular [186–189].

The device reported in [78] is a highly sensitive one-port reflective-mode phase-variation sensor able to detect the dielectric changes experienced by an edible oil when it is subjected to cumulative frying processes. The sensor consists of a microstrip transmission line terminated with a slot resonator, the sensing element, etched in the ground plane, and the working principle is the variation in the phase of the reflection coefficient at a fixed frequency (the output variable) generated by the changes in the dielectric properties of the subsequently fried oil. A 3D-printed container (holder) made of polylactic acid (PLA), where the oil is poured, was placed on top of the slot resonator, and conveniently sealed, so that the oil is in contact with the sensitive element (the slot resonator). Figure 21 depicts a generic layout, as well as a perspective sketch of the sensor. Note the presence of a via, used to connect the slot resonator to ground. The equivalent circuit model is also depicted in the figure, where the slot resonator is modelled by the capacitance C, the inductance L, and the conductance G, and Y_1 is the admittance of the inverter.[19] The frequency of operation must be tuned to the resonance frequency of the slot resonator loaded with the REF sample (i.e., fresh, or unused, oil). In practice, the via can be avoided by replacing it with a quarter-wavelength open-ended line. This increases somehow the overall size of the sensor, but not significantly (moreover, the size of the sensing region is

[19] In this prototype, the sensing resonator is cascaded to a single inverter stage, as a means to enhance the sensitivity. The admittance of this inverter was set in [78] to a small value, particularly, $Y_1 = 1/150$ S (corresponding to a high inverter impedance of $Z_1 = 150\ \Omega$) for the reasons that will be later justified in this subsection.

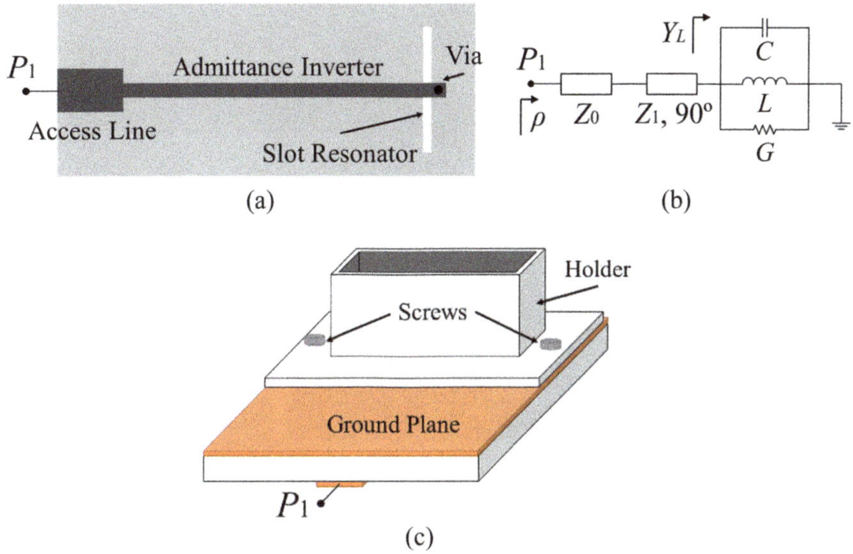

Fig. 21 Topology (**a**), circuit model (**b**), and perspective sketch seen from the back side (**c**) of the proposed reflective-mode sensor based on a slot resonator. In (**a**), the ground plane is depicted in grey. Reprinted with permission from [78]; copyright 2023 EuMA

unaffected). To enhance the sensitivity, it is convenient that the slot resonator exhibits a high-quality factor, i.e., a high *C*/*L* ratio. This explains the meandering in the slot resonator, as depicted in the fabricated prototype, shown in Fig. 22.

The operating frequency of the sensor was set to $f_0 = 1.27$ GHz. Thus, the slot resonator was designed in order to exhibit its fundamental resonance at that frequency, when it is loaded with REF sample, i.e., unused oil, as indicated before. The dielectric constant and the loss tangent of such oil was estimated to be $\varepsilon_{REF} = 4.0$ and $\tan \delta_{REF} = 0.092$, respectively, as it was inferred from an independent measurement of a transmission mode sensor, consisting of a matched line loaded with a slot resonator. By adjusting the resonance frequency and notch level, the above-cited dielectric parameters of the unused oil were obtained. The electromagnetically simulated response of the sensor by considering that it is loaded with unused (REF) oil (with a height of 7 mm) is depicted in Fig. 23 (both the magnitude and phase of the reflection coefficient are shown). The figure includes also the measured response, which is in good agreement with the electromagnetic simulation. From the simulated response, the reactive parameters of the slot resonator loaded with the REF sunflower oil, as well as the conductance, were obtained. For that purpose, the response of the bare sensor (i.e., with the slot resonator surrounded by air) was first obtained. For this case, losses were neglected (a good approximation), and the inductance and the capacitance of the resonator were inferred from the resonance frequency and phase slope at resonance. When the resonator was then loaded with oil, the inductance did not change. Thus, the capacitance was determined with accuracy from the new

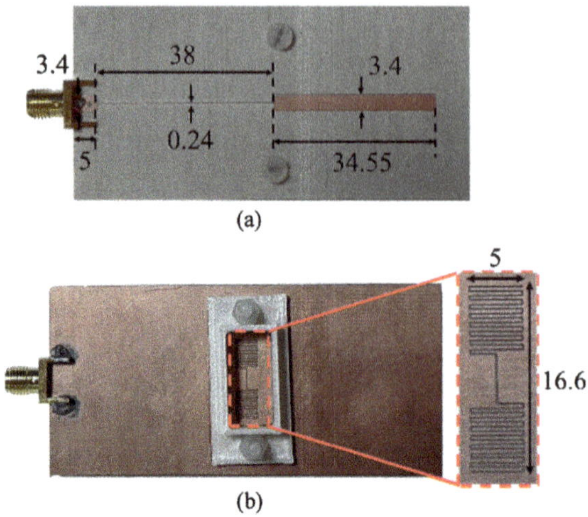

Fig. 22 Photograph of the fabricated sensor based on the topology of Fig. 21. **a** Top view; **b** bottom view. Dimensions are given in mm. The width of the slot is 0.2 mm and the separation between meanders is 0.2 mm. The height of the holder is 7 mm. The sensor was fabricated in the *Rogers 4003C* substrate with dielectric constant $\varepsilon_r = 3.55$, thickness $h = 1.524$ mm, and loss factor tan $\delta = 0.0022$. Note that the via in the circuit of Fig. 21a has been replaced with an open-ended quarter-wavelength matched line. Reprinted with permission from [78]; copyright 2023 EuMA

resonance frequency, whereas the conductance was obtained by curve fitting the response. With this parameter extraction procedure, the element values were found to be $L = 2.774$ nH, $C = 5.649$ pF, and $G = 2.444$ mS. The circuit response inferred by considering such circuit parameters is also included in Fig. 23, and the agreement is perfect.

The phase response of the sensor loaded with the REF oil indicates that losses are important (oils are lossy samples), as revealed by the positive slope at the operational frequency. Indeed, the sign of the slope depends on the relative weight of Y_0G against Y_1^2, where Y_0 is the characteristic admittance of the access line and reference admittance of the ports ($Y_0 = 1/Z_0 = 0.02$ S). Demonstrating this is as simple as obtaining the phase of the reflection coefficient, i.e., [78]

$$\phi_\rho = \arctan\left\{\frac{Y_0\left(C\omega - \frac{1}{L\omega}\right)}{Y_0G - Y_1^2}\right\} - \arctan\left\{\frac{Y_0\left(C\omega - \frac{1}{L\omega}\right)}{Y_0G + Y_1^2}\right\} \quad (2)$$

deriving it, i.e.,

$$\frac{d\phi_\rho}{d\omega} = \frac{\frac{Y_0}{Y_0G - Y_1^2}\left(C + \frac{1}{L\omega^2}\right)}{1 + \left[\frac{Y_0\left(C\omega - \frac{1}{L\omega}\right)}{Y_0G - Y_1^2}\right]^2} - \frac{\frac{Y_0}{Y_0G + Y_1^2}\left(C + \frac{1}{L\omega^2}\right)}{1 + \left[\frac{Y_0\left(C\omega - \frac{1}{L\omega}\right)}{Y_0G + Y_1^2}\right]^2} \quad (3)$$

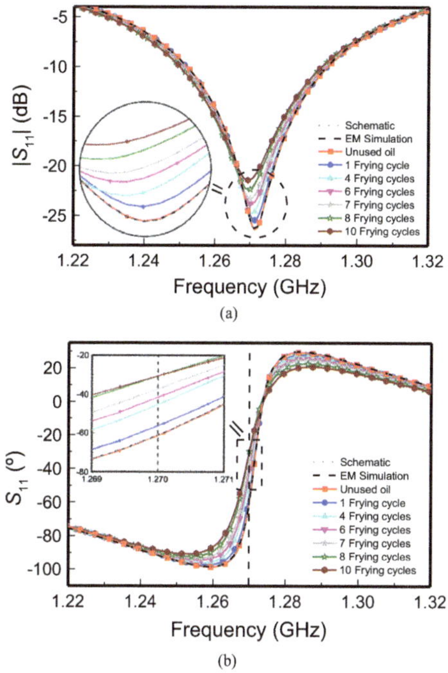

Fig. 23 Response of the sensor of Fig. 22 loaded with unused oil and with oils subjected to different frying cycles (indicated in the figure). **a** Magnitude response; **b** phase response. Note that $S_{11} \equiv \rho$, and hence, $|S_{11}| \equiv |\rho|$and $\angle S_{11} = \phi_\rho$. Reprinted with permission from [78]; copyright 2023 EuMA

and finally evaluating such derivative at the (angular) resonance frequency of the slot resonator, $\omega_0 = 2\pi f_0 = (LC)^{-1/2}$, the operating frequency. This gives

$$\left.\frac{d\phi_\rho}{d\omega}\right|_{\omega_0} = \frac{4Y_0CY_1^2}{Y_0^2G^2 - Y_1^4} \tag{4}$$

It is apparent from (4) that two operation regimes can be discerned. For low losses, i.e., with the conductance G satisfying $Y_0G < Y_1^2$, the phase slope is negative (similar to the lossless case, with $G = 0$). Moreover, the phase of the reflection coefficient at ω_0 is $\phi_\rho = \pm 180°$, since the reflection coefficient at that frequency is $\rho = (Y_0G - Y_1^2)/(Y_0G + Y_1^2)$, i.e., a negative real number. By contrast, for $Y_0G > Y_1^2$ (high-loss regime), the phase slope at ω_0 is positive, and the value of the phase of the reflection coefficient at that frequency is $\phi_\rho = 0°$, since the reflection coefficient is a positive real number. Nevertheless, regardless of the specific operation regime (either low-loss or high-loss), it is clear that if the admittance of the inverter, Y_1, is chosen so that $Y_0G \approx Y_1^2$, the derivative of the phase of the reflection coefficient with frequency at ω_0 is very high. This is a sufficient condition to achieve a strong dependence of the phase of the reflection coefficient at ω_0 (the output variable of the sensor) with the dielectric constant of the sample under test (edible oil in our case), i.e., a high sensitivity (this aspect will be later discussed in detail in Sect. 2.1.2). Nevertheless, let us mention that in the reported prototype example, it has been found that $Y_0G =$

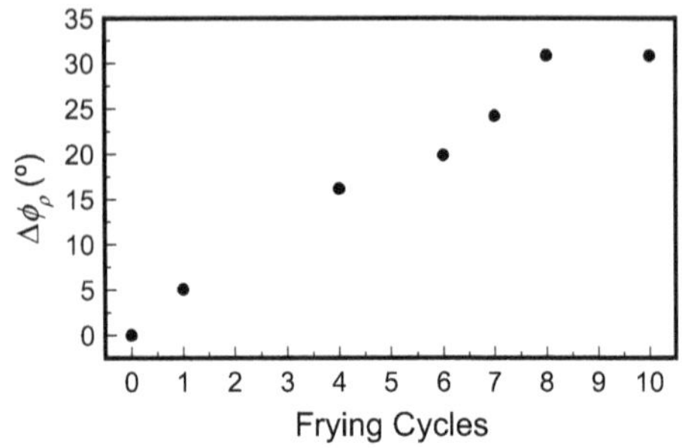

Fig. 24 Dependence of the phase of the reflection coefficient at f_0 with the number of frying cycles for the sensor of Fig. 22. Reprinted with permission from [78]; copyright 2023 EuMA

0.0000488 S^2 for the unused oil, whereas $Y_1^2 = (1/150\ S)^2 = 0.0000444\ S^2$. With such values, $Y_0 G$ is only slightly superior to Y_1^2, , a sufficient condition to achieve a high phase slope and hence a high sensitivity (this justifies the value of the admittance of the inverter).

A protocol for subjecting the oil to different frying cycles (using potatoes as fried element) was proposed in [78]. According to such protocol, the oil was utilized for frying 150 g of fresh potatoes every frying cycle, each one at 185 °C, during 10 min. An oil sample (of roughly 2 mL) was taken and cooled down for 30 min after the different frying cycles for characterization. The measured responses for the different oil samples are also depicted in Fig. 23, where it can be appreciated that the sensor is able to capture the changes experienced by the oil due to its usage. In particular, the phase response at the operating frequency varies significantly, as consequence of the significant phase slope. Figure 24 depicts the phase of the reflection coefficient at f_0, referred to the phase of the unused oil, as a function of the number of frying cycles. The phase varies roughly linearly, and the average sensitivity is found to be 3.22°/cycle. These results point out that the designed and fabricated sensor in [78] can be used to detect the deterioration-related dielectric changes experienced by edible oil when it is subjected to repeated frying cycles.

1.2.5 Differential-Mode Sensors

As indicated before, rather than a working principle, differential sensing is a mode of operation (examples of microwave differential sensors can be found in [89, 91, 92, 178, 190–204]). Since differential-mode sensors are based on a pair of independent (single-ended) sensors, such devices can operate under various working principles, e.g., frequency variation, phase variation, or coupling modulation. The output variables are the corresponding differential variables for each sensor type (differential frequency, phase, or magnitude), whereas the input variables are the differential variables (measurands) present at each independent sensor of the differential pair (e.g., a differential dielectric constant, a differential volume fraction, a differential temperature, or humidity, etc.). Nevertheless, many differential-mode sensors utilize the cross-mode transmission, or reflection, coefficient (depending on whether the sensor works in transmission or in reflection) as the differential output variable, since that variable is differential by nature. Note that if the two sensors of the differential pair are not coupled, the cross-mode transmission or reflection coefficient are simply given by [86, 205]

$$S_{ij}^{cd} = \frac{1}{2}\left(S_{ij,\mathrm{MUT}} - S_{ij,\mathrm{REF}}\right) \tag{5}$$

i.e., proportional to the difference between the transmission or reflection coefficients of each individual sensor. In expression (5), $i = 1, 2$ and $j = 1, 2$ denote the port number, and the sub-indexes MUT or REF are used to differentiate the sensing element (devoted either to the MUT or to the REF sample, or measurand). Usually the magnitude of the cross-mode transmission, or reflection, coefficient at a certain frequency, or at the frequency where that magnitude is a maximum, is the variable used for sensing [191–194]. Nevertheless, in applications where two output variables are needed (e.g., the determination of the complex permittivity of the MUT), the frequency where the cross-mode transmission, or reflection, coefficient is a maximum is also utilized.

In certain reported transmission-mode differential sensors, additional microwave components are introduced in order to reduce the number of accessible ports (from 4 to 2) and transform the differential output variable (e.g., the cross-mode transmission coefficient) to a simpler measurable variable (e.g., the magnitude of the transmission coefficient of the resulting two-port network). A clear example of a two-port transmission-mode differential permittivity sensor of this type can be found in [178]. The device is a phase-variation differential sensor based on a pair of meandered lines, and it uses a pair of race hybrid couplers, which act jointly as a differential-mode to common-mode conversion detector, able to detect imbalances between both sensing elements through the measurement of the transmission coefficient. Another two-port differential sensor that uses a similar principle, but based on CRLH artificial sensing lines is reported in [91]. In reflective-mode differential sensors, it is also possible to transform the typical differential output variable (the cross-mode reflection coefficient) to a transmission coefficient, by connecting the differential input port of the sensor (a composite port consisting of a pair of single-ended ports) to the isolated ports of a rat-race hybrid coupler. As demonstrated, e.g., in [196, 204], the transmission coefficient of the resulting two-port structure is proportional to the cross-mode reflection coefficient, and thus, such transmission coefficient can be used as output variable of the sensor.

As an illustrative example of a differential-mode sensor, let us report the microfluidic sensor presented in [178], based on a pair of meandered ordinary lines. Note that this sensor was included in the reference list of phase-variation sensors (see Sect. 1.2.4), since its working principle is the imbalance in the line pair caused by the difference between the phases of the lines, in turn generated by an asymmetric dielectric load. Differential phase to magnitude conversion was achieved by means of a pair of rat-race hybrid couplers, as anticipated in the preceding paragraph. The photograph of the sensor is depicted in Fig. 25, where the presence of fluidic channels on top of the meander (sensing) lines can be appreciated. It should be mentioned that, in order to avoid liquid absorption by the substrate, a dry film with thickness 50 μm and dielectric constant 3.6 was deposited on the upper side of the substrate. The sensor was applied in [178] to the measurement of the concentration of isopropanol in deionized (DI) water. The REF liquid is thus DI water, whereas the MUT is the mixture of isopropanol/DI water. As demonstrated in [178], with the configuration of Fig. 25, the transmission coefficient between the input and the output port ($S_{1'2}$, following the port designation of [178]), is proportional to the cross-mode transmission coefficient of the pair of sensing lines, i.e.,

$$S_{1'2} = -S_{21}^{cd} \tag{6}$$

In [178], the magnitude of such transmission coefficient at the operating frequency ($f_0 = 2$ GHz) was used as output variable. For measuring purposes, the REF liquid (pure DI water) was injected in the REF channel of line A, and subsequently different mixtures of isopropanol in DI water were injected in the MUT channel (line B), starting with a null volume fraction (corresponding to the REF liquid), and progressively increasing the volume fraction of isopropanol. The responses for the different mixtures are depicted in Fig. 25b. The sensor was able to resolve concentrations of isopropanol in DI water as small as 1%. The dependence of the transmission coefficient at f_0 as a function of the isopropanol content is depicted in Fig. 25c. In this sensor, the maximum sensitivity is roughly 12 dB/%. Let us mention that such high sensitivity was achieved by adjusting the electrical length of the pair of lines at the design frequency to a multiple of 180°, as specified in [178] (particularly, the length was set to 540° in the sensing lines of the device shown in Fig. 25).

2 Strategies for Sensor Performance Optimization

The objective of this section is to briefly report some methods, techniques, and approaches for performance optimization in planar microwave sensors, excluding those strategies based on coupled structures, since such strategies are the subject of the subsequent chapters of this book, and will be studied later in detail. Section 2.1 will focus on different approaches useful to boost up the sensitivity, probably the main sensor performance indicator, whereas Sect. 2.2 will discuss some techniques for resolution (also an important performance parameter) and precision improvement. Such performance optimization strategies are grouped in the surveys [206, 207] and exhaustively detailed in the references therein.

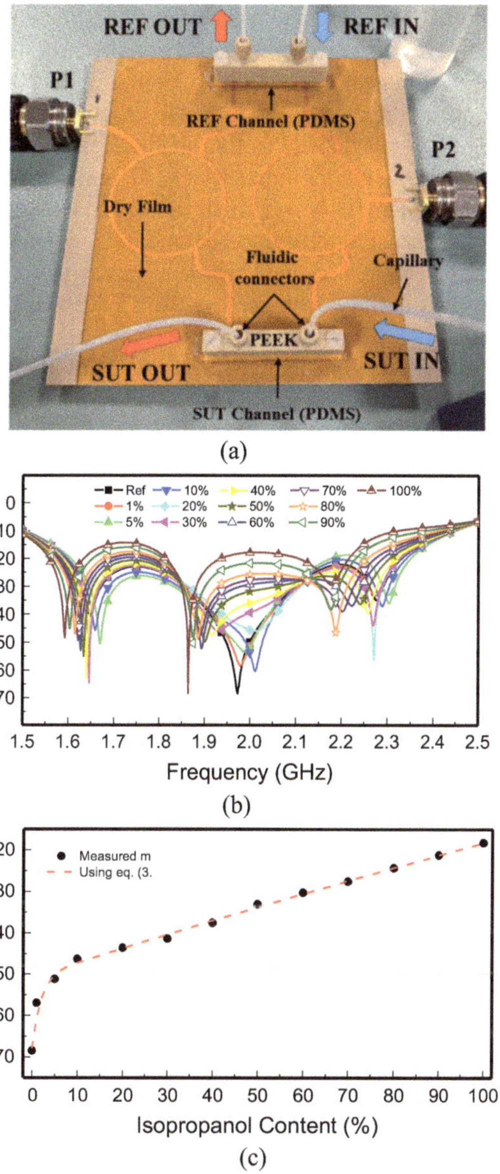

Fig. 25 **a** Photograph of the differential-mode microfluidic sensor based on a pair of meandered lines (present below the channels) and equipped with a pair of hybrid couplers for phase to magnitude conversion; **b** measured transmission coefficient that results by loading line A with the REF liquid (DI water) and line B with MUT samples corresponding to the indicated volume fractions of isopropanol in DI water; **c** value of the transmission coefficient (in dB) at f_0 as a function of the isopropanol content (the dashed line is the calibration curve). The sensor substrate is the *Rogers 4003C* with dielectric constant $\varepsilon_r = 3.55$, thickness $h = 0.8128$ mm, and dissipation factor tan $\delta = 0.0021$. The inner radius of the circularly-shaped couplers is 21.76 mm, whereas the width of the coupler lines is 0.98 mm, corresponding to a characteristic impedance of 70.71Ω. The width of the 50-Ω access lines and meandered lines is 1.81 mm, and the length of the meandered sensing lines is 131.77 mm. Reprinted from [178] with permission; copyright 2020 IEEE

2.1 *Sensitivity Optimization*

Sensitivity is usually considered to be the most relevant performance parameter in planar microwave sensors, or, at least, the main subject of research in planar microwave sensors, with dozens of works devoted to their optimization. The reason is that highly sensitive sensors, able to detect small changes in the measurand, are required in many practical applications. Ideally, a good sensor should exhibit not only a high sensitivity, but also a good resolution, linearity, dynamic range, accuracy, and repeatability. However, the simultaneous optimization of all sensor performance parameters is not possible, in general. For example, sensitivity and resolution, the later defined as the smallest variation of the input signal that can be reliably detected, are intimately related, and sensitivity optimization favors resolution improvement. However, highly sensitive sensors typically exhibit a limited linearity (i.e., a very nonlinear transfer function), which in turn reduces the range of measurable values of the input variable (input dynamic range). Thus, in applications where the measurand is expected to experience significant variations, sensitivity should be sacrificed in favor of linearity. Nevertheless, as indicated above, there are many situations that demand the detection of tiny changes of the measurand in the vicinity of a certain (reference, REF) value. In such cases, sensors exhibiting high sensitivity (and resolution) in the vicinity of that REF value are needed. For example, in the sensor depicted in Fig. 25, the sensitivity is very high in the vicinity of small volume fractions of isopropanol in DI water (the input variable), and it progressively degrades as such volume fraction increases. Such high sensitivity for small perturbations (small volume fractions of isopropanol in DI water) allows for the detection of small quantities of that substance in DI water (of the order of 1% of volume fraction of isopropanol). Similar highly sensitive sensors could be applied, for instance, to the detection of impurities in liquids, manifested as changes in the complex permittivity, or to the measurement of electrolyte content in diluted solutions of DI water [193], or in urine [194] (the latter of practical medical or veterinarian interest). Let us next review some strategies/approaches/techniques aimed to boost up the sensitivity in planar microwave sensors.

2.1.1 Sensitivity Optimization Through Resonator's Topology in Frequency-Variation Sensors

In frequency-variation resonant sensors, sensitivity optimization requires that the variation in the dielectric constant of the MUT generates the maximum possible change in the capacitance of the sensing resonant element. Let us consider frequency-variation sensors based on slotted resonators, for example, complementary split ring resonators (CSRRs), complementary spiral resonators (CSR), open complementary split ring resonators (OCSRRs), or dumbbell defect-ground-structure (DB-DGS) resonators, among others, see Fig. 26. For resonator's excitation, transmission lines are needed. Figure 26 depicts specific configurations for CSRRs, DB-DGS resonators, and OCSRRs, as well as the equivalent circuit models (the configuration and circuit model for the CSR are identical to those of the CSRR and are not shown in the figure). According to Fig. 26, the DB-DGS and the OCSRR are described by parallel resonant tanks, whereas the CSRR is modelled by a resonant tank series-connected to a capacitance (the coupling capacitance with the host line). The CSRR and the DB-DGS generate a notch in the transmission coefficient at the resonance frequency, whereas total transmission (by neglecting the effect of losses) is achieved for the OCSRR-loaded line at resonance. This means that the CSRR- and the DB-DGS-based sensors are stopband (notched) type structures, whereas the OCSRR sensor operates as a bandpass filter.

Although in frequency-variation sensors the magnitude of the resonance notch, or peak, can be used as a second output variable for sensing, e.g., to retrieve the loss factor of the MUT, the canonical sensitivity in such sensor is given by

$$S = \frac{df_0}{d\varepsilon_{MUT}} \tag{7}$$

where f_0 is the resonance frequency and $\varepsilon_{\mathrm{MUT}}$ is the dielectric constant of the MUT. It is obvious that the variation in the resonance frequency generated by a change in the dielectric constant of the MUT should be larger in high frequency sensors. Thus, for a faithful comparison of the sensitivity between different frequency-variation sensors, the relative frequency change is the one that should be considered. Therefore, the relative sensitivity, defined as

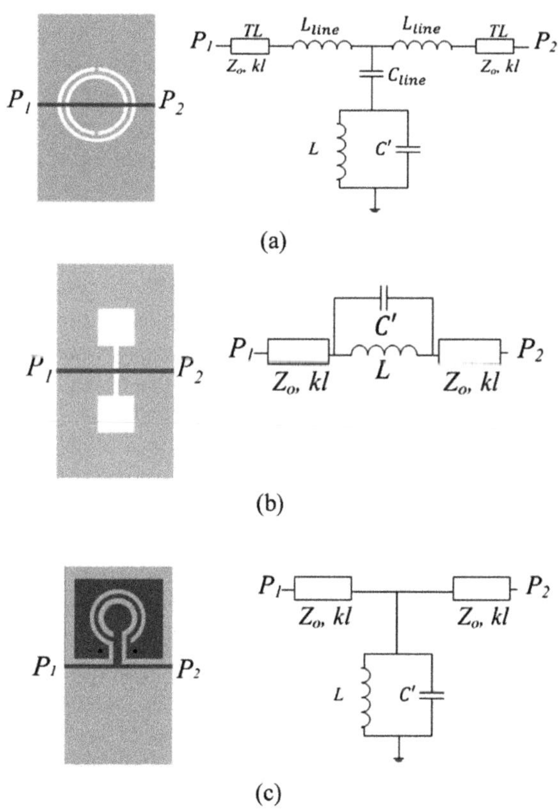

Fig. 26 Typical topology and circuit model of a microstrip line loaded with a CSRR (**a**), DB-DGS resonator (**b**), and OCSRR (**c**). Z_0 and kl are the characteristic impedance and electrical length, respectively, of the host lines at both sides of the resonator (k and l being the phase constant and length, respectively)

$$\overline{S} = \frac{1}{f_0} \cdot \frac{df_0}{d\varepsilon_{MUT}} \tag{8}$$

is the relevant parameter and figure of merit (useful for a realistic comparison) in these frequency-variation sensors. Assuming that the MUT is in the region of influence of the resonant element, a variation in the dielectric constant of the MUT alters the capacitance of such resonant element, and this in turn produces a shift in the resonance frequency. Thus, the relative sensitivity can be further developed and expressed as

$$\overline{S} = \frac{1}{f_0} \cdot \frac{df_0}{dC'} \cdot \frac{dC'}{d\varepsilon_{MUT}} \tag{9}$$

where *C'* is the capacitance of the resonant element loaded with the MUT.

The first derivative in (9) depends on the specific configuration of the excitation line and sensing resonator, i.e., it depends on the analytical dependence of f_0 with *C'*. The second derivative in the relative sensitivity can be analytically obtained provided several hypothesis and approximations are adopted [44]. The first assumption considers the analysis restricted to fully planar resonant sensing elements etched in a single metallic layer and, therefore, exhibiting an edge capacitance.[20] It is also considered that the metal is an ideal (perfect) conductor, with negligible resistivity and thickness (this approximation is very reasonable in the usual metallization present in common microwave substrates, typically made of a high conductivity material, such as Cu, and with thicknesses of the order of few tens of microns). The MUT is considered to be in contact with the resonant element and occupying the whole half-space. Namely, it is considered to be semi-infinite in the vertical direction (i.e., the direction orthogonal to the plane of the resonator), with transverse dimensions extending beyond the region occupied by the resonator. Moreover, the MUT is made of a uniform material. The substrate (also uniform) is sufficiently thick, so that the effects of any potential metallic pattern in the face opposite to the one where the resonator is etched (e.g., the line strip) can be neglected (i.e., the resonance frequency is not altered by the presence of such patterns).

Let us next consider that the sensor is implemented by a slot resonator, e.g., a CSRR, a DB-DGS, or an OCSRR, as depicted in Fig. 26, and let us neglect the effects of the line on the capacitance (and resonance) of the resonator. If the substrate is thick enough and it can also be considered to be semi-infinite in the vertical direction, it follows that the plane of the resonant element is a magnetic wall, and the electric field distribution at both half-spaces is a mirror image [121]. Under these conditions, the electric field in the plane of the resonator is tangential, and the capacitance of the resonator can be separated in two parts, i.e., the capacitances associated to the lower and upper half-spaces (representing the contributions of the MUT and substrate, respectively), both connected in parallel. If we designate by *C* the capacitance of the bare resonator, i.e., without MUT (or surrounded by air), the capacitance *C'* of the

[20] Namely, resonant elements implemented by means of two metallic layers, such as the broadside-coupled split ring resonator (BC-SRR), or the microstrip step-impedance shut stub (SISS) resonator, both exhibiting a broadside capacitance, are excluded in the present analysis.

resonator loaded with an arbitrary MUT can be expressed as [44, 77, 121, 122]

$$C' = C\frac{\varepsilon_r + \varepsilon_{MUT}}{\varepsilon_r + 1} \tag{10}$$

where ε_r is the dielectric constant of the substrate. According to (10), the second derivative in (9) is found to be

$$\frac{dC'}{d\varepsilon_{MUT}} = \frac{C}{\varepsilon_r + 1} \tag{11}$$

Concerning the first derivative in (9), as indicated, it depends on the specific sensor configuration and resonant element. According to the circuit models of Fig. 26, [77, 182, 208–210], for the OCSRR and DB-DGS resonators, the dependence of the resonance frequency on *C'* is simply

$$f_0 = \frac{1}{2\pi\sqrt{LC'}} \tag{12}$$

whereas for the CSRR, the following result is obtained'

$$f_0 = \frac{1}{2\pi\sqrt{L(C_{line} + C')}} \tag{13}$$

where L is the inductance of the resonator, not affected by the MUT, and C_{line} is the capacitance of the line section on top of the CSRR.

Using (13), it follows that

$$\frac{df_0}{dC'} = -\frac{1}{2}\cdot\frac{f_0}{C_{line} + C'} \tag{14}$$

and introducing (14) and (11) in (9), the relative sensitivity for the CSRR-loaded line is found to be

$$\overline{S} = -\frac{1}{2}\cdot\frac{C}{C_{line}(\varepsilon_r + 1) + C(\varepsilon_r + \varepsilon_{MUT})} \tag{15}$$

It can be easily deduced that the relative sensitivity for the OCSRR- and the DB-DGS-loaded microstrip lines is

$$\overline{S} = -\frac{1}{2}\cdot\frac{1}{\varepsilon_r + \varepsilon_{MUT}} \tag{16}$$

i.e., a result inferred from (15) by simply forcing $C_{line} = 0$.

From these results, it can be concluded that the relative sensitivity is superior in OCSRR- and DB-DGS-loaded microstrip lines, as compared to CSRR-loaded lines. Moreover, from (16), it follows that the relative sensitivity in sensors based on OCSRR and DB-DGS resonators does not depend on the geometry of the resonant element. However, note that the assumption of a semi-infinite MUT and substrate has been adopted. Concerning the MUT, it is possible to machine it in order to be thick enough, so that the semi-infinite approximation holds. By contrast, substrate thickness is sometimes determined by external factors that do not depend on the designer. This means that the semi-infinite substrate approximation cannot always be guaranteed. Despite that fact, even for relatively thin substrates, the plane of the resonant element is a quasi-magnetic wall [121]. Under these circumstances, the capacitance of the resonator can still be considered to be formed by the parallel connection of the capacitance of the MUT and the capacitance of the substrate. Such capacitance is given by an expression formally identical to (10), but replacing ε_r with $\varepsilon_{r,eq}$, the equivalent dielectric constant of the substrate, defined as the dielectric constant of a hypothetical semi-infinite substrate providing the same contribution to the resonator's capacitance [121, 211]. This means that, for a finite substrate, the relative sensitivities for the CSRR-loaded line and OCSRR-loaded (or DB-DGS-loaded) line should be rewritten as (15) and (16), but replacing ε_r with $\varepsilon_{r,\mathrm{eq}}$.

The equivalent dielectric constant depends on the ratio between the width of the capacitive slot and the thickness of the substrate [121]. Consequently, for finite substrates, i.e., for substrates not satisfying the semi-infinite approximation, the relative sensitivity in OCSRR- or DB-DGS-loaded microstrip lines, given by (16) with ε_r replaced with $\varepsilon_{r,\mathrm{eq}}$, does depend on the geometry. However, the dependence is soft, especially for relatively thick substrates, as it has been recently demonstrated [121] (note also that, as the substrate thickness increases, the equivalent dielectric constant tends to the nominal dielectric constant of the substrate, or $\varepsilon_{r,\mathrm{eq}} \rightarrow \varepsilon_r$). Nevertheless, it can be concluded that regardless of the specific structure, for sensitivity optimization it is convenient to deal with low dielectric constant substrates. In addition, it can be concluded from (15) and (16) that, as the dielectric constant of the MUT increases, the relative sensitivity decreases. Indeed, for high dielectric constant MUTs, the sensitivity scarcely depends on ε_r (or $\varepsilon_{r,\mathrm{eq}}$), since $\varepsilon_{\mathrm{MUT}}$ obscures the substrate dielectric constant (or equivalent dielectric constant).

Validation of the previous analysis was carried out in [121, 137, 206]. Particularly, in [137], it was demonstrated that the sensitivity in DB-DGS-based frequency-variation sensors is superior to that of the CSRR-based counterpart. Figure 27 depicts the layout of three sensing structures based on CSRR, DB-DGS and OCSRR (substrate characteristics and dimensions are included in the figure). The variation of the resonance frequency as a function of the dielectric constant of the MUT was obtained through electromagnetic simulation, and the relative sensitivity was then inferred from (8). The results, depicted in Fig. 28, reveal that the sensitivities of the DB-DGS- and OCSRR-based sensors are very similar and superior to that of the CSRR sensor, as expected from the previous analysis. The figure also includes the relative sensitivity as inferred from (15), for the CSRR-based sensor, and from (16), for the DB-DGS- and OCSRR-based sensors, and the agreement with the simulated values is good.

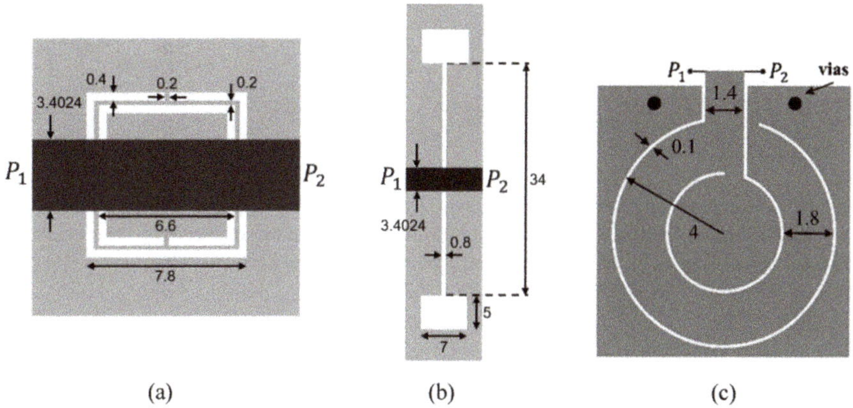

Fig. 27 Specific sensor topologies (layouts) used for validation of the sensitivity analysis. **a** CSRR-based sensor; **b** DB-DGS-based sensor; **c** OCSRR-based sensor. Dimensions are given in mm. The considered substrate is the *Rogers RO4003C* with dielectric constant $\varepsilon_r = 3.55$, thickness $h = 1.524$ mm, and loss tangent $\tan\delta = 0.0023$

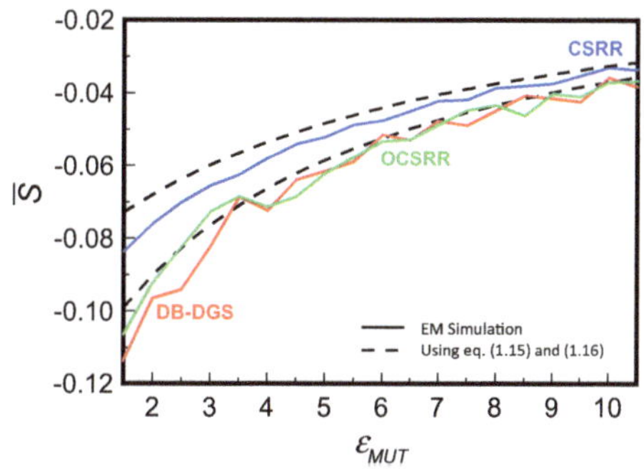

Fig. 28 Dependence of the relative sensitivity on the dielectric constant of the (semi-infinite) MUT for the sensors depicted in Fig. 27. Reprinted with permission from [206]; copyright 2020 MPDI

2.1.2 Sensitivity Optimization Through Cascaded High/Low Impedance Inverters in Reflective-Mode Phase-Variation Sensors

In reflective-mode phase-variation sensors based on transmission lines terminated with a sensing resonator, either distributed or semi-lumped, boosting up the sensitivity requires an adequate design of the resonant element. Moreover, the sensitivity can be significantly enhanced by cascading a set of impedance inverters with alternating high and low impedance to the sensing resonator. Let us first analyze the requirements of the resonator for sensitivity optimization, and then we will demonstrate that the cascade of inverters is beneficial for sensitivity improvement.

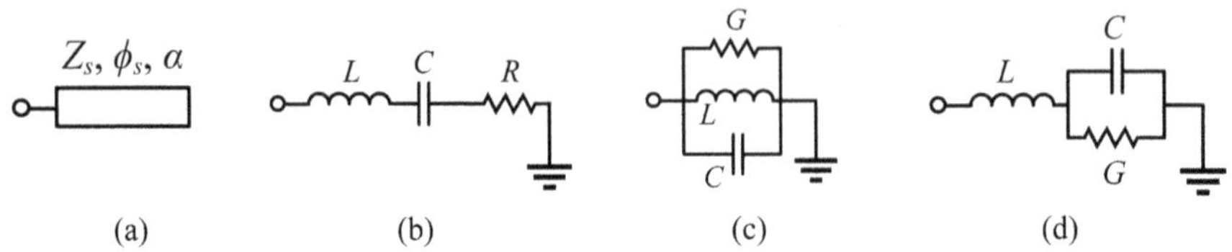

Fig. 29 Schematic of an open-ended line (**a**), lossy series resonator (**b**), lossy parallel resonator (**c**), and series resonator with losses accounted for by a conductance parallel connected to the capacitance (**d**). Losses are included in (**a**) through the attenuation constant, α, whereas in (**b**), (**c**) and (**d**) losses are accounted for by the resistance R or by the conductance G

Resonator Optimization

As it was anticipated in Sect. 1.2.4, the key aspect to achieve a high sensitivity in the limit of small perturbations in the vicinity of the dielectric constant of the REF sample, or material, is to generate a high phase slope in the phase response (phase of the reflection coefficient) at resonance (the operational frequency). To calculate such phase slope, and its dependence with resonator's parameters, let us distinguish between distributed and semi-lumped resonators.

Distributed Resonators

In distributed resonators, either an open-ended quarter-wavelength transmission line (equivalent to a grounded series resonator), or an open-ended half-wavelength transmission line (equivalent to a grounded parallel resonant tank), such phase slope at resonance is controlled by the characteristic impedance of the line, related to the quality factor of the resonator. Let us analyze this aspect in detail, first by excluding the effects of losses (in both the line and the MUT). Figure 29a depicts an open-ended line with characteristic impedance Z_s and electrical length ϕ_s (the sub-indexes denote that the line is used for sensing). The input impedance of such open-ended line by excluding losses is

$$Z_{in} = -jZ_s\cot\phi_s \tag{17}$$

and the reflection coefficient, referred to Z_0, the reference impedance of the port, is

$$\rho_{REF} = \frac{Z_{in} - Z_0}{Z_{in} + Z_0} = \frac{+Z_0 + jZ_s\cot\phi_s}{-Z_0 + jZ_s\cot\phi_s} \tag{18}$$

The phase of the reflection coefficient is thus

$$\phi_\rho = 2\arctan\left(\frac{Z_s\cot\phi_s}{Z_0}\right) + \pi \tag{19}$$

The derivative of the phase with the angular frequency (phase slope) can be expressed as

$$\frac{d\phi_\rho}{d\omega}=\frac{d\phi_\rho}{d\phi_s}\cdot\frac{d\phi_s}{d\omega}=-\frac{2}{\frac{Z_0}{Z_s}\sin^2\phi_s+\frac{Z_s}{Z_0}\cos^2\phi_s}\cdot\frac{l_s}{v_s} \tag{20}$$

where l_s and v_s are the physical length and phase velocity, respectively, of the sensing line [note that the last term in (20) is easily obtained, since $\phi_s=\omega\cdot l_s/v_s$]. It is apparent from (20) that for an open-ended quarter-wavelength (or odd multiple) resonator, with $\phi_s=(2n+1)\cdot\pi/2$ (and $n=0, 1, 2, 3,\ldots$), the phase slope (a negative number) is

$$\left.\frac{d\phi_\rho}{d\omega}\right|_{\phi_s=(2n+1)\cdot\frac{\pi}{2}}=-\frac{2Z_s}{Z_0}\cdot\frac{l_s}{v_s} \tag{21a}$$

and it increases with Z_s. Thus, to boost up the sensitivity, Z_s must be as high as possible, or at least $Z_s > Z_0$. By contrast, for an open-ended half-wavelength (or multiple) resonator, with $\phi_s=n\cdot\pi$, the phase slope is.

$$\left.\frac{d\phi_\rho}{d\omega}\right|_{\phi_s=n\cdot\pi}=-\frac{2Z_0}{Z_s}\cdot\frac{l_s}{v_s} \tag{21b}$$

and it follows that, in this case, Z_s should be low for phase slope and sensitivity enhancement. The previous line impedance requirements ($Z_s > Z_0$ for open-ended quarter-wavelength sensing lines, and $Z_s < Z_0$ for open-ended half-wavelength sensing lines) correspond to high-Q resonators.[21] Therefore, it can be concluded that high-Q distributed resonators are needed to boost up the sensitivity.

The previous analysis, with losses neglected, is valid in many situations. Nevertheless, there are cases where the loss factor of the MUT is significant (e.g., in most liquid samples), and losses must be accounted for. Therefore, let us next calculate the phase slope of the distributed open-ended quarter- and half-wavelength transmission line resonators at resonance by including the effects of losses. In such case, the input impedance (17) should be expressed as

$$Z_{in}=Z_s\coth\gamma l_s \tag{22}$$

where $\gamma=\alpha+j\beta$ is the complex propagation constant, and α and β are the attenuation constant and the phase constant, respectively, of the sensing line.[22] After some simple algebra, (22) can be expressed as

$$Z_{in}=Z_s\frac{\sinh 2\alpha l_s-j\sin 2\phi_s}{\cosh 2\alpha l_s-\cos 2\phi_s} \tag{23}$$

[21] Despite the fact that losses are neglected, the Q-factor is finite.

[22] Note that, for the lossless case, $\alpha=0$ and (22) rewrites as (17).

with $\phi_s = \beta l_s$. The reflection coefficient is thus

$$\rho_{\mathrm{REF}} = \frac{Z_s \frac{\sinh 2\alpha l_s - j \sin 2\phi_s}{\cosh 2\alpha l_s - \cos 2\phi_s} - Z_0}{Z_s \frac{\sinh 2\alpha l_s - j \sin 2\phi_s}{\cosh 2\alpha l_s - \cos 2\phi_s} + Z_0} \tag{24}$$

and it can be rewritten as

$$\rho_{\mathrm{REF}} = \frac{Z_s \sinh 2\alpha l_s - Z_0(\cosh 2\alpha l_s - \cos 2\phi_s) - j Z_s \sin 2\phi_s}{Z_s \sinh 2\alpha l_s + Z_0(\cosh 2\alpha l_s - \cos 2\phi_s) - j Z_s \sin 2\phi_s} \tag{25}$$

where the real and the imaginary parts in the numerator and the denominator have been separated.[23] The phase of the reflection coefficient is hence

$$\begin{aligned} \phi_\rho = {} & \arctan\left(\frac{Z_s \sin 2\phi_s}{Z_0(\cosh 2\alpha l_s - \cos 2\phi_s) - Z_s \sinh 2\alpha l_s}\right) \\ & + \arctan\left(\frac{Z_s \sin 2\phi_s}{Z_0(\cosh 2\alpha l_s - \cos 2\phi_s) + Z_s \sinh 2\alpha l_s}\right) \end{aligned} \tag{26}$$

Calculation of the phase slope (see expression 20, left), and evaluation at the key phases, $\phi_s = (2n + 1)\cdot\pi/2$ and $\phi_s = n\cdot\pi$, namely, corresponding to the two sets of possible resonance frequencies, gives:

$$\left.\frac{d\phi_\rho}{d\omega}\right|_{\phi_s=(2n+1)\cdot\frac{\pi}{2}} = -\frac{2 Z_s Z_0}{Z_0^2 \cosh^2 \alpha l_s - Z_s^2 \sinh^2 \alpha l_s} \cdot \frac{l_s}{v_s} \tag{27a}$$

$$\left.\frac{d\phi_\rho}{d\omega}\right|_{\phi_s=n\cdot\pi} = -\frac{2 Z_s Z_0}{Z_s^2 \cosh^2 \alpha l_s - Z_0^2 \sinh^2 \alpha l_s} \cdot \frac{l_s}{v_s} \tag{27b}$$

For the extreme case where $\alpha = 0$ (lossless line and MUT), (27a) and (27b) coincide with (21a) and (21b), respectively, as expected. In view of (27a, b), two operation regimes can be distinguished, depending on the sign of the denominator. For $\phi_s = (2n + 1)\cdot\pi/2$, the low-loss regime is given by those values of α satisfying $\tanh \alpha l_s < Z_0/Z_s$. . Under this regime, the phase slope is negative, and the magnitude of the slope (its absolute value) increases as Z_0/Z_s decreases. Note that the phase slope goes to infinite in the limiting case where $\tanh \alpha l_s = Z_0/Z_s$. For high losses, with $\tanh \alpha l_s > Z_0/Z_s$, the phase slope is positive, and the magnitude of the slope increases as Z_0/Z_s increases. For $\phi_s = n\cdot\pi$, the low loss regime is dictated by those values of α satisfying $\tanh \alpha l_s < Z_s/Z_0$, and the phase slope (negative) exhibits a magnitude that increases as Z_s/Z_0 decreases. By contrast, for $\tanh \alpha l_s > Z_s/Z_0$

[23] It has been assumed that the characteristic impedance, Z_s, is a real number, a reasonable approximation, despite the presence of losses, necessary to avoid an excessive complex formulation.

(high-loss regime), the phase slope is positive and its absolute value increases as Z_s/Z_0 increases.

Semi-lumped resonators

Let us next analyze the case of semi-lumped sensing resonators. Such resonators can be described by either a series or a parallel resonant tank.[24] Let us directly consider the lossy case, by including a resistance R in the series circuit and a conductance G in the parallel circuit, as depicted in Fig. 29b and c, respectively, the canonical circuits describing lossy resonators. In these circuits, L and C are the inductance and capacitance, respectively, of the semi-lumped resonator.

For the series sensing resonator, the impedance is given by

$$Z_{in} = R + j\left(L\omega - \frac{1}{C\omega}\right) \tag{28}$$

The reflection coefficient is hence

$$\rho_{REF} = \frac{R + j\left(L\omega - \frac{1}{C\omega}\right) - Z_0}{R + j\left(L\omega - \frac{1}{C\omega}\right) + Z_0} \tag{29}$$

and the phase of the reflection coefficient is

$$\phi_\rho = \arctan\left\{\frac{L\omega - \frac{1}{C\omega}}{R - Z_0}\right\} - \arctan\left\{\frac{L\omega - \frac{1}{C\omega}}{R + Z_0}\right\} \tag{30}$$

The phase slope is

$$\frac{d\phi_\rho}{d\omega} = \frac{\frac{1}{R-Z_0}\left(L + \frac{1}{C\omega^2}\right)}{1 + \left[\frac{L\omega - \frac{1}{C\omega}}{R-Z_0}\right]^2} - \frac{\frac{1}{R+Z_0}\left(L + \frac{1}{C\omega^2}\right)}{1 + \left[\frac{L\omega - \frac{1}{C\omega}}{R+Z_0}\right]^2} \tag{31}$$

and evaluation of the phase slope at resonance gives

$$\left.\frac{d\phi_\rho}{d\omega}\right|_{\omega_0} = \frac{4Z_0L}{R^2 - Z_0^2} \tag{32}$$

Similar to the distributed resonators, two operation regimes, depending on the level of losses, can be distinguished. For $R < Z_0$ (low-loss regime), the phase slope is negative, whereas for $R > Z_0$ (high-loss regime), the phase slope is positive. In

[24] A slot resonator, as the one depicted in Fig. 21, is described by a parallel resonant tank. A step-impedance resonator (SIR) is an example of a semi-lumped resonator that can be described by means of a series resonant tank.

both cases, the absolute value of the phase slope increases (proportionally) with L, corresponding to a high-Q resonator. Moreover, the magnitude of the phase slope at resonance increases as R approaches Z_0.

For the parallel resonator, circuit dual of the series resonator, it is convenient to deal with admittances. Thus, the input admittance is:

$$Y_{in} = G + j\left(C\omega - \frac{1}{L\omega}\right) \tag{33}$$

The reflection coefficient is thus

$$\rho_{REF} = \frac{Y_0 - G - j\left(C\omega - \frac{1}{L\omega}\right)}{Y_0 + G + j\left(C\omega - \frac{1}{L\omega}\right)} \tag{34}$$

and the phase of the reflection coefficient is

$$\phi_\rho = \arctan\left\{\frac{C\omega - \frac{1}{L\omega}}{G - Y_0}\right\} - \arctan\left\{\frac{C\omega - \frac{1}{L\omega}}{G + Y_0}\right\} \tag{35}$$

The phase slope is

$$\frac{d\phi_\rho}{d\omega} = \frac{\frac{1}{G-Y_0}\left(C + \frac{1}{L\omega^2}\right)}{1 + \left[\frac{C\omega - \frac{1}{L\omega}}{G-Y_0}\right]^2} - \frac{\frac{1}{G+Y_0}\left(C + \frac{1}{L\omega^2}\right)}{1 + \left[\frac{C\omega - \frac{1}{L\omega}}{G+Y_0}\right]^2} \tag{36}$$

and evaluation of the phase slope at resonance gives

$$\left.\frac{d\phi_\rho}{d\omega}\right|_{\omega_0} = \frac{4Y_0 C}{G^2 - Y_0^2} \tag{37}$$

For the parallel resonator, the low-loss regime occurs for $G < Y_0$, with a negative phase slope at resonance, whereas the sign of the phase slope is positive for $G > Y_0$ (high-loss regime). The phase slope is proportional to the capacitance of the resonator, which should be high for sensitivity optimization. Such high C value, for a fixed resonance frequency, provides a high-Q for the sensing (parallel) resonator. This analysis justifies the meandered topology of the slot resonator (modelled by a parallel resonant tank) in the prototype example reported in Sect. 1.2.4 (see Fig. 22), since such topology increases the resonator's capacitance and hence the Q-factor. Note also that losses tend to boost up the magnitude of the phase slope provided G approaches Y_0.

According to the previous analysis, both the series and the parallel resonator should exhibit a high-Q factor in order to obtain a high phase slope (in absolute value) at resonance, and hence a high sensitivity. It is worth mentioning that expressions (32) and (37) are formally identical, provided the pairs of elements L and C, and R and G, as well as Z_0 and Y_0, are interchanged, a consequence of duality [i.e., the structures of Fig. 29b and c are circuit duals]. Nevertheless, concerning the series resonators, though the usual form of including losses is as depicted in Fig. 29b, MUT losses are better accounted for by parallel connecting a conductance to the capacitance, as depicted in the circuit of Fig. 29d. For example, if an SIR is used as sensing resonant element, and a MUT is placed on top of it, the dielectric constant of the MUT will modify the capacitance of the SIR. However, the loss tangent of such material will not significantly affect the series resistance (mainly related to the ohmic losses of the inductive strip and expected to be very small). The loss factor of the MUT adds to the loss factor of the substrate (small if low-loss substrates are considered), and hence, losses in the MUT should be actually accounted for by parallel connecting a conductance to the SIR capacitance. In the circuit of Fig. 29d, the input impedance can be expressed as

$$Z_{in} = jL\omega + \frac{1}{jC\omega + G} = j\left(L\omega - \frac{C\omega}{G^2 + C^2\omega^2}\right) + \frac{G}{G^2 + C^2\omega^2} \tag{38}$$

The reflection coefficient is thus

$$\rho_{\text{REF}} = \frac{j\left(L\omega - \frac{C\omega}{G^2+C^2\omega^2}\right) + \frac{G}{G^2+C^2\omega^2} - Z_0}{j\left(L\omega - \frac{C\omega}{G^2+C^2\omega^2}\right) + \frac{G}{G^2+C^2\omega^2} + Z_0} \tag{39}$$

and the phase of the reflection coefficient is

$$\phi_\rho = \arctan\left\{\frac{L\omega - \frac{C\omega}{G^2+C^2\omega^2}}{\frac{G}{G^2+C^2\omega^2} - Z_0}\right\} - \arctan\left\{\frac{L\omega - \frac{C\omega}{G^2+C^2\omega^2}}{\frac{G}{G^2+C^2\omega^2} + Z_0}\right\} \tag{40}$$

The calculation of the phase slope provides a cumbersome expression (not shown). Nevertheless, evaluation of such phase slope at the resonance frequency (where the imaginary part of the input impedance nulls) gives:

$$\left.\frac{d\phi_\rho}{d\omega}\right|_{\omega_0} = \frac{\frac{4L}{Z_0}\left(1 - \frac{L}{C}G^2\right)}{\frac{G^2L^2}{Z_0^2C^2} - 1} \tag{41}$$

the resonance (angular) frequency being:

$$\omega_0 = \sqrt{\frac{1}{LC} - \frac{G^2}{C^2}} \tag{42}$$

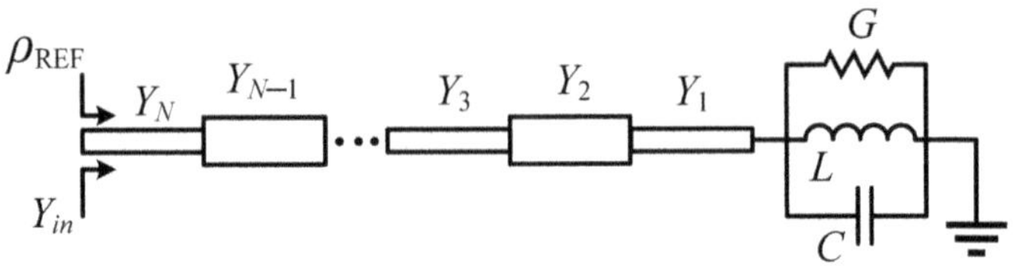

Fig. 30 Schematic of a cascade of high/low admittance inverters terminated with a parallel resonator

As expected, expression (41) is identical to expression (32) when losses are negligible ($R = 0$ and $G = 0$).

According to (41), the analysis of the effects of losses is not so simple, since that expression exhibits a pole and a zero at $G_p = Z_0C/L$ and $G_z = \sqrt{C/L}$, respectively. Nevertheless, it can be demonstrated that for conductances smaller than both the zero and the pole (or $G < \min\{G_p, G_z\}$), low-loss regime, the phase slope is negative, and its magnitude increases with L. Moreover, if $G_p < G_z$ (or, equivalently, $L/C > Z_0^2$), the magnitude of the phase slope in the low-loss regime increases with G. By contrast, for $G_p > G_z$ (or $L/C < Z_0^2$), the magnitude of the phase slope in the low-loss regime decreases with G. Let us mention that the usual situation in the lossy series resonator of Fig. 29d should be $L/C > Z_0^2$, since this enhances the phase slope and hence the sensitivity. Therefore, as far as the structure operates under low-loss regime, losses tend to optimize the phase slope and the sensitivity.

Sensitivity Optimization by Means of High/Low Impedance Inverters

The phase slope can be further enhanced by cascading a set of impedance inverters to the sensing resonator, either distributed or semi-lumped. Let us consider next the specific case of a sensing resonator that can be described by a parallel resonant tank, e.g., the slot resonator of the example of Fig. 22, and let us assume that the resonator is the termination of a cascade of N admittance inverters with admittance Y_i, where the sub-index i (with $i = 1, 2, \ldots N$) indicates the inverter position (see Fig. 30). The input admittance can be expressed as

$$Y_{in} = \left\{G + j\left(C\omega - \frac{1}{L\omega}\right)\right\}^{(-1)^N} \cdot \prod_{i=1}^{N} Y_i^{2(-1)^{i+N}} \tag{43}$$

For N even, the reflection coefficient can be expressed as

$$\rho_{REF} = \frac{Y_0 - \left[G + j\left(C\omega - \frac{1}{L\omega}\right)\right] \cdot \prod_{i=1}^{N} Y_i^{2(-1)^{i+N}}}{Y_0 + \left[G + j\left(C\omega - \frac{1}{L\omega}\right)\right] \cdot \prod_{i=1}^{N} Y_i^{2(-1)^{i+N}}} \tag{44}$$

and the phase of the reflection coefficient is thus

$$\phi_\rho = \arctan\left\{\frac{C\omega - \frac{1}{L\omega}}{G - \frac{Y_0}{\prod_{i=1}^{N} Y_i^{2(-1)^{i+N}}}}\right\} - \arctan\left\{\frac{C\omega - \frac{1}{L\omega}}{G + \frac{Y_0}{\prod_{i=1}^{N} Y_i^{2(-1)^{i+N}}}}\right\} \tag{45}$$

Note that (45) is formally identical to (35). Indeed, both expressions are identical if Y_0 is replaced with $Y_0/\prod_{i=1}^{N} Y_i^{2(-1)^{i+N}}$ in (35). Therefore, the phase slope evaluated at the resonance frequency should be given by (37), provided an identical substitution is made, namely

$$\left.\frac{d\phi_\rho}{d\omega}\right|_{\omega_0} = \frac{4C\frac{Y_0}{\prod_{i=1}^{N} Y_i^{2(-1)^{i+N}}}}{G^2 - \left(\frac{Y_0}{\prod_{i=1}^{N} Y_i^{2(-1)^{i+N}}}\right)^2} \tag{46}$$

For N odd, the reflection coefficient is

$$\rho_{REF} = \frac{Y_0 - \frac{\prod_{i=1}^{N} Y_i^{2(-1)^{i+N}}}{[G+j(C\omega - \frac{1}{L\omega})]}}{Y_0 + \frac{\prod_{i=1}^{N} Y_i^{2(-1)^{i+N}}}{[G+j(C\omega - \frac{1}{L\omega})]}} = \frac{Y_0 \cdot \left[G + j\left(C\omega - \frac{1}{L\omega}\right)\right] - \prod_{i=1}^{N} Y_i^{2(-1)^{i+N}}}{Y_0 \cdot \left[G + j\left(C\omega - \frac{1}{L\omega}\right)\right] + \prod_{i=1}^{N} Y_i^{2(-1)^{i+N}}} \tag{47}$$

and its phase is

$$\phi_\rho = \arctan\left\{\frac{C\omega - \frac{1}{L\omega}}{G - \frac{\prod_{i=1}^{N} Y_i^{2(-1)^{i+N}}}{Y_0}}\right\} - \arctan\left\{\frac{C\omega - \frac{1}{L\omega}}{G + \frac{\prod_{i=1}^{N} Y_i^{2(-1)^{i+N}}}{Y_0}}\right\} \tag{48}$$

i.e., an expression identical to (45) by replacing $Y_0/\prod_{i=1}^{N} Y_i^{2(-1)^{i+N}}$ with $\prod_{i=1}^{N} Y_i^{2(-1)^{i+N}}/Y_0$. Thus, the phase slope evaluated at the resonance frequency is simply

$$\left.\frac{d\phi_\rho}{d\omega}\right|_{\omega_0} = \frac{4C\frac{\prod_{i=1}^{N} Y_i^{2(-1)^{i+N}}}{Y_0}}{G^2 - \left(\frac{\prod_{i=1}^{N} Y_i^{2(-1)^{i+N}}}{Y_0}\right)^2} \tag{49}$$

The low-loss and the high-loss regime is given by those values of G providing a negative and positive sign, respectively, in the denominators of (46), for N even, and (49), for N odd. For the low-loss regime, the usual one (except for MUTs exhibiting significant loss factors, e.g., liquids), the absolute value of the phase slope (of negative sign) increases as $Y_0/\prod_{i=1}^{N} Y_i^{2(-1)^{i+N}}$ decreases, for N even, or increases, for N odd. In both cases, for phase slope (absolute value) enhancement, the admittance of the inverter adjacent to the sensing resonator must be low (i.e., $Y_1 < Y_0$), and then, the admittances of the subsequent cascaded inverters (if they are present) must alternate, i.e., $Y_2 > Y_0$, $Y_3 < Y_0$, and so on. For the high-loss regime, the absolute value of the phase slope (positive) increases as $Y_0/\prod_{i=1}^{N} Y_i^{2(-1)^{i+N}}$ increases, for N even, or decreases, for N odd. This means that, under this high-loss operational mode, the admittances of the inverters must alternate with $Y_1 > Y_0$, $Y_2 < Y_0$, $Y_3 > Y_0$, and so on (i.e., oppositely to the low-loss regime). Nevertheless, it is important to mention that when choosing the admittances of the inverters, we should avoid an excessively high slope, in order to prevent from a change of regime when the dielectric properties of the MUT change.

The prototype example of a reflective-mode phase-variation sensor reported in Sect. 1.2.4, devoted to detect the deterioration of sunflower edible oil by usage, constitutes a good example to clarify the effects of the admittance inverters on sensitivity optimization. In that example, it was shown that a single inverter ($N = 1$) suffices to achieve a high sensitivity, since the denominator of (49), an expression that coincides with (4) when $N = 1$, is very small (and positive, corresponding to sensor operation under the high-loss regime).

The calculation of the exact expression of the sensitivity in the limit of small perturbations for the structure of Fig. 30, with an arbitrary number of admittance inverters, is provided in [184], where further examples are reported. Note that the inverters in the device of Fig. 30, and those of the prototype examples that can be found in [184], have been implemented by means of quarter-wavelength transmission lines. This means that when the frequency deviates from the one corresponding to a quarter-wavelength line length (which is also the resonance frequency of the resonator), the reflection coefficients given by (44) and (47) are no longer valid, and neither are their phases. Consequently, the phase slopes are also invalid. Nevertheless, since the sensor operates at a single frequency, the one of the sensing resonator loaded with the REF sample (and the one that provides the inverter functionality to the line section/s), this aspect does not represent an issue.

To end this subsection, let us mention that an analysis similar to the previous one, focused on a cascade of admittance inverters terminated with a parallel resonant tank, can be applied to the sensor based on a cascade of admittance inverters terminated with a sensing resonant element describable by a series resonant tank. In this case, for the low-loss regime, the admittance of the inverter adjacent to the resonator should be high and then the admittances of the cascaded inverters should alternately be low and high, and so on, contrarily to the high-loss regime. Taking into account that an open-ended quarter-wavelength transmission line is equivalent to a series resonator, and an open-ended half-wavelength transmission line behaves as a parallel resonator, the same conclusions (in regard to the admittances of the inverters) that apply to the series and parallel semi-lumped resonators do also apply to their distributed counterparts. An exhaustive sensitivity analysis for the low-loss regime for distributed and semi-lumped sensing resonators can be found in [44, 179–183].

2.1.3 Sensitivity Optimization Through Embedded Channels

In the previous methods to enhance the sensitivity, detailed in Sects. 2.1.1 and 2.1.2, the key aspect was the resonator's topology and electrical characteristics (*Q*-factor). Moreover, in reflective-mode phase-variation sensors, introducing additional microwave circuitry (a cascade of high/low admittance inverters) contributes to boost up the sensitivity. Nevertheless, a critical aspect with direct influence on sensor sensitivity is the interaction of the MUT, or analyte, with the electromagnetic field generated by the sensing element (e.g., a planar resonator in resonant-type sensors, or a transmission line section in non-resonant sensors). In most planar sensors devoted to permittivity measurements (and related variables), the MUT is placed on top of the sensing element (either in contact or contactless, as discussed in Sect. 1.1.1). If the electric field generated by such sensing element is mostly circumscribed within the sensor's substrate, it follows that the MUT does not significantly perturb such field, resulting in a limited sensitivity. This occurs, for instance, in a microstrip transmission line section, or in a step-impedance resonator (SIR) loading a microstrip (host) line, where the MUT essentially alters the so-called fringing fields. By contrast, in other sensing elements, such as coplanar waveguide (CPW) transmission line sections, or slotted resonators (e.g., the CSRR or the DB-DGS), the plane of the line/resonator is a quasi-magnetic wall, and the field lines generated by such line/resonator are roughly a specular image with regard to that plane.[25] Under these conditions, the MUT significantly interacts with the overall electric field, and the sensitivity of the sensor can (potentially) be higher than that of microstrip-based sensing elements. Figure 31 depicts the top and cross-sectional views of a microstrip transmission line loaded with a folded DB-DGS resonator etched in the ground plane, where the electric field lines are sketched. According to such figure, it is apparent that the MUT interacts more strongly with the electric field of the folded slotted resonator than with the field generated by the line.

In closed lines, such as the strip line, or in substrate integrated waveguides (SIW), placement of the MUT on top (or bottom) of the line for sensing is reasonable as far as a slotted (sensing) resonator is etched in the metal layer in contact to (or in close proximity with) the MUT, so that part of the field lines is allowed to interact with the MUT. One solution to widespread the sensing possibilities in closed lines (strip lines or SIWs), or to improve the sensitivity in microstrip-based sensing structures, with most electric field lines circumscribed within the substrate, is embedding the MUT in the substrate. This is possible with modern micromachining technologies. Particularly, it is possible to implement planar sensors in microstrip, strip line, or SIW technology, devoted to liquid characterization, with embedded channels, or micropipes, containing the liquid under study [see Fig. 1b]. As compared to placement of those channels on top of the substrate [Fig. 1a], channel embedding represents a higher cost solution, but with improved sensitivity, since the MUT liquid can

[25] The higher the ratio between the thickness of the substrate and the width of the slots (of the resonant elements or CPW line), the closer the electric field distribution to a mirror image with regard to the plane of the sensing element [121].

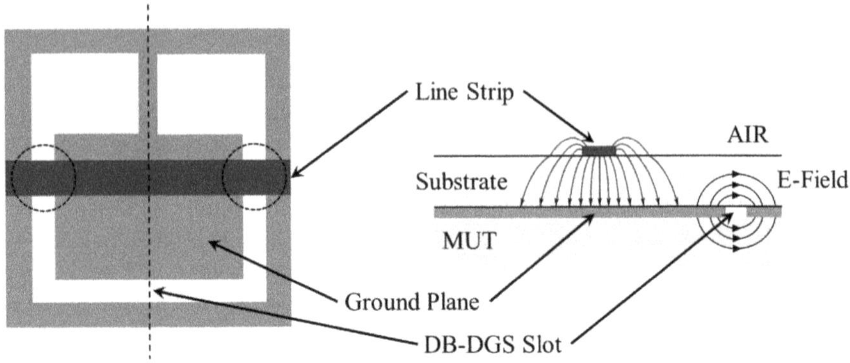

Fig. 31 Sketch of the top view and cross-sectional view (corresponding to the indicated dashed plane) of a microstrip line loaded with a folded DB-DGS resonator etched in the ground plane. The electric field lines are indicated. The upper-level metallization (strip of the line) is depicted in dark grey, whereas the light grey color corresponds to the lower-level metallization (ground plane). Reprinted with permission from [44], copyright 2023 Wiley

substantially interact with the electric field generated by the sensing element. Another possibility is the one depicted in Fig. 1c, with the pipe crossing vertically the substrate. Examples of planar sensors with embedded channel can be found in [84, 110, 212], whereas the works [76, 85, 213] present sensors with channels crossing vertically the substrate.

To illustrate the effects of channel embedding in the substrate on sensor sensitivity, Fig. 32 depicts a pair of transmission-mode frequency-variation sensors based on a SRR (the sensing element). In both sensors, the SRR is coupled through a narrow gap to the input and output access lines. Thus, the sensors operate as bandpass structures, and the output signal is the peak frequency in the frequency response (resonance frequency). The difference between both sensors concerns the location of the channel, i.e., on top of the sensing element in one case, and embedded in the substrate in the other case. In the response of both sensors to different (hypothetical) MUTs inside the channel (with dielectric constant variation in the range between $\varepsilon_{\text{MUT}} = 1$ and $\varepsilon_{\text{MUT}} = 2.3$), it can be appreciated that the resonance varies more strongly in the sensor with embedded channel. Hence, the sensitivity is superior in that sensor.

2.1.4 Sensitivity Optimization Using Functional Films

Many planar microwave sensors include a sensing material (in the form of a thin layer, or film) on top of the sensitive element, in order to enhance the sensitivity of the device to the input stimulus. The use of this type of films (called functional, or smart, films) is necessary for the measurement of certain variables that generate very small changes in the electrical characteristics of the sensing element (resonators or transmission line sections). For example, changes in ambient factors, such as

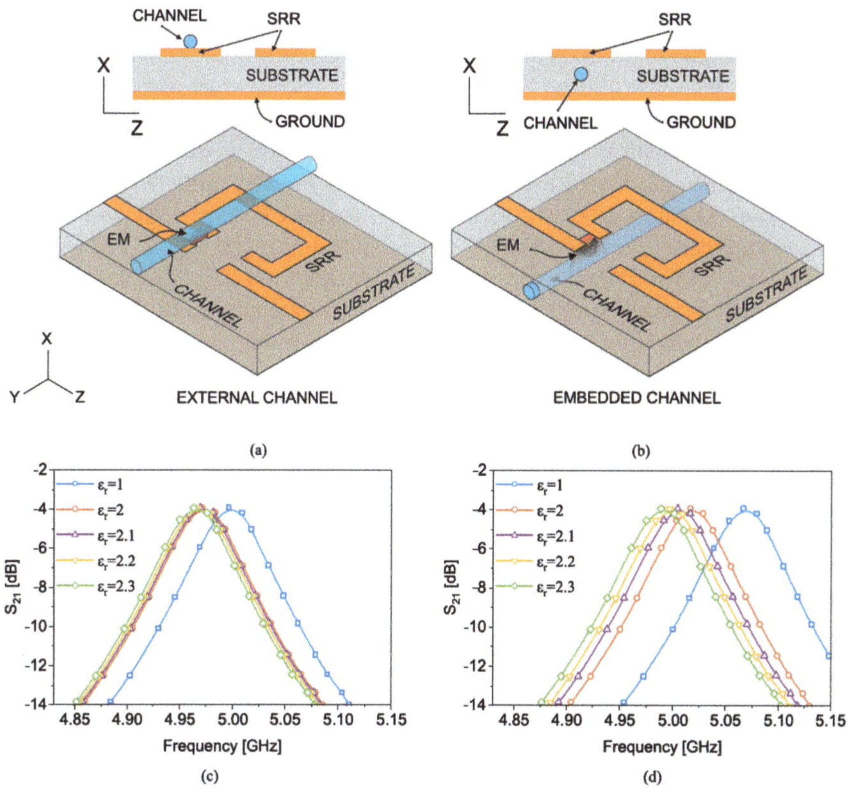

Fig. 32 Sensor with channel on top of the substrate (**a**) and embedded in the substrate (**b**), and frequency responses inferred from electromagnetic simulation (c) and (d), respectively, for various dielectric constant of the MUT inside the channel. The considered substrate is a 0.79 mm-thick *Rogers RO5880* with dielectric constant $\varepsilon_r = 2.2$ and loss factor tan $\delta = 0.0009$. Dimensions can be found in [110]. Figures (**c**) and (**d**) are reprinted with permission from [110]; copyright 2016 IEEE

temperature or relative humidity, do not significantly modify the resonance frequency of a planar resonator (either distributed or semi-lumped), or the electrical length of a transmission line (the reason is that such variables do not alter significantly the dielectric constant of the "environment" surrounding the sensor). However, there are materials that exhibit a high dependence of the permittivity, or conductivity, with such ambient variables. Therefore, coating the sensing element of the sensor with a certain specific material (or functional film), highly sensitive to the measurand (either temperature or humidity), provides a path for sensitivity optimization.

An exhaustive overview of the physical and electromagnetic properties of "smart" materials, as well as their transduction mechanisms involved in sensing, is out of the scope of this subsection (for that purpose, the authors recommend the sources [38, 214–216], and references therein). However, let us succinctly present some of the most common "smart" sensing materials useful for measuring environmental variables (as well as other measurands), indicating the stimuli to which they are sensitive and how such stimuli affect the electromagnetic properties of the materials.

Probably, the most utilized "smart" materials in planar microwave sensors are polymers [217, 218]. For example, polyamide thin films exhibit a highly dependent dielectric constant with temperature [219]. Thus, polyamide-based sensors have been applied to temperature measurements [51, 220]. However, most polymers, especially those exhibiting hygroscopic properties, such as Polyvinyl Alcohol (PVA), have been applied to the measurement of ambient humidity [50, 52, 221–224]. Other polymers useful for humidity measurements are Kapton [225, 226] and Poly (2-Hydroxyethyl Methacrylate) (PHEMA) [227]. There are also multi-sensitive polymers, i.e., materials whose properties are sensitive to various stimuli simultaneously, including not only ambient variables, but also chemical agents or gases. Examples include Polyetherurethane (PEUT) [228] (temperature and humidity), PEDOT:PSS (temperature, humidity, and pH), Polydimethylsiloxane (PDMS) (pH and pressure) [229, 230], Polymethyl Methacrylate (PMMA) (humidity and gas sensing) [231], or cellulose (humidity and chemical sensing) [232–236].

Nanostructures do also constitute a powerful sensing platform for many different applications, including bio-sensing, electrochemical sensing, and gas detection, among others [237–241]. The reason is their high sensitivity and selectivity, long-term reliability and stability, low-temperature operation, and capability for integration of arrays on a massive scale. There are many reported examples of microwave sensors, particularly RFID sensors, based on nanostructures, including nanostructured metallic oxides, for instance SnO_2 [243], silicon nanowires [244–247], and carbon nanotubes (CNTs) [22, 248–259]. Nevertheless, due to their very rich mechanical, electrical and dielectric properties, CNTs constitute one of the most exhaustively used nanostructures in microwave and RFID sensing. Both conductance and capacitance changes have been exploited in CNT-based sensors. Applications mainly include humidity, temperature, and mechanical strain sensing, as well as gas detection (in this case, selectivity to specific substances can be achieved by doping or by using functional groups). Further details on the underlying physics behind the transduction mechanisms and the specific processes for the preparation of the sensing structures are out of the scope of this chapter but can be found in [240, 260].

Graphene is another "smart" material that exhibits extraordinary chemical, physical and electrical properties [261]. Similar to CNTs, it can be functionalized or doped in diverse ways, in order to enhance its sensing properties. The reported sensing applications of graphene and its derivatives, i.e., graphene oxide (GO) and reduced graphene oxide (rGO), are innumerable, as reported in the review paper [262], and are out the scope of this subsection. Nevertheless, let us mention some examples of sensors based on graphene. For example, in [263], a RFID-sensor system that uses Pt-decorated reduced graphene oxide (Pt-rGO), spin-coated in the tag antenna, was applied to the detection of hydrogen gas. In [264, 265], the authors report a RFID GO sensor for the monitoring and discrimination of breath anomalies, utilizing the moisture emitted during inhalations and exhalations, and its effect on the DC resistance of a GO-based electrode. In [266], the dependence of the dielectric constant of GO with the humidity level was used to implement a battery-free RFID humidity sensor by coating a printed graphene antenna with a GO layer. An interesting survey of the up-to-date progress in graphene-based RF sensing platforms, including RFID sensors, can be found in [267] (see also the references therein for further details on the reported applications).

To end this subsection, let us indicate that further references relative to "smart" materials for sensing can be found in various review papers devoted to RFID sensors, e.g., [215, 268–270].

2.1.5 Sensitivity Enhancement Through Resonator Pattern Optimization

Even though in Sect. 2.1.1 it was shown that the geometry of certain slot resonators (e.g., OCSRRs or DB-DGS resonators) does not have influence on the relative sensitivity of frequency-variation sensors (an assertion valid under restrictive conditions discussed in that subsection), in general, the shape and dimensions of the sensing resonant element, or inclusion, have certain impact on sensitivity. There are well-known optimization methods able to provide the optimum geometry of a certain planar component (e.g., a semi-lumped resonator, or even a complex device, such as a filter) in order to achieve a certain required goal. Among such methods, space mapping and its variants (e.g., aggressive space mapping) [271–274], based on a quasi-Newton iterative approach, have been demonstrated to be effective in providing the designs that meet the target requirements (for example, a filter response with an in-band insertion loss below a certain predefined limit). In space mapping-based methods, a certain component topology, with specific dimensions, is considered as a seed of the algorithm, and optimization is based on iterative modifications of the geometry (dimensions) till the goal is achieved. The final (optimum) design might exhibit dimensions rather different than those of the initial pattern (seed), but the topology of the element under optimization is essentially the same. In other words, optimization represents a change in the dimensions of the component, but not in its shape and topology. In many applications, this geometry variation may suffice to achieve the desired goal. However, in many optimization problems, not necessarily the considered element topology is the best one in order to optimize a certain parameter. For example, the geometry of a certain planar resonator, e.g., a CSRR, can be optimized to maximize the sensitivity of a planar frequency-variation sensor based on it. However, the CSRR might not be the best choice for the implementation of the sensor with optimized sensitivity. Indeed, the optimum topology might not correspond to any identifiable element, but rather exhibit an arbitrary shape.

There are methods where the target is to optimize the shape of the element under consideration (rather than its dimensions departing from a certain topology) in order to obtain the best possible performance. In such methods, the optimization procedure is based on making a decision as to which parts of a certain pattering area are covered with metal and which parts are not (etched). Such pattern optimization methods have been applied to the design of a variety of planar components [275–285], and recently to the design of highly sensitive planar sensors [286–288]. The idea behind the pattern optimization method applied to planar sensors is to pixelate the sensing area, and then, by maximizing the sensitivity as a goal, apply a binary optimization algorithm to determine whether each pixel should be metalized or not. The outcome of the optimization is expected to be a pixelated pattern yielding the maximum possible sensitivity.

In [286], a highly sensitive sensor for the accurate characterization of dielectric materials was generated through pattern optimization. A square of size 10.2 mm × 10.2 mm in the middle of the ground plane of a microstrip line was considered as the sensing area (where the resonator to be optimized was etched) and was pixelated into 17 × 17 pixels. Then, a binary value of 1 or 0 was assigned to each pixel, so that each pixel was represented as a bit, where the value of 1 or 0 indicates the presence or absence, respectively, of metal on the area of the pixel. In such a way, a string of bits represents the shape of the resonator. The authors of [286] applied the binary version of the so-called particle swarm optimization (PSO), that is, the BPSO [289, 290], to this string (where each bit is an optimization parameter), to optimize the shape of the resonator for maximum sensitivity. The choice of the initial value of the optimization parameters strongly impacts the convergence of the optimization method. Thus, to further accelerate the convergence, the initial values of the bits were set such that the initial shape of the resonator was the one depicted in Fig. 33 (left inset). The process of optimizing the shape of the sensor was performed by writing a code in MATLAB and linking it to an electromagnetic solver (particularly *CST Microwave Studio*). After setting the initial parameters and generating binary strings, each string is converted to a pattern which can be simulated in CST. In CST, each sensor (with a specific pattern) is simulated by considered both cases, i.e., the bare sensor and the sensor loaded with a MUT with a predetermined dielectric constant. Then the results of the full-wave EM simulations are fed backed to MATLAB. Using the received transmission coefficients, the normalized resonance frequency shift of the loaded sensor (with respect to the unloaded one) is determined and the cost function is calculated as the inverse of that frequency shift. Figure 33 shows the values of the cost function during the iterations of the algorithm indicating the decrease in the cost function from 2.78 (for the initial shape of the sensor) to 2.17 (for the final shape), or equivalently, the resonance frequency shift was increased from 0.36 to 0.46. The pattern of the optimized sensor is shown in the right inset of Fig. 33. The details of the MATLAB program (BPSO algorithm) to generate the pixelated layout is out of the scope of this book (see [286] for further details).

Figure 34 demonstrates the sensitivity improvement when using the generated pixelated structure, as compared to the initial seed and the simple square shaped CSRR (additional traces on that figure correspond to other designs reported by other authors, included in [286], but not discussed in this section). The sensor based on the generated pixelated layout was experimentally validated in [286]. Other sensors based on pattern optimization were presented in [287, 288]. The authors recommend such references to those readers willing to gain further insight on these pattern optimization techniques.

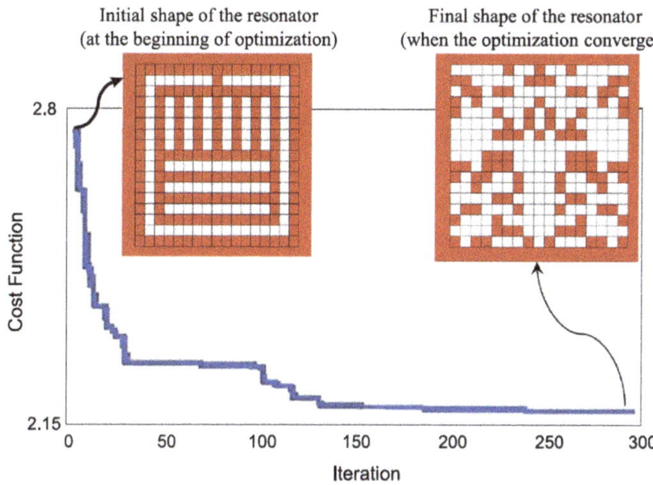

(a)

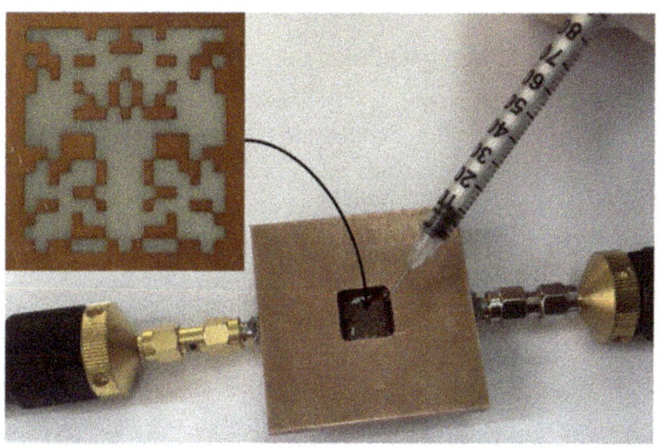

(b)

Fig. 33 Evolution of the cost function by application of the iterative BPSO algorithm in [286] (**a**) and photograph of the fabricated sensor (**b**). The considered substrate is the *Rogers RO4003C* laminate with dielectric constant $\varepsilon_r = 3.55$, loss tangent $\tan\delta = 0.0027$, and thickness $h = 0.5$ mm. The length and width of the host line (unaltered during the optimization process) are 50 mm and 1.1 mm, respectively. Reprinted with permission from [286]; copyright 2019 Springer Nature

2.1.6 Sensitivity Optimization Through Intermodulation Products

The nonlinearity of certain active components, such as transistors or diodes, can be applied to sensitivity enhancement in frequency-variation planar resonant sensors [291]. The idea behind this approach is based on the fact that, through the intermodulation products (IMP) of two tones, achieved thanks to transistor's nonlinearity (Fig. 35), the frequency variation caused by changes in the dielectric constant of the MUT can be expanded as compared to the one of a conventional sensor, thereby increasing the sensitivity. At least one of such tones should be due to the signal

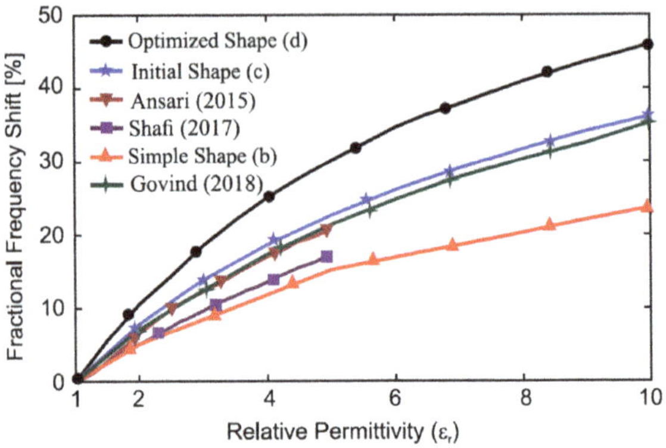

Fig. 34 Fractional frequency shift as a function of the dielectric constant of the MUT for the designed pixelated sensor in [286], and other sensors, including the one based on the initial seed (shape) of the sensing element, the one based on a simple square shaped CSRR, and other sensors reported by other authors. Reprinted with permission from [286]; copyright 2019 Springer Nature

coming from the sensing element, whereas the other can be arbitrarily fixed. Such intermodulation products can be realized by means of regenerative active resonators, which exhibit an inherently mixing capability. The mixing behavior is obtained as follows. One of the signals (tone), with frequency f_1, comes from a signal generator, whereas the second signal is generated from a feedback oscillator with the bandpass filter that determines the oscillation frequency f_2 (see Fig. 35). Due to the nonlinearity of the transistor, harmonics of the two tones, as well as intermodulation products, are generated as indicated in Fig. 35. Some of the IMP components are located at frequencies $2f_1 - f_2 = f_1 - \Delta f$, $3f_1 - 2f_2 = f_1 - 2\Delta f$, $4f_1 - 3f_2 = f_1 - 3\Delta f$,..., where $\Delta f = f_2 - f_1$. Thus, variations in the frequency of the sensing resonator, f_1, are magnified by the IMP index as $2\Delta f_1$, $3\Delta f_1$, $4\Delta f_1$, etc. Thus, the first stage (as sensing element) shifts only as Δf_1, while the 7th IMP changes as $4\Delta f_1$. Thanks to this mixing approach, variation in the original frequency f_1, or even at both frequencies, f_1 and f_2, can be multiplied by integer values when the intermodulation products are monitored.

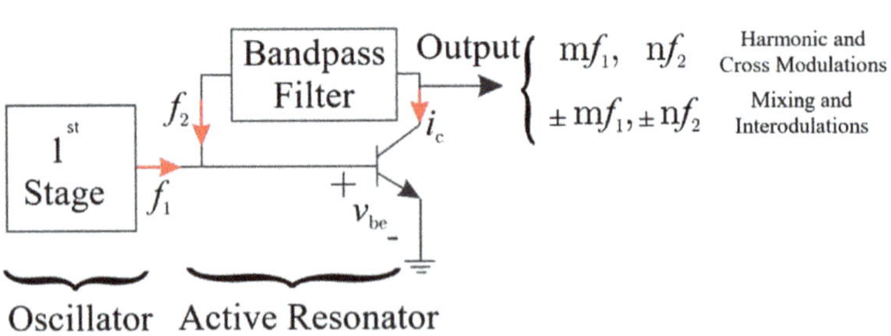

Fig. 35 Mixing/Intermodulation using double stage active resonator. Reprinted with permission from [291]; copyright 2020 IEEE

In [291], a prototype device sensor that considers the IMP frequencies as the output signals was reported. The frequencies where set to $f_2 = 5$ GHz and $f_1 = 4.75$ GHz. The resonators are double loop SRRs, designated as DSRRs in [291]. To generate a signal with the frequencies f_1 and f_2, the resonators were employed as the tank of oscillators. Each resonator was utilized in the feedback path of a regenerative amplifier, and its operation was transformed into a feedback oscillator once the resonator loss was fully compensated. In order to couple these two signals in a nonlinear component, the authors in [291] devised a mixing technique to couple the two individual signals as shown in Fig. 36a. In this design, the second sensing oscillator was used as both generator of f_2 and the mixing element for the signal f_1 being injected from the first stage. The second stage needs to operate at a nonlinear region for the bandwidth of interest to allow for mixing products being produced with reasonable gain at the output. In [291], the mixer was followed by a pair of wide-band high-gain linear antennas for transmission purposes, as a means to enable radio frequency identification (RFID) sensing. The photograph of the fabricated prototype, excluding the antennas, is depicted in Fig. 36b. This figure includes the measurement system, see Fig. 36c, the frequency of the individual oscillators, see Fig. 36d, and the intermodulation products, see Fig. 36e. The results of Fig. 36e were obtained via a pair of ultra-wideband high gain antennas separated 30 cm. The black curve, corresponding to the sensor in bare condition, transforms into the red curve with the water filling the tube on the left oscillator (f_1), while the second stage (f_2) for the right oscillator is intact. This proof-of-concept shows a multiplied downshift in the mixing products of 3rd, 5th, and 7th order that are proportional with their corresponding coefficients. Note that since the lower frequency is set for sensing, f_2 is left unaffected when f_1 is exposed to water. Also, the intermodulation modes higher than f_2 are shifted less than those lower than f_1.

2.2 Resolution and Precision Improvement

It is convenient to start this subsection by clearly defining, and differentiating, certain sensor performance indicators that sometimes are confused, particularly, resolution, accuracy, error, precision and repeatability [44]. Resolution is defined as the smallest variation of the input signal that can be reliably detected. For example, in electromagnetic encoders, the resolution is dictated by the period of the inclusions' chain. Accuracy, which is the opposite concept of error, is the degree of closeness of the measurand to the true value. If the difference between the measured and the actual value of the variable of interest is very small (small error), the sensor is said to be very accurate. Finally, precision and repeatability, synonymous concepts within the framework of sensors, describe the sensor's ability to provide identical output signals for the same input variable. Precision is related to the standard deviation of measured data, but not to the proximity between the mean value and the true value (or error). Thus, precision, accuracy, and resolution should not be confused. Indeed, a sensor

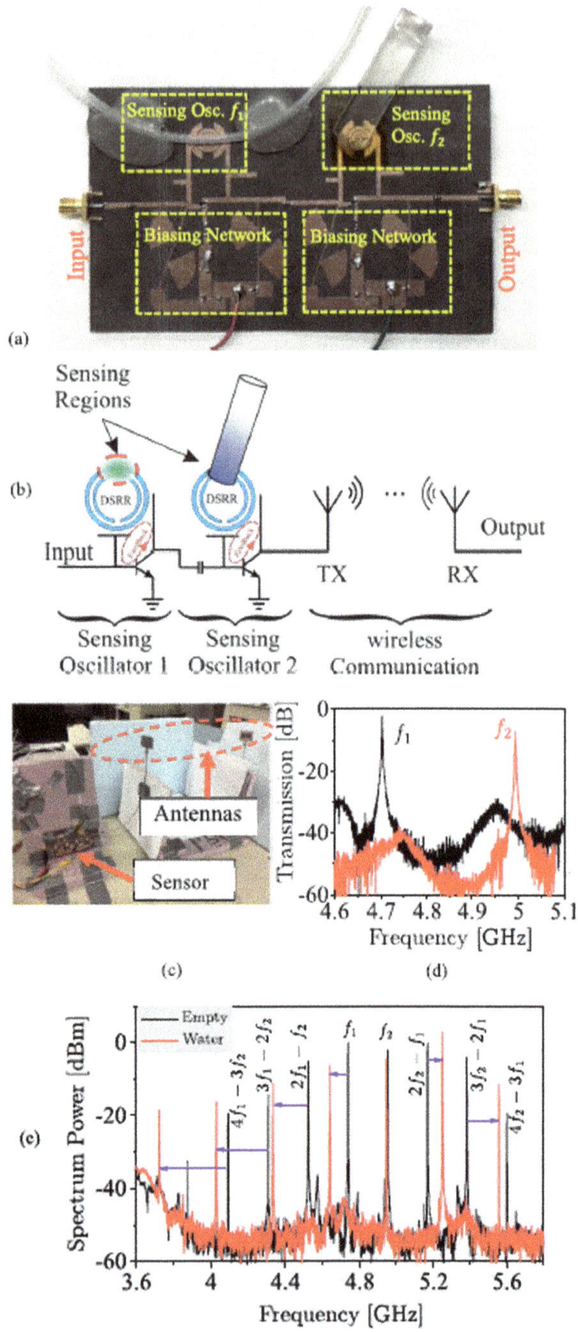

Fig. 36 Sensor based on IMPs. **a** Fabricated device; **b** schematic, including the antenna system; **c** measurement system; **d** response of the individual oscillators in vector network analyzer; (e) IMPs of f_1 and f_2 in spectrum analyzer. Reprinted with permission from [291]; copyright 2020 IEEE

might be very precise and exhibit a limited accuracy, or vice versa, and, in both cases, sensor resolution can be good or poor.

Another aspect that is worthy to be discussed concerns the link between resolution and sensitivity. Despite the fact that such performance parameters obey to different definitions (sensitivity is the variation caused in the output variable due to a change in the measurand), generally, a highly sensitive sensor exhibits good resolution. Note that, if the sensitivity is high, small variations in the input signal are expected to be discerned by the measuring device, thereby providing good resolution. Nevertheless, in analog sensors, resolution is ultimately dictated by the noise level. By contrast, in a sensor with a digital output, resolution is determined by the numerical step of the digital output.

Let us next briefly analyze some strategies or structures that have positive impact on sensor resolution and/or precision.

2.2.1 Resolution/Precision Improvement Through Loss Compensation in Frequency-Variation Sensors

As mentioned before in this chapter, frequency-variation sensors typically consist of a transmission line loaded with a planar resonator, e.g., an SRR, a CSRR, or a DB-DGS among many others. Depending on the specific sensor configuration, the device can operate either in transmission or in reflection. In the former case, two types of responses are possible, i.e., the notched-type or the bandpass-type. For example, the DB-DGS-based sensor of Fig. 6 exhibits a notched response, as it can be seen in Fig. 7. In both sensor types, the resonance frequency, dictated by the position of the notch (notched-type sensors) or peak (bandpass-type sensors), is the canonical output variable, as already indicated. However, the notch or peak magnitude can also be used as a secondary output variable, e.g., for the determination of the complex permittivity of the MUT, which includes two measurands, that is, the real and the imaginary part.[26] In other applications, the resonance frequency suffices as output variable (i.e., as far as the measurand is mainly related to the dielectric constant of the MUT), and sensor performance is mainly dictated by its sensitivity, resolution, and precision. In frequency-variation transmission-mode sensors solely based on the measurement of the resonance frequency (the unique output variable), the sharpness of the notch, or peak, intimately related to the Q-factor (and hence to the level of losses) of the sensing resonator, is a fundamental aspect with direct impact on sensor precision and resolution. Sharp responses provide good precision, a parameter related to the notch or peak bandwidth. Moreover, small changes in the measurand, and hence in the notch or peak frequency, are expected to be better detected if the sensor response exhibits an abrupt change at resonance. However, in many cases, the MUTs are lossy samples, and the resulting Q-factor of the sensing

[26] MUT losses have direct impact on the magnitude of the notch, or peak, of the frequency response of the sensor.

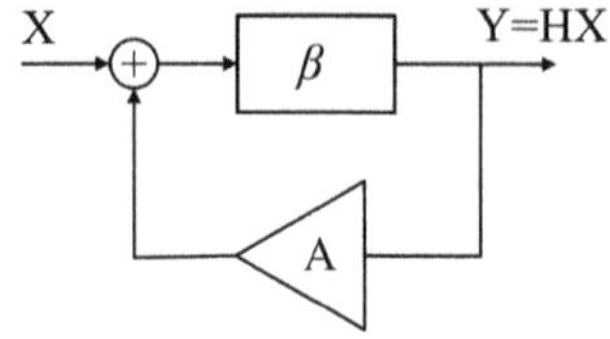

Fig. 37 Schematic of the sensor with active feedback loop

resonator is limited.[27] However, it is possible to enhance the Q-factor, and thereby generate sharp responses, through loss compensation by means of an active feedback loop, a technique reported in [292], to be discussed next.

The idea of loss compensation is very simple. It consists in providing active feedback (i.e., with gain) that partially restores the losses generated in the resonator, by adequately tuning the phase and gain of the open loop transfer function. The schematic of such an approach is depicted in Fig. 37, where A is the amplifier gain and β is the resonator transfer function.[28] According to this figure, the transfer function, H, of the closed loop system is

$$H = \frac{\beta}{1 - A\beta} \tag{50}$$

The optimum condition is obtained when the amplifier gain A partially compensates path losses from the resonator β at the frequency of interest, with an overall phase of the open loop transfer function $A\beta$ equal to 0°. Indeed, the level of loss power compensation is controllable with the gain of the amplifier, so that the dissipated power of the sensor (including the MUT on it) can be retrieved with proper amplifier gain and phase control. Nevertheless, as discussed in [292], special care must be taken in order to avoid full compensation ($A\beta = 1$) since, under these conditions, the circuit might start oscillating.[29]

In [292], an SRR-based transmission-mode frequency-variation sensor acting as a bandpass-type structure, that incorporates an active feedback loop, was designed, see Fig. 38. The sensor was implemented in the *Rogers RO5880* substrate with thickness $h = 0.8$ mm, dielectric constant $\varepsilon_r = 2.2$, and loss tangent tan $\delta = 0.0009$ [sensor dimensions are given in Fig. 38a]. Losses were compensated by means of a feedback amplifier based on a bipolar junction transistor (the *CEL NE68033*), and the phase was controlled with the help of transmission lines. The bare sensor resonates at 1 GHz, and the quality factor of the device without the regenerative feedback loop was estimated to be $Q_{\text{Passive}} = 190$ in [292]. The active circuit is triggered with a bias voltage that controls transistor's operation. Particularly, increasing the

[27] Nevertheless, it should be clarified that substrate, conductor, and radiation losses might also contribute to the overall losses of the sensing resonant element.

[28] Note that β has been used to designate the phase constant in previous sections. Nevertheless, in this section, β is identified with the resonator transfer function, following the nomenclature of the original source [292].

[29] Note that the circuit becomes unstable if $A\beta = 1$ (Barkhausen stability criterion).

bias voltage allows for higher transistor current that results in an increased peak in the transmission coefficient, as shown in Fig. 38c. This represents an increase of roughly 20 times in the quality factor for the optimum bias voltage (i.e., $Q_{\mathrm{Active}} \approx 3850$, as estimated in [292]), and, consequently, an important improvement in sensor resolution and precision. The high Q-factor ensures that variations in the measured resonance frequency due to changes in the effective dielectric constant of the MUT (caused by variations in glucose concentration in [292]), can be discerned with small steps.

The active sensor of [292] was used as a glucose-sensing device for noninvasive monitoring of glucose concentration in aqueous solutions. The intention was to be able to resolve glucose concentrations as small as 1 mMol/L, corresponding approximately to 18 mg/dL. A polytetrafluoroethylene (PTFE) tube covering the resonator gap as well as the SRR (to improve sensitivity) was used for liquid driving. Two syringe pumps with variable flow rates were utilized to combine DI water with a highly concentrated glucose solution (i.e., 1000 mMol/L) to achieve intermediate concentrations. Glucose concentration was varied from 100 mMol/L to 1000 mMol/L and monitored with the sensor resonance frequency continuously within a period of 75 s [see Fig. 38d]. Under the so-called passive mode, with $V_{\mathrm{bias}} = 0$ V, highly scattered measured resonance frequencies, due to a large 3-dB bandwidth, were found (a large 3-dB bandwidth allows for frequency span to undergo white noise). By contrast, at higher bias voltages (e.g., $V_{\mathrm{bias}} = 22$ V), the sensor loss compensation diminishes the white noise impact on the transmission profile, and subsequently, on the resonance frequency detection. Figure 38d clearly shows that the uncertainty in the frequency detection reduces from 312 kHz down to 3 kHz. This low frequency fluctuation reflects the high resolution of the sensor that helps discriminating smaller glucose concentration variations.

The results of Fig. 38, reported in [292], point out the potential of these active sensors based on loss compensation for resolution and precision improvement. Other similar sensors have been reported in [293–295]. In [296], a loss-compensated single-port reflective-mode active sensor was also presented.

2.2.2 High-Resolution Differential Sensors

Transmission-mode resonant-type differential sensors are quite sensitive to unbalances in the dielectric loading of the pair of sensing resonators and exhibit a high capability to resolve tiny differences between the MUT and the REF sample. By considering the magnitude of the cross-mode transmission coefficient as the differential output variable, it has been shown that these sensors are very adequate for the determination of electrolyte content in diluted aqueous solutions, and especially for the detection of very small concentrations of such components [191–194]. Since electrolytes are present as ions (e.g., Na^+, Ca^{2+} K^+, Cl^-, etc.), small concentrations of these elements in DI water suffice to generate a significant change in the conductivity of the solution. Therefore, it is expected that the loss tangent of the solution exhibits a high dependence with the concentration of electrolytes, and, for that reason, the

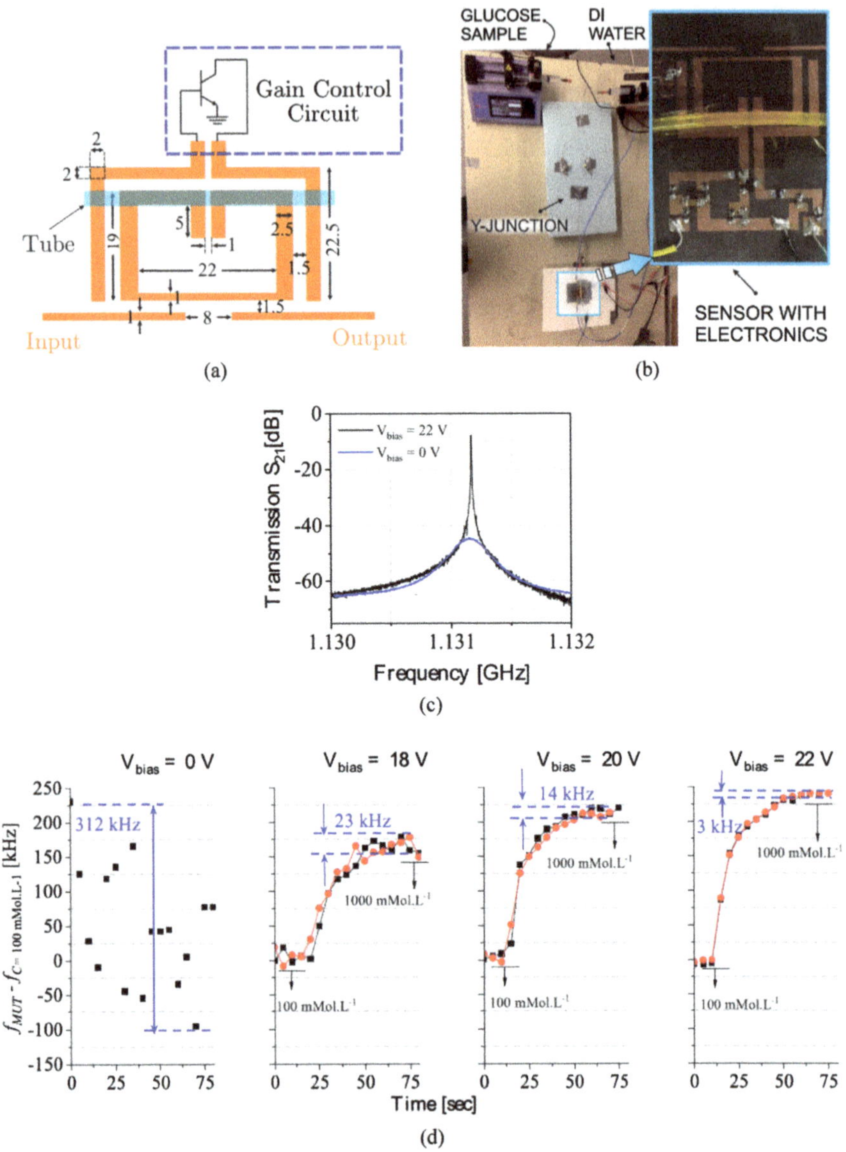

Fig. 38 Active sensor based on a SRR with active feedback for loss compensation, and results. **a** Topology of the sensor core (with indication of dimensions in mm) and schematic feedback loop; **b** sensing setup for glucose monitoring with two syringe pumps; **c** bias voltage impact on the transmission response; **d** monitoring glucose concentrations (in the range 100 mg/dL to 1000 mg/dL) with different states of the active sensor (V_{bias} = 0 V, 18 V, 20 V, 22 V) with two separate measurements (black and red lines). Reprinted with permission from [292]; copyright 2021 IEEE

main output variable in these sensors is the magnitude of the cross-mode transmission coefficient (particularly, measured at the frequency exhibiting the maximum value).[30] In these sensors, typically based on fluidic channels located on top of the sensing resonators, the REF liquid (pure DI water) is injected in one channel, whereas the other channel contains the MUT liquid, i.e., the electrolyte solution.

As an example, let us report the differential sensor presented in [193], based on a pair of SRR-loaded lines. The layout is depicted in Fig. 39a, where dimensions are indicated. The sensor was implemented on the *FR4* substrate with dielectric constant $\varepsilon_r = 4.4$ and thickness $h = 1.6$ mm. The details of the fluidic channels and mechanical parts are given in [193]. To avoid substrate absorption, a dry film of clear polyester, with an estimated thickness of 50 μm and dielectric constant of 3.5, was deposited on top of the SRRs. Figure 39b depicts the photograph of the whole sensing structure. The fluidic channels are located in such a way that the MUT liquid and REF liquids lye on top of the gaps of the SRRs, the most sensitive regions of the resonators, due to the high fringing fields present in those regions. This sensor was validated in [193] by considering DI water solutions with different types of electrolytes, particularly, NaCl, KCl and $CaCl_2$. Nevertheless, in this chapter, only the second set of experiments, carried out by injecting DI water solutions with different concentrations of KCl in the MUT channel, are reported. The REF liquid is, obviously, pure DI water, injected in the REF channel. The measured cross-mode transmission coefficient for the different KCl concentrations is depicted in Fig. 39c. The cross-mode transmission coefficient resulting when the MUT channel is filled with pure DI water, the REF sample (corresponding to a balanced loading), is also included. The maximum value of the modulus of the cross-mode transmission coefficient for this case is −38.95 dB. This is a small value, indicative of a good balance between the channels (and individual sensors) when both are loaded with pure DI water. Thanks to such small value, concentrations of KCl as small as 0.25 g/L can be resolved. By increasing the electrolyte concentration, the cross-mode transmission coefficient also increases, as expected. Figure 39d depicts the dependence of the maximum value of the cross-mode transmission coefficient, $|S_{21}^{cd}|_{max}$, with the concentration of KCl, designated as [KCl]. Actually, a pair of measurements was carried out in [193], in order to ensure that the results are repetitive [such measurements are distinguished by the labels 1 and 2 in Fig. 39d]. For small concentrations, the sensitivity is very high, and it progressively decreases as [KCl] increases. The maximum value of the sensitivity was found to be 0.032 $(g/L)^{-1}$, a high value.

[30] It has been shown that the magnitude of the transmission coefficient at resonance, or notch depth, in any of the individual (single-ended) sensors of the differential pair strongly depends on the electrolyte content, and this has been attributed to the high sensitivity of the loss tangent with the concentration of electrolytes. A strong variation in the notch depth of the MUT sensor in comparison with the notch depth of the REF sensor (loaded with DI water) is expected to cause an important variation in the cross-mode transmission coefficient (magnitude) in the vicinity of resonance.

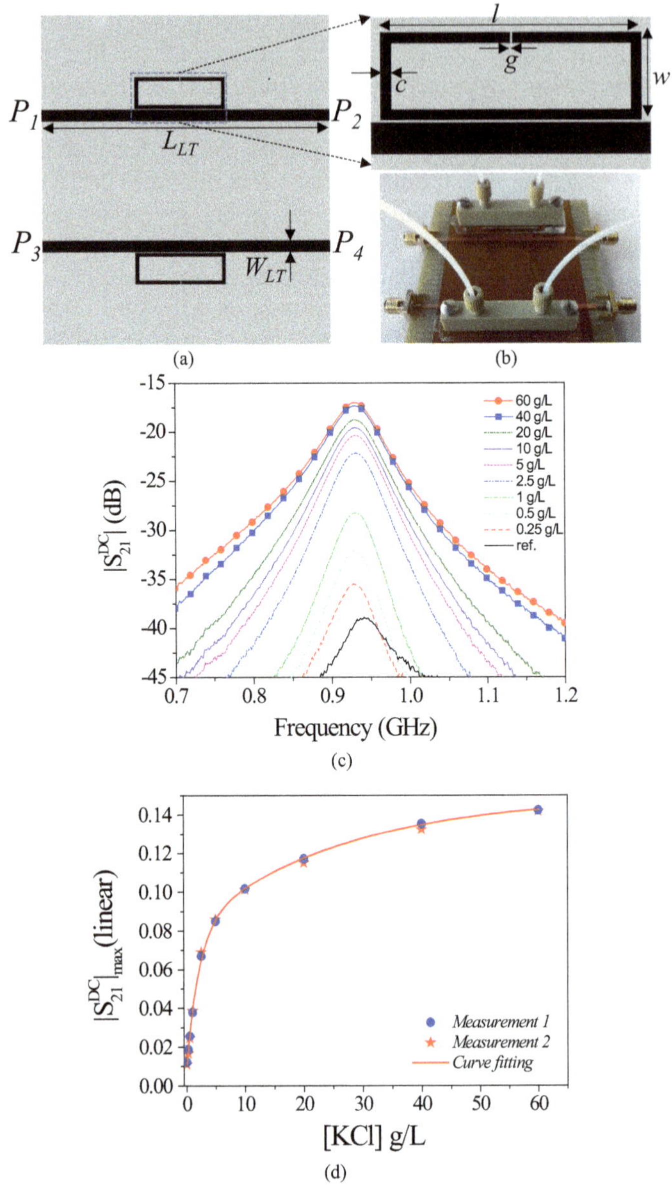

Fig. 39 Differential-mode SRR-based fluidic sensor for electrolyte concentration measurements. **a** Topology of the microwave structure; **b** photograph of the SRR-based differential sensor, including the microwave structure, the fluidic channels (plus mechanical accessories) and connectors; **c** cross-mode transmission coefficient for different values of KCl concentration; **d** variation of $|S_{21}{}^{DC}|_{max}$ with [KCl]. Dimensions (in mm) are: $L_{LT} = 76.5$, $W_{LT} = 2.79$, $l = 24$, $c = 1$, $w = 8.17$, $g = 0.2$. The ground plane in (**a**) is depicted in grey. Reprinted with permission from [193]; copyright 2019 IEEE

2.2.3 High-Resolution Electromagnetic Encoders

It was shown in Sect. 1.2.3 that the spatial resolution in electromagnetic encoders (either linear or angular) is dictated by the period of the inclusions' chain. In the specific linear electromagnetic encoder system of Fig. 15 (a quasi-absolute encoder with inclusions screen-printed in a rubber elevator belt), the period was set to $p =$ 4 mm. Such period, or resolution, suffices for many applications where the input dynamic range is in the range of several meters (e.g., conveyor belts, or elevators, among others). Indeed, it is reasonable to consider that an encoder period of roughly three orders of magnitude smaller than the maximum displacement span of the encoder is enough for the accurate determination of encoder position. Thus, for the measurement of linear displacements in the sub-meter range, as required in many motion control applications, reducing the period to values in the vicinity of 1 mm, or smaller, is necessary. In other words, linear electromagnetic encoder systems exhibiting sub-millimeter resolution are highly demanded. In rotary electromagnetic encoders, the number of pulses per revolution (PPR) is a figure of merit (FoM), similar to optical encoders, and such PPR is intimately related to the period of the (circular) chain. Nevertheless, in rotary encoders, the resolution is usually given as an angle, determined by $\theta = 2\pi/\text{PPR}$, and high-resolution (sub-degree) rotary encoders require small chain periods.[31]

The minimum achievable chain period in an electromagnetic encoder (linear or angular) is dictated by the stator, which should be able to detect the chain inclusions, not necessarily resonators, and to clearly discriminate two adjacent inclusions. In the linear encoder system of Fig. 15, the stator is based on a transmission line loaded with CSRRs, and the size of such CSRRs determines the dimensions of the inclusions (comparable to those of the CSRRs), as well as the period of the chain (of the same order than CSRR dimensions). In the rotary electromagnetic encoders reported in [163, 164], the chain inclusions are SRRs identical to those of the stator, but oppositely oriented. It is obvious that reducing (scaling) the overall dimensions of the stator and movable part (e.g., the rotor in rotary encoders) would contribute to improve resolution, but at the penalty of increasing the operational frequency of the system, and reducing the maximum air gap, or vertical space between the stator and the movable part. In order to avoid friction between the stator and the movable part of the system, and to provide robustness against vibrations and precession (in rotary encoders), it is necessary for the stator to be able to detect the chain inclusions by considering moderate (i.e., not extremely small) air gaps. Thus, dimensions (and frequency) scaling does not seem to be the preferred option for resolution optimization.

Probably, the optimum strategy in terms of spatial resolution is to consider an inclusions' chain based on transversely oriented narrow, and closely spaced, metallic strips. This strategy was adopted in the incremental-type linear electromagnetic encoders proposed in [167], with a period (and resolution) of $p = 0.6$ mm, and

[31] Note that the period is given by $p = r \cdot \theta$, where r is the encoder radius. Thus, in compact rotary encoders, i.e., with small radius, good resolution (θ small) necessarily requires a small period.

then applied to the synchronous quasi-absolute electromagnetic encoders proposed in [297], with a resolution, or period, of $p = 1.2$ mm. Let us briefly discuss next the system reported in [297], in particular the linear quasi-absolute electromagnetic encoder with motion direction detection (Fig. 40). Similar to the encoder of Fig. 15, in the system of [297], the stator contains three sensitive elements, particularly, three double stubs, each tuned to a different frequency (in Fig. 15, the sensitive elements of the stator are CSRRs). The encoder is based on three chains of transversely oriented strips, one that contains a certain ID code (position chain), and two chains with all the strips present at the predefined positions, separated a period p, see Fig. 40. Such chains are designated as clock chains and distinguished numerically as clock 1 chain and clock 2 chain. In the double stubs, the elements (stubs) are oriented face-to-face at their open-ended terminations. As far as the stubs of a pair are uncoupled, a single resonance (at the frequency that makes the electrical length of those stubs to be 90°) in the form of a notch arises. The distance between the open terminations of the stubs of any of the three stub pairs is enough to avoid inter-stub coupling, unless a strip of the chains is on top of such terminations, at small vertical distance (air gap). Such coupling generates a frequency splitting that significantly modifies the frequency response in the vicinity of resonance. Moreover, as justified in detail in [167, 297], the double stub generates a pole in the response (with roughly total transmission) to the left of the transmission zero, or zeros, whose position depends on the distance between the T-junctions of the stubs with the host transmission line. Thus, through a proper design, it is possible to force the first notch of the double stub loaded with a strip to coincide with the pole of the bare double stub. Under these circumstances, a significant excursion of the transmission coefficient at the frequency of the pole of the bare double stub is expected. Thus, by tuning the three carrier frequencies to the pole frequencies of the bare double stubs, strip detection when the encoder moves on top of the stator at short distance is possible. The carrier frequency for the detection of the strips of the position chain was set to $f_{c1} = 2.02$ GHz (double stub with intermedium stub lengths), whereas the one for the synchronism was set to $f_{c2} = 3.98$ GHz (smaller double stub). Finally, $f_{c3} = 0.959$ GHz (larger double stub) is the carrier frequency of the harmonic signal necessary to generate the redundant clock signal used for retrieving the motion direction. Note that the face-to-face open ends of the larger double stub are located in a different position (in the direction of encoder motion) to those of the smaller double stub, so that the corresponding clock signals are in relative lag or lead depending on the encoder motion direction.

Figure 41 depicts the photograph of the fabricated stator and a specific 16-bit encoder, where the position chain contains 8 strips at the indicated specific (predefined) positions. Since the carrier frequencies are tuned to the pole frequencies of the stub pairs when there is no strip on top of their open-end terminations, it follows that the envelope function of each AM modulated signal generated at the output port of the stator, caused by encoder motion, should exhibit a dip when a strip crosses the open end termination of the stub pair. Figure 42 depicts the three measured envelope functions of the fabricated encoder, by considering constant velocity, as well as constant acceleration of the encoder with regard to the sensitive part of the reader. For the experiment that conducted to the results depicted in Fig. 42a, the encoder

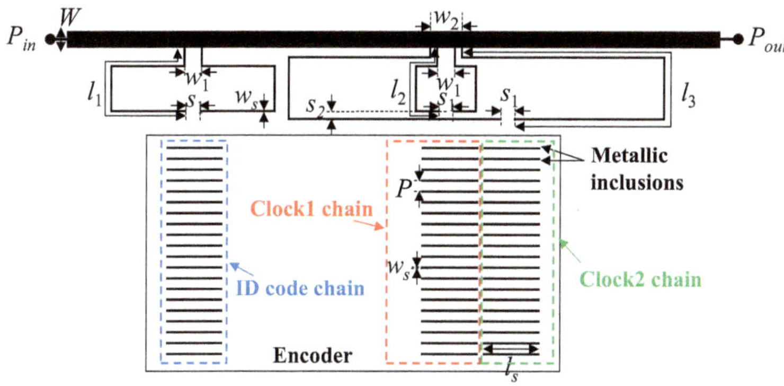

Fig. 40 Layout of the synchronous encoder system based on a microstrip transmission line loaded with three pairs of folded open-ended stubs, able to provide the direction of motion of the encoder. Dimensions (in mm) are: $W = 1.81$, $l_1 = 24.39$, $l_2 = 12.33$, $l_3 = 48.60$, $w_1 = 1.80$, $w_2 = 3.40$, $s_1 = 1.60$, $s_2 = 0.90$, $p = 1.20$, $w_s = 0.20$ and $l_s = 6.40$. Reprinted with permission from [297]; copyright 2021 IEEE

was displaced with a nominal constant velocity of 10 mm/s. The redundant clock signal (Clock 2) is advanced with regard to the signal Clock 1. Therefore, the motion direction of the encoder can be inferred. Moreover, from adjacent pulses, the encoder velocity is found to be 10 mm/s, in perfect agreement with the considered nominal value. With regard to Fig. 42b, the encoder was displaced over the sensitive part of the reader with a nominal constant acceleration of 3 mm/s^2. As can be seen, the encoder direction as well as the ID code can be inferred regardless of the encoder velocity (a constant velocity is not a requirement in these encoders). Further details on these quasi-absolute electromagnetic encoders, based on chains of transversally oriented metallic strips, can be found in [167, 297].

3 Selectivity and Robustness Improvement Assisted by Machine Learning (ML)

Microwave sensors, either planar or non-planar, are essentially permittivity sensors. Nevertheless, as mentioned previously, many physical, chemical, and biological variables, as well as multitude of analytes correlated with the permittivity, can be retrieved through microwaves (e.g., temperature, humidity, material composition, concentration of certain substances in mixtures, etc.). This versatility of microwaves for the measurement of many different types of variables might be detrimental in regard to two aspects: sensor robustness against environmental factors and sensor selectivity (or capacity to selectively determining the value of a certain variable in a multivariable measurement, or in the presence of uncontrollable stimuli, e.g., caused by variations in ambient factors). For example, since temperature affects

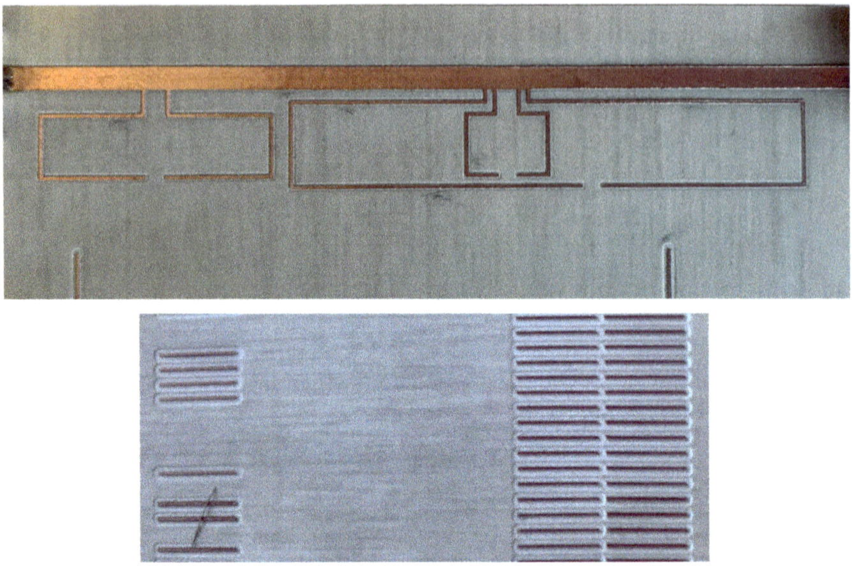

Fig. 41 Photograph of the fabricated sensitive part of the stator (upper part) and encoder (lower part). The stator was implemented in the *Rogers RO4003C* with dielectric constant $\varepsilon_r = 3.55$, thickness $h = 0.81$ mm, and loss tangent tan $\delta = 0.0021$. For the encoder, the substrate was the same, except the thickness, $h = 0.204$ mm. Reprinted with permission from [297]; copyright 2021 IEEE

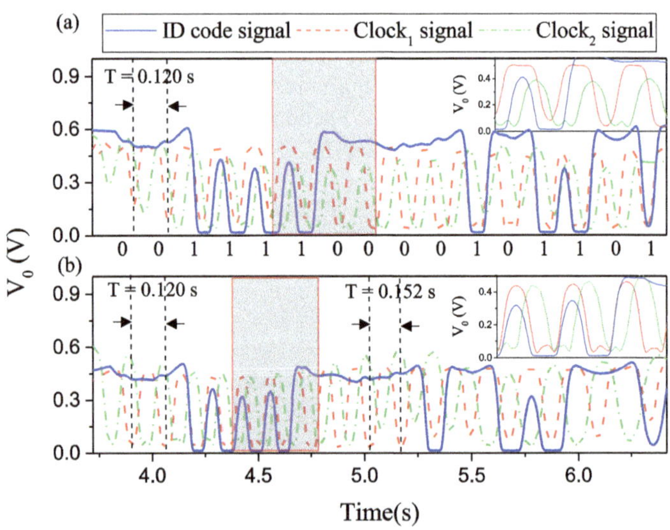

Fig. 42 Measured envelope function of the 16-bit encoder of Fig. 41 (with the indicated code) by considering encoder motion with (**a**) constant velocity and (**b**) constant acceleration. The considered air gap is 0.2 mm. Reprinted with permission from [297]; copyright 2021 IEEE

material's permittivity, identifying a certain MUT or variable related to material composition (e.g., the volume fraction of solute in a liquid solution) in environments subjected to substantial temperature variations might generate erroneous results. The reason is that different combinations of the measurand of interest (MUT, analyte, or input variable) and temperature may provide the same permittivity value, and hence the same sensor response. This is an example to illustrate the (potential) lack of robustness of microwave sensors against changes in ambient factors (temperature in the example).[32] Concerning selectivity, let us focus, as a representative case of interest, on the quantification of a certain substance (or substances) in a complex sample. In uniform and homogeneous samples consisting of two components (binary samples), retrieving material composition is simple, since there is a direct correlation between the permittivity and the relative concentration of the constitutive components. However, this does not occur in ternary or quaternary mixtures. In these cases, different combinations of the volume fractions of the various components might provide the same value of the permittivity, at least if the sensor operates at a single frequency. A possible solution to unequivocally determine material composition in complex mixtures (such as ternary and quaternary samples) is to obtain the sensor response at different frequencies (multi-frequency sensing). This approach, presented, e.g., in [123–125, 298, 299], is based on the different dispersive behavior (dependence of permittivity with frequency) of the various constitutive elements of the sample. For example, in ternary mixtures, retrieving the sensor information at two different frequencies suffices to determine sample composition, whereas three frequencies are needed in samples constituted by four components. Note that, in general, the number of required frequencies is equal to the number of sample components minus one, because the volume fractions must sum up one. Thus, as many conditions as components (and hence volume fractions) in the mixture are generated through this multi-frequency sensing strategy, thereby being possible the determination of sample composition.

Despite the fact that multi-frequency sensing suffices in certain applications, there are situations where the quantification of a certain component (or components) in very complex samples, sometimes subjected to external stimuli and conditions, is required. A paradigmatic example is the determination of glucose concentration in blood through low-cost non-invasive microwave sensors. Inferring glucose concentration in patients by simply contacting, e.g., the finger, to the sensitive area of the sensor, with the same level of accuracy to that of (high cost) biochemical analyzers, is of the highest interest within the Microwave Sensor's Community. However, blood is an extremely complex medium, and many other factors, such as the pressure exerted by the patient, the thickness of the different skin layers, the skin temperature and sweating, etc., may influence the sensor response. Thus, non-invasively obtaining glucose concentration in blood with planar microwave sensors does not seem an easy enterprise. Possibly, the best solution to deal with such complex systems is

[32] It has been previously indicated that differential measurements can alleviate the effects of cross-sensitivities against ambient factors. Nevertheless, if the variations of temperature, humidity, or pressure are severe, differential sensing does not provide a definite solution to this issue.

to resort to artificial intelligence (AI), and particularly to train the system under consideration and apply machine learning (ML) algorithms.

The idea behind ML is to generate a significant amount of data that includes the sensor response to a set of different values of the measurand, and for various conditions or external stimuli that also affect the sensor response, and use these data to train the ML algorithm. Thus, ML methods can be trained to recognize the complex relationships between input and output information. If a representative set of data is considered, it is potentially possible to identify the measurand of interest with a high level of accuracy. Various examples of microwave sensors assisted by ML can be found in the literature [300–307]. In this section, we will briefly report one of such examples, where ML is applied to the determination of volume fractions of solutions of acetone or methanol in DI water, disturbed by strong temperature variations [302]. The sensor is a transmission-mode device based on an SRR-loaded transmission line operating in bandpass configuration, and the liquid under test is driven to the sensitive part by means of a micro-tube. As mentioned, binary mixtures are typically identified through the resonance frequency,[33] directly correlated with the permittivity of the mixture. However, when different solutes and temperatures are considered, significant entanglement, or intertwining, between the input variables and sensor response (resonance frequency) is expected. Under these circumstances, ML-assisted sensing is a good option to properly discriminate the MUT regardless of the temperature. In [302], 10 different MUTs consisting of different concentrations of methanol/acetone in water within 0%–100%, incremented by 20%, were considered. Note that this gives a total of five classes for methanol in water and five classes for acetone in water, all combined into a total of 10 classes. Each class of data also contains the effect of the temperature cycle on the sensor, from 25 °C up to 50 °C. The first task for the machine learning system is to identify which class each data point belongs to. This allows to recognize the material regardless of the uncontrolled and variable environmental changes.

The frequency span used in the analysis of [302] was [1.188–1.21 GHz], and it contains 5001 points for each S_{21} profile that corresponds to a specific measurement temperature. Thus, the recorded transmission profiles consist of 5001 columns for S_{21}, with the added last column for material type in the form of a numerical label ranging from 0 to 9, and over 300 rows, corresponding to the different temperatures, as depicted in Fig. 43. In [302], two thirds of the rows for each class were randomly selected to train the network, and the remaining third was used as the test set. Several algorithms were employed in [302] to classify the input sensor data into corresponding classes. These classifiers include multilayer perceptron (MLP), decision tree (DT), K-nearest neighbors (KNN), support vector machine (SVM), random forest, Naive Bayes, AdaBoost (base estimator DT), AdaBoost (base estimator SVM), logistic regression, quadratic discriminant analysis (QDA), and linear discriminant analysis (LDA). An overview of the final accuracy of each method, as given in [302], indicates that MLP, SVM, DT, LDA, random forest, and KNN show

[33] Although not explicitly mentioned, it is assumed that a frequency-variation sensor is under consideration.

the best performance (above 96% accuracy). Upon completion of the training, the test set was used to verify whether the classifier maps each input to its corresponding class or not. The performance of the algorithm was visualized in [302] using a confusion matrix. The confusion matrix is populated based on the test set whose true outputs are known. Each row of this matrix (see Fig. 44) shows which classes are prediction results of the data within a known class. Fully correct mapping of the present data set can be distinguished with zeros at all elements except the one at the diagonal.

Although not discussed here, it was also shown in [302] that the temperature can also be retrieved through the use of regressors. Further details on the regressors for extracting temperature information, as well as on the limits of detection, out of the scope of this book, can be found in [302]. In this section, the main objective has been to point out the potential of ML-assisted microwave sensing to deal with complex systems, and this has been conducted by means of a specific, but representative, case example.

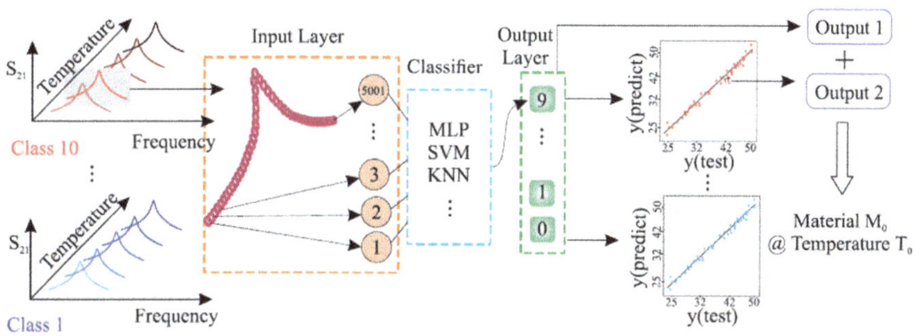

Fig. 43 Schematic of the ML classification and regression applied to the input measured transmission profiles. Reprinted with permission from [302]; copyright 2021 IEEE

Fig. 44 Classification results of randomly selected input data with confusion matrix to predict the material types (classes). Labels "0–4" denote 100%, 20%, 40%, 60%, and 80% acetone in water, while "5–9" represent similar concentrations of methanol in water. Reprinted with permission from [302]; copyright 2021 IEEE

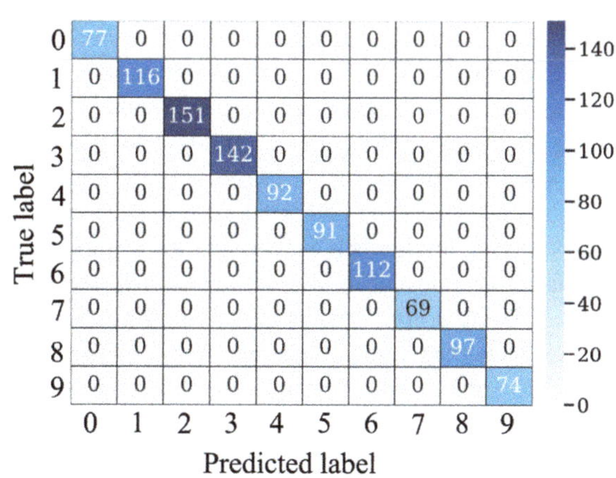

4 Key Aspects of Coupled Structures for Microwave Sensing

In Sect. 1.2, planar microwave sensors were classified according to their working principle, intimately related to the considered canonical output variable. The measurand (MUT's permittivity in most cases) has influence on the frequency response (magnitude and phase) of the sensor (either the reflection or the transmission coefficient). However, the considered output variable is dictated by the type of sensor response and/or by the parameter exhibiting an optimum sensitivity with the measured. For example, in resonant-type sensors, the canonical (and most used) output variable is the resonance frequency, very dependent on the permittivity of the MUT. Thus, most resonant microwave sensors are frequency-variation sensors. However, there are also resonant sensors that exploit the changes in the phase of the reflection coefficient at a fixed frequency, caused by variations in the permittivity of the MUT, and there are examples of extremely sensitive resonant-type phase-variation sensors [179, 182, 183]. Also, coupling-modulation sensors typically use (movable) resonant elements for sensing, though these sensors are mostly utilized for the measurement of spatial variables (displacements and velocities). In these sensors, the input variable (a linear or an angular displacement, for example) generates magnitude changes in the response, and the output variable is usually the amplitude of the transmission coefficient at resonance (which can be easily transformed to a voltage by means of an envelope detector). In non-resonant sensors, on the other hand, the phase of either the reflection or the transmission coefficient is the canonical output variable. Nevertheless, the magnitude can be used to retrieve the imaginary part of the complex permittivity, or the loss tangent.

As indicated previously, rather than a working principle, differential sensing is a mode of operation. Thus, differential-mode sensors can be implemented by means of pairs of sensors that individually exploit the resonance frequency variation, the phase variation, or even the magnitude variation in the transmission, or reflection, coefficient. Similarly, planar microwave sensors based on coupled structures, the subject of this book, can exploit the resonance frequency, the phase, the magnitude, a combination of these parameters, and even other parameters, such as the quality factor or the group delay. It is important to clarify, and highlight, that the sensors to be studied in this book are based, at least, on a pair of coupled resonators (semi-lumped or distributed) or coupled lines, the sensing elements. Thus, we exclude those sensors implemented by means of a transmission line coupled to a sensing resonator (or multiple resonators), exhaustively studied in other sources (e.g., in [44]). In other words, in the framework of this book, by coupled structures we mean (typically) pairs of coupled resonators, or multiple coupled resonators (e.g., the case of the sensors based on magneto-inductive wave –MIW– transmission lines), or pairs of coupled line sections (including coupled-line directional couplers). Nevertheless, we should clarify that in certain reported frequency-splitting sensors, a line is loaded with a pair of coupled (sensing) resonators [138–140]. In such sensors, inter-resonator coupling was caused by the necessary proximity of both resonant elements to the host (exciting) line, and such coupling was detrimental since it degrades the sensitivity, as it was demonstrated in [87]. Solutions to this issue, by separating the resonant elements by means of the so-called cascade configuration [87], or by means of the splitter/combiner configuration [141–143], were reported. Other solutions include schemes to minimize inter-resonator coupling [144, 145]. Nevertheless, inter-resonator coupling, or inter-line coupling is, in general, beneficial for sensing, provided the coupled elements and the associated microwave circuitry are conveniently designed, as it will be discussed throughout this book. Let us next very briefly comment those relevant aspects of coupled structures that make them useful for sensing (and for achieving high sensor performance), as a motivation for the reader to further delve into the following chapters (a detailed analysis of coupled structures is given in Chap. 2, and an in-depth study of planar microwave sensors based on different types of coupled structures is carried out in the subsequent chapters).

There is not a specific reason that explains why coupled structures are useful for microwave sensing. Moreover, the principles for sensor performance optimization vary, depending on the particular type of structure. Nevertheless, it is clear that coupled structures exhibit a rich phenomenology, as compared to that of their isolated counterparts. The reason is that, besides the features associated with the individual (uncoupled) elements (resonators or lines, typically), their interaction (through electromagnetic coupling) provides further degrees of freedom that can be exploited for sensing. For example, an isolated resonator (either distributed or semi-lumped) can be described (at least in the vicinity of resonance) by means of a series or a parallel resonant tank, with inductance L and capacitance C. However, if an identical resonator is coupled, let us say electrically, to it, then the mutual capacitance, C_M, should be introduced in the model describing the (coupled) structure, and this provides more possibilities for tailoring (reshaping) the frequency response of the structure, with (potentially) further options for sensor performance improvement.

Further deepening on coupled resonators, it is well known (as it will be further discussed in Chap. 2) that the fundamental resonance frequency of the uncoupled resonator splits into two resonance frequencies when coupling arises, and such frequencies are separated a distance that depends on the level of coupling. This aspect is very important for sensitivity and resolution optimization in reflective-mode phase-variation sensors, as well as in transmission-mode sensors based on the measurement of the resonance frequency (or frequencies) and/or quality factor, as it will be shown in dedicated chapters (Chaps. 5, 6, 8 and 9). Inter-resonator coupling substantially reshapes the magnitude and phase response of the sensor, which typically consists of a transmission line terminated with a pair of coupled resonators (reflective-mode sensors) or a transmission line loaded, or coupled, with a pair of coupled resonators (transmission-mode sensors). In the latter case (transmission-mode sensors), the structure can be either notched-type or band pass type. Regardless on the specific type of sensing structure, as a result of such sensor response reshaping, controllable through the level of coupling, it is potentially possible to tailor, and thus optimize, an important sensor performance indicator, i.e., the sensitivity, and, indirectly, the sensor resolution.

Concerning coupled lines, it is expected that the presence of a material (MUT) in the region between the lines (either in edge- or broadside-coupled lines) significantly perturbs the odd-mode characteristic impedance, and this aspect can be exploited for sensing, as it will be demonstrated in Chap. 3. There are also sensors based on coupled-line directional couplers, where the presence of a MUT in the coupling region alters the phase of the line pair for the even and odd propagating modes, and this aspect is instrumental to implement high sensitive sensors based on the variation of the notch frequency of the response corresponding to the coupling factor, or on the variation of the magnitude of the isolation (these sensors based on coupled-line directional couplers will be discussed in Chap. 4).

Finally, to end this section and this chapter, let us mention that, in Chap. 10, near-field position sensors and contactless resonant imaging systems based on the sensitivity of "meta-atoms" to the electromagnetic environment in magneto-inductive-wave (MIW) transmission lines are reported (MIW lines are a class of artificial transmission lines consisting of a chain of magnetically coupled resonators, typically SRRs [308]). This is a very specific type of position sensing and localization, where the considered object under test should be metallic and of a similar size to the inclusions (coupled resonators) of the MIW transmission line. The idea behind this approach relies on the fact that the presence of the object on top of a resonant element perturbs the impedance and therefore generates reflections giving rise to wave patterns, from which the position can be determined by means of specific developed algorithms. Electro-inductive-wave (EIW) transmission lines, where the constitutive resonators of the chain (typically CSRRs) are electrically coupled, have also been used for permittivity sensing [92], in this case exploiting the highly dispersive behavior of such artificial lines, achievable under weak inter-resonator coupling. This high dispersion provides strong phase variations when the EIW line is coated with a MUT with varying permittivity. Nevertheless, the EIW transmission line behaves as an effective medium, and despite the fact that (weak) coupling is a key aspect for generating a high level of dispersion, and hence high sensitivity, sensing is not based on specific features derived from coupling, such as frequency response reshaping. For this main reason, and because these phase-variation EIW-based sensors have been extensively studied in previous works [44, 92], we preclude from including them in this book.

References

1. Liao Y, Deschamps F, de Freitas Rocha Loures E, Pierin-Ramos LF (2017) Past, present and future of Industry 4.0—a systematic literature review and research agenda proposal. Int J Prod Res 55(12):3609–3629
2. Vaidyaa S, Ambadb P, Bhosle S (2018) Industry 4.0—a glimpse. Procedia Manuf 20:233–238
3. Popkova EG, Ragulina YV, Bogoviz (2019) Industry 4.0: industrial revolution of the 21st century, Springer
4. Galar-Pascual D, Daponte P, Kumar U (2019) Handbook of industry 4.0 and SMART systems. CRC Press, Taylor and Francis Group, Boca Raton, FL, USA

5. Wollschlaeger M, Sauter T, Jasperneite J (2017) The future of industrial communication: Automation networks in the era of the internet of things and industry 4.0. IEEE Ind Electron Mag 11:17–27
6. Yan L, Zhang Y, Yang LT, Ning H (2008) The Internet of things: from RFID to the next-generation pervasive networked systems: wireless networks and mobile communications. Auerbach Publications, Taylor & Francis Group, Boca Raton, FL, USA
7. Tsiatsis V, Karnouskos S, Holler J, Boyle D, Mulligan C (2018) Internet of things: technologies and applications for a new age of intelligence, 2nd edn. Academic Press
8. Davies J, Fortuna C (eds) (2020) The internet of things: from data to insight. John Wiley, Hoboken, NJ, USA
9. Hassan QF (2018) Internet of things A to Z: technologies and applications. John Wiley, Hoboken, NJ, USA
10. Yang LT, Di Martino B, Zhang Q (2017) Internet of everything. Mob Inf Syst 2017:3. Article ID 8035421
11. Sharma L (ed) (2021) Towards smart world: homes to cities using internet of things. Chapman and Hall/CRC Press, Taylor and Francis Group, Boca Raton, FL, USA
12. Kalansuriya P, Bhattacharyya R, Sarma S, Karmakar N (2012) Towards chipless RFID-based sensing for pervasive surface crack detection. In: 2012 IEEE international conference on RFID-technologies and applications (RFID-TA), Nice, France, Nov 2012, pp 46–51
13. Dey S, Kalansuriya P, Karmakar NC (2014) Chipless RFID based high resolution crack sensing through SWB technology. In: 2014 IEEE international microwave and RF conference (IMaRC), Bangalore, India, Dec 2014, pp 330–333
14. Ramos A, Girbau D, Lazaro A, Villarino R (2015) Wireless concrete mixture composition sensor based on time-coded UWB RFID. IEEE Microw Wireless Compon Lett 25(10):681–683
15. Ramos A, Lazaro A, Girbau D (2015) Time-coded chipless sensors to detect quality of materials in civil engineering. In: 2015 9th European conference on antennas and propagation (EuCAP), Lisbon, Portugal, Apr 2015
16. DiGiampaolo E, DiCarlofelice A, Gregori A (2017) An RFID-enabled wireless strain gauge sensor for static and dynamic structural monitoring. IEEE Sensors J 17(2):286–294
17. Caizzone S, DiGiampaolo E (2015) Wireless passive RFID crack width sensor for structural health monitoring. IEEE Sensors J 15(12):6767–6774
18. Lázaro A, Villarino R, Costa F, Genovesi S, Gentile A, Buoncristiani L, Girbau D (2018) Chipless dielectric constant sensor for structural health testing. IEEE Sensors J 18(13):5576–5585
19. Marindra AMJ, Tian GY (2018) Chipless RFID sensor tag for metal crack detection and characterization. IEEE Trans Microw Theory Techn 66(5):2452–2462
20. Marindra AMJ, Tian GY (2019) Multiresonance chipless RFID sensor tag for metal defect characterization using principal component analysis. IEEE Sensors J 19(18):8037–8046
21. Li MM, Wan GC, Yang M, Tong MS (2019) A chipless RFID sensor for metal crack detection based on notch characteristics. In: 2019 photonics & electromagnetics research symposium—fall (PIERS—Fall), Xiamen, China, Dec 2019, pp 217–221
22. Min SH, Kim HJ, Quan YJ, Kim HS, Lyu JH, Lee GY, Ahn SH (2020) Stretchable chipless RFID multi-strain sensors using direct printing of aerosolised nanocomposite. Sens Act A: Phys 313:112224
23. Marindra AMJ, Tian GY (2020) Chipless RFID sensor for corrosion characterization based on frequency selective surface and feature fusion. Smart Mater Struct 29:125010
24. Deif S, Daneshmand M (2020) Multiresonant chipless RFID array system for coating defect detection and corrosion prediction. IEEE Tran Ind Electron 67(10):8868–8877
25. Wan G, Kang W, Wang C, Li W, Li M, Xie L, Chen L (2020) Separating strain sensor based on dual-resonant circular patch antenna with chipless RFID tag. Smart Mater Struct 30:015007
26. Javed N, Azam MA, Amin Y (2021) Chipless RFID multisensor for temperature sensing and crack monitoring in an IoT environment. IEEE Sensors Lett 5(6):6001404

27. Dey S, Bhattacharyya R, Sarma SE, Karmakar NC (2021) A novel "smart skin" sensor for chipless RFID-based structural health monitoring applications. IEEE Internet Things J 8(5):3955–3971
28. Chambon H, Nicolay P, Floer C, Benjeddou A (2021) A package-less SAW RFID sensor concept for structural health monitoring. Mech Adv Mater Struct 28:648–655
29. Mc Gee K, Anandarajah P, Collins D (2021) Proof of concept novel configurable chipless RFID strain sensor. Sensors 21:6224
30. Pittella E, Schiavoni R, Monti G, Masciullo A, Scarpetta M, Cataldo A, Piuzzi E (2022) Split ring resonator network and diffused sensing element embedded in a concrete beam for structural health monitoring. Sensors 22(17):6398
31. Want R (2004) Enabling ubiquitous sensing with RFID. Computer 37:84–86
32. A. Rida, L. Yang, and M. Tentzeris, *RFID-enabled Sensor Design and Applications*, Artech House, Norwood, MA, USA, 2010.
33. Marrocco G (2010) Pervasive electromagnetics: sensing paradigms by passive RFID technology. IEEE Wireless Commun 17(6):10–17
34. Roselli L (ed) (2014) Green RFID systems. Cambridge University Press, Cambridge, United Kingdom
35. Perret E (2014) Radio frequency identification and sensors: from RFID to chipless RFID. Wiley, Hoboken, NJ, USA
36. Dey S, Saha JK, Karmakar NC (2015) Smart sensing: chipless RFID solutions for the internet of everything. IEEE Microw Mag 16(10):26–39
37. Forouzandeh M, Karmakar NC (2015) Chipless RFID tags and sensors: a review on time-domain techniques. Wirel. Power Transf. 2(2):62–77
38. Karmakar NC, Amin EM, Saha JK (2016) Chipless RFID sensors. Wiley, Hoboken, NJ, USA
39. Ramos A, Lazaro A, Girbau D, Villarino R (2016) RFID and wireless sensors using ultra-wideband technology. ISTE Press/Elsevier, London, UK
40. Cui L, Zhang Z, Gao N, Meng Z, Li Z (2019) Radio frequency identification and sensing techniques and their applications—a review of the state-of-the-art. Sensors 19:4012
41. Costa F, Genovesi S, Borgese M, Michel A, Dicandia FA, Manara G (2021) A review of RFID sensors, the new frontier of Internet of Things. Sensors 21:3138
42. Martín F, Herrojo C, Mata-Contreras J, Paredes F (2020) Time-Domain Signature Barcodes for Chipless-RFID and Sensing Applications. Springer, Cham, Switzerland
43. Khalid N, Mirzavand R, Iyer AK (2021) A survey on battery-less RFID-based wireless sensors. Micromachines 12:819
44. Martín F, Vélez P, Muñoz-Enano J, Su L (2022) Planar microwave sensors. Wiley/IEEE Press, Hoboken, NJ, USA
45. Kent G (1988) An evanescent-mode tester for ceramic dielectric substrates. IEEE Trans Microw Theory Techn 36(10):1451–1454
46. Kent G (1996) Nondestructive permittivity measurement of substrates. IEEE Trans Instrum Meas 45(1):102–106
47. Janezic MD, Baker-Jarvis J (1999) Full-wave analysis of a split-cylinder resonator for nondestructive permittivity measurements. IEEE Trans Microw Theory Techn 47(10):2014–2020
48. Janezic MD (2003) Nondestructive relative permittivity and loss tangent measurements using a split-cylinder resonator, Ph.D. Thesis, University of Colorado at Boulder
49. Janezic MD, Kuester EF, Baker-Jarvis J (2004) Broadband complex permittivity measurements of dielectric substrates using a split-cylinder resonator. In: IEEE MTT-S international microwave symposium digest, Fort Worth, TX, USA, Jun 2004, pp 1817–1820
50. Amin EM, Bhuiyan MS, Karmakar NC, Winther-Jensen B (2014) Development of a low cost printable chipless RFID humidity sensor. IEEE Sensors J 14(1):140–149
51. WM Abdulkawi, Sheta AFA (2020) Chipless RFID sensors based on multistate coupled line resonators. Sens Act A Phys 309:112025
52. Yeo J, Lee JI, Kwon Y (2021) Humidity-sensing chipless RFID tag with enhanced sensitivity using an interdigital capacitor structure. Sensors 21:6550

53. Yilmaz T, Foster R, Hao Y (2010) Detecting vital signs with wearable wireless sensors. Sensors 10(12):10837–10862
54. Elgeziry M, Costa F, Tognetti A, Genovesi S (2022) Wearable textile-based sensor tag for breath rate measurement. IEEE Sensors J 22(23):22610–22619
55. Puentes M, Maasch M, Schubler M, Jakoby R (2012) Frequency multiplexed 2-dimensional sensor array based on split-ring resonators for organic tissue analysis. IEEE Trans Microw Theory Techn 60(6):1720–1727
56. Puentes M, Maasch M, Schüssler M, Damm C, Jakoby R (2014) Analysis of resonant particles in a coplanar microwave sensor array for thermal ablation of organic tissue. In: 2014 IEEE MTT-S international microwave symposium (IMS'2014), Tampa, FL, USA
57. Jol HM (ed) (2009) Ground penetrating radar theory and applications. Elsevier, Amsterdam, The Netherlands
58. Zhao W, Forte E, Pipan M, Tian G (2013) Ground penetrating radar (GPR) attribute analysis for archaeological prospection. J Appl Geophys 97:107–117
59. Benedetto A, Pajewski L (2015) Civil engineering applications of ground penetrating radar. Springer
60. Rothwell EJ, Frasch JL, Ellison SM, Chahal P, Ouedraogo RO (2016) Analysis of the Nicolson-Ross-Weir method for characterizing the electromagnetic properties of engineered materials. Prog Electromagn Res 157:31–47
61. Costa F, Borgese M, Degiorgi M, Monorchio A (2017) Electromagnetic characterisation of materials by using transmission/reflection (T/R) devices. Electronics 6(4):95
62. Komarov SA, Komarov AS, Barber DG, Lemes MJL, Rysgaard S (2016) Open-ended coaxial probe technique for dielectric spectroscopy of artificially grown sea ice. IEEE Trans Geosci Remote Sens 54(8):4941–4951
63. La Gioia A, Porter E, Merunka I, Shahzad A, Salahuddin S, Jones M, O'Halloran M (2018) Open-ended coaxial probe technique for dielectric measurement of biological tissues: challenges and common practices. Diagnostics 8:40
64. Guihard V, Taillade F, Balayssac JP, Steck B, Sanahuja J, Deby F (2020) Permittivity measurement of cementitious materials with an open-ended coaxial probe. Constr Build Mater 230:116946
65. Aydinalp C, Joof S, Yilmaz T (2021) Towards non-invasive diagnosis of skin cancer: sensing depth investigation of open-ended coaxial probes. Sensors 21:1319
66. Keysight Technologies (2020) N1501A Dielectric Probe Kit: 10 MHz to 50 GHz. Technical Overview
67. Su L, Huang X, Guo W, Wu H (2020) A Flexible Microwave Sensor Based on Complementary Spiral Resonator for Material Dielectric Characterization. IEEE Sensors J 20(4):1893–1903
68. Elsheikh D, Eldamak AR (2021) Microwave textile sensors for breast cancer detection. In: Proc. 38th nat. radio sci. conf. (NRSC), Mansoura, Egypt, Jul 2021, pp 288–294
69. Martínez-Estrada M, Gil I, Fernández-García R (2021) An alternative method to develop embroidery textile strain sensors. Textiles 1(3):504–512
70. Nejad HR, Punjiya MP, Sonkusale S (2017) Washable thread based strain sensor for smart textile. In: Proc. 19th int. conf. solid-state sens., actuators microsyst. (TRANSDUCERS), Kaohsiung, Taiwan, Jun 2017, pp 1183–1186
71. Vélez P, Martín F, Fernández-García R, Gil I (2022) Embroidered Textile Frequency-Splitting Sensor Based on Stepped-Impedance Resonators. IEEE Sensors J 22(9):8596–8603
72. Moras M, Martínez-Domingo C, Escudé R, Herrojo C, Paredes F, Terés L, Martín F, Ramon E (2021) Programmable organic chipless RFID tags inkjet printed on paper substrates. Applied Sciences 11(17):7832
73. Grenier K, Dubuc D, Poleni P, Kumemura M, Toshiyoshi H, Fujii T, Fujita H (2009) New broadband and contact less RF, microfluidic sensor dedicated to bioengineering", 2009 IEEE MTT-S International Microwave Symposium Digest Boston, MA, USA, pp 1329–1332
74. Vélez P, Su L, Grenier K, Mata-Contreras J, Dubuc D, Martín F (2017) Microwave microfluidic sensor based on a microstrip splitter/combiner configuration and split ring resonators (SRRs) for dielectric characterization of liquids. IEEE Sensors J 17(20):6589–6598

75. Ebrahimi A, Scott J, Ghorbani K (2019) Ultrahigh-sensitivity microwave sensor for microfluidic complex permittivity measurement. IEEE Trans Microw Theory Techn 67(10):4269–4277
76. Chuma EL, Iano Y, Fontgalland G, Bravo Roger LL (2018) Microwave sensor for liquid dielectric characterization based on metamaterial complementary split ring resonator. IEEE Sensors J 18(24):9978–9983
77. Su L, Mata-Contreras J, Vélez P, Fernández-Prieto A, Martín F (2018) Analytical method to estimate the complex permittivity of oil samples. Sensors 18:984
78. Muñoz-Enano J, Vélez P, Casacuberta P, Su L, Martín F (2023) Characterization of the quality of edible oils subjected to industrial frying processes through high sensitivity microwave sensors. In: 53rd European microwave conference, Berlin, Germany, pp 17–22
79. Woodhouse IH (2006) Introduction to microwave remote sensing. CRC Press, Taylor & Francis Group, Boca Raton, FL, USA
80. Lillesand T, Kiefer RW, Chipman J (2015) Remote sensing and image interpretation, 7th edn. John Wiley, Hoboken, NJ, USA
81. Thenkabail PS (2015) Remote sensing handbook. CRC Press, Taylor & Francis Group, Boca Raton, FL, USA
82. Campbell JB, Wynne RH (2011) Introduction to remote sensing, 5th edn. The Guilford Press
83. Zarifi MH, Daneshmand M (2016) Liquid sensing in aquatic environment using high quality planar microwave resonator. Sens Actuators, B Chem 225:517–521
84. Wei Z, Huang J, Li J, Xu G, Ju Z, Liu X, Ni X (2018) A high-sensitivity microfluidic sensor based on a substrate integrated waveguide re-entrant cavity for complex permittivity measurement of liquids. Sensors 18(11):4005
85. Javed A, Arif A, Zubair M, Mehmood MQ, Riaz K (2020) A low-cost multiple complementary split-ring resonator-based microwave sensor for contactless dielectric characterization of liquids. IEEE Sensors J 20(19):11326–11334
86. Martín F, Medina F (2023) Balanced microwave transmission lines, circuits, and sensors. IEEE J Microw 3(1):398–440
87. Naqui J, Damm C, Wiens A, Jakoby R, Su L, Mata-Contreras J, Martín F (2016) Transmission Lines Loaded with Pairs of Stepped Impedance Resonators: Modeling and Application to Differential Permittivity Measurements. IEEE Trans Microw Theory Techn 64(11):3864–3877
88. Kozak R, Khorsand K, Zarifi T, Golovin K, Zarifi MH (2021) Patch antenna sensor for wireless ice and frost detection. Sci Rep 11:13707
89. Muñoz-Enano J, Vélez P, Gil M, Martín F (2020) An analytical method to implement high sensitivity transmission line differential sensors for dielectric constant measurements. IEEE Sensors J 20:178–184
90. Ebrahimi A, Coromina J, Muñoz-Enano J, Vélez P, Scott J, Ghorbani K, Martín F (2021) Highly sensitive phase-variation dielectric constant sensor based on a capacitively-loaded slow-wave transmission line. IEEE Trans Circ Syst I: Reg Pap 68(7):2787–2799
91. Damm C, Schüßler M, Puentes M, Maune H, Maasch M, Jakoby R (2009) Artificial transmission lines for high sensitive microwave sensors. In: 2009 IEEE sensors Christchurch, New Zeland, pp 755–758
92. Gil M, Vélez P, Aznar F, Muñoz-Enano J, Martín F (2020) Differential sensor based on electro-inductive wave (EIW) transmission lines for dielectric constant measurements and defect detection. IEEE Trans Ant Propag 68:1876–1886
93. Kremer F, Schönhals A (eds) (2003) Broadband Dielectric spectroscopy. Springer
94. Grenier K et al (2013) Recent advances in microwave-based dielectric spectroscopy at the cellular level for cancer investigations. IEEE Trans Microw Theory Techn 61(5):2023–2030
95. Vélez P, Muñoz-Enano J, Ebrahimi A, Herrojo C, Paredes F, Scott J, Ghorbani K, Martín F (2021) Single-frequency amplitude-modulation sensor for dielectric characterization of solids and microfluidics. IEEE Sensors J 21(10):12189–12201
96. Elgeziry M, Paredes F, Vélez P, Costa F, Genovesi S, Martín F (2022) A method to retrieve the output variables in reflective-mode phase-variation sensors. In: 52nd European microwave conference, Milan, Italy, pp 20–25

97. Vélez P, Paredes F, Casacuberta P, Elgeziry M, Su L, Muñoz-Enano J, Costa F, Genovesi S, Martín F (2023) Portable reflective-mode phase-variation microwave sensor based on a rat-race coupler pair and gain/phase detector for dielectric characterization. IEEE Sensors J 23(6):5745–5756
98. Puentes M, Weiss C, Schussler M, Jakoby R (2011) Sensor array based on split ring resonators for analysis of organic tissues. In: IEEE MTT-S int. microw. symp. dig., Baltimore, MD, USA
99. Mandel C, Kubina B, Schüßler M, Jakoby R (2011) Passive chipless wireless sensor for two-dimensional displacement measurement. In: 41st European microwave conference, Manchester, UK, pp 79–82
100. Ebrahimi A, Withayachumnankul W, Al-Sarawi S, Abbott D (2014) High-sensitivity metamaterial-inspired sensor for microfluidic dielectric characterization. IEEE Sensors J 14(5):1345–1351
101. Schueler M, Mandel C, Puentes M, Jakoby R (2012) Metamaterial inspired microwave sensors. IEEE Microw Mag 13(2):57–68
102. Boybay MS, Ramahi OM (2012) Material characterization using complementary split-ring resonators. IEEE Trans Instrum Meas 61(11):3039–3046
103. Lee CS, Yang CL (2014) Complementary split-ring resonators for measuring dielectric constants and loss tangents. IEEE Microw Wireless Compon Lett 24(8):563–565
104. Withayachumnankul W, Jaruwongrungsee K, Tuantranont A, Fumeaux C, Abbott D (2013) Metamaterial-based microfluidic sensor for dielectric characterization. Sens Actuators A 189:233–237
105. Salim A, Lim S (2016) Complementary split-ring resonator-loaded microfluidic ethanol chemical sensor. Sensors 16:1802
106. Yang CL, Lee CS, Chen KW, Chen KZ (2016) Noncontact measurement of complex permittivity and thickness by using planar resonators. IEEE Trans Microw Theory Techn 64(1):247–257
107. Su L, Mata-Contreras J, Velez P, Martin F (2017) Estimation of the complex permittivity of liquids by means of complementary Split ring resonator (CSRR) loaded transmission lines. In: IEEE MTT-S int. microw. Workshop series on advanced materials and processes (IMWS-AMP 2017), Pavia, Italy, pp 20–22
108. Jha AK, Delmonte N, Lamecki A, Mrozowski M, Bozzi M (2019) Design of microwave-based angular displacement sensor. IEEE Microw Wireless Compon Lett 29(4):306–308
109. Galindo-Romera G, Javier Herraiz-Martinez F, Gil M, Martinez-Martinez JJ, Segovia-Vargas D (2016) Submersible printed split-ring resonator-based sensor for thin-film detection and permittivity characterization. IEEE Sensors J 16(10):3587–3596
110. Abdolrazzaghi M, Zarifi MH, Daneshmand M (2016) Sensitivity enhancement of split ring resonator based liquid sensors. In: 2016 IEEE Sensors, Orlando, FL, USA
111. Abdolrazzaghi M, Zarifi MH, Pedrycz W, Daneshmand M (2016) Robust ultra-high resolution microwave planar sensor using fuzzy neural network approach. IEEE Sens J 17:323–332
112. Zarifi MH, Daneshmand M (2017) Monitoring solid particle deposition in lossy medium using planar resonator sensor. IEEE Sens J 17:7981–7989
113. Zarifi MH, Deif S, Abdolrazzaghi M, Chen B, Ramsawak D, Amyotte M, Vahabisani N, Hashisho Z, Chen W, Daneshmand M (2017) A microwave ring resonator sensor for early detection of breaches in pipeline coatings. IEEE Trans Ind Electron 65:1626–1635
114. Abdolrazzaghi M, Daneshmand M, Iyer AK (2018) Strongly enhanced sensitivity in planar microwave sensors based on metamaterial coupling. IEEE Trans Microw Theory Tech 66:1843–1855
115. Alahnomi RA, Zakaria Z, Ruslan E, Ab Rashid SR, Bahar AAM (2017) High-Q sensor based on symmetrical split ring resonator with spurlines for solids material detection," IEEE Sensors J., vol. 17, no. 9, pp. 2766–2775, May 2017.
116. Jankovic N, Radonic V (2017) A microwave microfluidic sensor based on a dual-mode resonator for dual-sensing applications. Sensors 17(12):2713
117. Zhou H, Hu D, Yang C, Chen C, Ji J, Chen M, Chen Y, Yang Y, Mu X (2018) Multi-band sensing for dielectric property of chemicals using metamaterial integrated microfluidic sensor. Sci Rep 8(1):14801

118. Ansari MAH, Jha AK, Akhter Z, Akhtar MJ (2018) Multi-band RF planar sensor using complementary split ring resonator for testing of dielectric materials. IEEE Sensors J 18(16):6596–6606
119. Lobato-Morales H, Choi JH, Lee H, Medina-Monroy JL (2019) Compact dielectric-permittivity sensors of liquid samples based on substrate integrated-waveguide with negative-order-resonance. IEEE Sensors J 19(19):8694–8699
120. Khanna Y, Awasthi YK (2020) Dual-band microwave sensor for investigation of liquid impurity concentration using a metamaterial complementary split-ring resonator. J Electron Mater 49(1):385–394
121. Muñoz-Enano J, Martel J, Vélez P, Medina F, Su L, Martín F (2021) Parametric analysis of the edge capacitance of uniform slots and application to frequency-variation permittivity sensors. Appl Sci 11(15):7000
122. Muñoz-Enano J, Vélez P, Gil M, Martín F (2022) Frequency variation sensors for permittivity measurements based on dumbbell-shaped defect ground structures (DB-DGS): analytical method and sensitivity analysis. IEEE Sensors J 22(10):9378–9386
123. Hosseini N, Baghelani M (2021) Selective real-time non-contact multi-variable water-alcohol-sugar concentration analysis during fermentation process using microwave split-ring resonator based sensor. Sens Act A: Phys 325:112695
124. Albishi AM, Mirjahanmardi SH, Ali AM, Nayyeri V, Wasly SM, Ramahi OM (2019) Intelligent sensing using multiple sensors for material characterization. Sensors 19(21):4766
125. Abduljabar AA, Hamzah H, Porch A (2021) Multi-resonators, microwave microfluidic sensor for liquid characterization. Microw Opt Technol Lett 63:1042–1047
126. Nuñez-Flores A, Castillo-Araníbar P, García-Lampérez A, Segovia-Vargas D (2018) Design and Implementation of a submersible split ring resonator based sensor for pisco concentration measurements. In: 2018 IEEE MTT-S Latin America microwave conference (LAMC 2018), Arequipa, Peru
127. Chakyar SP, Simon SK, Bindu C, Andrews J, Joseph VP (2017) Complex permittivity measurement using metamaterial split ring resonators. J Appl Phys 121:054101
128. Torun H, Cagri Top F, Dundar G, Yalcinkaya AD (2014) An antenna-coupled split-ring resonator for biosensing. J Appl Phys 116:124701
129. Verma A, Bhushan S, Tripathi PN, Goswami M, Singh BR (2017) A defected ground split ring resonator for an ultra-fast, selective sensing of glucose content in blood plasma. J Electromagn Waves Appl 31(10):1049–1061
130. Asad M, Neyadi SA, Aidaros OA, Khalil M, Hussein M (2016) Single port bio-sensor design using metamaterial split ring resonator. In: 2016 5th international conference on electronic devices, systems and applications (ICEDSA), Ras AI Khaimah, United Arab Emirates
131. Reyes-Vera E, Acevedo-Osorio G, Arias-Correa M, Senior DE (2019) A submersible printed sensor based on a monopole-coupled split ring resonator for permittivity characterization. Sensors 19:1936
132. Islam M, Ashraf F, Alam T, Misran N, Mat K (2018) A compact ultrawideband antenna based on hexagonal split-ring resonator for pH sensor application. Sensors 18(9):2959
133. Sujith R, Augustine R (2018) CPW fed antenna for nearfield sensor applications. In: 2018 IEEE conference on antenna measurements & applications (CAMA), Vasteras, Sweden
134. Elwi TA (2018) Metamaterial based a printed monopole antenna for sensing applications. Int J RF Microw Comput Aided Eng 28(7):e21470
135. Morshidi WHW, Zaharudin Z, Khan S, Nordin AN, Shaikh FA, Adam I, Kader KA (2018) Inter-digital sensor for non-invasive blood glucose monitoring. In: 2018 IEEE international conference on innovative research and development (ICIRD), Bangkok, Thailand, May 2018
136. Lu F, Wang H, Guo Y, Tan Q, Zhang W, Xiong J (2018) Microwave backscatter-based wireless temperature sensor fabricated by an alumina-backed Au slot radiation patch. Sensors 18:242
137. Muñoz-Enano J, Vélez P, Herrojo C, Gil M, Martín F (2019) On the sensitivity of microwave sensors based on slot resonators and frequency variation. In: 2019 int. conf. electromagn. advan. applications (ICEAA), Granada, Spain, Sept 2019, pp 112–115

138. Su L, Naqui J, Mata-Contreras J, Martín F (2015) Modeling metamaterial transmission lines loaded with pairs of coupled split-ring resonators. IEEE Ant Wireless Propag Lett 14:68–71
139. Su L, Naqui J, Mata-Contreras J, Martín F (2016) Modeling and applications of metamaterial transmission lines loaded with pairs of coupled complementary split ring resonators (CSRRs). IEEE Ant Wireless Propag Lett 15:154–157
140. Naqui J, Damm C, Wiens A, Jakoby R, Su L, Martín F (2014) "Transmission lines loaded with pairs of magnetically coupled stepped impedance resonators (SIRs): Modeling and application to microwave sensors. In: 2014 IEEE MTT-S int. microw. symp. (IMS2014), Tampa, FL, USA
141. Su L, Mata-Contreras J, Naqui J, Martín F (2016) Splitter/combiner microstrip sections loaded with pairs of complementary split ring resonators (CSRRs): modeling and optimization for differential sensing applications. IEEE Trans Microw Theory Techn 64(12):4362–4370
142. Su L, Mata-Contreras J, Martín F (2016) Configurations of splitter/combiner microstrip sections loaded with stepped impedance resonators (SIRs) for sensing applications. Sensors 16(12):2195
143. Su L, Naqui J, Mata-Contreras J, Martín F (2016) Cascaded splitter/combiner microstrip sections loaded with complementary split ring resonators (CSRRs): modeling, analysis and applications. In: 2016 IEEE MTT-S int. microw. symp. (IMS16), San Francisco, CA, USA
144. Ebrahimi A, Beziuk G, Scott J, Ghorbani K (2020) Microwave differential frequency splitting sensor using magnetic-LC resonators. Sensors 20:1066
145. Ebrahimi A, Scott J, Ghorbani K (2018) Differential sensors using microstrip lines loaded with two split-ring resonators. IEEE Sens J 18:5786–5793
146. Zarifi MH, Farsinezhad S, Wiltshire BD, Abdorrazaghi M, Mahdi N, Kar P, Daneshmand M, Shankar K (2016) Effect of phosphonate monolayer adsorbate on the microwave photoresponse of TiO_2 nanotube membranes mounted on a planar double ring resonator. Nanotechnology 27:375201
147. Horestani AK, Naqui J, Shaterian Z, Abbott D, Fumeaux C, Martín F (2014) Two-dimensional alignment and displacement sensor based on movable broadside-coupled split ring resonators. Sens Act A 210:18–24
148. Weiner O (1912) Die theorie des Mischkorpers fur das Feld der statonare Stromung i. die mittelwertsatze fur kraft, polarisation und energie. In: Der Abhandlungen der Mathematisch-Physischen Klasse der Konigl. Sachsischen Gesellschaft der Wissenschaften, vol 32, pp 509–604
149. Naqui J, Durán-Sindreu M, Martín F (2011) Novel sensors based on the symmetry properties of split ring resonators (SRRs). Sensors 11:7545–7553
150. Naqui J, Durán-Sindreu M, Martín F (2012) Alignment and position sensors based on split ring resonators. Sensors 12:11790–11797
151. Karami-Horestani A, Fumeaux C, Al-Sarawi SF, Abbott D (2013) Displacement sensor based on diamond-shaped tapered split ring resonator. IEEE Sensors J 13(4):1153–1160
152. Horestani AK, Naqui J, Abbott D, Fumeaux C, Martín F (2014) Two-dimensional displacement and alignment sensor based on reflection coefficients of open microstrip lines loaded with split ring resonators. Electron Lett 50(8):620–622
153. Horestani AK, Abbott D, Fumeaux C (2013) Rotation sensor based on horn-shaped split ring resonator. IEEE Sens J 13:3014–3015
154. Naqui J, Durán-Sindreu M, Martín F (2013) Transmission lines loaded with bisymmetric resonators and applications. In: IEEE MTT-S int. microw. symp. dig., Seattle, WA, USA
155. Naqui J, Martín F (2013) Transmission lines loaded with bisymmetric resonators and their application to angular displacement and velocity sensors. IEEE Trans Microw Theory Techn 61(12):4700–4713
156. Naqui J, Martín F (2014) Angular displacement and velocity sensors based on electric-LC (ELC) loaded microstrip lines. IEEE Sensors J 14:939–940
157. Naqui J, Coromina J, Karami-Horestani A, Fumeaux C, Martín F (2015) Angular displacement and velocity sensors based on coplanar waveguides (CPWs) loaded with S-shaped split ring resonator (S-SRR). Sensors 15:9628–9650

158. Naqui J (2016) Symmetry properties in transmission lines loaded with electrically small resonators: circuit modeling and applications. Springer, Heidelberg, Germany
159. Naqui J, Martín F (2016) Application of broadside-coupled split ring resonator (BC-SRR) loaded transmission lines to the design of rotary encoders for space applications. In: 2016 IEEE MTT-S int. microw. symp. (IMS'16), San Francisco, CA, USA
160. Herrojo C, Muela F, Mata-Contreras J, Paredes F, Martín F (2019) High-density microwave encoders for motion control and near-field chipless-RFID. IEEE Sensors J 19:3673–3682
161. Paredes F, C. Herrojo, and F. Martín, "Microwave encoders with synchronous reading and direction detection for motion control applications. In: 2020 IEEE-MTT-S int. microw. symp. (IMS'20), Los Angeles, CA, USA
162. Herrojo C, Vélez P, Paredes F, Mata-Contreras J, Martín F (2019) All-dielectric electromagnetic encoders based on permittivity contrast for displacement/velocity sensors and chipless-RFID tags. In: 2019 IEEE-MTT-S Int. Microw. Symp. (IMS'19), Boston, MA, USA
163. Mata-Contreras J, Herrojo C, Martín F (2017) Application of split ring resonator (SRR) loaded transmission lines to the design of angular displacement and velocity sensors for space applications. IEEE Trans Microw Theory Techn 65(11):4450–4460
164. Mata-Contreras J, Herrojo C, Martín F (2018) Detecting the rotation direction in contactless angular velocity sensors implemented with rotors loaded with multiple chains of split ring resonators (SRRs). IEEE Sensors J 18(17):7055–7065
165. Herrojo C, Paredes F, Mata-Contreras J, Ramon E, Núñez A, Martín F (2019) Time-domain signature barcodes: near-field chipless-RFID systems with high data capacity. IEEE Microw Mag 20(12):87–101
166. Mata-Contreras J, Herrojo C, Martín F (2018) Electromagnetic rotary encoders based on split ring resonators (SRR) loaded microstrip lines. In: 2018 IEEE MTT-S int. microw. symp. (IMS'18), Philadelphia, Pennsylvania, USA
167. Herrojo C, Paredes F, Martín F (2019) Double-stub loaded microstrip line reader for very high data density microwave encoders. IEEE Trans Microw Theory Techn 67(9):3527–3536
168. Herrojo C, Paredes F, Martín F (2020) 3D-printed high data-density electromagnetic encoders based on permittivity contrast for motion control and chipless-RFID. IEEE Trans Microw Theory Techn 68(5):1839–1850
169. Paredes F, Herrojo C, Mata-Contreras J, Martín F (2018) Near-field chipless-RFID sensing and identification system with switching reading. Sensors 18:1148
170. Paredes F, Herrojo C, Martín F (2021) Position sensors for industrial applications based on electromagnetic encoders. Sensors 21:2738 (28 pages)
171. Paredes F, Herrojo C, Moya A, Berenguel-Alonso M, Gonzalez D, Bruguera J, Delgado-Simao C, Martín F (2022) Electromagnetic encoders screen-printed on rubber belts for absolute measurement of position and velocity. Sensors 22:2044
172. Paredes F, Karami-Horestani A, Martín F (2022) Strategies to enhance the data density in synchronous electromagnetic encoders. Sensors 22:4356
173. Karami-Horestani A, Paredes F, Martín F (2022) Frequency-coded and programmable synchronous electromagǹetic encoders based on linear strips. IEEE Sensors Lett 6(8):1–4, Art no. 3501704
174. Karami-Horestani A, Paredes F, Martín F (2022) Near-field chipless-RFID system based on hybrid time/frequency domain encoding and power splitter reader. In: 52nd European Microwave Conference. Milan, Italy, pp 20–25
175. Karami-Horestani A, Paredes F, Martín F (2022) High data density absolute electromagǹetic encoders based on hybrid time/frequency domain encoding. IEEE Sensors J 22(24):23866–23876
176. Paredes F, Karami-Horestani A, Martín F (2023) Enhancing the bit density in linear electromagnetic encoders for chipless-RFID and motion sensing applications. In: European conference on antennas and propagation (EuCAP 2023), Florence, Italy, Mar. 27–31
177. de Bruijn NG (1975) Acknowledgement of priority to C. Flye Sainte-Marie on the counting of circular arrangements of 2n zeros and ones that show each n-letter word exactly once, T.H.-Report 75-WSK-06, Technological University Eindhoven

178. Muñoz-Enano J, Vélez P, Gil M, Mata-Contreras J, Martín F (2020) Differential-mode to common-mode conversion detector based on rat-race couplers: analysis and application to microwave sensors and comparators. IEEE Trans Microw Theory Techn 68:1312–1325
179. Muñoz-Enano J, Vélez P, Su L, Gil M, Casacuberta P, Martín F (2021) On the sensitivity of reflective-mode phase variation sensors based on open-ended stepped-impedance transmission lines: theoretical analysis and experimental validation. IEEE Trans Microw Theory Techn 69(1):308–324
180. Su L, Muñoz-Enano J, Velez P, Casacuberta P, Gil M, Martin F (2021) Highly sensitive phase variation sensors based on step-impedance coplanar waveguide (CPW) transmission lines. IEEE Sensors J 21(3):2864–2872
181. Casacuberta P, Muñoz-Enano J, Vélez P, Su L, Gil M, Martín F (2020) Highly sensitive reflective-mode detectors and dielectric constant sensors based on open-ended stepped-impedance transmission lines. Sensors 20:6236
182. Su L, Muñoz-Enano J, Vélez P, Gil M, Casacuberta P, Martín F (2021) Highly sensitive reflective-mode phase-variation permittivity sensor based on a coplanar waveguide (CPW) terminated with an open complementary split ring resonator (OCSRR). IEEE Access 9:27928–27944
183. Casacuberta P, Vélez P, Muñoz-Enano J, Su L, Gil M, Ebrahimi A, Martín F (2022) Circuit analysis of a Coplanar waveguide (CPW) terminated with a step-impedance resonator (SIR) for highly sensitive one-port permittivity sensing. IEEE Access 10:62597–62612
184. Muñoz-Enano J, Vélez P, Casacuberta P, Su L, Martín F, Reflective-mode phase-variation microwave sensor based on a step-impedance microstrip line terminated with a slot resonator for solid and liquid characterization. IEEE Trans Microw Theory Techn. https://doi.org/10.1109/TMTT.2023.3315832
185. Su L, Muñoz-Enano J, Vélez P, Casacuberta P, Gil M, Martín F (2021) Phase-Variation Microwave Sensor for Permittivity Measurements Based on a High-Impedance Half-Wavelength Transmission Line. IEEE Sensors J 21(9):10647–10656
186. Jha AK, Lamecki A, Mrozowski M, Bozzi M (2020) A highly sensitive planar microwave sensor for detecting direction and angle of rotation. IEEE Trans Microw Theory Techn 68(4):1598–1609
187. Horestani AK, Shaterian Z, Martín F (2020) Rotation sensor based on the cross-polarized excitation of split ring resonators (SRRs). IEEE Sensors J 20:9706–9714
188. Muñoz-Enano J, Vélez P, Su L, Gil M, Martín F (2020) A reflective-mode phase-variation displacement sensor. IEEE Access 8:189565–189575
189. Mehrjoo Z, Ebrahimi A, Beziuk G, Martín F, Ghorbani K (2023) Microwave rotation sensor based on reflection phase in transmission lines terminated with lumped resonators. IEEE Sensors J 23(7):6571–6580
190. Ferrández-Pastor FJ, García-Chamizo JM, Nieto-Hidalgo M (2017) Electromagnetic differential measuring method: application in microstrip sensors developing. Sensors 17(7):1650
191. Vélez P, Grenier K, Mata-Contreras J, Dubuc D, Martín F (2018) Highly-sensitive microwave sensors based on open complementary split ring resonators (OCSRRs) for dielectric characterization and solute concentration measurements in liquids. IEEE Access 6:48324–48338
192. Vélez P, Muñoz-Enano J, Gil M, Mata-Contreras J, Martín F (2019) Differential microfluidic sensors based on dumbbell-shaped defect ground structures in microstrip technology: analysis, optimization, and applications. Sensors 19(14):3189
193. Vélez P, Muñoz-Enano J, Grenier K, Mata-Contreras J, Dubuc D, Martín F (2019) Split ring resonator (SRR) based microwave fluidic sensor for electrolyte concentration measurements. IEEE Sensors J 19(7):2562–2569
194. Muñoz-Enano J, Vélez P, Gil M, Jose-Cunilleras E, Bassols A, Martín F (2020) Characterization of electrolyte content in urine samples through a differential microfluidic sensor based on dumbbell-shaped defect ground structures. Int J Microw Wireless Technol 12(9):817–824
195. Ebrahimi A, Tovar-López FJ, Scott J, Ghorbani K (2020) Differential microwave sensor for characterization of glycerol-water solutions. Sensors Actuators B Chem 321:128561

196. Muñoz-Enano J, Vélez P, Gil M, Martín F (2020) Microfluidic reflective-mode differential sensor based on open split ring resonators (OSRRs). Int J Microw Wireless Technol 12:588–597
197. Ebrahimi A, Scott J, Ghorbani K (2018) Transmission lines terminated with LC resonators for differential permittivity sensing. IEEE Microwave Wirel Compon Lett 28(12):1149–1151
198. Hallil H, Bahoumina P, Pieper K, Lachaud JL, Rebière D, Abdelghani A, Frigui K, Bila S, Baillargeat D, Zhang Q, Coquet P, Pichonat E, Happy H, Dejous C (2019) Differential passive microwave planar resonator-based sensor for chemical particle detection in polluted environments. IEEE Sens J 19(4):1346–1353
199. Vélez P, Muñoz-Enano J, Martín F (2019) Differential sensing based on quasi-microstrip-mode to slot-mode conversion. IEEE Microw Wireless Compon Lett 29:690–692
200. Camli B, Altinagac E, Kizil H, Torun H, Dundar G, Yalcinkaya AD (2020) Gold-on-glass microwave split-ring resonators with PDMS microchannels for differential measurement in microfluidic sensing. Biomicrofluidics14(5)
201. Gan HY, Zhao WS, Liu Q, Wang DW, Dong L, Wang G, Yin WY (2020) Differential microwave microfluidic sensor based on microstrip complementary split-ring resonator (MCSRR) structure. IEEE Sensors J 20(11):5876–5884
202. Jain MC, Nadaraja AV, Vizcaino BM, Roberts DJ, Zarifi MH (2020) Differential microwave resonator sensor reveals glucose-dependent growth profile of E. coli on solid agar. IEEE Microw Wireless Compon Lett 30(5):531−534, May 2020
203. Mohammadi S, Zarifi MH (2021) Differential microwave resonator sensor for real-time monitoring of volatile organic compounds. IEEE Sensors J 21(5):6105–6114
204. Herrojo C, Vélez P, Muñoz-Enano J, Su L, Casacuberta P, Gil M, Martín F (2021) Highly sensitive defect detectors and comparators exploiting port imbalance in rat-race couplers loaded with step-impedance open-ended transmission lines. IEEE Sensors J 21:26731–26745
205. Martín F, Zhu L, Hong J, Medina F (2018) Balanced microwave filters. Wiley/IEEE Press, Hoboken, NJ, USA
206. Muñoz-Enano J, Vélez P, Gil M, Martín F (2020) Planar microwave resonant sensors: a review and recent developments. Appl Sci 10:2615 (29 pages)
207. Abdolrazzaghi M, Nayyeri V, Martín F (2022) Techniques to improve the performance of planar microwave sensors: a review and recent developments. Sensors 22:6946
208. Ahn D, Park JS, Kim CS, Kim J, Qian Y, Itoh T (2001) A design of the low-pass filter using the novel microstrip defected ground structure. IEEE Trans Microw Theory Techn 49:86–93
209. Baena JD, Bonache J, Martín F, Marqués R, Falcone F, Lopetegi T, Laso MAG, García J, Gil I, Flores-Portillo M, Sorolla M (2005) Equivalent circuit models for split ring resonators and complementary split rings resonators coupled to planar transmission lines. IEEE Trans Microw Theory Techn 53:1451–1461
210. Su L, Muñoz-Enano J, Vélez P, Martel J, Medina F, Martín F (2021) On the modelling of microstrip lines loaded with dumbbell defect-ground-structure (DB-DGS) and folded DB-DGS resonators. IEEE Access 9:150878–150888
211. Muñoz-Enano J, Vélez P, Su L, Gil M, Casacuberta P, Martín F (2022) On the capacitance of slotted metamaterial resonators for frequency-variation permittivity sensing. In: 51st European Microwave Conference, London, UK, Apr 2022
212. Sharafadinzadeh N, Abdolrazzaghi M, Daneshmand M (2020) Investigation on planar microwave sensors with enhanced sensitivity from microfluidic integration. Sens Actuators A Phys 301:111752
213. Massoni E, Siciliano G, Bozzi M, Perregrini L (2018) Enhanced cavity sensor in SIW technology for material characterization. IEEE Microw Wireless Compon Lett 28(10):948–950
214. Amin EM, Saha JK, Karmakar NC (2014) Smart sensing materials for low-cost chipless RFID sensor. IEEE Sensors J 14:2198–2207
215. Gee KM, Anandarajah P, Collins D (2019) A review of chipless remote sensing solutions based on RFID technology. Sensors 19:4829
216. Mulloni V, Donelli M (2020) Chipless RFID sensors for the Internet of Things: challenges and opportunities. Sensors 20:2135

217. Harsányi G (2000) Polymer films in sensor applications: A review of present uses and future possibilities. Sens Rev 20:98–105
218. Pawlikowski G (2009) Effects of polymer material variations on high frequency dielectric properties. MRS Online Proc Libr (OPL) 1156:D02–D05
219. Kao KC (2004) Dielectric phenomena in solids, 1st edn. Academic Press, San Diego, CA, USA
220. Amin EM, Karmakar NC (2011) Development of a chipless RFID temperature sensor using cascaded spiral resonators. In: 2011 IEEE sensors, Limerick, Ireland, Oct 2011, pp 554–557
221. Chen YT, Kao HL (2006) Humidity sensors made on polyvinyl-alcohol film coated SAW devices. Electron Lett 42:948
222. Amin EM, Karmakar NC, Winther-Jensen B (2013) Polyvinyl-Alcohol (PVA)-based RF humidity sensor in microwave frequency. Prog Electromagn Res B 54:149–166
223. Amin EM, Bhuiyan S, Karmakar NC, Winther-Jensen B (2013) A novel EM barcode for humidity sensing. In: 2013 IEEE international conference on RFID (RFID), Penang, Malaysia, Apr–May 2013, pp 82–87
224. Abdolrazzaghi M, Hariri F, Chu M, Naguib H, Daneshmand M (2019) Relative humidity sensing using PANI/PVA integrated with feedback oscillator circuit. In: Proc. 2019 IEEE sensors, montreal, QC, Canada, 27–30 Oct 2019, pp 1–4
225. Virtanen J, Ukkonen L, Björninen T, Sydänheimo L (2010) Printed humidity sensor for UHF RFID systems. In: 2010 IEEE sensors applications symposium (SAS), Limerick, Ireland, Feb 2010, pp 269–272
226. Virtanen J, Ukkonen L, Bjorninen T, Elsherbeni AZ, Sydänheimo L (2011) Inkjet-printed humidity sensor for passive UHF RFID systems. IEEE Trans Instrum Measur 60(8):2768–2777
227. Reddy ASG, Narakathu BB, Atashbar MZ, Rebros M, Rebrosova E, Joyce MK (2011) Fully printed flexible humidity sensor. Procedia Eng 25:120–123
228. Potyrailo RA, Surman C (2013) A passive radio-frequency identification (RFID) gas sensor with self-correction against fluctuations of ambient temperature. Sens Act B: Chem 185:587–593
229. Manzari S, Occhiuzzi C, Nawale S, Catini A, di Natale C, Marrocco G (2012) Polymer-doped UHF RFID tag for wireless-sensing of humidity. In: 2012 IEEE international conference on RFID (RFID), Orlando, FL, USA, Apr 2012, pp 124–129
230. Bali C, Brandlmaier A, Ganster A, Raab O, Zapf J, Hübler A (2016) Fully inkjet-printed flexible temperature sensors based on carbon and PEDOT: PSS. Mater Today Proc 3:739–745
231. Zhang HD, Tang CC, Long YZ, Zhang JC, Huang R, Li JJ, Gu CZ (2014) High-sensitivity gas sensors based on arranged polyaniline/PMMA composite fibers. Sens Act A: Phys 219:123–127
232. Eyebe GA, Bideau B, Boubekeur N, Loranger E, Domingue F (2017) Environmentally-friendly cellulose nanofibre sheets for humidity sensing in microwave frequencies. Sens Act B: Chem 245:484–492
233. Yun S, Kim J (2010) Multi-walled carbon nanotubes–cellulose paper for a chemical vapor sensor. Sens Act B: Chem 150:308–313
234. Potyrail RA, Morris AWG (2007) Multianalyte chemical identification and quantitation using a single radio frequency identification sensor. ACS Anal Chem 79:45–51
235. Alimenti F, Mariotti C, Palazzi V, Virili M, Orecchini G, Mezzanotte P, Roselli L (2015) Communication and sensing circuits on cellulose. J Low Power Electron Appl 5:151–164
236. Fan J, Zhang S, Li F, Shi J (2020) Cellulose-based sensors for metal ions detection. Cellulose 27:5477–5507
237. Sun YF, Liu SB, Meng FL, Liu JY, Jin Z, Kong LT, Liu JH (2012) Metal oxide nanostructures and their gas sensing properties: a review. Sensors 12(3):2610–2631
238. Viter R, Iatsunskyi I (2019) Metal oxide nanostructures in sensing. In: Micro and nano technologies, nanomaterials design for sensing applications, Elsevier, pp 41–91
239. Ghafouri T, Tahmasebi R, Manavizadeh N, Nadimi E (2019) Gas sensing properties of silicon nanowires with different cross-sectional shapes toward ammonia: a first-principles study. J Nanoparticle Res 21:157

240. Zaporotskova IV, Boroznina NP, Parkhomenko YN, Kozhitov LV (2016) Carbon nanotubes: sensor properties. A review. Mod Electron Mater 2:95–105
241. Camilli L, Passacantando M (2018) Advances on sensors based on carbon nanotubes. Chemosensors 6:62
242. Guerin H, Le Poche H, Pohle R, Fernández-Bolaños M, Dijon J, Ionescu AM (2013) Carbon nanotube resistors as gas sensors: towards selective analyte detection with various metal-nanotube interfaces. In: European solid-state device research conference, Bucharest, Romania, Sep 2013, pp 326–329
243. Balachandran MD, Shrestha S, Agarwal M, Lvov Y, Varahramyan K (2008) SnO_2 capacitive sensor integrated with microstrip patch antenna for passive wireless detection of ethylene gas. Elect Lett 44:464–466
244. Nair RS, Perret E, Tedjini S, Baron T (2013) A group-delay-based chipless RFID humidity tag sensor using silicon nanowires. IEEE Ant Wireless Propag Lett 12:729–732
245. Nair R, Perret E, Tedjini S, Barron T (2012) A humidity sensor for passive chipless RFID applications. In: 2012 IEEE International conference on RFID-technologies and applications (RFID-TA), Nice, France, Nov 2012, pp 29–33
246. Vena A, Perret E, Tedjini S, Kaddour D, Potie A, Barron T (2012) A compact chipless RFID tag with environment sensing capability. In: 2012 IEEE/MTT-S int. microw. symp. digest, Montreal, QC, Canada
247. Ahoulou S, Perret E, Nedelec JM (2021) Functionalization and characterization of silicon nanowires for sensing applications: a review. Nanomaterials 11:999
248. Yang L, Zhang R, Staiculescu D, Wong CP, Tentzeris MM (2009) A novel conformal RFID-enabled module utilizing inkjet-printed antennas and carbon nanotubes for gas-detection applications. IEEE Ant Wirel Propag Lett 8:653–656
249. Yang L, Staiculescu D, Zhang R, Wong CP, Tentzeris MM (2009) A novel "green" fully-integrated ultrasensitive RFID-enabled gas sensor utilizing inkjet-printed antennas and carbon nanotubes. In: 2009 IEEE antennas and propagation society international symposium, North Charleston, SC, USA
250. Occhiuzzi C, Rida A, Marrocco G, Tentzeris MM (2011) Passive ammonia sensor: RFID tag integrating carbon nanotubes. In: 2011 IEEE international symposium on antennas and propagation (APSURSI), Spokane, WA, USA, pp 1413–1416
251. Occhiuzzi C, Rida A, Marrocco G, Tentzeris MM (2011) CNT-based RFID passive gas sensor. In: 2011 IEEE MTT-S int. microw. symp., Baltimore, MD, USA
252. Occhiuzzi C, Rida A, Marrocco G, Tentzeris M (2011) RFID passive gas sensor integrating carbon nanotubes. IEEE Trans Microw Theory Techn 59(10):2674–2684
253. Baccarelli R, Orecchini G, Alimenti F, Roselli L (2012) Feasibility study of a fully organic, CNT based, harmonic RFID gas sensor. In: 2012 IEEE international conference on RFID-technologies and applications (RFID-TA 2012), Nice, France, Nov 2012, pp 419–422
254. Vena A, Sydänheimo L, Tentzeris MM, Ukkonen L (2013) A novel inkjet printed carbon nanotube-based chipless RFID sensor for gas detection. In: 2013 European microwave conference, Nuremberg, Germany, Oct 2013, pp 9–12
255. Gou P, Kraut ND, Feigel IM, Bai H, Morga GJ, Chen Y, Tang Y, Bocan K, Stachel J, Berger L, Mickle M, Sejdic E, Star A (2014) Carbon nanotube chemiresistor for wireless pH sensing. Sci Rep 4:4468
256. Gammoudi I, Aissa B, Nedil M, Abdallah MM (2015) CNT-RFID passive tag antenna for gas sensing in underground mine. In: 2015 IEEE international symposium on antennas and propagation & USNC/URSI national radio science meeting, Vancouver, BC, Canada, Jul 2015, pp 1758–1759
257. Kutty AA, Björninen T, Sydänheimo L, Ukkonen L (2016) A novel carbon nanotube loaded passive UHF RFID sensor tag with built-in reference for wireless gas sensing. In: 2016 IEEE MTT-S Int. Microw. Symp. (IMS'16), San Francisco, CA, USA
258. Le NN, Fribourg-Blanc E, Phan HCT, Dang DMT, Dang CM (2018) A RFID-based wireless NH_3 gas detector using spin coated carbon nanotubes as sensitive layer. Int J Nanotechnol 15(1–3):3–13

259. Javed N, Azam MA, Qazi I, Amin Y, Tenhunen H (2021) A novel multi-parameter chipless RFID sensor for green networks. AEU-Int J Electron Commun 128:153512
260. Wang Y, Yeow JTW (2009) A review of carbon nanotubes-based gas sensors. J Sens 2009:1–24
261. Novoselov KS, Geim GK, Morozov SV, Jiang D, Zhang Y, Dubonos SV, Grigorieva IV, Firsov AA (2004) Electric field in atomically thin carbon films. Science 306:666–669
262. Nag A, Mitra A, Mukhopadhyay SC (2018) Graphene and its sensor-based applications: a review. Sens Act A: Phys 270:177–194
263. Lee JS, Oh J, Jun J, Jang J (2015) Wireless hydrogen smart sensor based on pt/graphene-immobilized radio-frequency identification tag. ACS Nano 9:7783–7790
264. Caccami MC, Mulla MYS, Occhiuzzi C, Di Natale C, Marrocco G (2018) Design and experimentation of a batteryless on-skin RFID graphene-oxide sensor for the monitoring and discrimination of breath anomalies. IEEE Sensors J 18(21):8893–8901
265. Caccami MC, Miozzi C, Mulla MYS, Di Natale C, Marrocco G (2017) An epidermal graphene oxide-based RFID sensor for the wireless analysis of human breath. In: 2017 IEEE international conference on RFID technology & application (RFID-TA), Warsaw, Poland, Sept 2017, pp 191–195
266. Huang X, Leng T, Georgiou T, Abraham J, Nair RR, Novoselov KS, Hu Z (2018) Graphene oxide dielectric permittivity at GHz and its applications for wireless humidity sensing. Sci Rep 8(1):1–7
267. Lee HJ (2021) Recent progress in radio-frequency sensing platforms with graphene/graphene oxide for wireless health care system. Appl Sci 11(5):2291
268. Costa F, Genovesi S, Monorchio A (2016) Normalization-free chipless RFIDs by using dual-polarized interrogation. IEEE Trans Microw Theory Techn 64(1):310–318
269. Singh R, Singh E, Singh Nalwa H (2017) Inkjet printed nanomaterial based flexible radio frequency identification (RFID) tag sensors for the internet of nano things. RSC Adv 7:48597–48630
270. Mezzanotte P, Palazzi V, Alimenti F, Roselli L (2021) Innovative RFID sensors for Internet of Things applications. IEEE J Microw 1(1):55–65
271. Bandler JW, Biernacki RM, Chen SH, Grobelny PA, Hemmers RH (1994) Space mapping technique for electromagnetic optimization. IEEE Trans Microw Theory Techn 42(12):2536–2544
272. Bandler JW, Biernacki RM, Chen SH, Hemmers RH, Madsen K (1995) Electromagnetic optimization exploiting aggressive space mapping. IEEE Trans Microw Theory Techn 41(12):2874–2882
273. Koziel S, Cheng QS, Bandler JW (2008) Space mapping. IEEE Microw Mag 9(6):105–122
274. Rayas-Sánchez JE, Koziel S, Bandler JW (2021) Advanced RF and microwave design optimization: a journey and a vision of future trends. IEEE J Microw 1(1):481–493
275. Haji-Ahmadi MJ, Nayyeri V, Soleimani M, Ramahi OM (2017) Pixelated checkerboard metasurface for ultra-wideband radar cross section reduction. Sci Rep 7:11437
276. Han X, Xu H, Chang Y, Lin M, Wenyuan Z, Wu X, Wei X (2020) Multiple diffuse coding metasurface of independent polarization for RCS reduction. IEEE Access 8:162313–162321
277. Shan H, Jiang K, Xing J, Jiang T (2022) BPSO and Staggered triangle layout optimization for wideband RCS reduction of pixelate checkerboard metasurface. IEEE Trans Microw Theory Tech 70:3406–3414
278. Kern DJ, Werner DH, Monorchio A, Lanuzza L, Wilhelm MJ (2005) The design synthesis of multiband artificial magnetic conductors using high impedance frequency selective surfaces. IEEE Trans Antennas Propag 53:8–17
279. Bayraktar Z, Gregory MD, Wang X, Werner DH (2012) A versatile design strategy for thin composite planar double-sided high-impedance surfaces. IEEE Trans Antennas Propag 60:2770–2780
280. Ghaderi B, Nayyeri V, Soleimani M, Ramahi OM (2018) Pixelated metasurface for dual-band and multi-polarization electromagnetic energy harvesting. Sci Rep 8:13227
281. Ghadimi A, Nayyeri V, Khanjarian M, Soleimani M, Ramahi OM (2020) Design and simulation of a wideband, wide-angle and polarization-insensitive microwave absorber based on pattern optimization of resistive films. J Phys D: Appl Phys 54:055102

282. Ghadimi A, Nayyeri V, Khanjarian M, Soleimani M, Ramahi OM (2020) A systematic approach for mutual coupling reduction between microstrip antennas using pixelization and binary optimization. IEEE Antennas Wirel Propag Lett 19:2048–2052
283. Borgese M, Costa F, Genovesi S, Monorchio A, Manara G (2018) Optimal design of miniaturized reflecting metasurfaces for ultra-wideband and angularly stable polarization conversion. Sci Rep 8:7651
284. Hojjati A, Soleimani M, Nayyeri V, Ramahi OM (2021) Ternary optimization for designing metasurfaces. Sci Rep 11:17110
285. Genovesi S, Mittra R, Monorchio A, Manara G (2006) Particle swarm optimization for the design of frequency selective surfaces. IEEE Antennas Wirel Propag Lett 5:277–279
286. Saadat-Safa M, Nayyeri V, Ghadimi A, Soleimani M, Ramahi OM (2018) A pixelated microwave near-field sensor for precise characterization of dielectric materials. Sci Rep 9:13310
287. Wadhwani K, Hussaini S, Mazumder A, Syed A (2022) Solvent-Based optimization of CSRR and IDC RF bio-sensors. IEEE Sensors J 22:5651–5661
288. Wang BX, Zhao WS, Wang DW, Wang J, Li W, Liu J (2021) Optimal design of planar microwave microfluidic sensors based on deep reinforcement learning. IEEE Sensors J 21:27441–27449
289. Jin N, Rahmat-Samii Y (2007) Advances in particle swarm optimization for antenna designs: Real-number, binary, single-objective and multiobjective implementations. IEEE Trans Antennas Propag 55:556–567
290. Jun X, Chang H (2009) The discrete binary version of the improved particle swarm optimization algorithm. In: Proc. 2009 IEEE int. conf. management and service science, Beijing, China, 20–22 September 2009, pp 1–6
291. Abdolrazzaghi M, Daneshmand M (2020) Exploiting sensitivity enhancement in micro-wave planar sensors using intermodulation products with phase noise analysis. IEEE Trans Circuits Syst I Regul Pap 67:4382–4395
292. Abdolrazzaghi M, Katchinskiy N, Elezzabi AY, Light PE, Daneshmand M (2021) Noninvasive glucose sensing in aqueous solutions using an active split-ring resonator. IEEE Sensors J 21(17):18742–18755
293. Abdolrazzaghi M, Daneshmand M (2017) Dual active resonator for dispersion coefficient measurement of asphaltene nano-particles. IEEE Sensors J 17:7248–7256
294. Abdolrazzaghi M, Daneshmand M (2018) A phase-noise reduced microwave oscillator sensor with enhanced limit of detection using active filter. IEEE Microw Wirel Compon Lett 28:837–839
295. Abdolrazzaghi M, Zarifi MH, Floquet CF, Daneshmand M (2017) Contactless asphaltene detection using an active planar microwave resonator sensor. Energy Fuels 31:8784–8791
296. Kazemi N, Schofield K, Musilek P (2021) A high-resolution reflective microwave planar sensor for sensing of vanadium electrolyte. Sensors 21:3759
297. Herrojo C, Paredes F, Martín F (2021) Synchronism and direction detection in high-resolution/high-density electromagnetic encoders. IEEE Sensors J 21(3):2873–2882
298. Hosseini N, Baghelani M, Daneshmand M (2020) Selective volume fraction sensing using resonant-based microwave sensor and its harmonics. IEEE Trans Microw Theory Techn 68(9):3958–3968
299. Mosavirik T, Soleimani M, Nayyeri V, Mirjahanmardi SH, Ramahi OM (2021) Permittivity characterization of dispersive materials using power measurements. IEEE Trans Instrum Meas 70:1–8, 6005508
300. Harrison L, Ravan EAM (2020) Material identification using a microwave sensor array and machine learning. Electronics 9(2):288
301. Abdolrazzaghi M, Kazemi N, Daneshmand M (2020) Machine learning to immune microwave sensors from temperature impact. In: 2020 IEEE international symposium on antennas and propagation and North American radio science meeting, Montreal, QC, Canada, pp 843–844
302. Kazemi N, Abdolrazzaghi M, Musilek P (2021) Comparative analysis of machine learning techniques for temperature compensation in microwave sensors. IEEE Trans Microw Theory Techn 69(9):4223–4236

303. Mosavirik T, Hashemi M, Soleimani M, Nayyeri V, Ramahi OM (2021) Accuracy-improved and low-cost material characterization using power measurement and artificial neural network. IEEE Trans Instrum Meas 70:1–9
304. Ricci M et al (2021) Machine-learning-based microwave sensing: a case study for the food industry. IEEE J Emerg Sel Top Circuits Syst 11(3):503–514
305. Aydinalp C, Joof S, Akinci MN, Akduman I, Yilmaz T (2022) Microwave dielectric property retrieval from open-ended coaxial probe response with deep learning. IEEE Access 10:1216–1227
306. Kazemi N, Gholizadeh N, Musilek P (2022) Selective microwave zeroth-order resonator sensor aided by machine learning. Sensors 22:5362
307. Gocen C, Palandoken M (2022) Machine learning assisted novel microwave sensor design for dielectric parameter characterization of water-ethanol mixture. IEEE Sensors J 22(3):2119–2127
308. Shamonina E, Kalinin VA, Ringhofer KH, Solymar L (2002) Magneto-inductive waveguide. Elect Lett 38(8):371–373

Coupled Planar Microwave Resonators and Transmission-Line Structures

Enrique Bronchalo

Abstract This chapter presents the fundamental concepts on coupled microwave structures, including coupled resonators, coupled transmission lines and coupled-line structures. Although many concepts and results here presented have a general validity in the microwave field, the chapter focuses on planar coupled structures, which are the building blocks of the sensors described in the book. The chapter introduces the fundamentals of microwave resonators (lumped and distributed) and coupled planar transmission lines, with special emphasis on coupled identical microstrip lines. Coupled resonators are addressed from the field and circuit approaches, introducing the concepts of electric and magnetic coupling, and resonance-frequency splitting. The last section of the chapter presents the coupled-line directional couplers and two-port coupled-line networks.

In a broad sense, the term electromagnetic coupling is used to mean a non-radiative electromagnetic interaction between two or more resonating or guiding structures. The non-radiative character of the coupling implies that the distance between the coupled structures is small in terms of the involved wavelengths. The electromagnetic fields associated to the coupling between resonating or guiding structures, commonly referred to as near fields, are very sensitive to the environment conditions, making coupled structures good candidates for sensing at local scale.

The purpose of this chapter is to present the basic concepts about the coupled structures that the reader will find in the following chapters. Although several textbooks in Microwave Engineering cover different parts of the chapter contents [1–5], our aim is to provide a compact and unified view of the subject, avoiding complex theoretical developments and focused on planar structures. The first two sections of the chapter present the fundamentals on microwave resonators and coupled transmission lines required to expose the more specific contents of the last two sections. These

E. Bronchalo (✉)
Instituto de Investigación en Ingeniería, Universidad Miguel Hernández de Elche, 03202 Elche, Spain
e-mail: ebronchalo@umh.es

F. Martín and E. Bronchalo (eds.), *Coupled Structures for Microwave Sensing*,
Lecture Notes in Electrical Engineering 1150,
https://doi.org/10.1007/978-3-031-53861-2_2

are dedicated to coupled resonators and coupled-line couplers and transmission-line sections, which are the basic components of the sensing devices presented in the rest of the book.

1 Microwave Resonators: Basic Concepts

An electromagnetic resonator is a physical system where electromagnetic fields can be confined, oscillating at a single resonance frequency, or at any frequency of a particular resonance spectrum. For the purposes of this book, the involved resonant frequencies belong to the microwave bands, and therefore we will use the term microwave resonator to denote this kind of oscillatory systems. There are several types of microwave resonators: purely dielectric, metallic cavities, or based on microwave circuit elements, lumped or distributed. Although many of the definitions and results of this section are valid for all kind of resonators, we will focus on planar transmission-line resonators.

If an isolated microwave resonator had no loss mechanism (ohmic, dielectric or radiative), its electromagnetic energy would remain constant, and the electric and magnetic fields would remain oscillating at the resonance frequency. Real resonators have, to some extent, losses, and therefore the energy stored in a resonator is dissipated at a rate that depends on its storage capacity and the mechanisms of losses. On the other hand, real resonators are not isolated, so that they interchange electromagnetic energy with the environment, which, in practical applications, means sources and loads. In this section, we will pay a special attention to the relations between energy storage, dissipation and interchange with external sources and loads. The interchange of energy between resonators will be addressed in Sect. 3.

1.1 Basic Lumped-Element Resonators

The simpler examples of microwave resonator are the lumped-element series and parallel RCL circuits (Fig. 1). As these circuits are analyzed in many textbook on electromagnetics [1, 4], we just present the basic definitions and results, which can easily be generalized to more complex resonators.

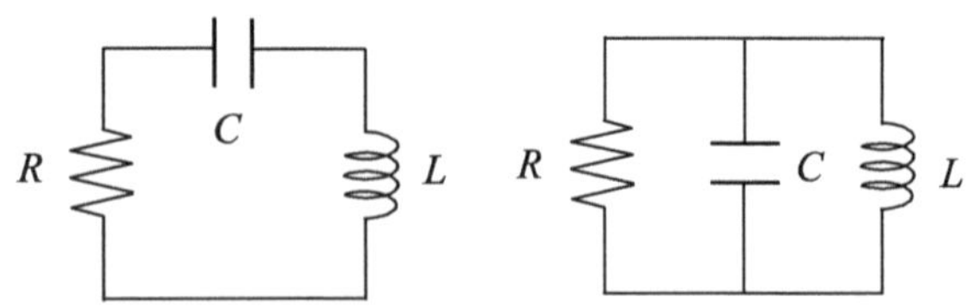

Fig. 1 Left: series RLC resonator. Right: parallel RLC resonator

A simple analysis of these circuits shows that, as long as the energy dissipation is slow enough, the time evolution of currents and voltages has the form of damped oscillations with angular resonance frequency

$$\omega_r = \omega_0 \sqrt{1 - \left(\frac{\alpha}{\omega_0}\right)^2} \tag{1}$$

where

$$\omega_0 = \frac{1}{\sqrt{LC}} \tag{2}$$

and α is the attenuation constant.[1] For the series resonator

$$\alpha = \frac{R}{2L} \tag{3}$$

and for the parallel resonator

$$\alpha = \frac{G}{2C} \tag{4}$$

where $G = 1/R$ is the resistor conductance. While the condition $\alpha < \omega_0$ guarantees the existence of free damped oscillations, for practical resonators $\alpha << \omega_0$, and therefore $\omega_r \cong \omega_0$. On what follows, we will take ω_0 as the angular resonance frequency, assuming implicitly that the attenuation is low enough.

The magnetic energy stored in the inductance and the electric energy stored in the capacitor can be easily calculated. The time-averaged magnetic and electric energies, $< U_M >$ and $< U_E >$, for the series resonator are

$$\langle U_M \rangle = \langle U_E \rangle = \frac{1}{4} L I_0^2 e^{-2\alpha t} = \frac{1}{2} U_0 e^{-2\alpha t} \tag{5}$$

where I_0 is the current amplitude and U_0 the total energy at $t = 0$. A similar result holds for a parallel resonator:

$$\langle U_M \rangle = \langle U_E \rangle = \frac{1}{4} C V_0^2 e^{-2\alpha t} = \frac{1}{2} U_0 e^{-2\alpha t} \tag{6}$$

where V_0 is the voltage amplitude at $t = 0$. Thus, the average magnetic and electric energies are equal at resonance. The instantaneous values of the magnetic and electric energies oscillate around these average values with a 90° phase-shift, that is, if U_M varies as $\cos^2 \omega_0 t$, then U_E varies as $\sin^2 \omega_0 t$. The total energy decreases with time as

[1] This attenuation constant, with units of nepers/second, is not to be confused with the attenuation constant of a transmission line, with units of nepers/meter, that appears later in this chapter. Although we will employ the same symbol for both, they rarely appear in the same context.

$$U = U_0 e^{-2\alpha t} \tag{7}$$

A measure of the relative rate of energy dissipation of an isolated resonator is the unloaded quality factor, Q_u, a dimensionless parameter defined as

$$Q_u = \omega_0 \frac{U}{P} \tag{8}$$

where P is the power dissipated in the resonator. It can be readily shown that, for the series resonator

$$Q_u = \frac{\omega_0 L}{R} \tag{9}$$

and for the parallel resonator

$$Q_u = \frac{\omega_0 C}{G} \tag{10}$$

For both types of resonators,

$$Q_u = \frac{\omega_0}{2\alpha} \tag{11}$$

The magnitudes and results presented in the previous paragraphs refer to isolated resonators, but real resonators interact with other systems, or they would be useless. To continue with simple examples, consider the two resonant circuits of the previous section connected to an external voltage or current source that oscillates at an arbitrary angular frequency ω (Fig. 2).

Usually, the frequency range of interest when using resonators is a more or less narrow band around the resonance frequency. In that case, the input impedance of the series circuit seen by the source can be approximated by

$$Z_{in} = R + j\omega_0 L\left(\frac{\omega}{\omega_0} - \frac{\omega_0}{\omega}\right) \cong R\left(1 + 2jQ_u\frac{\omega - \omega_0}{\omega_0}\right) \tag{12}$$

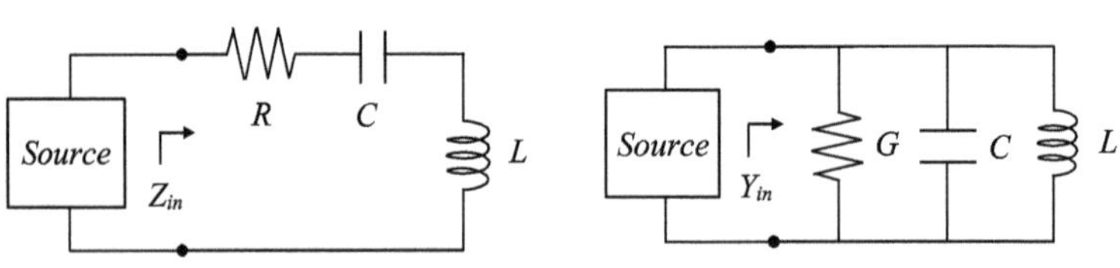

Fig. 2 Series and parallel resonators excited by external sources. The source box represents not only a generator, but also any circuitry connecting it to the resonator

At the resonance frequency, $Z_{in} = R$ and the reactance of the resonator vanishes. The 3 dB bandwidth of $|Z_{in}(\omega)|$ around the minimum $|Z_{in}(\omega_0)| = R$ results

$$\frac{\Delta\omega_{3dB}}{\omega_0} = \frac{1}{Q_u} \tag{13}$$

Similarly, the input admittance of the parallel resonator can be approximated by

$$Y_{in} = G + j\omega_0 C\left(\frac{\omega}{\omega_0} - \frac{\omega_0}{\omega}\right) \cong G\left(1 + 2jQ_u\frac{\omega - \omega_0}{\omega_0}\right) \tag{14}$$

At the resonance frequency, $Y_{in} = G$ and the susceptance of the resonator vanishes. The 3 dB bandwidth of $|Y_{in}(\omega)|$ around the minimum $|Y_{in}(\omega_0)| = G$ is given by (13) as well.

An arbitrary resonator, more complex in general than the simple resonant circuits presented above, has a reactance or susceptance null at the resonance frequency. In the first case, it can be locally described by a RLC series circuit, and in the second by a RLC parallel circuit. Therefore, as the simple RLC resonators, a microwave resonator at frequencies around the resonance can be described by three parameters: its resonance frequency, its quality factor and its equivalent resistance or conductance.

Alternatively, the series resonator impedance around the resonance can be described as

$$Z_{in}(\omega) \cong R + 2jx\frac{\omega - \omega_0}{\omega_0} \tag{15}$$

where x is the reactance slope parameter, defined as

$$x \equiv \frac{\omega_0}{2}\frac{dX}{d\omega}\bigg|_{\omega=\omega_0} \tag{16}$$

being $X(\omega)$ the resonator reactance. In the same way, for a parallel resonator around the resonance

$$Y_{in}(\omega) \cong G + 2jb\frac{\omega - \omega_0}{\omega_0} \tag{17}$$

where b is the susceptance slope parameter defined as

$$b \equiv \frac{\omega_0}{2}\frac{dB}{d\omega}\bigg|_{\omega=\omega_0} \tag{18}$$

being $B(\omega)$ the resonator susceptance.

1.2 Transmission-Line Resonators

At microwave frequencies, owing to parasitic effects or to size issues, resonators are rarely made-up of lumped elements as capacitors and inductors. More frequently, they are made-up of transmission-line sections, metallic cavities or dielectrics. For the purposes of this book, we will focus on transmission-line resonators as the more representative resonators of planar microwave circuits.

Real transmission lines have ohmic losses, dielectric losses (unless the dielectric is the vacuum), and, if the fields are not confined in the plane transversal to the propagation axis, radiative losses as well. Nevertheless, for many purposes losses can be neglected without affecting the results too much.

To confine spatially the electromagnetic energy, the extremes of a resonating transmission line have to be loaded with reactive loads providing reflection coefficients equal to 1 in magnitude. As a simple example of transmission-line resonator, consider the line section with opens at both ends represented in Fig. 3a. The voltage and current waves propagating forward and backward in the line of Fig. 3a have to fulfill the open-end conditions at the extremes of the line. If the losses in the line can be considered negligible, the null-current condition in $x = 0$ leads to the following voltage and current distributions:

$$V(x) = V_0 cos(\beta x) \tag{19}$$

$$I(x) = -j\frac{V_0}{Z_0}sin(\beta x) \tag{20}$$

where the voltage amplitude V_0 is arbitrary, Z_0 is the line impedance and $\beta = \omega/v$ is the phase constant, and v is the propagation velocity. It is important to note that voltage and current waves in formulas (19) and (20) are 90° out of phase. Therefore, no power is dissipated in the resonator, in coherence with the assumption about the lossless character of the line.

As the line is supposed to be lossless,

$$Z_0 = \sqrt{L/C} \tag{21}$$

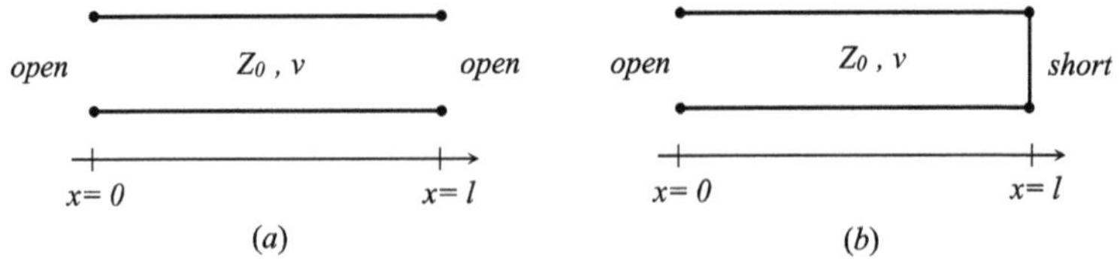

Fig. 3 *a*: Half-wave transmission-line resonator. *b*: Quarter-wave transmission-line resonator. Z_0 is the characteristic impedance, and v the propagation velocity

$$v = \frac{1}{\sqrt{LC}} \tag{22}$$

where L and C are respectively the line per-unit-length inductance and capacitance. The null-current condition in $x = l$ is fulfilled only for a discrete set of angular frequencies ω_n ($n = 1, 2\ldots$), which are the resonant frequencies of the line:

$$f_n = n\frac{v}{2l} \tag{23}$$

Thus, the transmission-line resonator has an infinite set of resonant frequencies. This is a general result, valid for resonators made-up of distributed capacitances and inductances. If condition (23) is expressed in terms of wavelengths, the results is

$$\lambda_n = \frac{2l}{n} \tag{24}$$

For the fundamental resonant mode ($n = 1$), the resonator length is half a wavelength, and for this reason the line of Fig. 3a is referred to as half-wave resonator. If the lossless half-wave resonator is connected to a source, the resonator input admittance seen by the source is

$$Y_{in} = jY_0 tg(\beta l)$$

where $Y_0 = 1/Z_0$ is the line characteristic admittance. Real transmission lines always have losses, and therefore the input admittance of the half-wave resonator at resonance has a finite real value. If the attenuation constant α of the transmission line is small enough ($\alpha l < < 1$), the input admittance of the half-wave resonator can be approximated by

$$Y_{in} \cong Y_0(\alpha l + j \cdot tg(\beta l)) \tag{25}$$

At the resonant frequencies (23), the input susceptance vanishes, and $Y_{in} \cong Y_0 \alpha l$. Therefore, the half-wave resonator behaves as a parallel RLC resonator.

Another common transmission-line resonator is the quarter-wave resonator shown in Fig. 3b. The conditions of null-current at $x = 0$ and null-voltage at $x = l$ leads to the following set of resonant frequencies and wavelengths:

$$f_n = \frac{2n - 1}{4}\frac{v}{l} \tag{26}$$

$$\lambda_n = \frac{4}{2n - 1}l \tag{27}$$

The fundamental resonant mode correspond to a line length of a quarter-wave. The voltage and current distributions (19) and (20) are also valid for the quarter-wave line. The input admittance at frequencies near the fundamental frequency is, for low losses ($\alpha l << 1$),

$$Y_{in} \cong Y_0(\alpha l - j \cdot ctg(\beta l)) \tag{28}$$

Again, the input susceptance vanishes at resonance, and thus the half-wave resonator behaves as a parallel resonator.

Both the half-wave and the quarter-wave resonators are examples of resonators that can oscillate freely (without external sources) at the resonance frequencies. Other transmission-line circuits are considered resonating lines, as long as their input reactances or susceptances vanish at specific frequencies, even if they cannot support free oscillations at those frequencies. For instance, the short-circuited line of Fig. 3b presents a purely real input impedance (zero, in the lossless case) when its length is a half wavelength, thus behaving as a series resonator for an external source. However, the null-current conditions at the left end, required if the resonator were isolated, cannot be fulfilled for $l = \lambda/2$.

The unloaded quality factor of a transmission-line resonator can be calculated by finding the equivalent lumped-circuit resonator, determined by three parameters (R, L and C), with the conditions that both resonators have the same resonance frequency, resistance or conductance, and reactance or susceptance parameter. Then, (9) or (10) can be used to find Q_u.

Another possibility consists in evaluating the energy stored in the resonator and the power dissipated by the different mechanisms of losses. We will follow this way to calculate, as an example, the unloaded quality factor of a half-wave resonator. For simplicity, we will consider only dielectric and ohmic losses, neglecting possible radiative losses. To account for the distributed losses in the transmission line, we introduce a per-unit-length conductance G for the dielectrics and a per-unit-length resistance R for the conductors. If their values are small enough ($G << \omega C$, $R << \omega L$ at the frequencies of interest), the voltage and current distributions in the line do not differ very much from those of the lossless case, (19) and (20). Therefore, we can reasonably use these distributions to calculate the dissipated power. The power dissipated in the conductors is

$$P_c = \int_0^l \frac{1}{2} R |I(x)|^2 dx = Rl \left(\frac{V_0}{2Z_0} \right)^2 \tag{29}$$

and in the dielectrics

$$P_d = \int_0^l \frac{1}{2} G |V(x)|^2 dx = Gl \left(\frac{V_0}{2} \right)^2 \tag{30}$$

It can easily be shown that, at resonance, the time-averaged electric and magnetic stored energies are equal, as for the lumped resonant circuits of Sect. 1.1. Then, the total stored energy can be calculated simply as the maximum stored electric energy, or twice the average electric energy:

$$U = \int_0^l \frac{1}{2} C|V(x)|^2 dx = Cl\left(\frac{V_0}{2}\right)^2 \tag{31}$$

The resulting unloaded quality factor is

$$Q_u = \omega_0 \frac{U}{P_c + P_d} = \frac{\beta_0}{2\alpha} \tag{32}$$

where ω_0 is the angular resonance frequency, β_0 the phase constant at resonance and α the attenuation constant of the transmission line. Here we have used the usual approximations for low-loss transmission lines, (21), (22) and

$$\alpha \cong \frac{1}{2}\left(\frac{R}{Z_0} + \frac{G}{Y_0}\right) \tag{33}$$

Applying the above procedure to the quarter-wave resonator of Fig. 3b leads also to (33). However, this result cannot be generalized to more complex transmission-line resonators.

Looking at (32), we can see that different dissipative mechanisms contribute to the unloaded quality factor in the form of an inverse-sum rule,

$$\frac{1}{Q_u} = \frac{1}{Q_c} + \frac{1}{Q_d} \tag{34}$$

where the conductor (Q_c) and dielectric (Q_d) contributions to quality factor are given by

$$Q_c = \frac{\omega_0 L}{R} \tag{35}$$

$$Q_d = \frac{\omega_0 C}{G} \tag{36}$$

The dielectric loss contribution to the quality factor can be expressed in a more compact way, taking into account that, for a transmission line with homogeneous dielectric like the one consider here,

$$\frac{G}{\omega C} = tan\delta \tag{37}$$

where $\tan\delta$ is the loss-tangent of the dielectric. Then, for transmission-line resonators with homogeneous dielectric,

$$Q_d = \frac{1}{tan\delta} \tag{38}$$

The sum-rule (34) can be generalized to the case where N dissipative mechanism (eventually including radiative losses) are present,

$$\frac{1}{Q_u} = \sum_{k=1}^{N} \frac{1}{Q_k} \tag{39}$$

This is a general result, valid for any kind of resonator.

1.3 Resonators in One-Port Configuration

For practical uses, resonators are connected to external sources and loads in one- or two-port configurations. The connections are realized in the form of coupling networks, usually (but not necessarily) lossless, that regulate the exchange of electromagnetic power between the resonator and the sources or loads. Figure 4 shows a lumped-element parallel resonator in one-port, or reflection, configuration. The resonator is connected to a source of internal impedance Z_g through a lossless coupling network that, in planar circuits, is integrated with the resonator in a single one-port network.

The measurement of the resonator response is performed by a VNA or equivalent equipment connected to the input of the coupling network. Thus, the measured magnitude is the reflection coefficient S_{11} of the one-port network composed of the coupling network and the resonator. However, the analysis of the resonator response is simpler if referred to the connection plane between the coupling network and the resonator. In most practical cases, the reflection coefficient in that plane, Γ_{in}, differs slightly from the measured S_{11}, being the main difference a phase shift [6]. In the case where the coupling network is an inverter, a useful case for filter design, $S_{11} = -\Gamma_{\text{in}}$ [7, 8]. More detailed analysis of the relation between S_{11} and Γ_{in} and the effects of the coupling network can be found in [9, 10].

The whole network connected to the resonator can be substituted by an equivalent current source in parallel with an admittance $G_{ex} + jB_{ex}$, where G_{ex} and B_{ex} depend on the coupling network (Fig. 5). In practical cases, and especially if the resonator has a high quality factor, the susceptance B_{ex} is too low around the resonance frequency

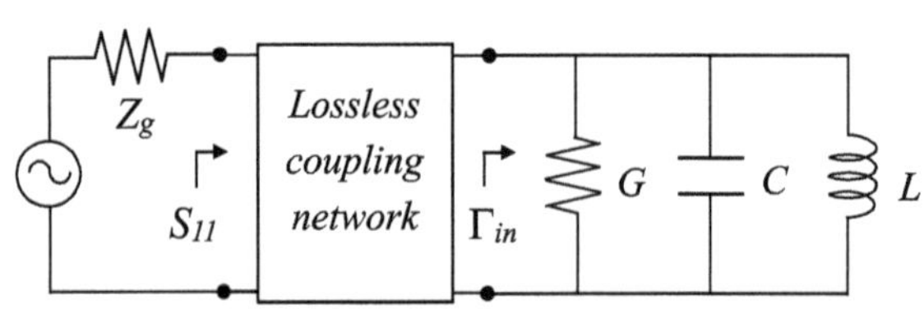

Fig. 4 A parallel resonant circuit connected to a source by means of a lossless coupling network

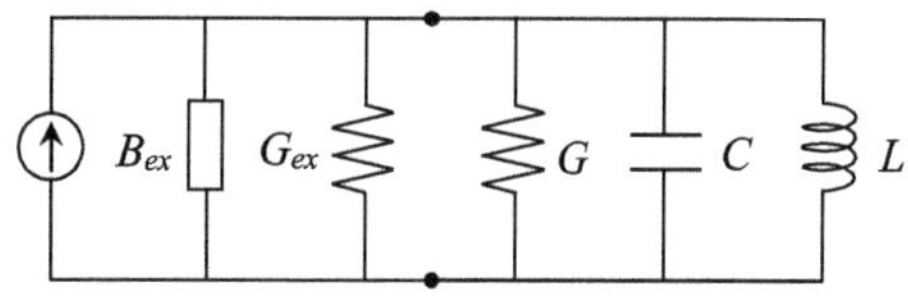

Fig. 5 Parallel circuit equivalent to the circuit of Fig. 4

to play a relevant role, besides a very small frequency shift in the resonance [11, 12]. For the sake of simplicity, we will neglect it on what follows. The conductance G_{ex}, however, adds to G and reduces the quality factor. Using (10) for a parallel resonator, the quality factor of the resonator connected to the source, or *loaded* quality factor, Q_L, is now

$$Q_L = \frac{\omega_0 C}{G + G_{ex}} \tag{40}$$

As in the precedent section, the loaded quality factor can be expressed by adding inversely the different contributions, in this case the resonator and source components:

$$\frac{1}{Q_L} = \frac{1}{Q_u} + \frac{1}{Q_{ex}} \tag{41}$$

where Q_{ex} is the external quality factor:

$$Q_e = \frac{\omega_0 C}{G_{ex}} \tag{42}$$

A measure of the relative magnitude of G_{ex} compared to G is the coupling factor between the resonator and the source, κ, defined as the ratio of the power dissipated in the source internal impedance, P_g, to the power dissipated in the resonator, P_r. Considering that in the circuit of Fig. 5 the voltage across G_{ex} and G is the same,

$$\kappa \equiv \frac{P_g}{P_r} = \frac{G_{ex}}{G} = \frac{Q_u}{Q_{ex}} \tag{43}$$

The relation between the loaded and unloaded quality factors results

$$Q_L = \frac{Q_u}{1 + \kappa} \tag{44}$$

The coupling factor κ affects to the reflection coefficient that the resonator presents to the source. If B_{ex} is neglected in the circuit of Fig. 5, the input reflection coefficient of the resonator is

$$\Gamma_{in}(\omega) = \frac{G_{ex} - Y_r}{G_{ex} + Y_r} = \frac{G_{ex} - G - j\omega_0 C\Omega}{G_{ex} + G + j\omega_0 C\Omega} \tag{45}$$

where

$$\Omega = \frac{\omega}{\omega_0} - \frac{\omega_0}{\omega}$$

Using the coupling factor κ and the unloaded quality factor, Γ_{in} results

$$\Gamma_{in}(\omega) = \frac{\kappa - 1 - jQ_u\Omega}{\kappa + 1 + jQ_u\Omega} \tag{46}$$

The magnitude response

$$|\Gamma_{in}(\omega)| = \sqrt{\frac{(\kappa - 1)^2 + (Q_u\Omega)^2}{(\kappa + 1)^2 + (Q_u\Omega)^2}} \tag{47}$$

is used to find the unloaded quality factor and the coupling factor. At resonance, $\Omega = 0$, and the magnitude of $|\Gamma_{in}|$ reaches its minimum value Γ_0:

$$\Gamma_0 \equiv |\Gamma_{in}(\omega_0)| = \left|\frac{\kappa - 1}{\kappa + 1}\right| \tag{48}$$

If $\kappa = 1$, the reflection coefficient at resonance is 0, that is, the resonator is matched to the source. In this situation, it is said that the resonator is critically coupled to the source.

Expression (48) contains an ambiguity, as two solutions for κ exist for a particular value of Γ_0:

$$\kappa_1 = \frac{1 + \Gamma_0}{1 - \Gamma_0} > 1 \tag{49a}$$

$$\kappa_2 = \frac{1 - \Gamma_0}{1 + \Gamma_0} < 1 \tag{49b}$$

The first solution corresponds to the case $G_{ex} > G$, and in this case the resonator is said to be over-coupled. For the second solution, $G_{ex} < G$, and the resonator is said to be under-coupled. As example, the graph of Fig. 6 shows $|\Gamma_{in}(\omega)|$ for a parallel resonator with $Q_u = 50$ and three values of κ (0, 1/2 and 2), the last two corresponding to $\Gamma_0 = 1/3$.

Practical microwave resonator frequently have high quality factors, and therefore the resonance is narrow, so that only frequencies in a narrow range around ω_0 are relevant. In that range,

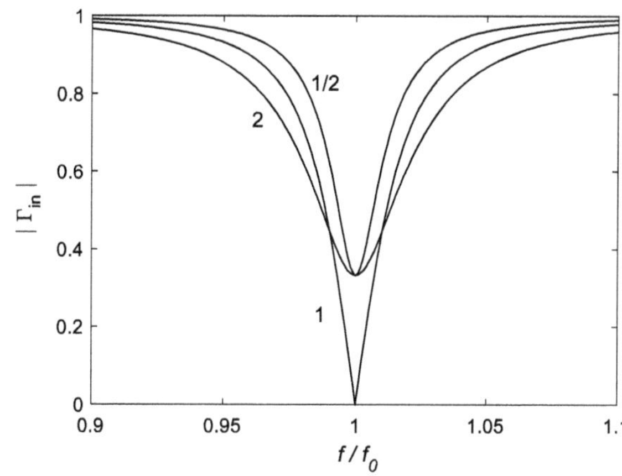

Fig. 6 Magnitude of the reflection coefficient of a parallel resonator with $Q_u = 50$ for three values of the coupling factor. These curves are valid for a series resonator with the same values of Q_u and κ (see text)

$$\Omega = \frac{\omega}{\omega_0} - \frac{\omega_0}{\omega} \cong 2\frac{\omega - \omega_0}{\omega_0} \tag{50}$$

Introducing this approximation in (47), the 3 dB-bandwidth of the resonance is found to be

$$\frac{\Delta f_{3dB}}{f_0} = \frac{1}{Q_u}\frac{2\sqrt{1 - 2\Gamma_0^2}}{1 \pm \Gamma_0} \tag{51}$$

where the sign (+) correspond to the under-coupled and the sign (−) to the over-coupled case. Therefore, the minimum of the reflection response and the 3 dB bandwidth are *almost* enough to obtain the unloaded quality factor, except for the sign in the denominator: it is necessary to know as well whether the resonator is under- or over-coupled. Plotting $\Gamma_{in}(\omega)$ in the complex plane (i.e., in the Smith chart) removes this ambiguity. For $\kappa = 1$ (critical coupling), the curve passes through the origin of the complex plane, it encircles the origin for $\kappa > 1$ (over-coupling), and it does not reach the origin for $\kappa < 1$ (under-coupling). The graph of Fig. 7 shows $\Gamma_{in}(\omega)$ plotted in the complex plane for the example of Fig. 6.

An alternative to using (51) is to measure the resonance bandwidth for a power fraction different from 3 dB, that is, for a particular $|\Gamma_{in}|$ value [13]. Solving (47) with the approximation (50) (unless the quality factor is too low), and measuring Γ_0 at the resonance minimum provides the relation between the bandwidth and Q_u at the required power fraction. Indeed, it is necessary to know whether the resonator is over- or under-coupled to remove the ambiguity about κ.

As the magnitude of the reflection coefficient, the phase response provides relevant information about the resonance. For a parallel resonator, the phase of $\Gamma_{in}(\omega)$ can be obtained directly from (46). Figure 9 shows the phase response for a parallel resonator and the same κ and Q_u values of Fig. 6 and Fig. 7.

The analysis of the series resonator follows the same lines as the previous analysis of the parallel resonator. The equivalent circuit analogous to the parallel circuit of

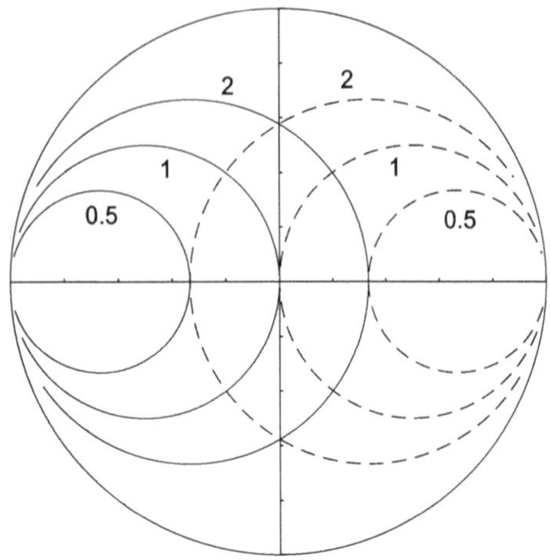

Fig. 7 Reflection coefficients of a parallel resonator (solid line) and a series resonator (dashed line) with $Q_u = 50$ and three values of the coupling factor, represented in the complex plane (Smith chart)

Fig. 5 is the series circuit of Fig. 8, where the source impedance, transformed by the coupling network, is $R_{ex} + jX_{ex}$. The reactance X_{ex} is considered to have very small influence in the resonance, and the coupling factor is defined as

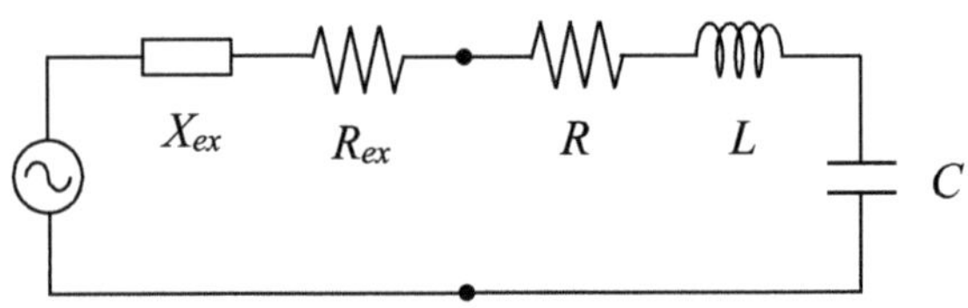

Fig. 8 Circuit model of a series resonator coupled to a generator by a lossless coupling network. The impedance $R_{ex} + jX_{ex}$ represents the generator impedance transformed by the coupling network

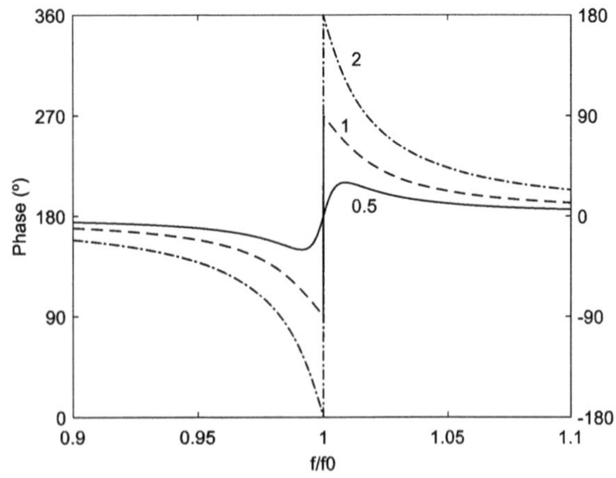

Fig. 9 Phase response of a parallel (left axis) and a series resonator (right axis) for $Q_u = 50$ and three values of the coupling factor

$$\kappa \equiv \frac{P_g}{P_r} = \frac{R_{ex}}{R} \tag{52}$$

After the substitutions $G_{ex} \leftrightarrow R_{ex}$, $G \leftrightarrow R$ and $C \leftrightarrow L$, expressions (40) to (44), (47) to (49) and (51) hold for the series resonator. Neglecting X_{ex}, the input reflection coefficient of the resonator results

$$\Gamma_{in}(\omega) = \frac{Z_r - R_{ex}}{Z_r + R_{ex}} = \frac{1 - \kappa + jQ_u\Omega}{1 + \kappa + jQ_u\Omega} \tag{53}$$

which is, except for a phase shift of 180°, equivalent to (46). The phase responses of the parallel and series resonators are represented in Fig. 9.

It has to be emphasized that the plots of Figs. 6, 7 and Fig. 9 refer to the input port of the resonator, that is not, in the general case, the measurement reference plane of a VNA or equivalent device. Among other effects of lesser importance, the transformation from Γ_{in} to S_{11} implies displacement of the circles of Fig. 7 and phase shifts in the curves of Fig. 9 [14].

The practical realization of coupling networks to achieve the required κ or Q_{ex} value can follow different approaches. Figure 10 shows some examples of coupling networks for microstrip open-loop resonators [2] and coplanar quarter-wave resonators [15]. The indicated geometrical parameters serve to control the coupling strength.

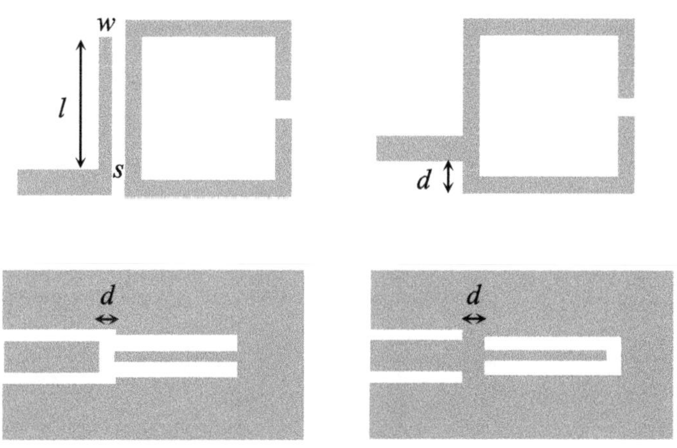

Fig. 10 Top: Open-loop microstrip resonators coupled to an input line through a coupled-line section (left) and a tapped-line (right). Bottom: capacitive (left) and inductive (right) coupling between an input line and a quarter-wave coplanar resonator. Letters indicate the geometrical parameters that control the coupling strength

1.4 Resonators in Two-Port Configuration

Resonators in two-port configuration can be connected to sources and loads in two different manners: transmission or absorption [13]. In the transmission mode, a series resonator is connected in series, or a parallel resonator in parallel, with the source and load, thus having a passband response. In the absorption mode, a series resonator is connected in parallel, or a parallel resonator in series, with the source and load, resulting a stopband response. The four possibilities are plotted in Fig. 11, along with the equivalent circuits at resonance and at frequencies very far from resonance, and the qualitative transmission response. Z_r represents the impedance of a series resonator, and Y_r the admittance of a parallel resonator. To simplify the drawing, the coupling networks have not been represented.

In the transmission mode, the coupling networks at input and output ports can be different, in which case there are different input and output coupling factors, κ_1 and κ_2. In the absorption mode, the resonator is coupled directly to the line connecting the source and the load, and therefore only a coupling factor κ is needed to describe the coupling. Figure 12 shows one example of each connection mode. In the absorption mode, a series resonator is magnetically coupled to a transmission line, and the coupling transforms the series RCL circuit into a parallel RCL resonator. This coupling mechanism is frequently used to couple a dielectric resonator to a transmission line [16, 17].

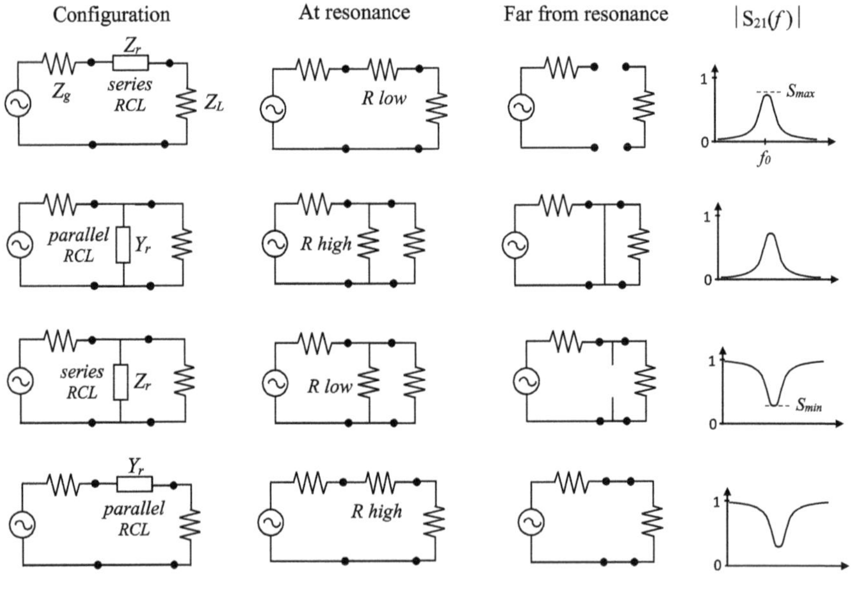

Fig. 11 Resonators in two-port configurations. Upper rows: transmission configurations for series and parallel resonators. Lower rows: absorption configuration for series and parallel resonators. The coupling networks are not represented

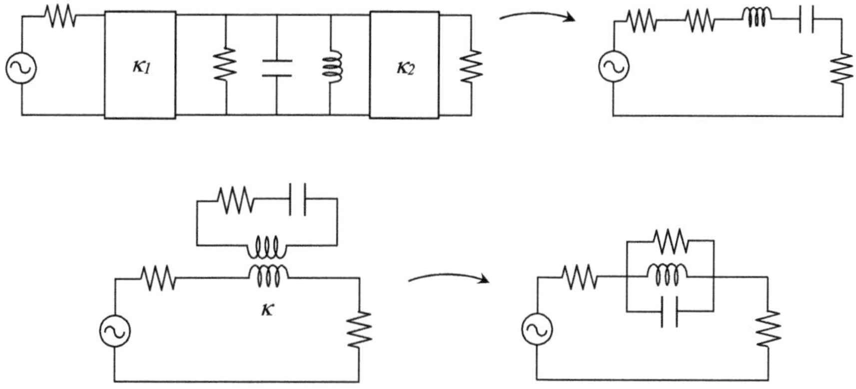

Fig. 12 Top: parallel resonator in transmission mode, with input and output coupling networks. Bottom: series resonator magnetically coupled to a transmission line (left) and its equivalent parallel resonator in absorption mode (right)

The relations between the unloaded, loaded and external quality factors are similar to those of the reflection configuration. For the transmission mode

$$\kappa_{1,2} = \frac{Q_u}{Q_{ex1,2}} \tag{54}$$

$$Q_L = \frac{Q_u}{1 + \kappa_1 + \kappa_2} \tag{55}$$

where $Q_{ex,i}$ is the external quality factor of port i. Frequently the coupling networks equal in the input and output ports, in which case

$$Q_L = \frac{Q_u}{1 + 2\kappa} \tag{56}$$

For the absorption mode

$$Q_L = \frac{Q_u}{1 + \kappa} \tag{57}$$

$$\kappa = \frac{Q_u}{Q_{ex}} \tag{58}$$

where Q_{ex} includes the dissipation in the source and in the load.

The transmission (S_{21}) or reflection (S_{11}) responses of the different connection modes can be analyzed in a similar way to that of the one-port configuration of the precedent section [18, 19]. The analysis is simpler if the generator and load impedances, Z_g and Z_L, are real and equal. Although this is a particular case, it supposes the most frequent practical situation. For the transmission mode, the S_{21}

parameter, referred to ports placed at the input and output of the coupling networks-resonator cascade, is [14, 19, 20]

$$S_{21} = e^{j\phi} \frac{S_{max}}{1 + jQ_L\Omega} \tag{59}$$

where the phase constant ϕ depends on the coupling networks and S_{max} is the maximum transmission magnitude, which occurs at resonance:

$$S_{max} \equiv |S_{21}(f_0)| = \frac{2\sqrt{\kappa_1 \kappa_2}}{1 + \kappa_1 + \kappa_2} \tag{60}$$

For equal input and output couplings, expressions (55) and (60) lead to the following relation between loaded and unloaded quality factors:

$$Q_u = \frac{Q_L}{1 - S_{max}} \tag{61}$$

Substituting the approximation (50) in (59), it is straightforward to obtain the relation between the loaded quality factor and the 3 dB bandwidth of the transmission peak:

$$Q_L = \frac{f_0}{\Delta f_{3dB}} \tag{62}$$

From (59), the phase of the transmission parameter results

$$\phi_{21} = \phi - arctg(Q_L\Omega) \cong \phi - arctg\left(\frac{2(f - f_0)}{\Delta f_{3dB}}\right) \tag{63}$$

It is easily seen that the phase-shift between the 3 dB frequencies is 90°. As example, the magnitude and phase responses for a resonator with $Q_u = 100$ and $\kappa_1 = \kappa_2 = 2$ in transmission mode are plotted in Fig. 13.

The analysis of the S_{21} response for the absorption mode leads to [17]

$$S_{21} = e^{j\phi} \frac{S_{min} + jQ_L\Omega}{1 + jQ_L\Omega} \tag{64}$$

where the phase ϕ depends, as in the precedent case, on the details of the coupling network, and where the minimum transmission, which occurs at resonance, is

$$S_{min} \equiv |S_{21}(f_0)| = \frac{\kappa}{1 + \kappa} \tag{65}$$

The relation (57) between Q_L and Q_u results

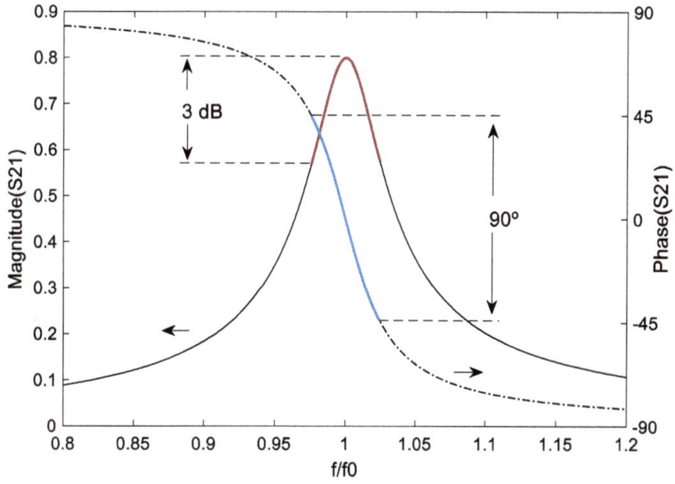

Fig. 13 Transmission response in magnitude (left axis) and phase (right axis) of a resonator with $Q_u = 100$ and $\kappa_1 = \kappa_2 = 2$ in transmission mode. The phase ϕ in (63) has been taken as 0. The colored parts of the curves correspond to the 3 dB bandwidth around the resonance frequency

$$Q_u = \frac{Q_L}{1 - S_{min}} \tag{66}$$

From (64), the loaded quality factor can be obtained from the 3 dB bandwidth of the transmission valley (see Fig. 14):

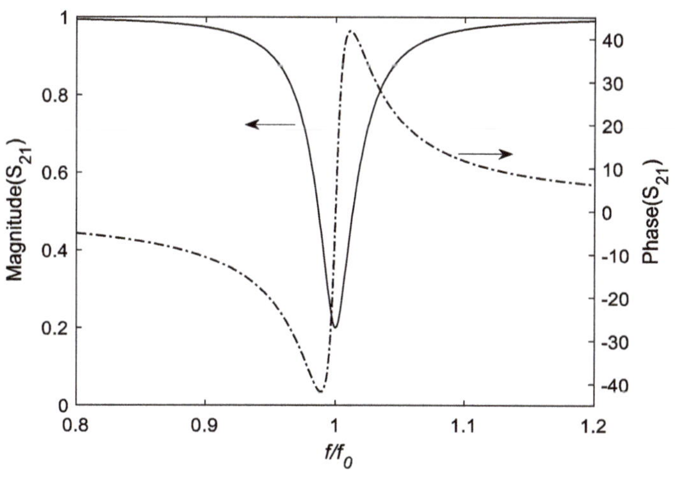

Fig. 14 Transmission response in magnitude (left axis) and phase (right axis) of a resonator in absorption mode with $Q_u = 100$ and $\kappa = 0.25$. The phase ϕ in (68) has been take as 0

$$Q_L = \frac{f_0}{\Delta f_{3dB}}\sqrt{1 - 2S_{min}^2} \tag{67}$$

The phase response for the absorption mode is

$$\phi_{21} = \phi + arctg\left(\left(\frac{\kappa + 1}{\kappa}\right)Q_L\Omega\right) - arctg(Q_L\Omega) \tag{68}$$

Figure 14 shows the magnitude and phase transmission responses of a resonator in absorption mode with $Q_u = 100$ and $\kappa = 0.25$.

The reflection coefficients S_{11} for absorption and transmission modes can be calculated from the corresponding S_{21} ((59) or (64)). Firstly, the cascade input coupling network-resonator-output coupling network is transformed into an equivalent resonator series- or shunt-connected to the generator and the load (see two examples in Fig. 12), resulting in one of the four two-port networks of Fig. 11. In the particular but frequent case of equal source and generator loads that we are considering, it can be easily shown [1] that, for the shunt-connected networks,

$$S_{11} = S_{21} - 1 \tag{69}$$

and for the series-connected networks,

$$S_{11} = 1 - S_{21} \tag{70}$$

2 Coupled Planar Transmission Lines

Coupled transmission lines are constitutive elements of many microwave structures, like directional couplers or coupled-line filters. For the purposes of this book, they have interest specifically in relation to coupled planar resonators and directional couplers, which are the basic components of the sensors addressed in the following chapters.

Planar transmission lines, like microstrip, coplanar or slotline are open lines that involve at least two dielectrics, typically a low-loss substrate and air. The exception is the one-dielectric stripline, scarcely used for sensor applications as the strip is buried in the dielectric, what complicates the contact with the medium to be sensed. The fundamental propagation mode in a multidielectric transmission line is a quasi-TEM mode, which is intrinsically dispersive. For narrow or moderate bandwidth applications, the variation of the propagation properties with frequency can be neglected, and the propagation mode behaves essentially like a TEM mode in a dielectric medium characterized by an effective permittivity. However, when two or more planar lines propagating quasi-TEM modes couple, the propagation modes of

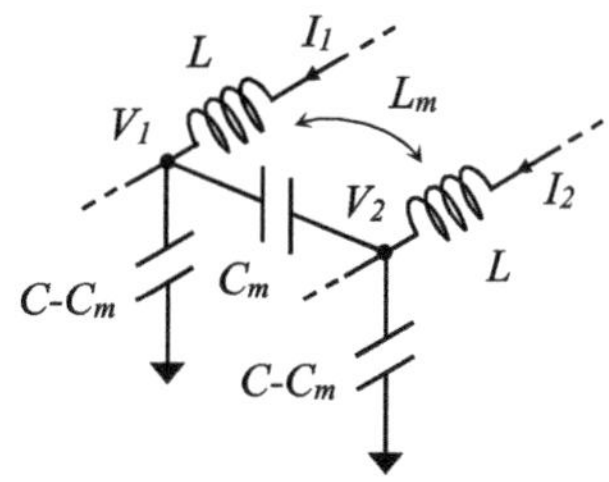

Fig. 15 Distributed element model of a pair of coupled identical lossless transmission lines

the coupled system have different effective permittivities, even although the variation of these permittivities with frequency in the bandwidth of interest is negligible.

The technical literature dedicated to propagation in coupled transmission lines is very wide and extends back many decades [21–30]. It covers a variety of coupled-line propagating structures with two or more conductors in dielectric media with different properties (inhomogeneous in the cross section or along the propagation axis, lossy, dispersive, anisotropic, nonlinear…). For the purposes of this book, it is enough to present the basic theory for the simple case of two symmetric coupled planar lines on a lossless substrate, along with a short introduction to the propagation in asymmetric coupled lines. The effects of losses [31–34] and dispersion [33–35] can be incorporated to the theory if needed.

2.1 *Distributed Circuit Model of Coupled Transmission Lines*

The classical distributed-parameter model of a single transmission line can be extended to a coupled-line pair by adding mutual capacitances and inductances. Figure 15 shows the distributed circuit model[2] of two identical coupled lines of length dx, where the self-inductance L, the mutual inductance L_m, the capacitance to ground $C\text{-}C_m$, and the mutual capacitance C_m are per-unit-length magnitudes (these magnitudes, multiplied by dx, give the absolute electrical parameters of the element). If line 2 is grounded, the per-unit-length charge in line 1 is CV_1, indicating that C is the self-capacitance of each uncoupled line in presence of the other line. As the separation between the lines is increased, C approaches the capacitance of the isolated line.

Assuming harmonic time-dependence $exp(j\omega t)$, the analysis of voltages and currents leads to the following system of coupled differential equations:

$$\frac{d}{dx}\begin{pmatrix} V_1 \\ V_2 \end{pmatrix} = -j\omega \begin{pmatrix} L & L_m \\ L_m & L \end{pmatrix} \begin{pmatrix} I_1 \\ I_2 \end{pmatrix} \tag{71}$$

[2] Note that the two lines share a common ground, a feature of unbalanced or single-ended planar lines like microstrip or coplanar waveguide transmission lines.

$$\frac{d}{dx}\begin{pmatrix} I_1 \\ I_2 \end{pmatrix} = -j\omega\begin{pmatrix} C & -C_m \\ -C_m & C \end{pmatrix}\begin{pmatrix} V_1 \\ V_2 \end{pmatrix} \tag{72}$$

These equations can be combined to give a pair of second order differential equation for the voltages:

$$\frac{d^2}{dx^2}\begin{pmatrix} V_1 \\ V_2 \end{pmatrix} = -\omega^2\begin{pmatrix} L & L_m \\ L_m & L \end{pmatrix}\begin{pmatrix} C & -C_m \\ -C_m & C \end{pmatrix}\begin{pmatrix} V_1 \\ V_2 \end{pmatrix} \tag{73}$$

An identical system of equations is obtained for the currents. There are two independent solutions of (73), or propagation modes, in the form of sinusoidal waves, i.e., with a dependence $exp(\pm j\beta x)$. As the coupled system is symmetric, the modes have a specific symmetry, even ($V_1 = V_2$, $I_1 = I_2$) or odd ($V_1 = -V_2$, $I_1 = -I_2$). For the even (odd) mode, the voltage and current waves in each line evolve as if it were an isolated line with equivalent even (odd) capacitance and inductance. For the even mode, these equivalent parameters are

$$C_e = C - C_m \tag{74a}$$

$$L_e = L + L_m \tag{74b}$$

and for the odd mode

$$C_o = C + C_m \tag{75a}$$

$$L_o = L - L_m \tag{75b}$$

These even and odd parameters can be inserted in the usual formulas of transmission line theory to obtain the propagation velocity and characteristic impedance of each mode:

$$Z_{e,o} = \sqrt{\frac{L_{e,o}}{C_{e,o}}} \tag{76}$$

$$v_{e,o} = \frac{1}{\sqrt{L_{e,o}C_{e,o}}} \tag{77}$$

The preceding formal analysis is equivalent to an even–odd analysis of the coupled structure, allowed by its symmetry. Placing a magnetic wall (open) in the symmetry plane for the even voltage distribution and an electric wall (short) for the odd distribution leads to (74) and (75).

The mutual electric and magnetic influence between the lines can be characterized by two dimensionless parameters, the capacitive (k_C) and inductive (k_L) coupling factors, defined as

$$k_C = \frac{C_m}{C} \tag{78}$$

$$k_L = \frac{L_m}{L} \tag{79}$$

Later in this chapter, in the context of inter-resonator coupling, we will find two parameters that have a very similar definition and meaning. We would like to emphasize that the just defined coupling parameters k_C and k_L, involving per-unit-length capacitances and inductances, have a local nature, while the analogous coupling parameters that will be defined in Sect. 3.2 refer globally to a coupled resonating system.

In terms of k_C and k_L, the properties of the even and odd modes can be expressed in the following form:

$$Z_{e,o} = Z\sqrt{\frac{1 \pm k_L}{1 \mp k_C}} \tag{80}$$

$$v_{e,o} = \frac{v}{\sqrt{(1 \pm k_L)(1 \mp k_C)}} \tag{81}$$

where the upper sign is for even and the lower is for odd, and where

$$Z = \sqrt{L/C} \tag{82}$$

$$v = \frac{1}{\sqrt{LC}} \tag{83}$$

It is worthy to mention that Z and v do not coincide with the impedance and velocity of the uncoupled lines, although they tend to the uncoupled values as the line separation increases, i.e., as the coupling weakens. Formula (80) shows that $Z_e > Z > Z_o$.

For pure TEM propagation, like in the case of coupled striplines in a homogeneous dielectric, the velocities of the even and odd modes are equal to the speed of light in the dielectric. This implies that $k_C = k_L = k$ and $Z_e Z_o = Z^2$. The coupling coefficient k is then unique, and it relates with the even and odd impedances as

$$k = \frac{Z_e - Z_o}{Z_e + Z_o} \tag{84}$$

For quasi-TEM propagation, k_C and k_L are, in general, different, and so they are the velocities of the even and odd modes. In the same way as for quasi-TEM propagation in a single line, an effective permittivity can be defined for each propagating mode in terms of the per-unit-length capacitances of the mode in the actual medium and in the vacuum, with the same conductor geometry:

$$\varepsilon_{re,o} = \frac{C_{e,o}}{C_{0e,o}} \tag{85}$$

where the sub-index 0 indicates the vacuum. In the very common case of non-magnetic substrate, the inductances L and L_m does not differ from those of a coupling structure with the same conductor geometry in a vacuum dielectric background. Therefore, the mode velocities can be expressed as

$$v_{e,o} = \frac{1}{\sqrt{L_{e,o}C_{e,o}}} = \frac{1}{\sqrt{L_{0e,o}C_{0e,o}\varepsilon_{re,o}}} = \frac{c}{\sqrt{\varepsilon_{re,o}}} \tag{86}$$

where c is the speed of light in vacuum.

The preceding considerations show that, for coupled planar lines on non-magnetic substrates, k_L is equal to k_C for the same conductor geometry in vacuum. Then, (78) and (79) indicate that the calculation of k_C and k_L for a given geometry reduces to an electrostatic problem: calculating the capacitances C and C_m with dielectric substrate and in vacuum. For microstrip coupled lines, the numerical and experimental results indicate that $k_L > k_C$ (see Sect. 2.2). This implies that $\varepsilon_{re} > \varepsilon_{ro}$, and thus the odd mode is faster than the even mode.

If the coupling is sufficiently weak so that the condition $k_L k_C << 1$ is fulfilled, then

$$\frac{Z_e - Z_o}{Z_e + Z_o} \cong \frac{k_L + k_C}{2} \tag{87}$$

Therefore, by analogy with the expression (84) for a homogeneous medium, a single coupling factor k can be defined as the mean of the inductive and capacitive coupling factors. In addition,

$$\left(\frac{c}{v}\right)^2 \cong \frac{\varepsilon_{re} + \varepsilon_{ro}}{2} \tag{88}$$

Thus, in conditions of weak coupling, and for devices where the two modes propagate in a short distance, so that the phase shift between the modes is small (for instance, in a typical quarter-wave directional coupler), the mean of the mode effective permittivities can be taken as the effective permittivity for the joint propagation of the two modes.

The analysis of asymmetrical coupled lines, while being more complex than the symmetrical case, can be carried out in the same manner. For lossless lines, after imposing a *exp*($\pm$jβx) dependence with x, Eq. (73) takes the form

$$\frac{d^2}{dx^2}\begin{pmatrix} V_1 \\ V_2 \end{pmatrix} = -\omega^2 \begin{pmatrix} L_1 & L_m \\ L_m & L_2 \end{pmatrix}\begin{pmatrix} C_1 & -C_m \\ -C_m & C_2 \end{pmatrix}\begin{pmatrix} V_1 \\ V_2 \end{pmatrix}$$

where C_i and L_i are the self-capacitances and self-inductances of the unequal coupled lines. The system can be uncoupled into two propagation modes with non-symmetric distribution of voltages and currents, and which are commonly denoted as c-mode and π-mode [26–28]. The capacitive and inductive coupling factors are defined as

$$k_C = \frac{C_m}{\sqrt{C_1 C_2}} \tag{89}$$

$$k_L = \frac{L_m}{\sqrt{L_1 L_2}} \tag{90}$$

The mode velocities are

$$\frac{1}{v_{c,\pi}^2} = \frac{1}{2}\left(\frac{1}{v_1^2} + \frac{1}{v_2^2}\right) - \frac{k_L k_C}{v_1 v_2} \pm \frac{1}{v_1 v_2}\sqrt{\left(\frac{v_1^2 - v_2^2}{2 v_1 v_2}\right)^2 + k_C^2 + k_L^2 - k_L k_C \left(\frac{v_1^2 + v_2^2}{v_1 v_2}\right)} \tag{91}$$

where $v_{1,2} = (L_{1,2} C_{1,2})^{-1/2}$, and where the signs + and – correspond, respectively, to the modes c and π. This expression coincides with (81) if $v_1 = v_2$. Moreover, if the asymmetry is suppressed in (89) and (90) by forcing $C_1 = C_2$ and $L_1 = L_2$, the velocity of the c-mode in (91) corresponds to the even mode of a symmetric line, and that of the π-mode corresponds to the odd mode.

2.2 Coupling Parameters of Coupled Microstrip Lines

Among the different coupled planar transmission lines, coupled microstrip lines have received a particular interest over decades. This is probably due to its widespread application in passband filters and directional couplers, along with the technical simplicity of manufacturing edge-coupled microstrip lines. For this reason, a number of theoretical or semiempirical models have been developed to calculate the coupling factors between two edge-coupled microstrip lines [33–42]. Figure 16 shows the cross section of two edge-coupled identical microstrip lines on a substrate of height h and permittivity ε_r. The strip width is w, the gap between lines is s and the conductor thickness is t.

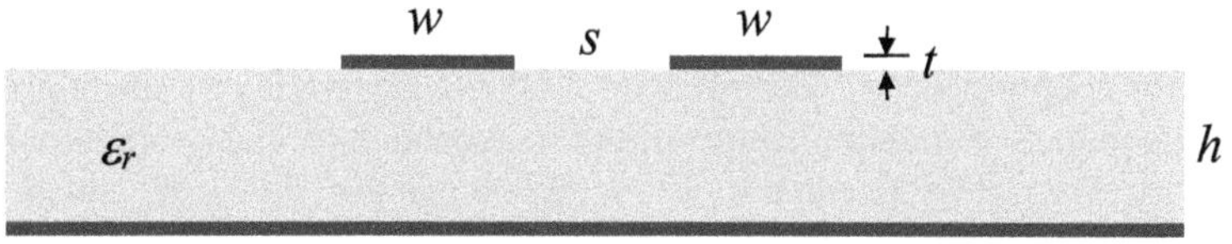

Fig. 16 Cross section of two symmetrical coupled microstrip lines

Most models express the even- and odd-mode impedances and effective permittivities in terms of the normalized dimensions $u = w/h$ and $g = s/h$. Expressions (80) and (81) allows obtaining k_C and k_L from impedances and velocities directly:

$$k_C = \frac{Z_e v_e - Z_o v_o}{Z_e v_e + Z_o v_o} \tag{92}$$

$$k_L = \frac{Z_e v_o - Z_o v_e}{Z_e v_o + Z_o v_e} \tag{93}$$

The more accurate parameterizations are those of Hammerstand and Jensen [39] and Kirschning and Jansen [41, 42]. The last includes dispersion corrections and is valid in the parameter range.

$$0.1 \leq u \leq 10, \ \ 0.1 \leq g \leq 10, \ \ 1 \leq \epsilon_r \leq 18$$

Within these limits, the accuracy of the model, according to the authors, is better than 0.7% for the static parameters (low-frequency impedances and effective permitivities). As the frequency increases, the correction terms for dispersion allows to extend the model up to a frequency of 20 GHz/h(mm) within an error of 2.5% in the specified range of u, g and ε_r. If the permittivity range is limited to 12.9, the model is valid with the same accuracy up to 25 GHz/h(mm). The model admits additional corrections for losses [33, 34] or for the finite conductor thickness (t in Fig. 16) [38]. This model has been widely applied in microwave-circuit simulation software packages.

As the formulations of both Hammerstand-Jensen model and Kirsching-Jansen model involve a number of expressions containing many parameters, we will not include them here, referring the reader to the original articles. It is, however, illustrative to show graphically the dependence of the different parameters related to the coupling with the separation between lines and the substrate permittivity. The left-side graph of Fig. 17 shows the impedance of the even and odd modes of two symmetric coupled microstrip lines as a function of the gap between the strips. Also shown are the impedance of the uncoupled microstrips and Z (82). As can be seen, the even mode has higher impedance, and the odd mode lower, than the uncoupled mode, to which value tend Z_e and Z_o as s increases. The right-side graph of the figure shows, for the same coupled-lines, the velocities of the coupled modes normalized to the uncoupled value.

In Fig. 18, the left-side graph shows the dependence of k_L and k_C with the gap width s for the w and h values of Fig. 17 and three substrate permittivities. It can be seen that the coupling factors decrease with s following a roughly exponential dependence. The inductive coupling factors does not depend appreciably on the substrate permittivity. The right-side graph shows, for a 500 μm gap, the k_C dependence with ε_r (k_L is almost independent of ε_r), along with the mean of k_C and k_L (dashed lined) and the coupling factor k calculated with (84), as for a homogeneous dielectric. The difference between the mean of k_C and k_L and k is hardly visible at the scale of the

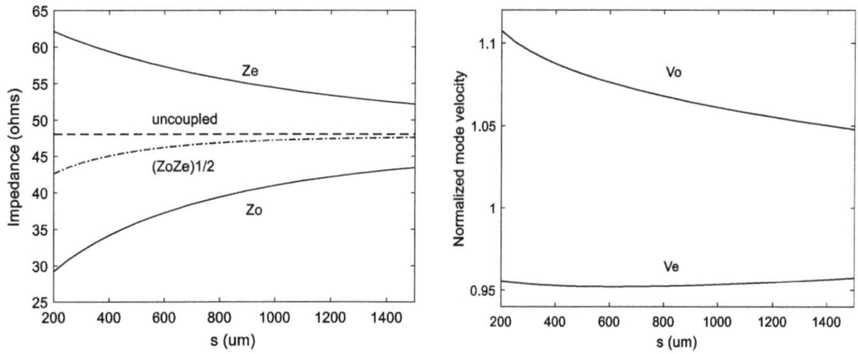

Fig. 17 Left: Even and odd mode impedances for two coupled microstrip lines ($w = 1$ mm, $h = 1$ mm, $\varepsilon_r = 10$) as a function of the gap between the lines. Dash line: impedance of the uncoupled microstrip lines. Dash-dot line: impedance (82). Right: for the same substrate and line width, mode velocities normalized to the velocity of a microstrip line of the same width in the same substrate. All magnitudes calculated at 1 GHz with the Kirsching-Jansen formulas

graph, indicating that (87) is a good approximation for weak or moderate coupling. The graph shows that increasing the permittivity reduces the coupling strength.

The coupling parameters of asymmetrical coupled microstrip lines have received a limited attention, compared with those of symmetrical coupled microstrips. Tripathi and Chang [43] have applied the Green's function integral equation method, along with suitable charge distributions in the strips, to calculate the propagation parameters of c- and π-modes by numerical techniques. Bedair [44] divided the mutual and

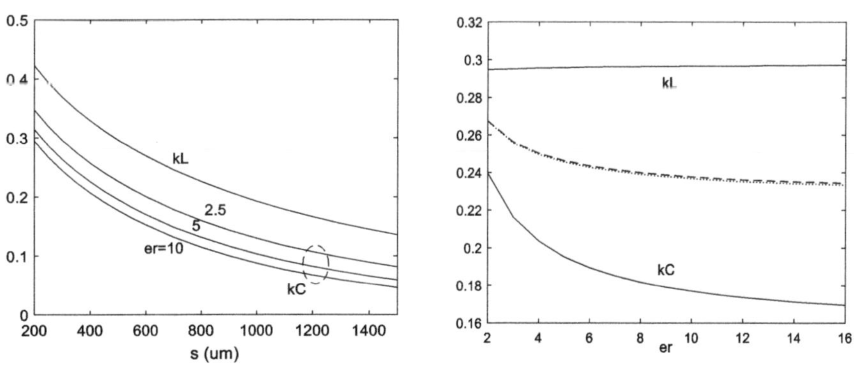

Fig. 18 Right: capacitive and inductive coupling factors dependence with line gap, for the substrate height and line width of Fig. 17 and three values of the substrate permittivity. Left: dependence of k_L and k_C with substrate permittivity for the same substrate height and line width as in Fig. 17, and a gap $s = 500$ um. The dashed line is the mean of k_L and k_C, and the dotted line is the coupling factor k calculated with (84), as for an homogeneous dielectric. All magnitudes calculated at 1 GHz with the Kirsching-Jansen formulas

self-capacitances of the asymmetric coupled lines into different parts (air, parallel-plate capacitance to ground, fringing fields…) providing explicit, although somewhat complex, expressions for each contribution.

2.3 Other Coupled Planar Transmission Lines

Bhat and Koul [45] have applied a variational approach for the strips capacitances, along with suitable charge distributions, to calculate the coupling parameters of a variety of coupled symmetric microstrip-like transmission lines. Figure 19 shows the cross section of two of these structures: double-layer coupled microstrip (*a*) and sandwiched coupled microstrip (*b*). The structure *b* has a specific interest in relation to coupling-based sensors, as it is equivalent to a uniform layer of the medium to be sensed placed over two coupled microstrip lines. The graphs of Fig. 20 show the dependence of the coupling factor between the lines on the permittivity of the top layer (left), the permittivity of the bottom layer (center) and the thickness of the top layer (right), calculated with the method of [45]. These graphs indicate that, in order to obtain high coupling factors, a low permittivity substrate and a thick top layer are convenient.

Coplanar waveguide is probably, after microstrip, the second most frequently used planar line. The coupling parameters of edge-coupled coplanar waveguides (with and without lower ground plane) and broadside-coupled coplanar waveguides (Fig. 21) have been calculated by conformal mapping techniques. Simons ([46] and references therein) gives closed-form expressions for the even- and odd-mode impedances, which allow to calculate the coupling factor by using (84) or (87). Broadside-coupled coplanar waveguides have the advantage of presenting a very low dispersion in both even and odd modes [47].

3 Coupled Microwave Resonators

The common approach to study coupled resonators is to suppose that the field distribution in the proximity of each resonator can be approximated by the field distribution of the isolated resonator plus a perturbation due to the other resonator. The

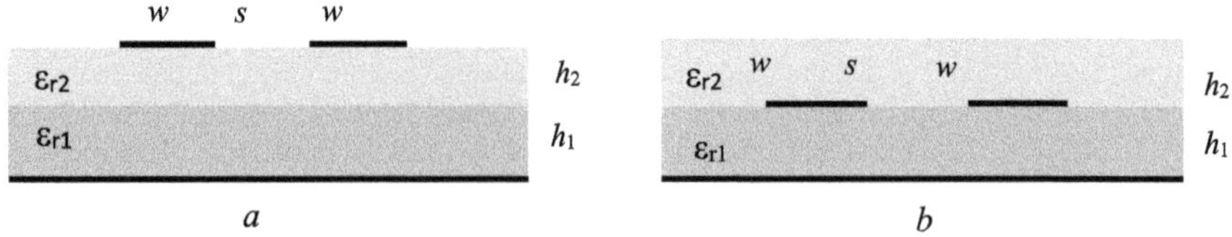

Fig. 19 (*a*) Double-layer coupled microstrip. (*b*) Sandwiched double-layer coupled microstrip

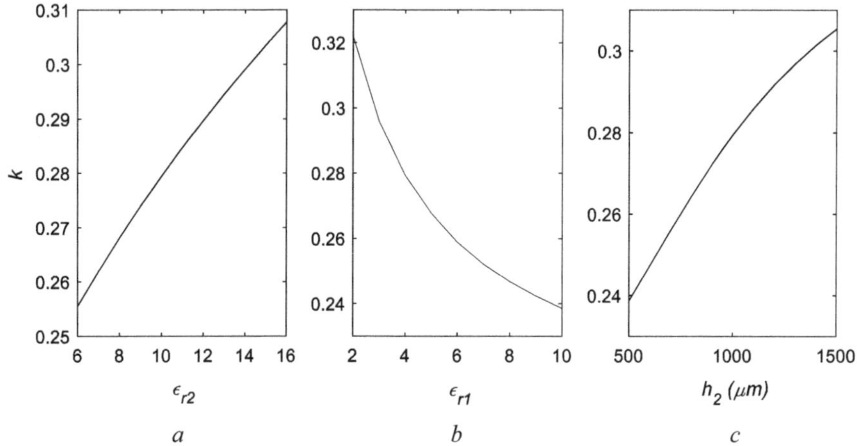

Fig. 20 Coupling factor for the coupled lines of Fig. 19b for the following parameters: $w = 1000\ \mu m$, $s = 500\ \mu m$, $h_1 = 1000\ \mu m$. (*a*) k dependence on ε_{r2} for $\varepsilon_{r1} = 4$ and $h_2 = 1000$ um. (*b*) k dependence on ε_{r1} for $\varepsilon_{r2} = 10$ and $h_2 = 1000$ um. (*c*) k dependence on h_2 for $\varepsilon_{r1} = 4$ and $\varepsilon_{r2} = 10$

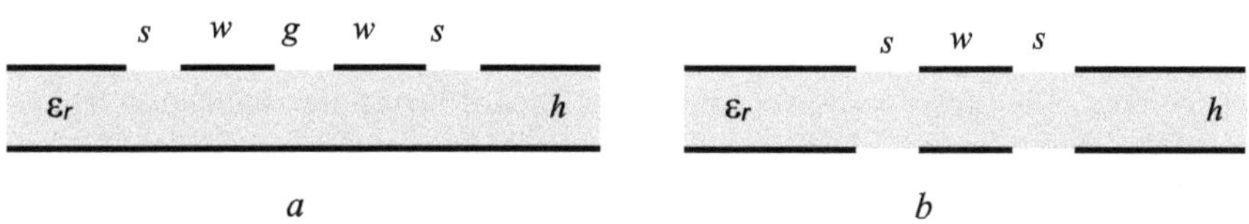

Fig. 21 Left: Edge-coupled coplanar waveguide with ground plane. Right: broadside coupled coplanar waveguide

perturbing fields are supposed to be comparatively weak. For most practical cases, this approximation provides a realistic picture of the coupled system.

To measure the strength of the mutual perturbation between two coupled resonators, a dimensionless coupling factor is used. As the mutual influence can be due predominantly to electric fields, magnetic fields or both, the coupling factor contains contributions of magnetic and electric origin. The way in which these contributions combine in a unique magnitude has given rise to a number of works in the technical literature [2, 48–56]. Due to the different theoretical approaches (circuital or electromagnetic), the kind of resonators studied (dielectric, transmission-line or metallic cavities) or the application (microwave filters, metamaterials…), some differences in definitions exists between different works. The purpose of this section is to provide a homogeneous view.

The fundamental concepts about coupling between resonators can be explained by using simplified models, as, for instance, symmetric coupled structures. The analysis of more complex structures, like unequal coupled resonators with different self-resonant frequencies, is more complex, but the underlying physical mechanisms are essentially the same. Thus, we will use symmetric structures in this section to expose the theory, and provide references to the relevant works devoted to asymmetrical coupled resonators [2, 49, 55, 56].

Apart from being simpler than the field approach, the circuit approach is generally accurate enough in the context of microwave planar resonators. Nevertheless, for completeness, we will briefly outline the field approach to the coupling between microwave resonators.

3.1 Field Approach to Inter-Resonator Coupling

In the context of electromagnetic theory, the study of coupled resonators has been addressed by several authors [51–53, 55, 56] through a formalism of coupling between resonant modes of the uncoupled resonators (Coupled Mode Theory). The fields of the coupled system are expanded as linear combinations of the uncoupled-mode fields. The uncoupled fields of one resonator do not fulfil, in general, the boundary conditions at the surface of another resonator, but, unless they are very close (strong coupling conditions), these fields are weak enough as to disregard this effect. The theory leads to an eigenvalue problem, where the eigenvalues are the resonant frequencies of the coupled system, and the eigenvectors are the expansion coefficients. The theory is conveniently simplified if losses are assumed to be low, i.e., if the quality factors of the involved resonators are not too low.

As an application of the Coupled Mode Theory to the kind of resonators discussed in this book, let us consider the coupling between two lossless, identical planar microwave resonators in the same substrate, i.e., edge-coupled. In this simple case, following the formalism of [55] and taking into account only the fundamental mode of each resonator, the eigenvalue problem takes the following form:

$$\omega_u^2 \mathcal{M}^{-1} \mathcal{E} \begin{pmatrix} a_1 \\ a_2 \end{pmatrix} = \omega^2 \begin{pmatrix} a_1 \\ a_2 \end{pmatrix} \tag{94}$$

Here, ω_u and ω are, respectively, the angular resonance frequencies of the uncoupled resonators and of the coupled system, the sub-indexes 1,2 denote the two resonant modes, and a_i are the electric field amplitudes of each uncoupled mode in the expansion. The matrices $\mathcal{M}$ and $\mathcal{E}$ contain the following magnetic and electric energy terms:

$$\mathcal{M}_{ij} = \int_V \boldsymbol{H}_i{}^* \mu \boldsymbol{H}_j dV \tag{95}$$

$$\mathcal{E}_{ij} = \int_V (\varepsilon \boldsymbol{E}_i)^* \boldsymbol{E}_j dV \tag{96}$$

where the integration volume V is large enough to cover the near fields of both resonators. For usual substrates, the magnetic permeability is very close to the vacuum permeability μ_0, and the permittivity ε is, in this case, the same for the uncoupled and coupled resonators in all points of V (the conductors, supposed very thin, are considered as surfaces).

Before proceeding with the eigenvalue problem (94), let us define the electric coupling factor between the two resonant modes as

$$k_E \equiv \frac{\int_V (\varepsilon \boldsymbol{E}_1)^* \boldsymbol{E}_2 dV}{\left(\int_V (\varepsilon \boldsymbol{E}_1)^* \boldsymbol{E}_1 dV \cdot \int_V (\varepsilon \boldsymbol{E}_2)^* \boldsymbol{E}_2 dV\right)^{1/2}} \tag{97}$$

and the magnetic coupling factor as

$$k_M \equiv \frac{\int_V \boldsymbol{H}_1{}^* \mu \boldsymbol{H}_2 dV}{\left(\int_V \boldsymbol{H}_1{}^* \mu \boldsymbol{H}_1 dV \cdot \int_V \boldsymbol{H}_2{}^* \mu \boldsymbol{H}_2 dV\right)^{1/2}} \tag{98}$$

Note that these coupling factors can be interpreted as ratios of mutual energies to self-energies. Note also that the denominators in these expressions are identical due to the equality of magnetic and electric average energies at resonance. The electric fields of the uncoupled modes can be chosen as real, implying imaginary magnetic fields, so that the resulting k_E and k_M are real.

In terms of the coupling factors just defined, the eigenvalue Eq. (94) leads to the following resonance frequencies for the coupled system:

$$\omega_1 = \omega_u \sqrt{\frac{1 + k_E}{1 + k_M}} \tag{99}$$

$$\omega_2 = \omega_u \sqrt{\frac{1 - k_E}{1 - k_M}} \tag{100}$$

Thus, the resonance frequency of the uncoupled resonators splits in two owing to the coupling. For the coupled mode with frequency ω_1, $a_1 = a_2$, while for ω_2, it results $a_1 = -a_2$. If the uncoupled electric fields are chosen in phase, ω_1 would correspond to an even and ω_2 to and odd resonance (if they are chosen with opposite phases, the signs of k_E and k_M would change, ω_1 and ω_2 would switch their values). It is worthy to note that there is no frequency splitting, i.e., no coupling, if $k_E = k_M$. A result analogous to (99, 100) has been obtained by Powell and coworkers [57] for two lossless split-ring resonators coaxially aligned in a homogeneous dielectric background. Both coupled systems are similar, as in both cases the permittivity distributions for the coupled and uncoupled resonators are the same, as long as the conductors are very thin.

The preceding results indicate that the electric (magnetic) coupling strength depends on the degree of overlap between the electric (magnetic) field distributions of the uncoupled resonant modes when the two resonators are coupled.

In the context of microwave filters, several authors [48, 49] have defined a coupling factor k for identical resonators coupled by electric and magnetic coupling (mixed coupling) in the following way:

$$k \equiv \left| \frac{\omega_1^2 - \omega_2^2}{\omega_1^2 + \omega_2^2} \right| \tag{101}$$

where we have explicitly taken the absolute value to keep k as a positive magnitude. Applying the expressions (99) and (100) to this definition results in

$$k = \left| \frac{k_M - k_E}{1 - k_E k_M} \right| \tag{102}$$

We will see in next section that the same expression can be obtained by analyzing two coupled lumped resonators in the context of circuit theory. Under conditions of weak or moderate coupling strength, this expression can be approximated by

$$k \cong |k_M - k_E| \tag{103}$$

Thus, we can see that magnetic and electric coupling combine in function of their signs to reduce or reinforce the total coupling. Eventually, in a particular coupled system, they could compensate mutually to give a null coupling [58, 59].

3.2 *Lumped-Element Approach to Inter-Resonator Coupling*

Although the field approach provides a general framework to study inter-resonator coupling, the circuit approach is much simpler, and, in many practical cases, it offers a very good approximation to practical microwave coupled resonators. This is especially true for transmission-line resonators like the planar microwave resonators considered in this book.

To start with a simple coupled system, consider the two parallel-circuit resonators of Fig. 22, coupled through a mutual capacitance C_m and a mutual inductance L_m. For simplicity, we neglect any loss mechanism in the model. The capacitance of each coupled resonator to ground is C-C_m, where C is the self-capacitance of each resonator. Being symmetric, the coupled system can be analyzed as formed by two halves with the same (even) or opposite (odd) distribution of currents and voltages (Fig. 23), each one leading to a resonant mode. For the even resonance, the symmetry plane is a magnetic wall (open), the capacitance to ground is C-C_m and the total inductance affecting the current I_L is $L + L_m$. For the odd resonance, the symmetry plane is an electric wall (short), the capacitance to ground is $C + C_m$ and the total

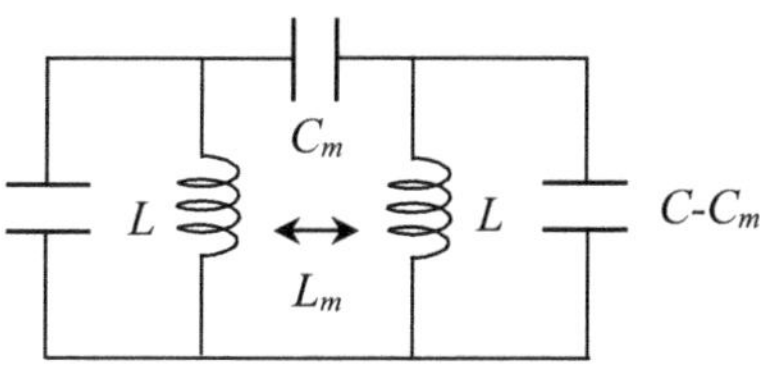

Fig. 22 Two identical lumped-element resonant circuits coupled by a mutual capacitance C_m and a mutual inductance L_m

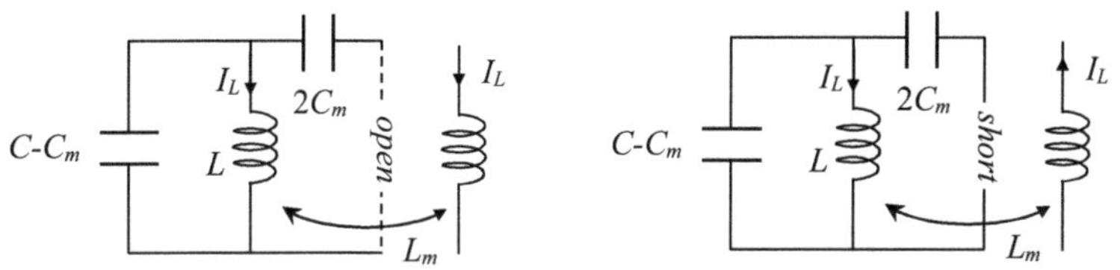

Fig. 23 Left half of the symmetric circuit of Fig. 21. Left: even resonance (equal currents in the inductances, open in the symmetry plane). Right: odd resonance (opposite currents in the inductances, short in the symmetry plane)

inductance affecting I_L is L-L_m. Therefore, the resonant frequencies for the even and odd modes are, respectively,

$$\omega_e = \frac{1}{\sqrt{(L + L_m)(C - C_m)}} \tag{104}$$

$$\omega_o = \frac{1}{\sqrt{(L - L_m)(C + C_m)}} \tag{105}$$

In a similar way as in Sect. 1.3 the coupling between a resonant circuit and an external load was introduced in terms of a ratio of conductances, now we can define a capacitive (or electric) coupling factor as a measure of the mutual electrical influence between the interacting resonators:

$$k_C = \frac{C_m}{C} \tag{106}$$

Analogously, the inductive (or magnetic) coupling factor between the resonators is defined as

$$k_L = \frac{L_m}{L} \tag{107}$$

The resonant frequencies of the even and odd modes can be expressed in terms of k_C and k_L as

$$\omega_e = \frac{\omega_0}{\sqrt{(1+k_L)(1-k_C)}} \tag{108}$$

$$\omega_o = \frac{\omega_0}{\sqrt{(1-k_L)(1+k_C)}} \tag{109}$$

where

$$\omega_0 = \frac{1}{\sqrt{LC}}$$

It is straightforward to verify that a relation analog to (102), obtained in the context of Coupled Mode Theory, holds with identical lumped resonators, with k_C and k_L instead of of k_E and k_M:

$$\frac{\omega_e^2 - \omega_o^2}{\omega_e^2 + \omega_o^2} = \frac{k_C - k_L}{1 - k_C k_L}$$

Therefore, the coupling between the two identical resonators can be defined as a positive magnitude as

$$k = \left|\frac{k_L - k_C}{1 - k_C k_L}\right| = \left|\frac{\omega_e^2 - \omega_o^2}{\omega_e^2 + \omega_o^2}\right| \tag{110}$$

Expression (110) suggests that k_C and k_L play for lumped resonant circuits an analogous role to that of k_E and k_M for general electromagnetic resonators. Moreover, under weak-coupling conditions, (99) and (100) can be approximated by expressions similar to (108) and (109) by using $1 + x \approx (1\text{-}x)^{-1}$ for small x, and with ω_u instead of ω_0.[3] This analogy will be explored with more detail in Sect. 3.3 for transmission-line coupled resonators.

If the two resonating circuits have different circuit parameters, k_C and k_L are defined as [48, 50]

$$k_C = \frac{C_m}{\sqrt{C_1 C_2}} \tag{111}$$

$$k_L = \frac{L_m}{\sqrt{L_1 L_2}} \tag{112}$$

Expression (110) is still valid if the two unequal resonators, when uncoupled, have the same resonance frequency. In the general case of asynchronous resonators with electric and magnetic coupling, the resonant frequencies of the coupled system

[3] The angular frequency ω_u introduced in Sect. 3.1 refers to identical uncoupled resonators (isolated from each other), while ω_0, is the resonance frequency of each coupled resonator when the other one is shorted to ground (see Fig. 22). Under weak-coupling conditions, the difference between ω_0 and ω_u is small.

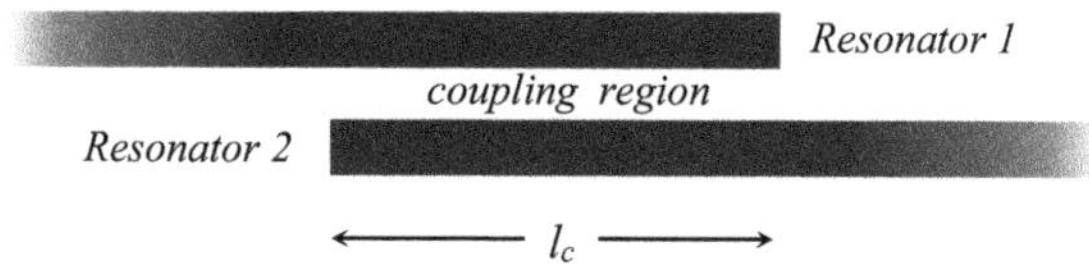

Fig. 24 Two microstrip resonators coupled in a length l_c

are more complicated functions of the electrical parameters [2, 50]. Under weak coupling conditions ($k_L, k_C << 1$), these expressions leads to

$$|k_L - k_C| = \frac{1}{2}\left(\frac{\omega_{02}}{\omega_{01}} + \frac{\omega_{01}}{\omega_{02}}\right)\sqrt{\left(\frac{\omega_2^2 - \omega_1^2}{\omega_2^2 + \omega_1^2}\right)^2 - \left(\frac{\omega_{02}^2 - \omega_{01}^2}{\omega_{02}^2 + \omega_{01}^2}\right)^2} \tag{113}$$

where ω_{0i} are the resonant angular frequencies of the uncoupled resonators and ω_i those of the coupled resonant modes.

3.3 Distributed Element Approach to Planar Resonator Coupling

Planar-line coupled resonators can be analyzed by using the distributed element models of single and coupled transmission lines. In the definitions (2.97–2.98) of k_E and k_M, the mutual and self- electric and magnetic energies can be calculated in terms of voltages and currents, instead of electric and magnetic fields. Let us consider, for illustration purposes, two microstrip resonators coupled in a region of length l_c (Fig. 24). The details of the uncoupled regions or the type of terminations are not of interest for the definitions to follow.

The time-averaged electric energy per unit length in the coupling region results

$$\frac{dU_{E12}}{dx} = \frac{1}{4}\left\{C_1|V_1(x)|^2 + C_2|V_2(x)|^2 - 2C_m Re(V_1(x)V_2^*(x))\right\}$$

In the field approach of Sect. 3.1, the average electric energy per unit volume in the coupling region is[4]

$$\frac{dU_{E12}}{dV} = \frac{1}{4}\varepsilon\left(|\boldsymbol{E}_1|^2 + |\boldsymbol{E}_2|^2 + 2Re(\boldsymbol{E}_1\boldsymbol{E}_2{}^*)\right)$$

Therefore, the term $-C_m V_1 V_2^*$ plays the role of the term $\varepsilon \boldsymbol{E}_1\boldsymbol{E}_2{}^*$ in the field approach. Thus, the definition of k_E (97) takes the following form in the distributed element model:

[4] In this expression, the permittivity is supposed to be real (lossless substrate). For the general case, the different energy terms are $(\varepsilon \boldsymbol{E}_i)^* \boldsymbol{E}_j$ instead of ε; $\boldsymbol{E}_i{}^* \boldsymbol{E}_j$.

$$k_E = \frac{-\int_{l_c} C_m V_1 V_2^* dx}{\sqrt{\int_{l_1} C_1 |V_1|^2 dx \int_{l_2} C_2 |V_2|^2 dx}} \tag{114}$$

where l_i are the total lengths of the resonators. Note that the capacitances C_i in each resonator could be different in the coupled and uncoupled regions, or even depend on the position if the line geometry changes along the line. This expression takes a simpler form if the capacitances C_i in the coupled and uncoupled regions are approximately the same. In that case, taking real voltage distributions $V_i(x)$ in the resonators,

$$k_E = -k_C \frac{\int_{l_c} V_1 V_2 dx}{\sqrt{\int_{l_1} {V_1}^2 dx \int_{l_2} {V_2}^2 dx}} \tag{115}$$

where k_C is the capacitive coupling factor of two coupled lines defined in Sect. 2.1. Analogous considerations lead to the following definition for k_M:

$$k_M = \frac{\int_{l_c} L_m I_1 I_2^* dx}{\sqrt{\int_{l_1} L_1 |V_1|^2 dx \int_{l_2} L_2 |V_2|^2 dx}} \tag{116}$$

If the inductances L_i are approximately the same in the coupled and uncoupled regions,

$$k_M = k_L \frac{\int_{l_c} I_1 I_2^* dx}{\sqrt{\int_{l_1} |I_1|^2 dx \int_{l_2} |I_2|^2 dx}} \tag{117}$$

where k_L is the inductive coupling factor of two coupled lines defined in Sect. 2.1.

Expression (115) indicate that k_E depends on two factors: the electric coupling at local scale (k_C) and the degree of overlapping between the voltage distributions in the coupling region. The same considerations apply to k_M, k_L and the current distributions.

The formulation of coupling between transmission-line resonators in terms of voltage and current distributions has been applied in filter design for purposes like suppression of harmonics of the center-bandpass frequency [60–62], or to generate transmission zeros without cross-coupling [58, 59].

As an example of application of the present approach to coupled planar microwave resonators, let us consider the coupling between two identical square open-loop resonators of Fig. 25. This problem has been studied in [49], where the coupling was obtained by means of an electromagnetic simulation software. For the sake of simplicity, we will suppose that the gap g is wide enough to neglect the capacitive effect in the open ends of each resonator, and that the separation s in the coupling

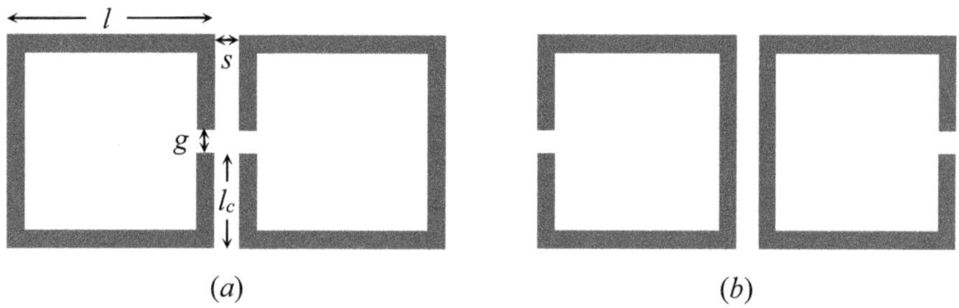

Fig. 25 Square open-loop microstrip resonators coupled (*a*) electrically and (*b*) magnetically

region is enough for the coupling to be considered weak. Therefore, the current and voltage distributions in each resonator, at the fundamental resonance, are approximately those of a half-wave resonator,[5] given by (19) and (20). Under weak-coupling conditions, k_E and k_M can be calculated respectively by (115) and (117). For the configuration (*a*), where the coupling region involves line-sections with high voltage values,

$$k_E = -k_C \frac{2\int_0^{l_c} cos^2(\beta x)dx}{\int_0^{4l-g} cos^2(\beta x)dx} \cong \frac{-k_C}{\pi}(2\beta l_c + sin(2\beta l_c))$$

$$k_M = k_L \frac{2\int_0^{l_c} sin^2(\beta x)dx}{\int_0^{4l-g} sin^2(\beta x)dx} \cong \frac{k_L}{\pi}(2\beta l_c - sin(2\beta l_c))$$

where β is the phase constant in the line at the resonance frequency. As the coupling is supposed to be weak, $k \cong |k_M - k_E|$, and it results

$$k \cong \frac{1}{\pi}|k_L(2\beta l_c - sin(2\beta l_c)) + k_C(2\beta l_c + sin(2\beta l_c))| \tag{118}$$

For the fundamental, half-wavelenght resonance, l_c is close to λ/16. Substituting in (118), the factor multiplying k_C results around 20 times greater that the factor multiplying k_L. Therefore, the coupling factor is essentially determined by k_C, i.e., the coupling has an electric character.

For the configuration (*b*), where the coupling region involves line-sections with high current magnitudes, the same procedure leads to

[5] If the coupling strength is high, or if the line width changes in the coupled region respect to the uncoupled region, the current and voltage distributions in both regions are different, and fulfill continuity conditions in the frontiers.

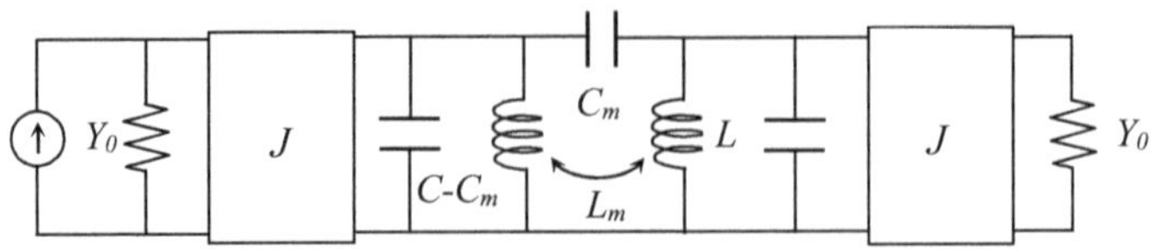

Fig. 26 Two identical lumped-element lossless resonators coupled electrically and magnetically in a two-port configuration. The coupling networks are lossless admittance inverters

$$k \cong \frac{1}{\pi}\left|k_L\left(\beta l + \frac{sin(2\beta l_c) + cos(2\beta l_c)}{2}\right) + k_C\left(\beta l - \frac{sin(2\beta l_c) + cos(2\beta l_c)}{2}\right)\right| \tag{119}$$

Now, the factor of k_L results around 20 times greater than the factor of k_C. Thus, the coupling factor is essentially determined by k_L, so that the coupling has a magnetic character.

3.4 Coupled Resonators as Two-Port Networks

Coupled planar microwave resonators are the basic building blocks of standard planar microwave band-pass filters [2, 5]. As we will see in Chaps. 5 to 9, they are used as well as coupling-based sensors. For sensing purposes, the number of resonators in the device is small, frequently two [63–69], as the aim of a sensing device is to obtain high sensitivity respect the physical property to be sensed, rather than obtaining a sharp frequency response. To show the more significant features of a two-port coupled-resonator network, let us consider the lumped-element symmetrical lossless network of Fig. 26, where the ports have characteristic admittances Y_0. The symmetry of the network and its lossless character simplify the analysis, yet keeping the model not far from real networks, as long as the Q_u of the resonators is high enough. Although built up with lumped elements, it can describe a network of transmission-line resonators in a frequency range around the resonance. The input/output coupling networks can be represented by identical lossless admittance inverters.[6] Each port-admittance inverter combination is equivalent, in the resonator side, to a conductance

$$G_{ex} = \frac{J^2}{Y_0} \tag{120}$$

The external quality factors of both resonators are (see Sect. 1.3)

[6] Using inverters as coupling networks does not suppose a serious limitation, as any lossless reciprocal two-port network can be converted into an inverter simply by adding suitable line sections at its ports [82]. Therefore, the only implication in the S parameters is a phase-shift.

$$Q_{ex} = \frac{\omega_0 C}{G_{ex}} \tag{121}$$

where ω_0 is the unloaded resonance frequency. Properly speaking, as the coupled system has two resonant frequencies, the external quality factor at each one can be different. But, as long as the coupling is not too high, these frequencies are close, and (121) provides a good approximation.

The response of the lossless coupled network of Fig. 26 can be calculated readily by using the coupling matrix formalism presented in [2]. The result is

$$S_{21} = \frac{2jkQ_{ex}}{(1 + 2jQ_{ex}u)^2 + (kQ_{ex})^2} \tag{122}$$

$$S_{11} = \frac{2(1 + 2jQ_{ex}u)}{(1 + 2jQ_{ex}u)^2 + (kQ_{ex})^2} - 1 \tag{123}$$

where k is the coupling factor and u is the normalized frequency deviation respect the unloaded resonance frequency

$$u = \frac{\omega - \omega_0}{\omega_0}$$

The maximum transmission is $|S_{21}| = 1$ and occurs at the resonant frequencies

$$\omega_{1,2} = \omega_0 \left(1 \mp \frac{1}{2}\sqrt{k^2 - \frac{1}{{Q_{ex}}^2}}\right) \tag{124}$$

The condition $Q_{ex}k >> 1$ implies a low external coupling compared with the inter-resonator coupling. If this condition is fulfilled, ω_1, ω_2 and k meet the relation (110). For moderate external coupling, the effect of Q_{ex} is to reduce slightly the interval between the S_{21} peaks.

It has to be mentioned that the coupling matrix approach leading to the results above has a narrowband character, besides considering the coupled system as lossless. A lossy version of the symmetric network of Fig. 26 was studied by Cohn and Shimizu [7, 70], also in narrowband approximation, obtaining the following transmission power ratio:

$$|S_{21}|^2 = \frac{1}{A_0 + A_1 u^2 + A_2 u^4} \tag{125}$$

where

$$A_0 = \left(\frac{\left(1 + \frac{Q_{ex}}{Q_u}\right)^2}{2kQ_{ex}} + \frac{kQ_{ex}}{2}\right)^2$$

$$A_1 = 2\left(\frac{\left(1+\frac{Q_{ex}}{Q_u}\right)^2}{k^2} - Q_{ex}{}^2\right)$$

$$A_2 = 4\left(\frac{Q_{ex}}{k}\right)^2$$

and where Q_u is the unloaded quality factor of the resonators. The transmission poles for the pair of lossy resonators are obtained maximizing $|S_{21}|^2$ respect u, resulting

$$\omega_{1,2} = \omega_0\left(1 \pm \frac{1}{2}\sqrt{k^2 - \left(\frac{1}{Q_{ex}} + \frac{1}{Q_u}\right)^2}\right) \tag{126}$$

which coincides with (124) for $Q_u = \infty$.

The influence of Q_{ex}, k and Q_u in the transmission response can be seen in the graphs of Figs. 26, 27, 28. In Fig. 27, the coupling factor k is varied while keeping constant Q_{ex} and Q_u. For the moderate k values of the graph,

$$k = \frac{f_2^2 - f_1^2}{f_2^2 + f_1^2} \cong \frac{f_2 - f_1}{f_0}$$

i.e., the coupling factor gives a good approximation of the fractional bandwidth between the transmission peaks, a result that can be checked by inspection in the graph. As the resonators are lossy ($Q_u = 400$), the magnitude of S_{21} at the resonant frequencies is lower than 1.

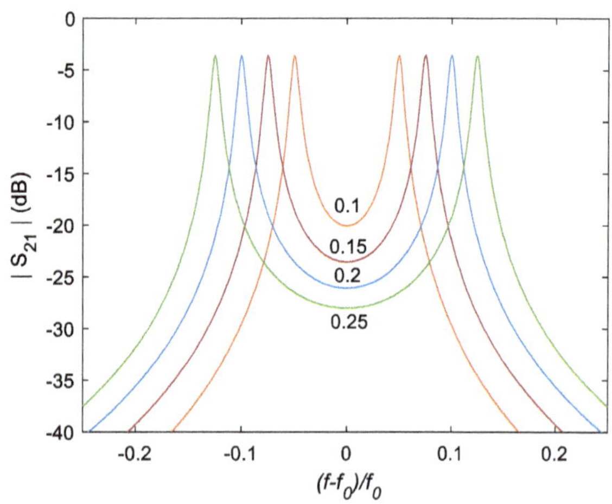

Fig. 27 Transmission magnitude response of two coupled symmetric resonators with $Q_u = 400$, $Q_{ex} = 200$ and different values of k

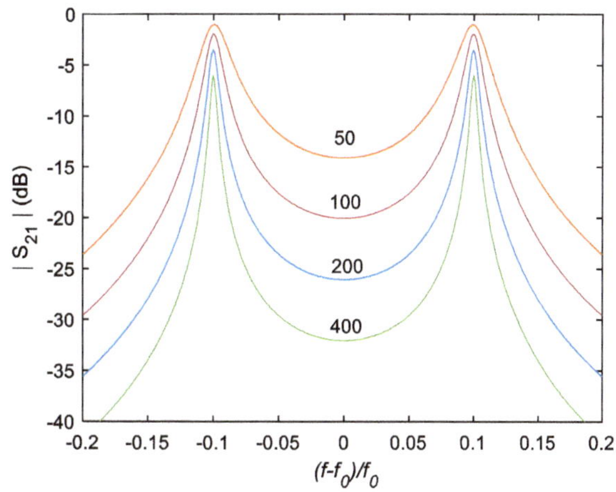

Fig. 28 Transmission magnitude response of two coupled symmetric resonators with $Q_u = 400$, $k = 0.2$, and different values of Q_{ex}

Figure 28 shows the influence of the external quality factor. As the interaction between the coupled resonant system and the source/load increases, Q_{ex} decreases and the resonant peaks widens. In addition, the losses associated to the resonators at resonance reduce as Q_{ex} decreases, because the equivalent conductance of the load (G_{ex}, see (120)) increasingly dominates over the resonators conductances. It is worthy to note also the strong influence of Q_{ex} in the minimum S_{21}, placed at the uncoupled resonance frequency.

Figure 29 shows how the transmission at the resonant peaks approaches 0 dB as the unloaded quality factor of the resonators increases.

The phase response in the lossless case is obtained directly from (122):

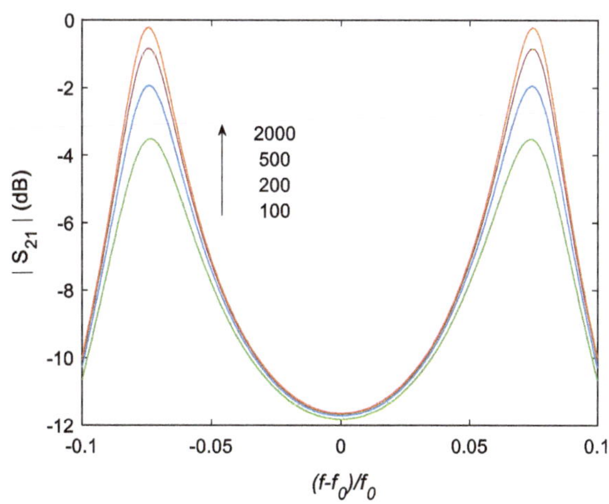

Fig. 29 Transmission magnitude response of two coupled symmetric resonators with $k = 0.15$, $Q_{ex} = 50$, and different values of Q_u

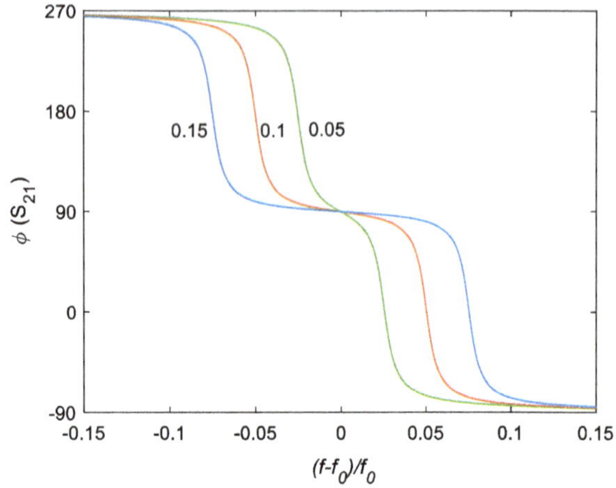

Fig. 30 Transmission phase response of two coupled symmetric resonators with $Q_u = \infty$, $Q_{ex} = 100$ and different values of k

$$\phi_{21} = \frac{\pi}{2} - arctg\left(\frac{4Q_{ex}u}{1 + Q_{ex}^2\left(k^2 - 4u^2\right)}\right) \tag{127}$$

Figure 30 shows the phase response for a lossless coupled-resonator pair with $Q_{ex} = 100$ and three values of k. It can be seen a jump of near $-180°$ associated with each resonance, so that the total phase shift from low to high frequencies is –360°. The slope of these jumps increases with Q_{ex}.

4 Coupled Transmission-Line Sections and Directional Couplers

Coupled-line sections are routinely used in many microwave devices, like filters [2, 5 and references therein], directional couplers [3 and references therein], baluns [71] and matching networks [72]. As the reader can see on Chaps. 3 and 4, they find application also as basic components of microwave sensors. Although coupled-line sections made up of 3 [73] or more coupled lines can be found in technical literature, the majority of published works deal with coupled-line structures involving two lines. These are the type of structures considered in this section.

Depending on the number of terminated ports, a coupled-line section can conform a network or sub-network of 1 to 4 ports. Here we are mainly interested in 2- and 4-port coupled-line networks, which are the most frequently used in sensor applications. We will start by presenting the coupled-line directional coupler, a 4-port network consisting simply in two coupled lines whose ends are connected to external ports.

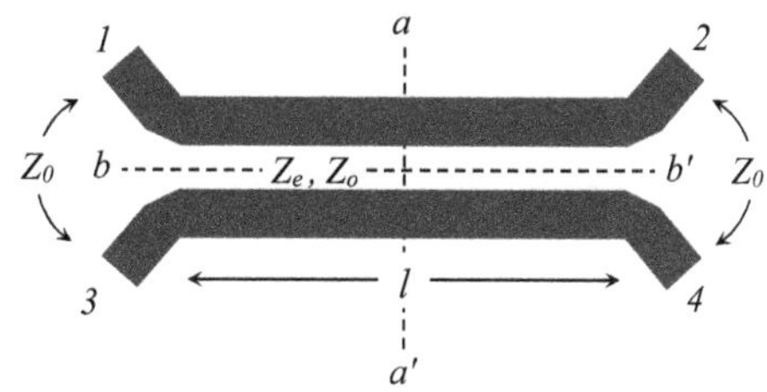

Fig. 31 Layout of a symmetrical edge-coupled microstrip directional coupler

4.1 Planar Coupled-Line Directional Couplers

Figure 31 shows the layout of a directional coupler composed of two edge-coupled microstrip lines, as an illustrative example of a variety of coupled planar lines. Although directional couplers made up of asymmetrical coupled-lines have been studied and used [3, 74, 75], for the purposes of this book it is enough, and much simpler, to consider the analysis of symmetric directional couplers. The extremes of each line are connected to external ports of impedance Z_0. In the coupling region, of length *l*, the even- and odd-mode impedances are, respectively, Z_e and Z_o. This network has two perpendicular symmetry planes, indicated by dashed lines in the figure. Thus, even–odd analysis can be applied to obtain the network S-parameters. When the excitations in ports 1 and 3 have even (odd) symmetry, only the even (odd) mode propagates in the left half of the structure, and the same is valid for the right half. When the excitations in the left and right halves have even (odd) symmetry, there is a magnetic (electric) wall in the plane *a-a*'. The S matrix can be obtained by superposing excitations with even and odd symmetries respect to the planes *a-a*'and *b-b*' [1, 3, 4]. Before proceeding with the analysis, let us expose some definitions, design requirements and approximations.

The proper operation of the directional coupler requires that the directional coupler is matched ($S_{ii} = 0$) at the design frequency. In TEM lines, the power entering from port 1 is partially coupled to port 3 (from 2 to 4, and so on) [3], while the remaining power exits through port 2. Then, it is said that 2 is the direct port, 3 the coupled port and 4 the isolated port respect to the input port 1. This kind of coupler is called backward directional coupler, and it is also the common type of coupler in quasi-TEM lines, like microstrip or coplanar.[7] The matching condition in the input port is fulfilled if [3]

$$Z_e Z_o = Z_0^2 \tag{128}$$

In quasi-TEM lines, the electrical length of the coupling region is different from the even and odd modes, owing to different mode velocities:

[7] In the type of couplers called forward directional couplers, a portion of the power entering through port 1 couples to port 4, while port 3 remains isolated [3]. This type of coupler, although possible, is rarely based on quasi-TEM coupled lines, being commonly based on other kinds of coupled structures, like metallic or dielectric waveguides.

$$\theta_{e,o} = \frac{\omega l}{v_{e,o}} \tag{129}$$

Although the mode impedances may be quite different, the mode velocities do not differ so much. Therefore, it is reasonable to approximate both mode velocities by a single velocity v calculated with (88), i.e., with the mean mode effective permittivity. The effective electrical length of the coupler can be taken as:

$$\theta = \frac{\omega l}{c}\sqrt{\frac{\varepsilon_{re} + \varepsilon_{ro}}{2}} \tag{130}$$

Alternatively, θ can be approximated as the mean of θ_e and θ_o. In most practical cases, the difference between both choices is quite small.

With this approximation and the matching condition (128), the S parameters of the coupler result

$$S_{11} = 0 \tag{131a}$$

$$S_{21} = \frac{\sqrt{1-k^2}}{\sqrt{1-k^2}cos\theta + j \cdot sin\theta} \tag{131b}$$

$$S_{31} = \frac{jk \cdot sin\theta}{\sqrt{1-k^2}cos\theta + j \cdot sin\theta} \tag{131c}$$

$$S_{41} = 0 \tag{131d}$$

where the coupling factor k is defined, as in Sect. 2.2, by (2.84). In fact, as propagation is quasi-TEM, this is equivalent to approximate k by (87). The remaining S parameters can be obtained from (131) by symmetry considerations ($S_{42} = S_{31}$, $S_{32} = S_{41}$, …).

The characteristics of a directional coupler are expressed by means of three power ratios, expressed in dB:

- Coupling:

$$C = -20log_{10}|S_{31}| \tag{132}$$

- Isolation:

$$I = -20log_{10}|S_{41}| \tag{133}$$

- Directivity:

$$D = 20log_{10}|S_{31}/S_{41}| \tag{134}$$

The three parameters are not independent, as $I = C + D$. Both directivity and isolation are infinite for a perfect coupler, although they have finite values for real couplers.

The maximum coupling is achieved when the coupler electrical length is 90°, i.e., when the coupling length is a quarter wavelength. For quasi-TEM coupled lines, an effective wavelength is calculated at the design frequency with the velocity (88). For $\theta = 90°$,

$$S_{21} = -j\sqrt{1-k^2} \tag{135a}$$

$$S_{31} = k \tag{135b}$$

and therefore

$$C = -20log_{10}k$$

The difference between even and odd mode velocities degrades the response of a quasi-TEM directional coupler respect to a TEM one [76–78]. The graph of Fig. 32 shows the coupling, isolation and matching of a microstrip coupled-line directional coupler with $C = 10$ dB on a substrate 1 mm thick and permittivity $\varepsilon_r = 9$, simulated with Keysight ADS software. The S_{21} parameter is higher than -1 dB and is not shown. Also shown with dashed line is the coupling for an ideal (TEM) coupler with the same central frequency and maximum coupling level (the S_{11} and S_{41} parameters are $-\infty$ dB for an ideal coupler). Several approaches have been proposed to improve the response of quasi-TEM directional couplers. A frequently used technique is to place a dielectric overlay on the coupled lines to reduce the difference between even- and odd-effective permittivities [76–78]. As discussed in Sect. 2.2.1, $\varepsilon_{re} > \varepsilon_{ro}$ in microstrip lines. A dielectric layer over the strips increases the mutual capacitance, thus reducing the difference between ε_{re} and ε_{ro} (see 74a, 75a). Although with different purposes, this technique relays on the same principles of the sensors presented in Chaps. 3 and 4.

4.2 *Two-Port Coupled-Line Networks*

Terminating two ports of a coupled-line section with suitable loads leads to a two-port coupled-line network. Figure 33 shows some examples of two-port coupled-line sections: open-ended (*a*), short-ended (*b*), ends connected by a load (*c*), and coupled-line section with opposite open-ends (*d*) typically used, with a length of a quarter-wavelength, in bandpass filters. In the figure, only two conducting strips are shown, being the third conductor the reference ground plane. The two strips are supposed to be identical on what follows, so that the coupled modes are the even and odd modes described in Sect. 2.1.

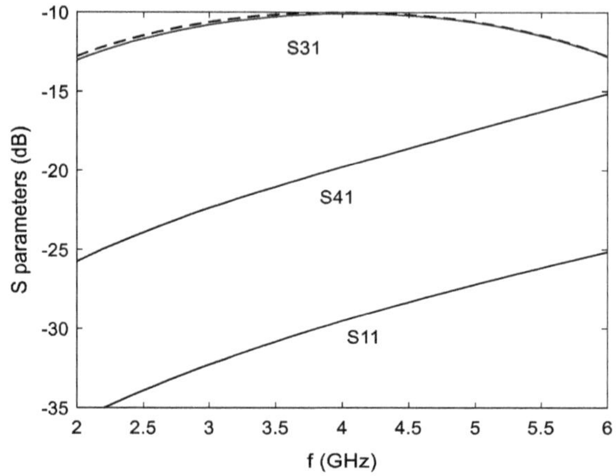

Fig. 32 Response of a microstrip coupled-line directional coupler on a substrate with $\varepsilon_r = 9$ and 1 mm thick. The design data are: $Z_0 = 50\ \Omega$, $C = 10$ dB, electrical length 90° at $f_0 = 4$ GHz. The synthetized line width and separation are $w = 852.4\ \mu$m and $s = 323.6\ \mu$m, and the physical length $l = 7.665$ mm. The dashed line shows the S_{31} parameter of an ideal coupler with the same parameters k, Z_0 and f_0

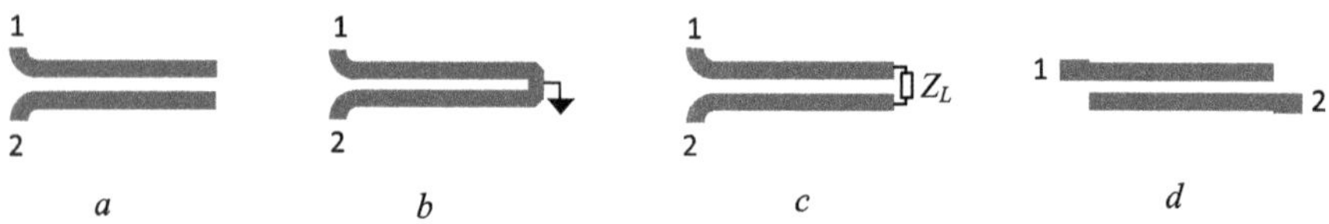

Fig. 33 Different configurations of two-port coupled-line sections

The S parameters of these 2-port networks are readily calculated from the symmetrical directional coupler S-parameters (2.130*a-d*) by considering the reflection coefficients at the line terminations. As an example, Fig. 34 shows the signals entering and exiting the internal coupler of network (*a*) of Fig. 33 when it is excited from port 1. Although this method is simple, here we will adopt a different approach, based on differential signals.

Networks of the types *a-c* in Fig. 33 have been used as basic component of permittivity sensors [79]. For sensing purposes, it is convenient to excite the network

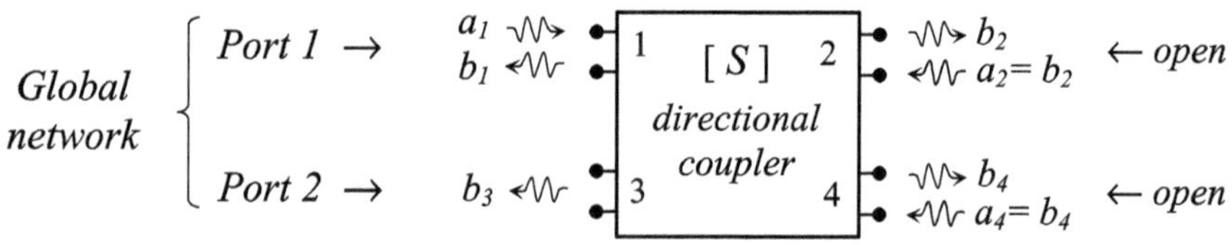

Fig. 34 Signals entering and exiting the internal directional coupler of coupled-line section (*a*) of Fig. 33 when the generator is connected to port 1

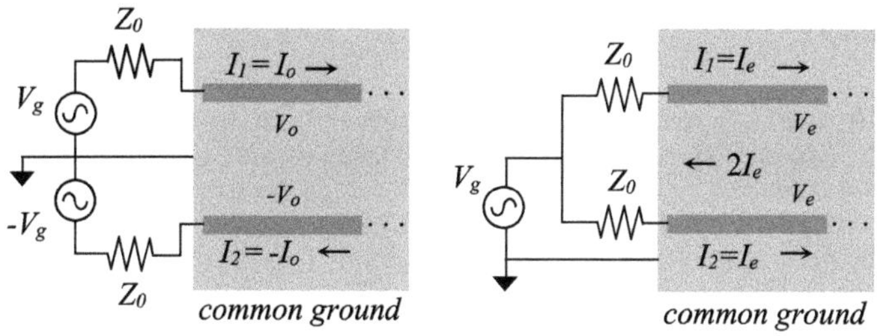

Fig. 35 Differential- (left) and common-mode (right) excitation of two symmetrical coupled lines

only with the odd-mode, as the reader can see in Chap. 3. This requires out-of-phase input signals in ports 1 and 2, and, in this situation, the coupled-line behaves as a differential or balanced line. The differential-mode and common-mode description of propagation in symmetrical coupled-lines is alternative to the even–odd mode description [80, 81]. Figure 35 shows schematically the source configuration to excite the differential mode (left) and common mode (right). In the differential mode, the current flowing in the lines is I_o and the voltage between the lines is $2V_o$. In the common mode, the current flowing is $2I_e$ and the voltage V_e. Therefore, the characteristic impedances of the differential- and common-mode are

$$Z_d = 2Z_o \tag{136}$$

$$Z_c = \frac{Z_e}{2} \tag{137}$$

A coupled-line section excited by a differential source can be analyzed as a single line of impedance Z_d, and, if it is excited by the common-mode, it behaves like a single line of impedance Z_e. As an example, the input impedance of the open-ended coupled-line section (*a*) of Fig. 33, if the lines are considered lossless, is

$$Z_{in} = -jZ_d ctg\theta_o$$

where θ_o is the electrical length of the lines for the odd-mode. This input impedance refers to a single differential port connected to the coupled-line section like in the left side of Fig. 35. Analogously, for the short-ended coupled-line section (*b*) of Fig. 33,

$$Z_{in} = jZ_d tg\theta_o$$

Chapter 3 provides more details and specific applications of coupled-line sections terminated in different loads.

References

1. Collin RE (2001) Foundations for microwave engineering, Wiley-IEEE Press
2. Hong JS, Lancaster MJ (2001) Microstrip filters for RF/Microwave applications. John Wiley & Sons, New York
3. Mongia RK, Barthia P, Hong S, Bahl IJ (2007) RF and microwave coupled-line circuits, Norwood, MA: Artech House Microwave Library
4. Pozar DM (2011) Microwave engineering. John Wiley & Sons, Hoboken, New Jersey
5. Cameron RJ, Kudsia CM, Mansour RR (2018) Microwave filters for communication systems: fundamentals. John Wiley & Sons, Design and Applications
6. Shahid S, Ball JAR, Wells CG, Wen P (2011) Reflection type Q-factor measurement using standard least squares methods. IET Microw. Antennas Propag. 4(4):426–432
7. Matthaei GL, Young y L, Jones EMT (1964) Microwave filters, impedance-matching networks and coupling structures, New York: McGraw-Hill
8. Kwok RS, Liang J-F (1999) Characterization of High-Resonators for Microwave-Filter applications. IEEE Trans Microw Theory Techn 47(1):111–114
9. Kajfez D (1994) Q-Factor. MS, Vector Fields, Oxford
10. Sun E-Y, Chao S-H (1995) Unloaded Q measurement—The critical points method. IEEE Trans Microw Theory Techn 43(8):1983–1986
11. Kajfez D, Hwan EJ (1984) Q-Factor measurement with network analyzer. IEEE Trans Microw Theory Techn 32(7):666–670
12. Leong K, Mazierska J (2002) Precise measurements of the Q factor of dielectric resonators in the transmission Mode—Accounting for noise, crosstalk, delay of uncalibrated lines, coupling loss, and coupling reactance. IEEE Trans Microw Theory Techn 50(9):2115–2127
13. Bray JR, Roy L (2004) Measuring the unloaded, loaded, and external quality factors of one- and two-port resonators using scattering-parameter magnitudes at fractional power levels. IEE Proc Microw Antennas Propag 151(4):345–350
14. Gregory AP (2021) Q-factor Measurement by using a Vector Network Analyser. NPL Report MAT 58
15. Sanada A, Takehara H, Yamamoto T, Awai I (2002) Lambda//4 stepped-impedance resonator bandpass filters fabricated on coplanar waveguide. In: 2002 IEEE MTT-S International Microwave Symposium Digest, pp. 385–388, Seattle, USA
16. Khanna APS, Garault Y (1983) Determination of loaded, unloaded and external quality factors of a dielectric resonator coupled to a microstrip line. IEEE Trans Microw Theory Techn 31(3):261–264
17. Probst S, Song FB, Bushev PA, Ustinov AV, Weides M (2015) Efficient and robust analysis of complex scattering data under noise in microwave resonators, Rev Sci Instr, 86(2), paper 024706
18. Sánchez M, Martín E, Zamarro J (1990) Unified and simplified treatment of techniques for characterizing transmission reflection, or absorption resonators. IEE Proceedings 137(4):209–212
19. Kajfez D, Chebolu S, Abdul-Gaffoor MR, Kishk AA (1999) Uncertainty Analysis of the Transmission-Type Measurement of–Factor. IEEE Trans Microw Theory Techn 47(3):367–371
20. Petersan P, Anlage S (1998) Measurement of resonant frequency and quality factor of microwave resonators: comparison of methods. J Appl Phys 84(6):3392–3402
21. Pipes LA (1937) Matrix theory of multiconductor transmission lines. Phil Mag 24:97–100
22. Oliver BM (1954) Directional electromagnetic couplers. Proc IRE 42:1686–1692
23. Pierce JR (1954) Coupling of modes of propagation. J Appl Phys 25(2):179–183
24. Schelkunoff A (1955) Conversion of Maxwell's equations into generalized telegraphist's equations. Bell Syst Tech J 34(5):995–1043
25. Adair JE, Haddad GI (1969) Coupled-mode analysis of nonuniform coupled transmission lines. IEEE Trans Microw Theory Techn 17(10):746–752

26. Krage MK, Haddad GI (1970) Characteristics of coupled microstrip transmission Lines-I: Coupled-Mode formulation of inhomogeneous lines. IEEE Trans Microw Theory Techn 18(4):217–222
27. Marx KD (1973) Propagation modes, equivalent circuits, and characteristic terminations for multiconductor transmission lines with inhomogeneous dielectrics. IEEE Trans Microw Theory Techn 21(7):450–457
28. Tripathi VK (1975) Asymmetric coupled transmission lines in an inhomogeneous medium. IEEE Trans Microw Theory Techn 23(9):734–739
29. Speciale RA (1975) Even- and Odd-Mode waves for nonsymmetrical coupled lines in nonhomogeneous media. IEEE Trans Microw Theory Techn 23(11):897–908
30. Wei C, Harrington RF, Mautz JR, Sarkar TK (1984) Multiconductor transmission lines in multilayered dielectric media. IEEE Trans Microw Theory Techn 32(4):439–450
31. Bedair S, Sobhy M (1980) Approximate solution for propagation modes on lossy multiconductor transmission lines in inhomogeneous media. Electron Lett 16(24):914–915
32. Denlinger E (1980) Losses of microstrip lines. IEEE Trans Microw Theory Techn 28(6):513–522
33. Garg R, Bahl I (1979) Characteristics of coupled microstriplines. IEEE Trans Microw Theory Techn 27(7):700–705
34. Garg R, Bahl I (1980) Correction to characteristics of coupled microstriplines. IEEE Trans Microw Theory Tech 28(3):272–272
35. Getsinger WJ (1973) Dispersion of Parallel-Coupled Microstrip. IEEE Trans Microw Theory Tech 21(3):144–145
36. Bryant TG, Weiss JA (1968) Parameters of microstrip transmission lines and of coupled pairs of microstrip lines. IEEE Trans Microw Theory Techn 16(12):1021–1027
37. Krage MK, Haddad GI (1970) Characteristics of coupled microstrip transmission Lines-II: Evaluation of Coupled-Line Parameters. IEEE Trans Microw Theory Techn 18(4):222–228
38. Jansen RH (1978) High-Speed computation of single and coupled microstrip parameters including dispersion, High-Order modes, loss and finite strip thickness. IEEE Trans Microw Theory Techn 26(2):75–82
39. Hammerstad E, Jensen O (1980) Accurate Models for Microstrip Computer-Aided Design. In: 1980 IEEE MTT-S International Microwave Symposium Digest, pp 407–409., Washington, DC, USA
40. Kal S, Bhattacharya D, Chakraborti NB (1981) Empirical relations for capacitive and inductive coupling coefficients of coupled microstrip lines. IEEE Trans Microw Theory Techn 29(4):386–388
41. Kirschning M, Jansen R (1984) Accurate Wide-Range design equations for the Frequency-Dependent characteristic of parallel coupled microstrip lines. IEEE Trans Microw Theory Techn 32(1):83–90
42. Kirschning M, Jansen R (1985) Corrections to accurate Wide-Range design equations for the Frequency-Dependent characteristic of parallel coupled microstrip lines. IEEE IEEE Trans. Microw. Theory Techn. 33(3):288
43. Tripathi VK, Chang CL (1978) Quasi-TEM parameters of non-symmetrical coupled microstrip lines. Int J Electron 45(2):215–223
44. Bedair SS (1984) Characterization of some asymmetric coupled transmission lines. IEEE Trans Microw Theory Techn 32(1):108–110
45. Bhat B, Koul SK (1982) Unified approach to solve a class of strip and Microstrip-Like transmission lines. IEEE Trans Microw Theory Techn 30(5):679–686
46. Simons RN (2001) Coplanar waveguide circuits, components, and systems. John Wiley and Sons, New York
47. Nguyen C (1993) Dispersion characteristics of the Broadside-Coupled coplanar waveguide. IEEE Trans Microw Theory Tech 41(9):1630–1633
48. Tyurnev V, Belyaev BA (1990) Interaction of parallel microstrip resonators (in Russian). Elektronnaya tekhnika. Ser. Elektronika SVCh 4:25–30

49. Hong J-S, Lancaster MJ (1996) Couplings of microstrip square Open-Loop resonators for Cross-Coupled planar microwave filters. IEEE Trans Microw Theory Techn 44(11):2099–2019
50. Hong JS (2000) Couplings of asynchronously tuned coupled microwave Resonators. IEE Proc Microw Antennas Propag 147(5):354–358
51. Awai I, Zhang Y (2007) Coupling coefficient of Resonators—An intuitive way of its understanding. Electron Commun Jpn 90(9):962–968
52. Awai I, Zhang Y (2008) Separation of coupling coefficient between resonators into magnetic and electric components toward its application to BPF development. In 2008 China-Japan Joint Microwave Conference, pp 61–65, Shanghai, China
53. Awai I, Iwamura S, Kubo H, Sanada A (2006) Separation of coupling coefficient between resonators into electric and magnetic contributions. Electron Commun Jpn 89:1033–1039
54. Tyurnev VV (2010) Coupling coefficients of resonators of resonators in microwave filter theory. Prog Electromagn Res B 21:47–67
55. Elnaggar SY, Tervo RJ, Mattar SM (2015) Energy coupled mode theory for electromagnetic resonators. IEEE Trans Microw Theory Techn 63(7):2115–2123
56. Elnaggar S, Tervo R, MMS (2015) General expressions and physical origin of the coupling coefficient of arbitrary tuned coupled electromagnetic resonators. J Appl Phys, 118, p 194901
57. Powell DA, Hannam K, Shadrivov IV, Kivshar YS, Powell DA, Hannam K, Shadrivov IV, Kivshar YS (2011) Near-field interaction of twisted split-ring resonators. Phys Rev B, 83, paper 235420
58. Wang H, Chu Q-X (2007) Generation of transmission zero through electric and magnetic mixed coupling. In 2007 International Conference on Microwave and Millimeter Wave Technology, pp 1–3, Guilin, China
59. Tang S-C, Yu C-H, Chiou Y-C, Kuo J-T (2009) Extraction of electric and magnetic coupling for coupled symmetric microstrip resonator bandpass filter with tunable transmission zero. In: 2009 Asia Pacific Microwave Conference, pp 206, Singapore
60. Belyaev BA, Serzhantov AM, Laletin NV (2001) Coupling factors of regular microstrip resonators. In 2001 Microwave Electronics: Measurement, Identification, Applications. Conference Proceedings. MEMIA'2001, pp 73–76, Novosibirsk, Russia
61. Dai GL, Zhang XY, Chan C, Xue Q, Xia MY (2009) An investigation of open- and short-ended resonators and their applications to bandpass filters. IEEE Trans Microw Theory Techn 57(9):2203–2210
62. Sánchez-Soriano MA, Torregrosa-Penalva G, Bronchalo E (2012) Multispurious suppression in parallel-coupled line filters by means of coupling control. IET Microw. Antennas Propag. 6(11):1269–1276
63. Zarifi MH, Daneshmand M (2016) Wide dynamic range microwave planar coupled ring resonator for sensing applications. Appl Phys Lett, 108, paper 232906
64. Juan G, Potelon B, Quendo C, Bronchalo E, Sabater-Navarro J (2019) Highly-Sensitive glucose concentration sensor exploiting inter-resonators couplings. In 2019 49th European Microwave Conference (EuMC), pp 662–665, Paris, France
65. Herrera-Sepulveda LV, Olvera-Cervantes JL, Corona-Chavez A, Kaur T (2019) Sensor and methodology for determining dielectric constant using electrically coupled resonators. IEEE Microw Wireless Compon Lett 29(9):626–628
66. Morales-Lovera HN, Olvera-Cervantes JL, Corona-Chavez A, Kaur Kataria T (2020) Dielectric anisotropy sensor using coupled resonators. IEEE Trans Microw Theory Techn, 68(4), pp 1610–1616
67. Herrera Sepulveda LV, Olvera Cervantes J, Saavedra CE (2021) Multifrequency Coupled-Resonator sensor for dielectric characterization of liquids. IEEE Trans Instrum Meas, 70, pp 1–7
68. Nguenouho OB, Chevalier A, Potelon B, Quendo C (2022) Dielectric characterization and modelling of aqueous solutions involving sodium chloride and sucrose and application to the design of a bi-parameter RF-sensor. Sci Rep 12:7209
69. Luckasavitch K, Kozak R, Golovin K, ZMH (2022) Magnetically coupled planar microwave resonators for real-time saltwater ice detection. Sensors and Actuators: A. Physical, 333, paper 113245

70. Cohn SB, Shimizu JK (1955) Strip transmission lines and components: second quarterly progress report. In: Stanford Research Institute, SRI Project 1114. Stanford Research Institute, SRI Project 1114
71. Ang K, Robertson ID (2001) Analysis and design of Impedance-Transforming planar marchand baluns. IEEE Trans Microw Theory Techn 49(2):402–406
72. Jensen T, Zhurbenko V, Krozer V, Meincke P (2007) Coupled transmission lines as impedance transformer. IEEE Trans Microw Theory Techn 55(12):2957–2965
73. Sun S, Zhu L (2006) Capacitive-Ended interdigital coupled lines for UWB bandpass filters with improved Out-of-Band performances. IEEE Microw Wireless Compon Lett. 16(8):440–442
74. Tsai C-M, Gupta KC (1992) A Generalized model for coupled lines and its applications to Two-Layer planar circuits. IEEE Trans Microw Theory Techn 40(12):2190–2199
75. Cristal E (1966) Coupled-Transmission-Line directional couplers with coupled lines of unequal characteristic impedances. IEEE Trans Microw Theory Techn 14(7):337–346
76. Sheleg B, Spielman E (1974) Microstrip Broad-Band directional couplers using with dielectric 'overlays. IEEE Trans Microw Theory Techn 22(12):1216–1220
77. Su L, Itoh TT, Rivera J (1983) Design of an overlay directional coupler by a Full-Wave analysis. IEEE Trans Microw Theory Techn 31(12):1017–1022
78. Klein y J, Chang K (1990) Optimum dielectric overlay thickness for equal even- and odd-mode phase velocities in coupled microstrip circuits. Electron Lett., 26(5), pp 274–276
79. Piekarz I, Sorocki J, Wincza K, Gruszczynski S (2017) Microwave sensors for dielectric sample measurement based on Coupled-Line Section. IEEE Trans Microw Theory Techn 65(5):1615–1631
80. Bockelman D, Eisenstadt W (1995) Combined differential and Common-Mode scattering parameters: theory and simulation. IEEE Trans Microw Theory Techn 43(7):1530–1539
81. Martín F, ZL, Hong J, Medina F (2018) Balanced microwave filters, Hoboken, NJ, USA: Wiley/IEEE Press
82. Levy R (1973) A generalized design technique for practical distributed reciprocal ladder networks. IEEE Trans Microw Theory Techn 21(8):519–526

Permittivity Sensors Based on Coupled Line Sections

Ilona Piekarz and Jakub Sorocki

Abstract This chapter focuses on coupled-line section sensors operating under differential excitation utilized for wideband characterization of dielectric materials. Various realizations of sensors designed for the determination of complex permittivity of solid, as well as liquid materials, are presented together with the theoretical analysis, calibration methods, and measurement results. It is shown that utilization of differential excitation in coupled-line-based sensors allows to increase the sensitivity to small-sized materials under test (i.e., having a size comparable with the distance between the sensors' strips). Thus, these sensors can be utilized for the characterization of small amounts of liquids placed in a microfluidic channel. A comparison between the coupled-line-based sensors and other sensors described in the literature, allowing for the characterization of dielectric materials in a wide frequency range, is also provided.

Coupled transmission lines are utilized for the realization of various microwave circuits, such as directional couplers [1–9], filters [10–14], phase shifters [5, 6, 15], etc. Recently, it has been shown that coupled lines can also be used for the realization of sensors allowing for dielectric material characterization and detection of material parameter changes in a wide frequency range [16–18]. The major advantage of such a technique is its high sensitivity to the sample-under-test when out-of-phase excitation is provided to the coupled-line section. This allows the sample to be as narrow as the spacing between the coupled strips. The coupled-line section sensors can be realized in two configurations i.e., edge- [16, 17] and broadside-coupled [18]. The edge coupled-line sensor configuration [16, 17] is preferable for the characterization of solid materials and larger amounts of liquids since a minimal volume of the liquid sample is required to ensure that the electric field is entirely confined within

I. Piekarz (✉) · J. Sorocki
Institute of Electronics, AGH University of Krakow, 30-059, Kraków, Poland
e-mail: ilona.piekarz@agh.edu.pl

J. Sorocki
e-mail: jakub.sorocki@agh.edu.pl

F. Martín and E. Bronchalo (eds.), *Coupled Structures for Microwave Sensing*, Lecture Notes in Electrical Engineering 1150,
https://doi.org/10.1007/978-3-031-53861-2_3

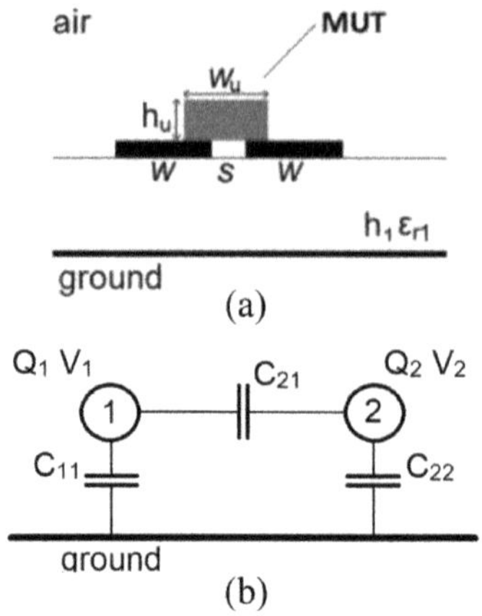

Fig. 1 Coupled-line system composed of three conductors, i.e. two coupled conductors and the third as a ground plane. Cross-sectional view (**a**) and an equivalent capacitance network (**b**). In this case $C_{11} = C_{22}$ due to the identical geometrical dimensions of the strips. Reprinted with permission from Piekarz et al. [16]; copyright 2016 IEEE

the sample. On the other hand, the broadside-coupled configuration allows for the realization of a microfluidic channel in between the coupled strips. As a result, when differential excitation is applied, the electromagnetic field is fully confined within the liquid-under test. Therefore, a high sensitivity is obtained while requiring a low sample volume.

In a coupled-line system composed of n conductors, $n - 1$ propagation modes can be distinguished. Due to this fact, when the structure consists of three conductors (see Fig. 1a), i.e., two signal conductors placed over a common ground plane, the circuit can be described by a per-unit-length capacitance system, composed by C_{11} and C_{22} (capacitances of the first and second line, respectively), and C_{21} (the mutual capacitance between the lines), as it is shown in Fig. 1b [19]. Q_1 and Q_2 denote charges gathered on conductors 1 and 2, respectively, whereas V_1 and V_2 denote the voltages of 1st and 2nd conductor (their potential referenced to the ground conductor). Assuming that conductors 1 and 2 have identical geometrical dimensions, as shown in Fig. 1a, the self-capacitances C_{11} and C_{22} are equal and the capacitance matrix can be expressed in terms of even- and odd-mode excitations, as shown in Fig. 2a–c [19]. For the even-mode excitation, both conductors are excited with equal amplitude and in-phase voltages, V_e (the charges gathered on both conductors are equal, Q_e). In such a case, there is no voltage between lines and the system from Fig. 2a simplifies to the circuit of Fig. 2b. For the odd-mode excitation, both conductors are excited with equal amplitude and out-of-phase voltages, V_o (charges gathered on both conductors are equal, Q_o). In such a case there is a doubled voltage between the lines, and the system of Fig. 2a simplifies to the circuit of Fig. 2c. Based on the calculated equivalent circuits (see Fig. 2b and c), the capacitances of the coupled transmission lines for even- and odd-mode excitations in the presence of uniform dielectric medium having a relative dielectric constant of unity are equal [19]:

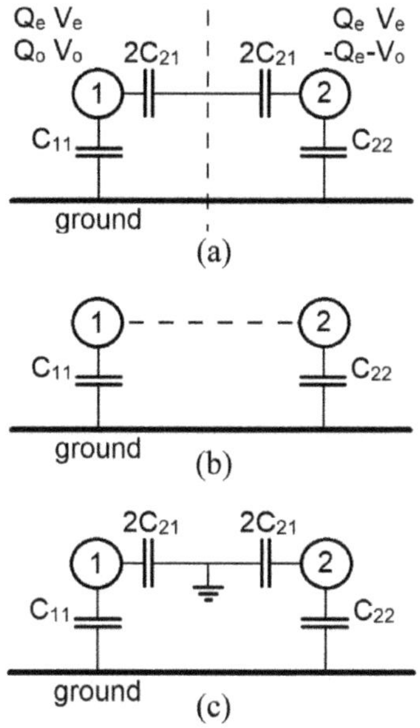

Fig. 2 Cross-section of two symmetrical coupled transmission lines for the even- and odd-mode excitations (**a**), an equivalent capacitance network for even-mode excitation (**b**), and an equivalent capacitance network for odd-mode excitation (**c**)

$$C_{oe} = C_{11} \tag{1a}$$

$$C_{oo} = C_{11} + 2C_{21} \tag{1b}$$

Having found the line capacitances for the even- and odd-mode excitations, the even- and odd-mode impedances of the coupled-line section in the presence of an inhomogeneous dielectric medium can be calculated as [19]:

$$Z_{oe} = \frac{1}{c\sqrt{C_e C_{oe}}} \tag{2a}$$

$$Z_{oo} = \frac{1}{c\sqrt{C_o C_{oo}}} \tag{2b}$$

where c is the free-space velocity of light and C_e, C_o denote the even- and odd-mode per-unit-length capacitances in the presence of inhomogeneous dielectric medium, i.e.,

$$C_e = C_{oe} \cdot \varepsilon_{effe} \tag{2c}$$

$$C_o = C_{oo} \cdot \varepsilon_{effo} \tag{2d}$$

where ε_{effe} and ε_{effo} are the even and odd mode effective dielectric constants.

Covering the coupled-line section with dielectric material-under-test (MUT) having thickness h_u and permittivity $\varepsilon_{MUT}^{comp} \neq 1$ in the way presented in Fig. 1a, the C_{11} and C_{21} are replaced with C_I and C_m. The presence of MUT causes a slight change of the capacitance C_1 between the line and a ground plane and significantly influences the mutual capacitance C_m between lines. Electric field distribution related to capacitance C_m is partially within the volume of the dielectric sample while most of the field for even excitation is within the volume of the substrate as can be seen in Fig. 3. Therefore, the coupled lines' odd-mode impedance (see Eq. (2b)) exhibits much higher sensitivity on the dielectric cover than the even-mode impedance, as it can be seen in Fig. 4 for different dielectric covers (different thickness and permittivity). Moreover, it can be assessed that the tested sample can be as small as the distance between the utilized coupled lines due to the above-described phenomena. Sensitivity s_o of the effective odd-mode permittivity to the presence of MUT can be expressed as:

$$s_o = \frac{d\varepsilon_{effo}}{d\varepsilon_{MUT}} \approx \frac{2C_{21}}{C_{11} + 2C_{21}} \tag{2e}$$

assuming relative permittivity of the MUT ε_{MUT} affects fully and only C_{21}, what indicates that either low C_{11} and/or high C_{21} is required to maximize it.

In the following sections, both edge- and broadside-coupled line section-based sensors operating in either differential reflection or differential reflection/transmission mode, respectively, in a broad frequency range, are elaborated. The usability of the developed class of sensors is experimentally validated through the characterization of solid and liquid materials with known dielectric properties. Moreover, a measurement setup is proposed to connect the differentially fed sensor with one-port single-ended measurement instruments using an integrated balanced-to-unbalanced circuit for the cost of an extra calibration step.

1 One-Port Differential Edge Coupled-Line Section Sensors

The unknown dielectric material covering the sensing element can be characterized based on the measurement of the differential S-parameters of the coupled-line sensor. General schematic diagrams of the proposed sensors composed of such odd-mode short-ended and open-ended coupled-line sections are presented in Fig. 5. The complex permittivity of the MUT is determined based on the measured complex odd-mode effective permittivity of the sensor and working backward using electromagnetic (EM) simulation tool. The complex odd-mode effective permittivity is calculated from the differential input impedance of the sensor obtained from the measured one-port differential S-parameter S_{dd} [21, 22] or two-port single-ended S-parameters ($S_{11}, S_{12}, S_{21}, S_{22}$) and their mathematical conversion into the differential one [23]:

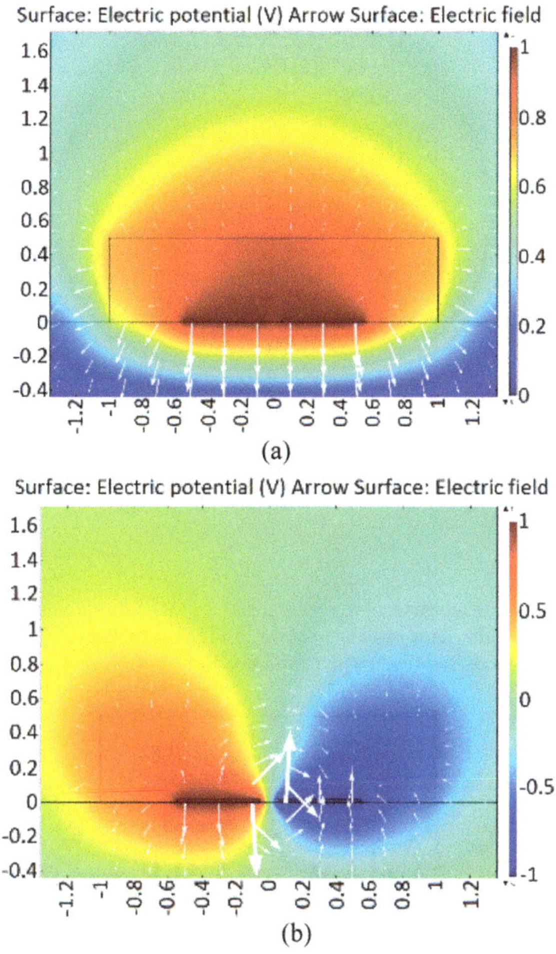

Fig. 3 Electric potential and electric field distribution for even—(**a**) and odd-mode (**b**) excitations. It can be clearly seen that in the case of odd-mode excitation, most of the field is within the dielectric sample covering coupled lines (dielectric structure as shown in Fig. 1, the dielectric sample is covering coupled strips). Reprinted with permission from Piekarz et al. [16]; copyright 2016 IEEE

$$S_{dd} = \frac{1}{2}(S_{11} + S_{22} - S_{12} - S_{21}) \tag{3}$$

$$Z_{in} = Z_T \frac{1 + S_{dd}}{1 - S_{dd}} \tag{4}$$

where Z_T is the reference impedance of the S-parameters measurement system (typically 50 Ω). The obtained Z_{in} (4) needs to be compared to the input impedance of the coupled-line sensor model (see Fig. 5c and d) to determine the odd-mode impedance Z_{oo} and odd-mode electrical length θ_o.

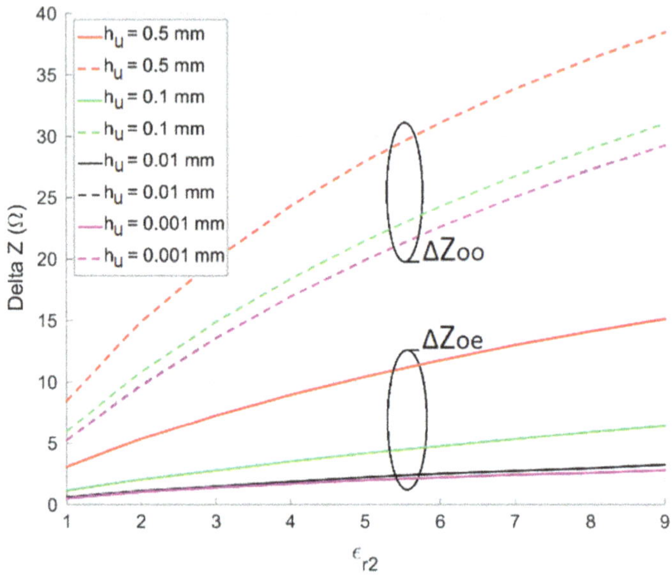

Fig. 4 Percentage change of Z_{oe} and Z_{oo}. Results obtained with the use of *Linpar* software [20] for the dielectric structure shown in Fig. 1, where $w = 1$ mm, $s = 0.08$ mm, $\varepsilon_{r1} = 3.38$, $h_1 = 1.524$ mm. Reprinted with permission from Piekarz et al. [16]; copyright 2016 IEEE

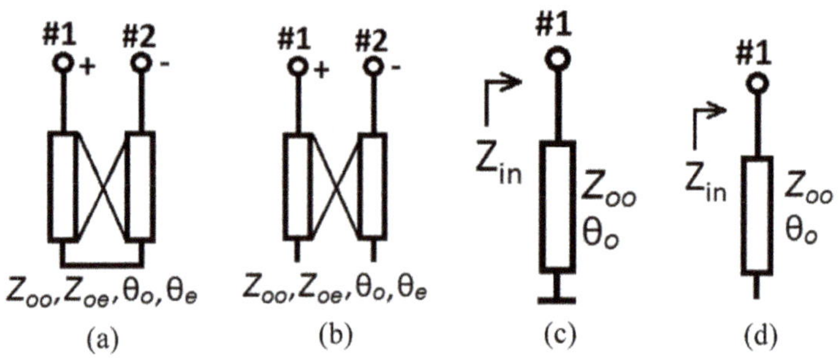

Fig. 5 The end-shorted (**a**) and open-ended coupled-line sensors (**b**) and their simplification into single transmission lines (**c–d**), when differential excitation is applied (at ports #1 and #2). The sensors are described by even- and odd-mode impedances (Z_{oe}, Z_{oo}) and even- and odd-mode electrical lengths (θ_e, θ_o). Reprinted with permission from Piekarz et al. [17]; copyright 2017 IEEE

1.1 Odd-Mode Short-Ended Coupled-Line Sensors

A schematic diagram of the proposed end-shorted coupled-line sensor is presented in Fig. 6a. For such a circuit the odd-mode parameters are calculated assuming differential excitation. Thus, the sensor's equivalent circuit simplifies to a single shorted transmission line having characteristic impedance Z_{oo} and electrical length θ_o, as shown in Fig. 5c. In general, two approaches to the realization of such a

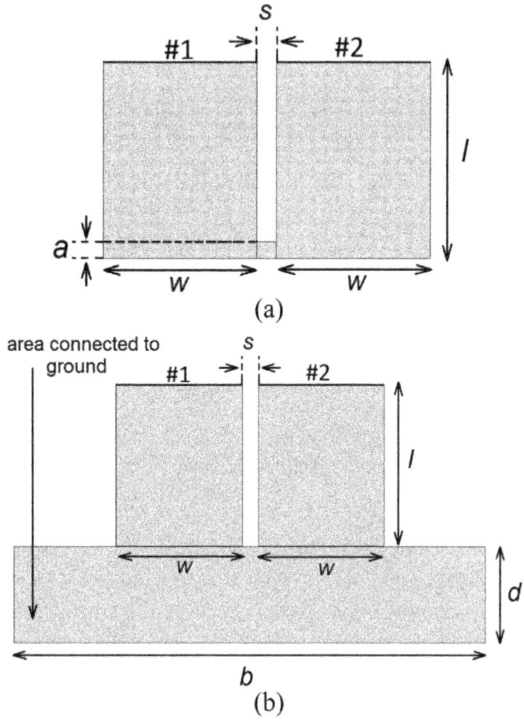

Fig. 6 Two different physical realizations of an odd-mode end-shorted coupled-line sensor: **a** small connecting segment—S^{I} and **b** grounding pad—S^{II}. Reprinted with permission from Piekarz et al. [17]; copyright 2017 IEEE

coupled-line sensor can be proposed since the principle of operation requires the odd-mode equivalent circuit to be shorted to ground. In the first approach (S^{I}) a small connecting segment at the end is used (see Fig. 6a) and in the second one (S^{II}) a metallization pad connected to the ground plane is introduced (see Fig. 6b).

In the case when S^{I} is considered, a physically small connecting segment can be modeled as an inductor (see Fig. 7). Thus, when differential excitation is applied, the sensor simplifies into the ground shorted circuit as shown in Fig. 7b. The input impedance Z_{in} of the circuit of Fig. 7b is

$$Z_{in} = Z_{oo}\frac{j\omega L + Z_{oo}\tanh(\gamma l)}{Z_{oo} + j\omega L\tanh(\gamma l)} \tag{5}$$

$$\gamma = \alpha + j\beta = \sqrt{(k_c^2 - \frac{\omega^2}{v^2}\varepsilon_{eff}^{comp}} \tag{6}$$

$$\varepsilon_{eff}^{comp} = \varepsilon_{eff}\left(1 - j\tan(\delta)_{eff}\right) \tag{7}$$

where γ is the propagation constant, α and β are the attenuation and phase constants, respectively, $\omega = 2\pi f$, where f is the frequency (in Hz) at which the input impedance

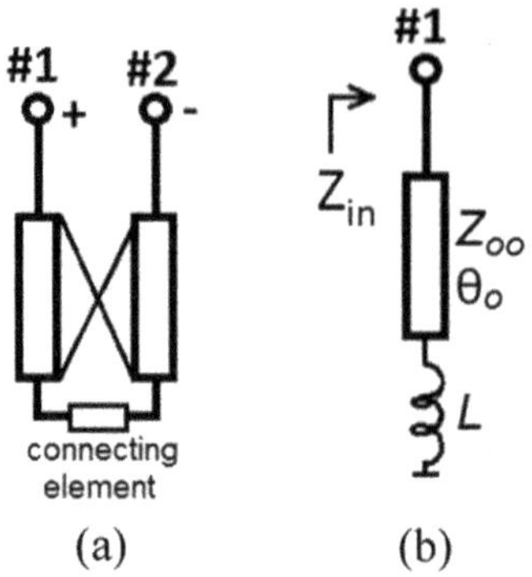

Fig. 7 Schematic diagram of **a** the physically realized end-shorted coupled-line sensor S[I] with connecting element and **b** simplification of the circuit in case of differential excitation, where the connecting element is modeled by inductor having inductance *L*. Reprinted with permission from Piekarz et al. [17]; copyright 2017 IEEE

is calculated, l is the physical length of the coupled-line section (in meters), v is the free-space velocity of light, k_c is the cut-off wave number, ε_{eff} is the real part of the odd-mode effective permittivity of the sensor, $\tan(\delta)_{eff}$ is the odd-mode effective loss tangent and ε_{eff}^{comp} is the complex odd-mode effective permittivity of the sensor. Assuming that TEM (Transverse Electromagnetic) type wave is propagating, $k_c = 0$ and the propagation constant γ is equal to:

$$\gamma = j\frac{\omega}{v}\sqrt{\varepsilon_{eff}\left(1 - j\tan(\delta)_{eff}\right)} \tag{8}$$

Thus, Z_{in} can be written as:

$$Z_{in} = \frac{A}{\sqrt{\varepsilon_{eff}^{comp}}} \frac{j\omega L + \frac{A}{\sqrt{\varepsilon_{eff}^{comp}}}\tanh\left(jY\sqrt{\varepsilon_{eff}^{comp}}\right)}{\frac{A}{\sqrt{\varepsilon_{eff}^{comp}}} + j\omega L\tanh\left(jY\sqrt{\varepsilon_{eff}^{comp}}\right)} \tag{9}$$

$$A = \frac{1}{vC_0} \tag{10}$$

$$Y = \frac{\omega l}{v} \tag{11}$$

where C_o denotes the per-unit-length odd-mode capacitance of strips with a uniform dielectric medium having a relative dielectric constant equal to 1. Next, using the Maclaurin series expansion of $\tanh(x)$ function:

$$\tanh(x) \approx x - \frac{x^3}{3} + \frac{2x^5}{15} + O\left(x^7\right) \tag{12}$$

one can calculate ε_{eff}^{comp} based on (3)–(4) and (9)–(11). Assuming approximation of $\tanh(x)$ with the first two elements (for $|x| < \pi/3$ the approximation error does not exceed 0.1128), one can rewrite (9) as a quadratic equation with respect to ε_{eff}^{comp}:

$$m_1\left(\varepsilon_{eff}^{comp}\right)^2 + m_2\left(\varepsilon_{eff}^{comp}\right) + m_3 = 0 \tag{13}$$

where:

$$m_1 = -\frac{\omega L Z_{in} Y^3}{3} \tag{14}$$

$$m_2 = -Z_{in}\omega L Y - j\frac{A^2Y^3}{3} \tag{15}$$

$$m_3 = Z_{in}A - jA\omega L - jA^2Y \tag{16}$$

As a result, ε_{eff}^{comp} is found to be a root of (13):

$$\varepsilon_{eff}^{comp} = \frac{-m_2 \pm \sqrt{m_2^2 - 4m_1m_3}}{2m_1} \tag{17}$$

The solution (either with a + or – sign) that yields a positive value within the roughly expected range for the real part of complex permittivity is selected as a physically meaningful one at a given frequency point. It must be noted that such an approximation of tanh(x) is only valid for $|\tanh(x)| << \pi/2$ and the approximation error increases with increasing x.

On the other hand, when S^{II} is considered, the utilized metallization pad connected to the ground eliminates the appearance of inductance L present in the S^{I} sensor, and therefore, the input impedance of the odd-mode equivalent circuit is equal to:

$$Z_{in} = Z_{oo}\tanh(\gamma l) \tag{18}$$

Using (3) and (4), (6)–(8), (10) and (11) the input impedance Z_{in} can be derived as:

$$Z_{in} = \frac{A}{\sqrt{\varepsilon_{eff}^{comp}}}\tanh\left(jY\sqrt{\varepsilon_{eff}^{comp}}\right) \tag{19}$$

Assuming approximation of $\tanh(x)$ with its first two elements of Maclaurin series expansion (see Eq. 12), ε_{eff}^{comp} can be calculated as:

$$\varepsilon_{eff}^{comp} = \frac{-j3AY + 3Z_{in}}{jAY^3} \tag{20}$$

S^I sensor is smaller than S^{II} since it requires only a short connection between the coupled-line sections. Nevertheless, such a connection introduces small inductance L which needs to be determined during calibration and can lead to the propagation of errors related to the calculation of L.

1.2 Odd-Mode Open-Ended Coupled-Line Sensors

A schematic diagram of the proposed open-ended coupled-line sensor (S^{III}) is presented in Fig. 5b. The circuit can be simplified in the ideal case to a single open-ended transmission line, as shown in Fig. 5d when differential excitation is assumed. However, in a physically realized sensor (see Fig. 8a), at the end of the open coupled-line section a fringing field [24–27] occurs which can be modeled for the odd-mode equivalent circuit with a transmission line section in series with capacitance C, as shown in Fig. 8b. The input impedance of such a circuit equals:

$$Z_{in} = Z_{oo}\frac{\frac{-j}{\omega C} + Z_{oo}\tanh(\gamma l)}{Z_{oo} - \frac{j}{\omega C}\tanh(\gamma l)} \tag{21}$$

Using (1a), (1b) and (2a)–(2e), (4)–(6), (8) and (9) the input impedance Z_{in} can be rewritten as:

$$Z_{in} = \frac{A}{\sqrt{\varepsilon_{eff}^{comp}}}\frac{\frac{-j}{\omega C} + \frac{A}{\sqrt{\varepsilon_{eff}^{comp}}}\tanh\left(jY\sqrt{\varepsilon_{eff}^{comp}}\right)}{\frac{A}{\sqrt{\varepsilon_{eff}^{comp}}} - \frac{j}{\omega C}\tanh\left(jY\sqrt{\varepsilon_{eff}^{comp}}\right)} \tag{22}$$

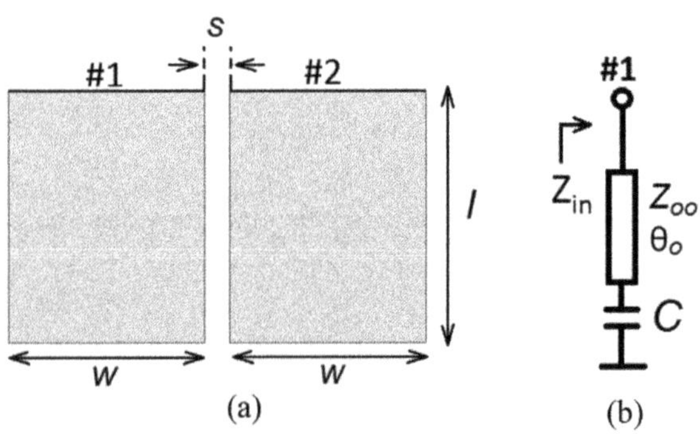

Fig. 8 An open-ended coupled-line sensor: **a** physical realization and **b** an equivalent circuit under differential excitation (at ports #1, #2) including fringing field at the open end modeled with a capacitor. Reprinted with permission from Piekarz et al. [17]; copyright 2017 IEEE

As in the previous cases, assuming an approximation of $\tanh(x)$ with its first two elements of Maclaurin series expansion, (22) can be rewritten as a quadratic equation with respect to ε_{eff}^{comp}:

$$p_1\left(\varepsilon_{eff}^{comp}\right)^2 + p_2\left(\varepsilon_{eff}^{comp}\right) + p_3 = 0 \tag{23}$$

where:

$$p_1 = \frac{Z_{in}Y^3}{3\omega C} \tag{24}$$

$$p_2 = \frac{Z_{in}Y}{\omega C} - j\frac{A^2Y^3}{3} \tag{25}$$

$$p_3 = Z_{in}A + j\frac{A}{\omega C} - jA^2Y \tag{26}$$

As a result, ε_{eff}^{comp} is found to be a root of (23):

$$\varepsilon_{eff}^{comp} = \frac{-p_2 \pm \sqrt{p_2^2 - 4p_1p_3}}{2p_1} \tag{27}$$

Similarly, as in the previous case, the solution (either with a + or – sign) that yields a positive value within the roughly expected range for the real part of complex permittivity is selected as a physically meaningful one at a given frequency point.

S^{III} occupies a similar area as S^{I} and requires the determination of end effect lumped-element mode value during the calibration procedure, which can negatively affect the determination error of the MUT's complex permittivity. Moreover, the value of C varies with the permittivity of MUT, since this capacitance models fringing fields partially enclosed within the volume of MUT and such a relation cannot be easily described. This effect influences the accuracy of determination of the effective permittivity of the sensor, hence the complex permittivity of MUT. In addition, the phenomenon of fringing fields is more time- and resource-demanding to investigate using an EM simulation tool, which is used for the calculation of MUT permittivity since it requires 3D instead of 2.5D field analysis.

1.3 Calculation of Complex Permittivity of MUT

Having obtained the complex odd-mode effective permittivity ε_{eff}^{comp} of the sensor, i.e., the end-shorted/open-ended coupled-line section covered with MUT, the complex permittivity of MUT ε_{MUT}^{comp} (see Fig. 1) itself can be determined using an appropriate iterative procedure, similar to the one described in [28]. The permittivity of MUT is found as follows:

1. First, one-port differential S-parameter $S_{dd}^{meas\_cov}$ or two-port single-ended S-parameters ($S_{11,}$ $S_{12,}$ $S_{21,}$ S_{22}) of the sensor covered with MUT need to be measured. In the case of measured single-ended S-parameters, mathematical conversion into the differential one (3) is required.
2. Next, determine the odd-mode effective permittivity of the sensor with MUT $\left(\varepsilon_{eff}^{comp}\right)_{meas_{cov}}$ using appropriate analytical equations provided in Sects. 1.1 and 1.2.
3. Create an EM model of the utilized coupled-line sensor with cover and calculate $\left(\varepsilon_{eff}^{comp}\right)_{sim_{cov}}$ based on EM-calculated differential S-parameters.
4. Compute $\Delta\varepsilon_{eff}$ and $\Delta\tan(\delta)_{eff}$:

$$\Delta\varepsilon_{eff} = \left| Re\left(\left(\varepsilon_{eff}^{comp}\right)_{sim_{cov}}\right) - Re\left(\left(\varepsilon_{eff}^{comp}\right)_{meas_{cov}}\right) \right| \tag{28}$$

$$\Delta\tan(\delta)_{eff} = \left| \frac{Im\left(\left(\varepsilon_{eff}^{comp}\right)_{sim_{cov}}\right)}{Re\left(\left(\varepsilon_{eff}^{comp}\right)_{sim_{cov}}\right)} - \frac{Im\left(\left(\varepsilon_{eff}^{comp}\right)_{meas_{cov}}\right)}{Re\left(\left(\varepsilon_{eff}^{comp}\right)_{meas_{cov}}\right)} \right| \tag{29}$$

5. Having known the approximate thickness and width of the dielectric sample, iterate steps (3)–(4) with different complex permittivity values ε_{MUT}^{comp} of MUT until the measured and simulated results match the assumed error.

As seen, the complex permittivity of MUT ε_{MUT}^{comp} can be obtained from measured differential S-parameters of the coupled-line sensor covered by the MUT sample and then working backward using the EM simulation program. The value of MUT complex permittivity ε_{MUT}^{comp} changes until measured $\left(\varepsilon_{eff}^{comp}\right)_{meas_{cov}}$ and calculated $\left(\varepsilon_{eff}^{comp}\right)_{sim_{cov}}$ agree with the assumed error.

1.4 Sensor Calibration

The physically realized sensors are composed of the end-shorted/open-ended coupled-line section and 50-Ω input signal lines for practical purposes. Hence, the proposed method requires a two-step calibration procedure:

1. First, 50-Ω input signal lines need to be de-embedded from the setup using an open-ended 50-Ω transmission line calibration circuit to shift the reference planes and provide measurements of the S-parameters of the sensor only.

2. Next, the measurement of one calibration dielectric sample with known permittivity $\left(\varepsilon_{MUT}^{comp}\right)_{cal}$ is required to determine one constant parameter C_0, in the case of the S^{II} sensor, and two constant parameters (C_0, L) and (C_0, C) in the case of S^{I} and S^{III} sensors, respectively.

As seen, the proposed calibration technique is timesaving since only de-embedding and one dielectric sample of known permittivity are required. As a reference, a very low-loss sample is preferred for ease of calculation.

Permittivity $\left(\varepsilon_{MUT}^{comp}\right)_{cal}$ of the calibration sample needs to be selected considering the physical realization of the sensor, i.e., the chosen substrate and its permittivity ε_{r1}. As there are two regions in the test setup cross-section (see Fig. 1) that have different permittivity, two components of the overall mutual capacitance between the sensor's coupled strips can be distinguished. One is related to the electrical field enclosed within the sensor's substrate and is constant, and the second one is related to the electrical field enclosed within MUT which changes with different samples and affects the overall capacitance value, hence, the measured odd-mode impedance. In general, the higher the $\left(\varepsilon_{MUT}^{comp}\right)_{cal}$ with respect to ε_{r1} the higher the effective permittivity value change. However, the higher the MUTs permittivity, the larger the hyperbolic tangent function argument, and thus the approximation error becomes at an unacceptable level from lower frequencies upwards. This is because the value of tanh at a given frequency is dependent on MUT permittivity and the approximation error rises non-linearly with the increase of tanh argument. To illustrate the issue, values of C_0 have been calculated for the coupled-line sensor S^{II} (w = 0.8 mm, $s = 0.1$ mm, $h_1 = 0.76$ mm, $\varepsilon_{r1} = 3.38$) covered with various dielectric materials and presented in Fig. 9. As seen at low frequencies all calibration samples yield comparable value of C_0. However, starting from roughly 3 GHz values start to deviate from each other with increasing rates aligned with increasing sample permittivity, hence impact of using tanh function approximation becomes visible. Therefore, the calibration sample needs to be selected as a trade-off between sensitivity maximization and measurement error minimization. Considering the above, the use of calibration sample featuring relative permittivity in the vicinity of sensor's substrate permittivity is favorable.

1.4.1 Calibration of the End-Shorted Coupled-Line Section Sensor S^{I}

The calibration procedure of S^{I} is as follows: after de-embedding of input signal lines, one needs to calculate the input impedance Z_{in}^{cal} delivered from the measured S-parameters (3) and (4) of the sensor covered with the calibration dielectric sample. The measured odd-mode input impedance Z_{in}^{cal} is then compared against the impedance Z_{in}^{EM} obtained from the EM model of the sensor covered with the calibration sample. Parameter L is determined by selecting its value in such a way that Z_{in}^{EM} matches Z_{in}^{cal} with the lowest possible error. Once L is determined, the C_0 value can be obtained by finding the roots of the following quadratic equation:

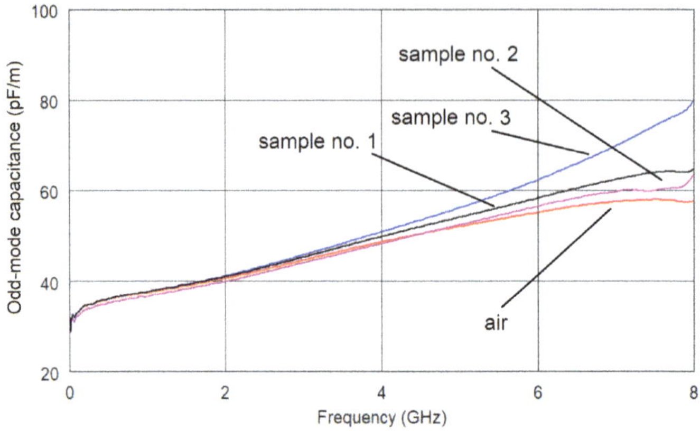

Fig. 9 Odd-mode capacitance C_O calculated based on measured *S*-parameters of sensor S[II] ($w = 0.8$ mm, $s = 0.1$ mm, $h_1 = 0.76$ mm, $\varepsilon_{r1} = 3.38$) covered with various dielectric materials: sample no. 1 – $Re\left(\varepsilon_{MUT}^{comp}\right) = 4.15$, $h_{MUT} = 0.76$ mm, sample no. 2 – $Re\left(\varepsilon_{MUT}^{comp}\right) = 3.38$, $h_{MUT} = 0.76$ mm, sample no. 3 – $Re\left(\varepsilon_{MUT}^{comp}\right) = 6.15$, $h_{MUT} = 1.57$ mm

$$n_1 C_0^2 + n_2 C_0 + n_3 = 0 \tag{30}$$

where:

$$n_1 = jZ_{in}^{cal}\omega L \tanh\left(jY\sqrt{\left(\varepsilon_{eff}^{comp}\right)_{cal}}\right) \tag{31}$$

$$n_2 = \frac{-j\omega L + Z_{in}^{cal}}{v\sqrt{\left(\varepsilon_{eff}^{comp}\right)_{cal}}} \tag{32}$$

$$n_3 = \frac{-\tanh\left(jY\sqrt{\left(\varepsilon_{eff}^{comp}\right)_{cal}}\right)}{v^2\left(\varepsilon_{eff}^{comp}\right)_{cal}} \tag{33}$$

Therefore, C_0 equals:

$$C_0 = \frac{-n_2 \pm \sqrt{n_2^2 - 4n_1 n_3}}{2n_1} \tag{34}$$

The solution (either with a + or – sign) that yields a positive value of capacitance within the roughly expected range is selected as a physically meaningful one at a given frequency point.

1.4.2 Calibration of the End-Shorted Coupled-Line Section Sensor S^{II}

The calibration procedure of S^{II} is as follows: after de-embedding of input signal lines, the value of C_0 needs to be determined from the measured S-parameters (3) and (4) of the sensor covered with the calibration dielectric sample having $\left(\varepsilon_{MUT}^{comp}\right)_{cal}$. In such a case, the odd-mode effective permittivity $\left(\varepsilon_{eff}^{comp}\right)_{cal}$ can then be determined. The value C_0 is calculated using (18) as:

$$C_0 = \frac{\tanh\left(jY\sqrt{\left(\varepsilon_{eff}^{comp}\right)_{cal}}\right)}{Z_{in}^{cal}v\sqrt{\left(\varepsilon_{eff}^{comp}\right)_{cal}}} \tag{35}$$

where Z_{in}^{cal} is the measured odd-mode input impedance of the sensor covered with the calibration sample.

1.4.3 Calibration of the Open-Ended Coupled-Line Section Sensor S^{III}

The calibration procedure of S^{III} is like the one of S^{I}. After the de-embedding of input signal lines, one needs to calculate the input impedance Z_{in}^{cal} delivered from the measured S-parameters (3), and (4) of the sensor covered with the calibration dielectric sample. The obtained Z_{in}^{cal} is then compared against the input impedance Z_{in}^{EM} determined from the EM model of the sensor covered with the calibration sample. Parameter C is determined by selecting its value in such a way that Z_{in}^{EM} best matches Z_{in}^{cal}. Once C is determined, the C_0 value can be obtained by finding the roots of the following quadratic equation:

$$o_1C_0^2 + o_2C_0 + o_3 = 0 \tag{36}$$

where:

$$o_1 = \frac{-jZ_{in}^{cal}\tanh\left(jY\sqrt{\left(\varepsilon_{eff}^{comp}\right)_{cal}}\right)}{\omega C} \tag{37}$$

$$o_2 = \frac{Z_{in}^{cal} + j\frac{1}{\omega C}}{v\sqrt{\left(\varepsilon_{eff}^{comp}\right)_{cal}}} \tag{38}$$

$$o_3 = -\frac{\tanh\left(jY\sqrt{\left(\varepsilon_{eff}^{comp}\right)_{cal}}\right)}{v^2\left(\varepsilon_{eff}^{comp}\right)_{cal}} \tag{39}$$

Therefore, C_0 equals:

$$C_0 = \frac{-o_2 \pm \sqrt{o_2^2 - 4o_1o_3}}{2o_1} \tag{40}$$

The solution (either with a + or – sign) that yields a positive value of capacitance within the roughly expected range o is selected as a physically meaningful one at a given frequency point.

1.5 Measurements of Low-Dispersive Solid Dielectric Materials

To validate the proposed sensor class, exemplary end-shorted and open-ended coupled-line sensors were developed. In all cases the microstrip dielectric structure shown in Fig. 1, where $h_1 = 0.51$ mm and $\varepsilon_{r1} = 3.38$ (*Arlon 25N* laminate) was used. The strips' widths and separation between lines have been chosen to obtain high sensitivity on the relatively narrow sample (minimum spacing between lines is limited by the available manufacturing technology), see Table 1. The geometry of the sensor is of high importance for the sensitivity. The closer the coupled strips, the higher their mutual capacitance C_m while keeping the self-capacitance C_1 at the same level. As it was shown, for odd-mode excitation the capacitive matrix of the coupled line section reduces to a single capacitance being $C_{odd} = 2C_m + C_1$. Since only mutual capacitance is affected by MUT, it is desired to maximize the contribution of this capacitance to the C_{odd} to maximize the sensitivity of the sensor. Additionally, note that the sensors have different physical lengths, as listed in Table 1, which restricts their band of operation, i.e., the longer the sensor the lower the maximum frequency of operation.

The developed sensors were used to characterize low-dispersive solid dielectric materials. Microwave dielectric laminates were used as exemplary test materials, whose electrical parameters were taken from datasheets and are listed in Table 2.

Table 1 The geometry of the realized coupled-line sensors

Parameter	Dimension (mm)
w	0.8
s	0.1
l[b]	4
l[a]	2.5
d	3
b	15

[a]Length of sensors S^{II} and S^{III}
[b]Length of sensor S^{I}

Table 2 Parameters of the solid dielectric samples used as MUT

No.	Material	h_u (mm)	ε_{MUT}^{comp} datasheet		ε_{MUT}^{comp} measured [29]	
			ε_{MUT}	$\tan(\delta)_{MUT}$	ε_{MUT}	$\tan(\delta)_{MUT}$
1	Arlon AD450	0.76	4.5	0.0030	4.55	0.0061
2	Arlon 25N	1.52	3.38	0.0025	3.2	0.0059
3	Arlon AD600	1.57	6.15	0.0030	6.15	0.0051
4	FR4	0.80	4.2–4.8	0.0200	4.17	0.0178

To confirm the information specified in datasheets, the samples were measured at 7 GHz using a resonant cavity, and the resulting complex permittivity values were determined through the cavity perturbation method [29]. The batch variation is also listed in Table 2. In the next step, all three manufactured sensors were exposed to the test materials, and respective S-parameters were captured. From that, the resulting complex permittivity values were determined using procedures elaborated in Sects. 1.1 and 1.2. During measurement, the MUT samples were pressed against the sensor to minimize the influence of air gaps between the sensor and MUT which affects the resulting effective permittivity of the stack-up.

All *S*-parameters were measured using vector network analyzer *Agilent 5224A* having IF filter bandwidth set to 500 Hz. Subsequently, *AWR Microwave Office* software was used for EM simulations required to determine the complex permittivity of MUT as described in the provided algorithm (see Sect. 1.2). Since *AWR AXIEM*® and *AWR EMSight*® solvers available in the *AWR Microwave Office* are 2.5D type solvers, it has been assumed that the MUT sample covers the entire coupled-line sensor (solid MUT width equals at least $s + 2w$). All the obtained results are summarized in Table 3 to provide a direct comparison of the performance of the proposed sensors.

1.5.1 The End-Shorted Coupled-Line Sensor S^I

An exemplary sensor S^I having geometry shown in Fig. 6a and dimensions as listed in Table 1 was manufactured and measured. A photograph of the fabricated sensor S^I unit itself and covered with an exemplary MUT are shown in Fig. 10a and b, respectively. Additionally, to apply the described in Sect. 1.3 calibration procedure, the de-embedding circuit composed of a single 50-Ω input signal line was manufactured as well (see Fig. 10c) to shift the reference plane to the very edge of the couple-line section. For the calibration purpose, sample no. 1 was used and the required L and C_0 values were determined following the procedure presented in Sect. 1.4.1: $L =$ 0.065 nH while C_0 versus f is shown in Fig. 11.

Having found the sensor‘s calibration constants, the proposed sensor was exposed to test materials no. 2–4 (see Table 2). Based on the measured differential *S*-parameters, the complex odd-mode effective permittivity values $\left(\varepsilon_{eff}^{comp}\right)_{meas_{cov}}$, i.e., ε_{eff}^{meas} and $\tan(\delta)_{eff}^{meas}$ were determined and are shown in Figs. 12 and 13 (black solid

Table 3 Parameters of the dielectric samples used as MUT—comparison of permittivity measurements

Sample	$\varepsilon_{MUT} \pm \Delta\varepsilon_{MUT}$, $\tan(\delta)_{MUT} \pm \Delta\tan(\delta)_{MUT}$, frequency range (GHz)														
	Datasheet			Resonant[a]			S^{I}			S^{II}			S^{II}		
25N	3.38 + 0.03	0.002 + 0.0005	1–10	3.2	0.006	7	3.45	0–0.003	2.5–7.0	3.40	0.005–0.010	2.3–7.5	3.3	0.010–0.014	2.0–5.8
AD600	6.15 + 0.06	0.003 + 0.0007	1–10	6.15	0.005	7	5.90	0.015	3.0–7.0	5.80	0.003–0.010	2.0–7.5	5.8	0.035–0.045	3.0–5.8
FR4	4.34	0.02	1	4.17	0.018	7	4.10	0.015	2.5–6.0	4.30	0.007–0.015	2.5–7.5	3.5	0.08	3.0–5.5

[a]Measured using a resonant cavity and calculated based on the cavity perturbation method [29]

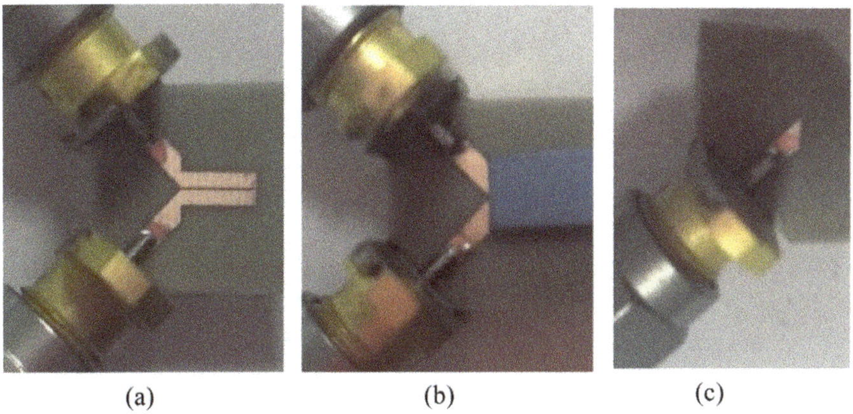

Fig. 10 Photographs of **a** manufactured S^I, **b** proposed sensor covered with MUT, and **c** open-ended 50-Ω signal line manufactured for de-embedding purposes. Reprinted with permission from Piekarz et al. [17]; copyright 2017 IEEE

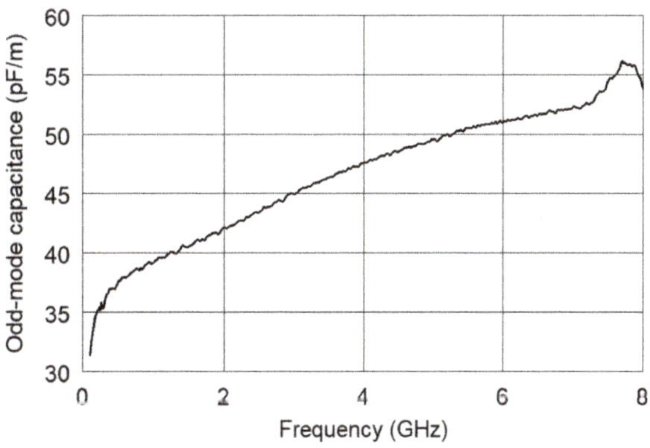

Fig. 11 Value of C_0 obtained from measured differential *S*-parameters of sensor S^I covered with calibration sample no. 1 (see Table 2)

lines). Note, that the measured odd-mode permittivity becomes relatively noisy below approximately 0.5 GHz which is the result of a finite dynamic range of the measurement equipment and the need for measuring S_{11} and S_{22} at the level below or near the noise floor. In the next step, the utilized end-shorted sensor S^I has been modeled in *AWR Microwave Office* software together with MUT. The complex permittivity of MUT ε_{MUT}^{comp} in EM simulations was swept in such a manner that the calculated ε_{eff}^{sim}, $\tan(\delta)_{eff}^{sim}$ of the EM model and measured ε_{eff}^{meas}, $\tan(\delta)_{eff}^{meas}$ values are in agreement with the assumed error at each frequency within the sensors‘ operation band. One or more electromagnetically calculated effective parameters ε_{eff}^{sim}, $\tan(\delta)_{eff}^{sim}$ were overlayed in Figs. 12 and 13 (grey lines) along with the corresponding complex permittivity values of covering samples ε_{MUT}^{comp} i.e., their real part ε_{MUT}, and loss tangent $\tan(\delta)_{MUT}$ to fit the measured curves. Note, that the values for some traces exceed the *y*-axis range of the plot, however, the plot range was chosen in such a way as

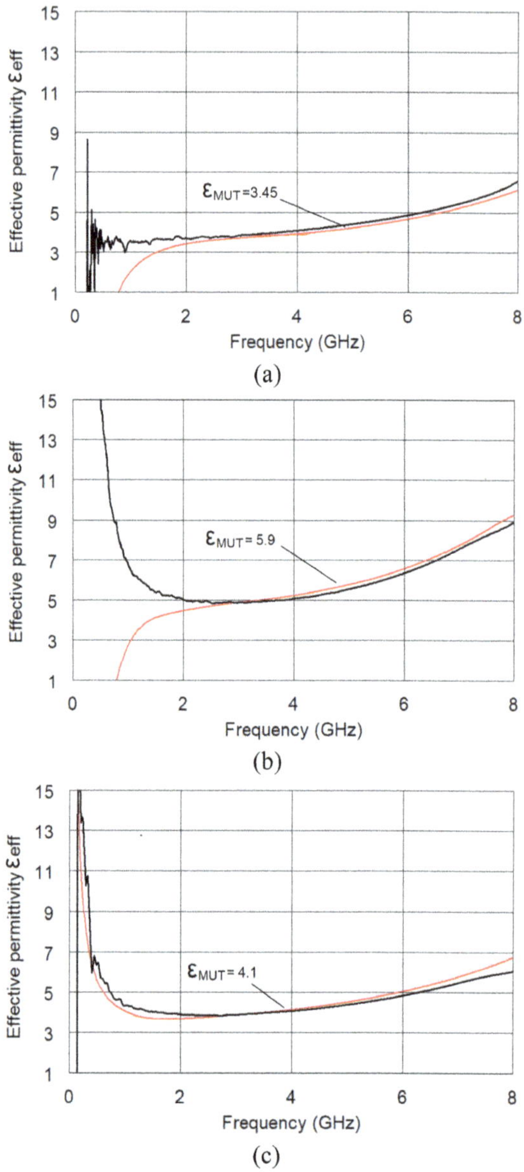

Fig. 12 Measured (thick solid line) and simulated (thin solid line with symbols) real odd-mode permittivity of samples: **a** no. 2, **b** no. 3 and **c** no. 4, respectively. Results obtained for sensor S^I

to show clearly the usable data. Moreover, a rapid value change is the result of the discontinuity of the tanh function used in the sensors' model or numerical calculations and the selection of appropriate solutions. The assumed maximum differences between the measured and simulated real parts of complex odd-mode permittivity values $\Delta\varepsilon_{eff}$ and $\Delta\tan(\delta)_{eff}$ was set as:

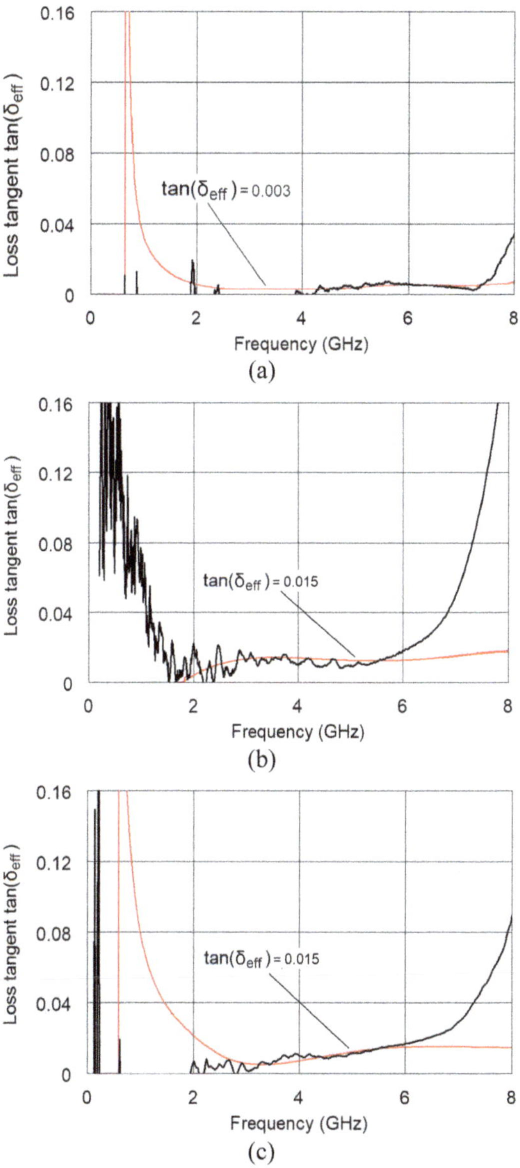

Fig. 13 Measured (thick solid line) and simulated (thin solid line with symbols) loss tangent of samples: **a** no. 2, **b** no. 3, and **c** no. 4, respectively. Results obtained for sensor S^I

$$\Delta\varepsilon_{eff} = \left|\varepsilon_{eff}^{sim} - \varepsilon_{eff}^{meas}\right| = 0.2 \tag{41}$$

$$\Delta\tan(\delta)_{eff} = \left|\tan(\delta)_{eff}^{sim} - \tan(\delta)_{eff}^{meas}\right| = 0.01 \tag{42}$$

Additionally, the measured effective permittivity values were averaged within lower frequencies (<2 GHz) for the fitting process to fulfill (41), and (42).

The percentage error of the determined permittivity value ε_{MUT} of sample no. 2 does not exceed 2% in the frequency range of 2.5–7.5 GHz, whereas in the case of sample no. 3 the percentage error of the calculated ε_{MUT} is not worse than 4% within frequencies 3–7 GHz. The percentage error of the determined permittivity value ε_{MUT} of sample no. 4 does not exceed 5.5% in the frequency range of 2–6 GHz. The determined values of loss tangent of MUTs $\tan(\delta)_{MUT}$ vary from 0.003 to 0.015 (see Fig. 13a–d) and for samples no. 2 and no. 3 are larger than the values provided by the materials manufacturer (see Table 2). Such an error is related to the very low $\tan(\delta)_{MUT}$ value of the measured sample, hence a relatively high error in value determination. As was pointed out in Krupka [30], for transmission line methods of permittivity measurements, a typical error of $\tan(\delta)_{MUT}$ determination is ±0.01. However, when the measured sample is more lossy, the error decreases as can be seen in the case when FR4 laminate is used as MUT for which $\tan(\delta)_{MUT}$ is assumed to be 0.02 (see Fig. 13d).

1.5.2 The End-Shorted Coupled-Line Sensor S^{II}

Similarly, an exemplary sensor having geometry shown in Fig. 6b and dimensions as listed in Table 1 was manufactured and measured. A photograph of the sensor is shown in Fig. 14a and the measurement setup composed of the manufactured sensor covered with an exemplary MUT sample is presented in Fig. 14b. A de-embedding circuit was manufactured as well, see Fig. 14c. Again, sample no. 1 was used to determine the C_0 and its value vs frequency is provided in Fig. 15.

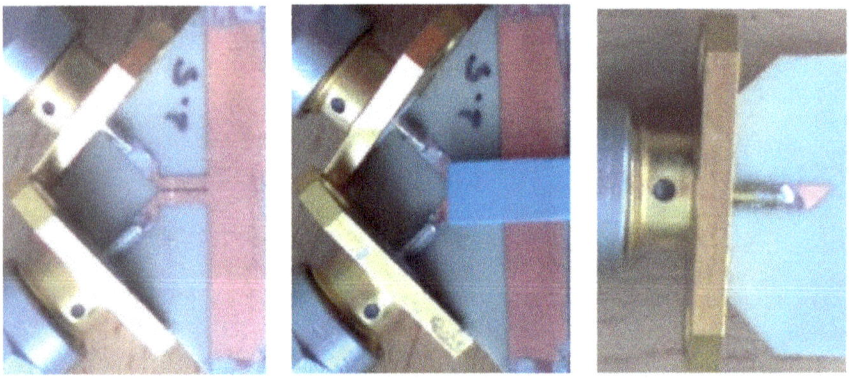

(a) (b) (c)

Fig. 14 Photographs of **a** manufactured S^{II}, **b** measurement setup composed of the proposed sensor covered with an exemplary dielectric MUT, and **c** open-ended 50-Ω signal line manufactured for de-embedding purposes. Reprinted with permission from Piekarz et al. [17]; copyright 2017 IEEE

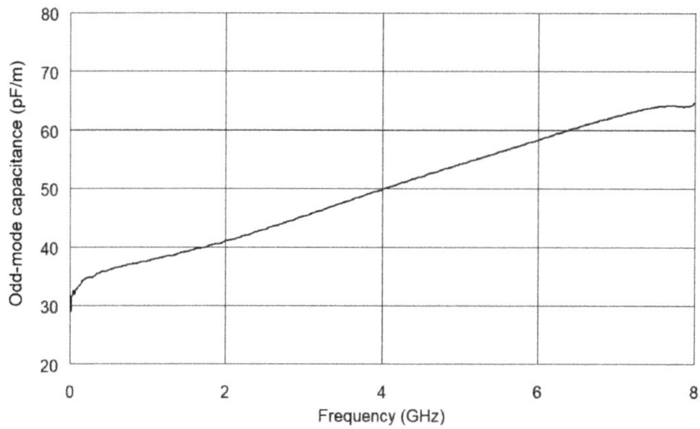

Fig. 15 The value of C_0 obtained from measured differential S-parameters of sensor S[II] covered with the calibration sample no. 1 (see Table 2)

The developed demonstrator was exposed to dielectric materials no. 2–4 (see Table 2) and based on the measured differential S-parameters, the complex odd-mode effective permittivity ε_{eff}^{meas} and $\tan(\delta)_{eff}^{meas}$ were calculated and are shown in Figs. 16 and 17. An analogous procedure was performed as in Sect. 1.4.3 to determine the complex permittivity of the measured MUTs ε_{MUT}^{comp}. In this case, the maximum differences defined in (41) and (42) are also held. Similarly, as in the case of sensor S[I], for lower frequencies (<2 GHz), the measured effective permittivity values have been averaged for the fitting process to fulfill (41), (42) and to determine the corresponding MUT permittivity values ε_{MUT}, as it is presented in Figs. 16 and 17.

The percentage error of the determined permittivity value ε_{MUT} of sample no. 2 does not exceed 0.6% in the frequency range of 2.3–7.5 GHz, whereas in the case of sample no. 3 does not exceed 5.7% in the frequency range of 1.5–8 GHz. The percentage error of the determined permittivity value ε_{MUT} of sample no. 4 does not exceed 0.9% in the frequency range of 2–8 GHz. The determined values of loss tangent of MUTs $\tan(\delta)_{MUT}$ vary from 0.003 to 0.015 (see Fig. 15a–d) which is greater than the assumed values (see Table 2). As mentioned in Sect. 1.4.3, such an error is related to very low $\tan(\delta)_{MUT}$. The measurement of more lossy FR4 laminate ($\tan(\delta)_{MUT} = 0.02$) yields a smaller error for loss tangent (see Fig. 17d), as in the case of the end-shorted sensor S[I].

1.5.3 The Open-Ended Coupled-Line Sensor S[III]

Finally, an exemplary sensor having geometry shown in Fig. 8a and dimensions as listed in Table 1 has been manufactured and measured. A photograph of the sensor is shown in Fig. 18a and the measurement setup composed of the manufactured sensor covered with an exemplary MUT is presented in Fig. 18b. The de-embedding circuit shown in Fig. 18c was manufactured alongside. Again, sample no. 1 was used

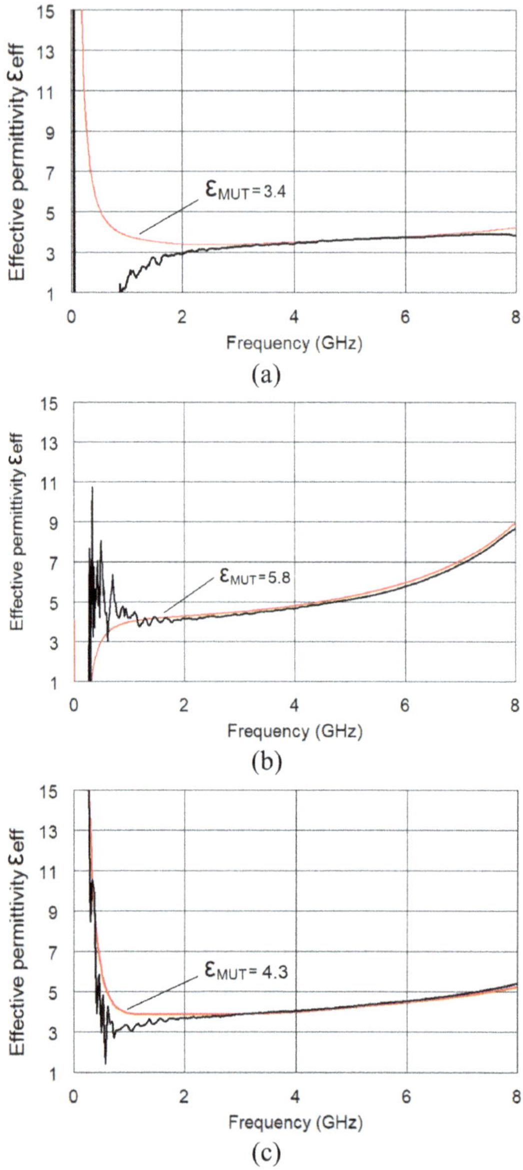

Fig. 16 Measured (thick solid line) and simulated (thin solid line with symbols) real odd-mode permittivity of samples: **a** no. 2, **b** no. 3, and **c** no. 4, respectively. Results obtained for sensor S^{II}

for calibration and the unit-specific C and C_0 values were determined following the procedure presented in Sect. 2.3: $C = 0.001$ pF while C_0 versus f is shown in Fig. 19.

The S^{III} sensor demonstrator was exposed to materials no. 2–4 (see Table 2) and based on the measured differential S-parameters the complex odd-mode effective permittivity values (ε_{eff}^{meas}, $\tan(\delta)_{eff}^{meas}$) were calculated and are shown in Figs. 20 and 21. An analogous procedure as in previous cases was carried out to determine

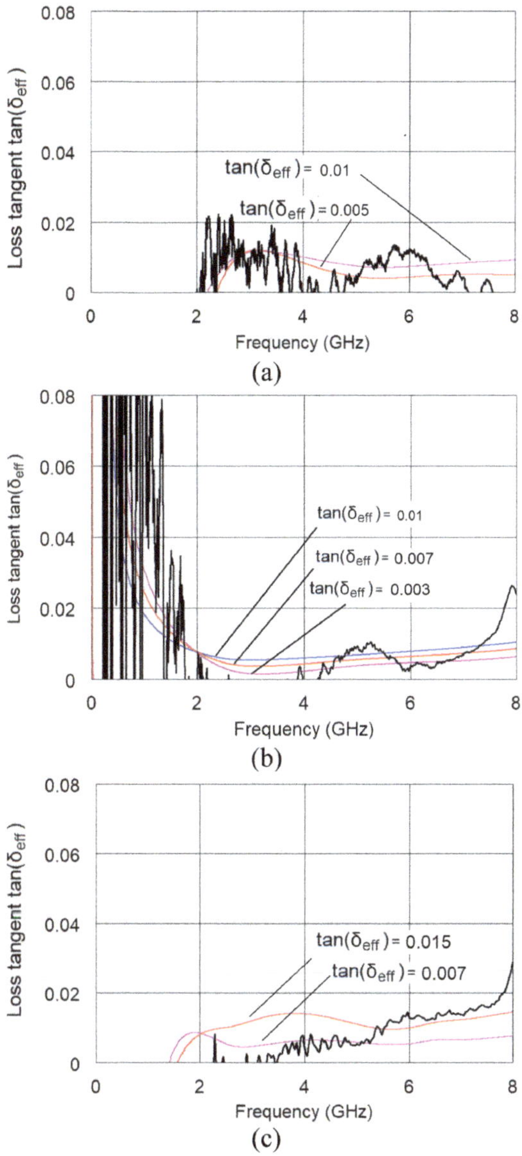

Fig. 17 Measured (thick solid line) and simulated (thin solid line with symbols) loss tangent of samples: **a** no. 2, **b** no. 3, **c** and no. 4, respectively. Results obtained for sensor S^{II}

the complex permittivity of the measured MUTs ε_{MUT}^{comp}. The maximum differences defined in (41) and (42) hold. Similarly, the measured effective permittivity values at lower frequencies were averaged for the fitting process to fulfill (41) and (42) and to determine the corresponding MUT permittivity values, as it was presented in Figs. 20 and 21.

The percentage error of the determined permittivity value ε_{MUT} of sample no. 2 does not exceed 2.4% in the frequency range 10 MHz–8 GHz, whereas for the sample

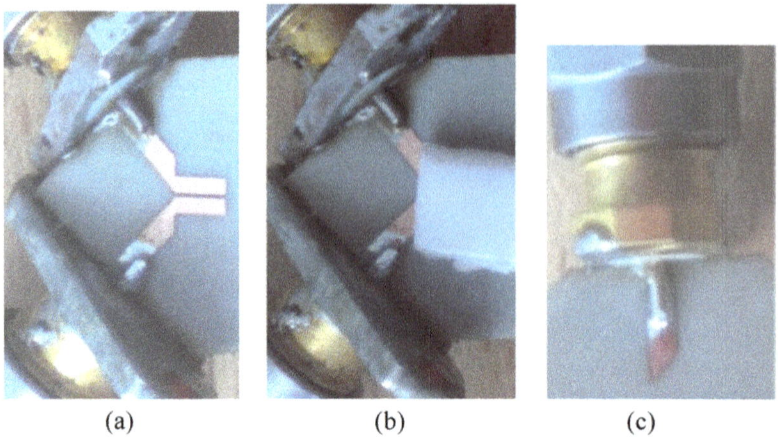

Fig. 18 Photographs of **a** manufactured S^{III}, **b** measurement setup composed of the sensor covered with MUT, and **c** open-ended input signal line manufactured for calibration purposes. Reprinted with permission from Piekarz et al. [17]; copyright 2017 IEEE

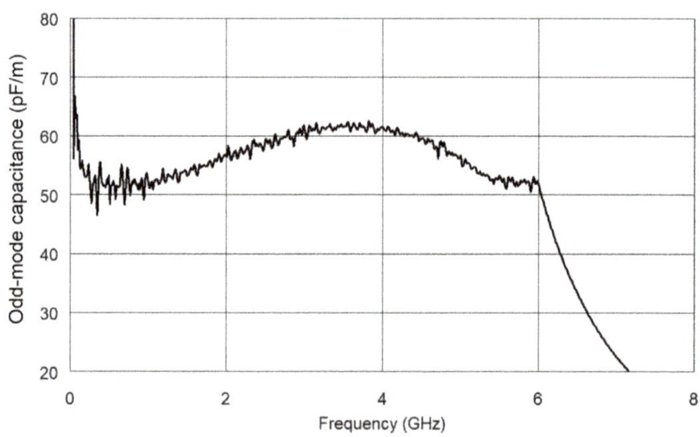

Fig. 19 C_0 value obtained from measurement of differential *S*-parameters of sensor S^{III} covered with calibration sample no. 1 (see Table 2)

no. 3 it is not worse than 5.7% in the frequency range 3–8 GHz. The percentage error of the determined permittivity value ε_{MUT} of sample no. 4 does not exceed 19.3% in the frequency range of 3–8 GHz. The established values of $\tan(\delta)_{MUT}$ of sample no. 2–4 vary between 0.01 and 0.08 and feature generally larger values than the expected ones as well as ones determined using S^{I} and S^{II}.

As can be seen, the open-ended coupled-line sensor features similar selectivity as the end-shorted sensors, i.e., the samples having lower permittivity can be easily distinguished from the samples having higher permittivity. However, the end-shorted

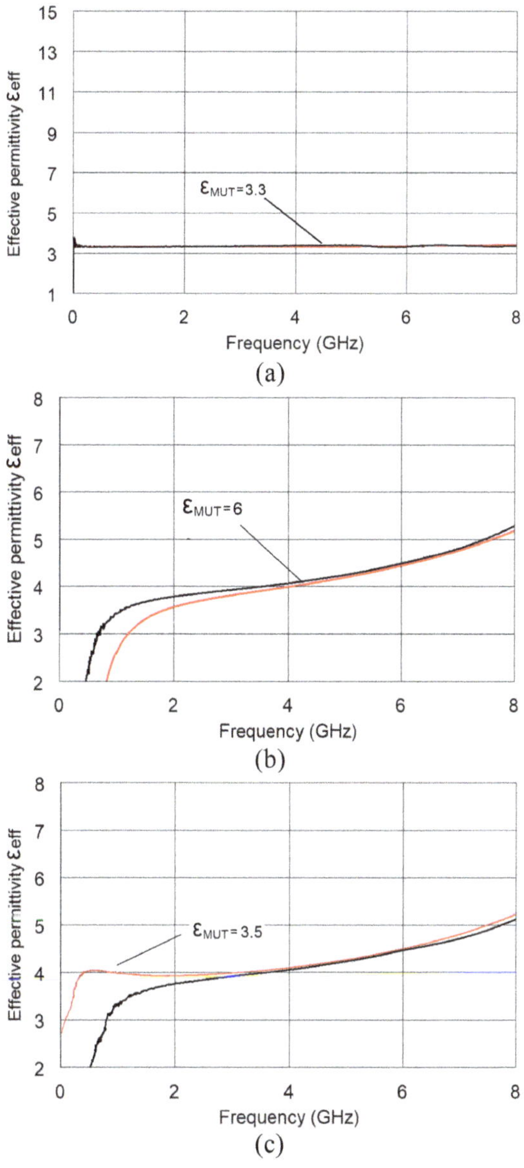

Fig. 20 Measured (thick solid line) and simulated (thin solid line with symbols) real odd-mode permittivity of samples: **a** no. 2, **b** no. 3, and **c** no. 4, respectively. Results obtained for sensor S^{III}

coupled-line sensor features higher accuracy of determination of complex permittivity of MUT than the open-ended coupled-line sensor, which is related to the existence of the fringe capacitance at the end of the open-ended coupled strips. This capacitance varies with the permittivity of MUT [31], which results in the change of the C value with respect to one established through the calibration process. Therefore, the open-ended coupled-line sensor can be above all used for the detection of

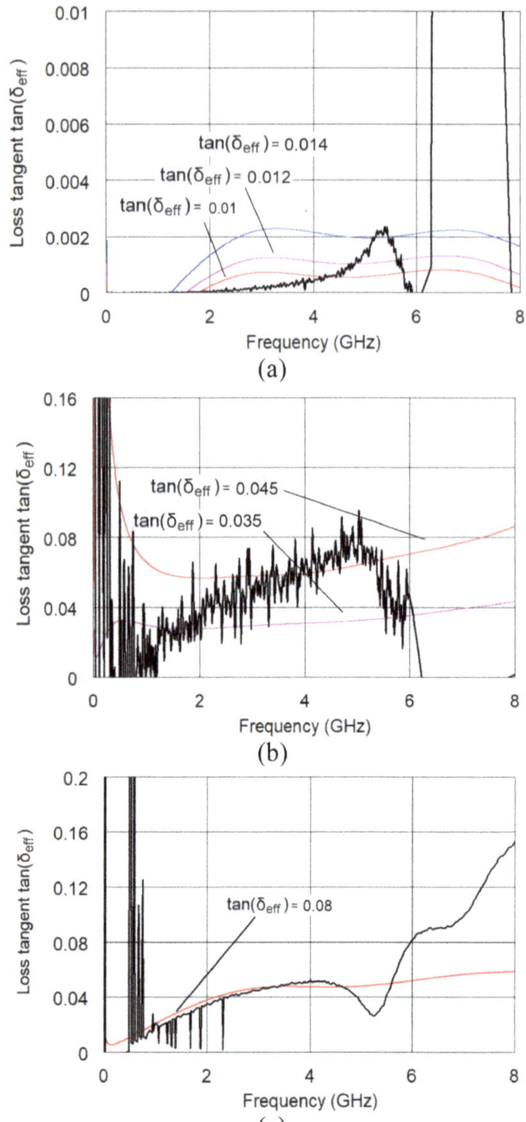

Fig. 21 Measured (thick solid line) and simulated (thin solid line with symbols) loss tangent of samples: **a** no. 2, **b** no. 3, **c** no. 4, and **d** no. 5, respectively. Results obtained for sensor S^{III}

dielectric sample permittivity change, contrary to the end-shorted sensors, which allows for the calculation of the MUT's complex permittivity value.

1.6 Measurement of Liquid Dielectric Materials Featuring Frequency-Dependent Permittivity

The presented sensor class not only allows for the non-resonant characterization of solid materials but also is suitable for the characterization of liquid dielectric materials having frequency-dependent complex permittivity. In general, each of the coupled-line sensors, e.g., S^{I}, S^{II}, S^{III}, can be used for that purpose. However, it is worth considering which one is most suitable for a particular application taking into account material properties and sensor features. There are three main approaches for the realization of a liquid measurement system utilizing the proposed coupled-line sensors. The first one is a fluid-flow system with a liquid channel in-between the coupled strips e.g., a microfluidic channel (for example Polydimethylsiloxane - PDMS channel [32] or quartz capillary [33, 34]) allowing for the electromagnetic field to be strongly affected by the measured material. Thus, a small volume of liquid-under-test is required, which is advantageous, since many of today's methods of biological materials characterization require a significant quantity of a test specimen [31, 35, 36]. Moreover, continuous measurement is possible. The second one is a steady-fluid system where, e.g., a droplet or larger volume of liquid-under-test is being held on top of the coupled-line sensor in a special cell. The third one does not require additional containers or channels, however, requires a relatively large volume of liquid-under-test since the sensor needs to be immersed in the measured liquid to cover the entire coupled-line section.

To experimentally verify the applicability of the proposed coupled-line sensors for the characterization of liquid samples, three liquid test dielectrics were measured i.e., propan-2-ol (isopropanol), butan-1-ol, and ethanol. The measurements were performed at a temperature of 30 °C. Since the utilized liquid samples feature relatively large permittivity change for frequencies lower than 1 GHz [37], sensor S^{III} was utilized due to its possibility of characterization of dielectric samples at lower frequencies (as shown in Sect. 1.4.3). The previously presented microstrip sensor S^{III} of Sect. 1.4.3 was adopted. Since the sensor is to be immersed in the liquid-under-test, the sensors' inputs were realized in the stripline technique to avoid the liquid from covering and influencing the de-embedded transmission lines. The photograph of the manufactured sensor is shown in Fig. 22. Since the capacitance C at the end of S^{III} varies with the permittivity of MUT deteriorating the measurement accuracy, three calibration samples, two having larger permittivity AD1000 ($\varepsilon = 10.6$, $h = 1.27$ mm), LTCC ($\varepsilon = 7.7$, $h = 1.13$ mm), and one having lower permittivity 25N ($\varepsilon = 3.38$, $h = 1.52$ mm) were used to calculate C for frequencies lower than 1 GHz and higher than 1 GHz, respectively. In the next step, the sensor was immersed in a container with liquids-under-test using a holding fixture. The measured complex odd-mode effective permittivities were compared with their electromagnetic counterparts obtained based on the sensor's model in *AWR Microwave Office* software. The complex permittivity of the measured liquids was determined using the algorithm presented in Sect. 1.2. The comparison of real and imaginary parts of permittivity

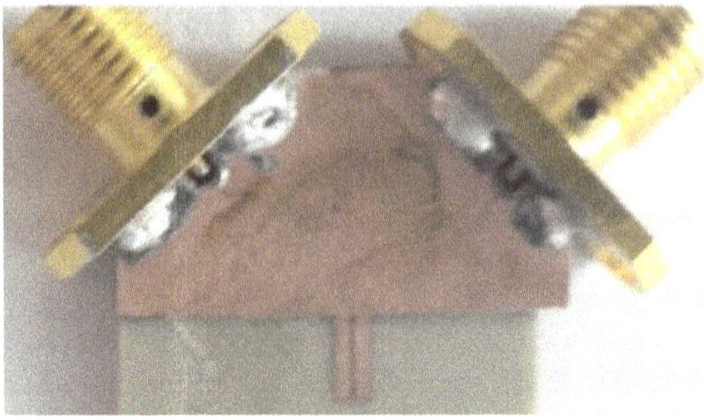

Fig. 22 Photograph of the manufactured sensor S^{III} utilized for measurement of liquids where only the coupled-line section is exposed on liquid-under-test. Reprinted with permission from Piekarz et al. [17]; copyright 2017 IEEE

measured using the proposed sensor and values provided in Gregory and Clarke [37] is shown in Fig. 23.

The percentage error of the determined real part of the complex permittivity value of ethanol does not exceed 2.5% in the frequency range of 20 MHz–2 GHz, whereas in the case of propan-2-ol and butan-2-ol, the percentage error is not worse than ±6%

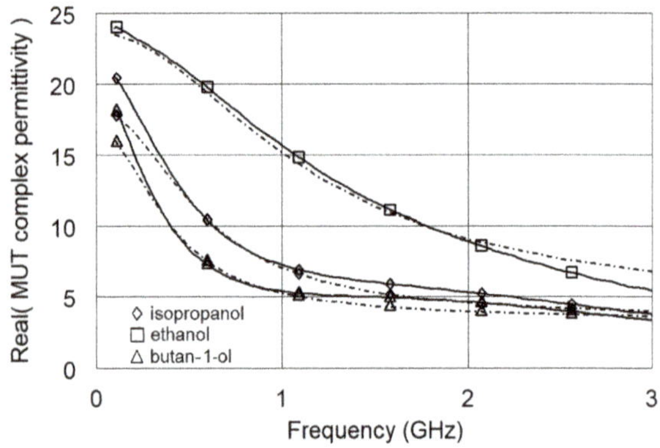

(a)

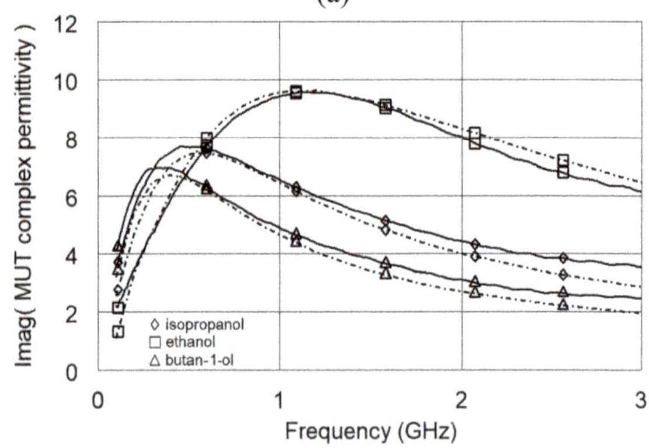

(b)

Fig. 23 Measured (solid lines) and provided in Gregory and Clarke [37] (dotted lines) **a** real and **b** imaginary part of complex permittivity of liquid dielectric samples: isopropanol, butan-1-ol and ethanol. Results obtained for sensor S^{III}

within 20 MHz–3 GHz band. The percentage error of the determined imaginary part of the complex permittivity value of ethanol does not exceed 3.8% in the frequency range of 20 MHz–3 GHz, whereas in the case of propan-2-ol and butan-2-ol the percent error of the calculated permittivity is not worse than ±4.5% within 20 MHz–3 GHz band. As seen, the use of three calibration samples allowed for improvement in the accuracy of sensor S^{III} having known roughly the expected permittivity.

1.7 Error Analysis

It is required for the newly developed sensors to carry out the sensitivity analysis and to observe the influence of various parameters on the accuracy of the performed measurements. In the presented approach, the complex permittivity of MUT is determined based on the measured complex effective permittivity of the sensor covered with MUT samples. Thus, the error sources were identified the main group of which are:

(a) Uncertainty of sensor *S*-parameters' measurements related to Vector Network Analyzer accuracy.
(b) Uncertainty of sensor system *S*-parameters' measurements related to the influence of air gaps between the sensor and MUT.
(c) Uncertainty of sensor system *S*-parameters' measurements related to the influence of incorrect placement of the MUT sample.
(d) Uncertainty of system calibration.

The first component of error sources is related to the uncertainty of VNA scattering parameters' measurement. The measurement accuracy of magnitude and phase of single-ended *S*-parameters is given by the manufacturer of the measurement equipment. The conducted analysis revealed that the error related to VNA measurement uncertainty is usually lower by an order of magnitude with respect to the other two components and therefore can be neglected.

The second part of error sources depends on the existence of air gaps between the MUT sample and the sensor. The quality of contact between the coupled-line sensor and measured sample has been investigated, since even a small air gap between the two causes changes in the measured complex effective permittivity, based on which the permittivity of MUT is calculated. Due to this fact, the sample needs to be sufficiently pressed against the sensor to minimize the influence of the air layer on the measured complex odd-mode effective permittivity. It has to be noted, that even light pressure allows to minimize this error, which has been empirically evaluated. Nevertheless, the roughness of the sensors' metal electrodes, as well as the surface roughness and hardness (softness) of the MUT sample, might introduce an air layer in-between MUT and the sensor as well. Therefore, the influence of an air layer thickness versus the resulting complex effective permittivity was calculated electromagnetically using *AWR Microwave Office* for sensor S^{II} covered with dielectric sample no. 2, assuming, that the air layer thickness varies between 5 and 17 μm, which is a typical thickness of metallization layer of microwave laminates. Such an air layer thickness has no significant influence on the determination of loss tangent of MUT, however, affects the real part of the odd-mode effective permittivity, which results in less than 5.5% lower value, assuming (41). However, the effect of air gaps on the calculation of complex permittivity of MUT can be minimized by including in the EM simulation model an air layer of typical thickness. Such a thickness can be calculated by taking a set of measurements of the calibration sample and using the algorithm presented in Sect. 1.2. Based on the permittivity obtained from EM calculation for various air layer thicknesses, a mean value of air layer thickness can be found. It has to be noted, that based on the performed measurements, the maximum air layer thickness does not exceed 10 μm (the sensor's metal thickness is approximately 17 μm).

The third element of uncertainty are the errors related to the inaccurate placement of the MUT sample in such a way that the coupled-line section is not completely covered. The accuracy of the human hand in positioning was analyzed in the literature [38, 39] and according to [39] the root mean square amplitude of hand-positioning error is equal to 0.5 mm. Therefore, the influence of positioning accuracy of the sample vs the resulting complex permittivity was determined electromagnetically using *AWR Microwave Office* for S^I sensor covered with dielectric sample no. 2, assuming that the sample is shifted by 0.5 mm towards the sensor. This source of error does not significantly influence the determination of loss tangent and results in a decrease of the real part of MUT complex permittivity by up to 8%, assuming (41). However, this error related to MUT positioning can be minimized either by increasing the coupled-line section length (at the expense of the sensor's bandwidth) or adding a sample fixing point. This could be either realization of sensors' connecting lines in the stripline technique (see Fig. 22) or the sensor itself in the microstrip technique, which allows the edge of the stripline to be the guide for sample positioning (at the expense of more complex fabrication and assembly) or introduction of on-PCB sample positioning pins or screws (at the expense of MUT sample pre-processing). It has to be noted that based on the performed measurements, the maximum error in hand-positioning of the sample does not exceed 0.2 mm.

The fourth element of error sources are the errors related to the system calibration, i.e., input lines de-embedding and calibration sample measurement. The basic de-embedding procedure includes only phase shift compensation (assuming uniform system impedance); however, such a procedure is the source of an increased measurement error, especially when it comes to low-loss material measurement. To minimize such an error, one needs to extend the procedure and additionally include conductor losses during de-embedding.

The measurement of the calibration sample, hence, the determination of sensors' constants are affected by the previously discussed error sources (a)–(c). It must be noted, however, that such a calibration sample needs to be measured only once for a particular sensor unit and this source of total uncertainty can be minimized by utilization of an accurate positioning method, e.g., motor positioning.

To compare the worst-case measurement errors of the proposed sensors against other techniques described in the literature, an overview is provided in Table 4.

Table 4 Performance comparison of various dielectric materials permittivity determination methods

	Percent error—permittivity determination (worst case) (%)	Bandwidth of operation (GHz)
[40]	8.7	Single frequency
[41]	10	0–11
[42]	20	0–14
[43]	5.7	Single frequency
S^{I}	±2.8	2.5–7[a]
S^{II}	±2.9	2.3–8[a]
S^{III}	19.3	3–8[a]

[a]The bandwidth of operation can be increased by shortening the coupled-line section constituting the sensor

The second group of error sources of MUT complex permittivity determination are the errors related to the realization of sensors, which cannot be easily generalized:

(a) The assumed maximum error between the measured and simulated complex permittivity allows for the determination of MUT complex permittivity, Eqs. (41) and (42).
(b) The error related to the selection of the calibration sample and determination of the sensor‘s calibration constants C_0, L, and C based on the measured input impedance when exposed to the calibration dielectric sample.
(c) The number of elements used for the Maclaurin series expansion of a hyperbolic tangent function, utilized to simplify (5), (18), and (21) and allow for the determination of ε_{eff}^{comp} based on the measured data.

Firstly, it can be stated that the error of tanh function approximation is of lesser importance since the same calculation procedure is used for measured and EM-calculated data when determining the complex permittivity of MUT. On the other hand, one can note that the lower the relative permittivity of the sensor‘s substrate the lower the effective permittivity of the entire measurement setup, hence a wider range of MUTs effective permittivity can be measured with relatively low error. Therefore, to characterize samples featuring higher complex permittivity values, their thicknesses need to be respectively thinner resulting in the complex effective permittivity within the measurement range of the selected sensor. It is, however, important to provide very good contact between the sensor substrate and the measured material (especially for solid materials). Therefore, to minimize the measurement error, the MUT should be pressed against the sensor with a force of a certain magnitude that considers the stiffness of both the sample and substrate (softer materials require less force to mate well). Considering the above, the thickness of the coupled-line section conductor should be low (e.g., 0.5 oz/ft^2—17 μm, or less) to minimize the air gaps between the sensor and the measured sample. In the case of very thin MUTs, laminate technology might not be sufficient to ensure little to no air gaps and thus microfabrication with thin metal e.g., silicon wafer, should be considered for sensor implementation.

1.8 Summary and the Selection Guide for the Edge Coupled-Line Section Sensors

The proposed coupled-line sensors are suitable for the non-resonant characterization of solid as well as liquid dielectric materials. However, due to different constructions, they exhibit different advantages and limitations which are discussed here.

The end-shorted coupled-line sensor S^{I} features simple construction since it requires only a short connection between the coupled strips at their ends. Nevertheless, such a connection introduces small inductance which needs to be determined during the calibration process. Therefore, the sensor is prone to errors related to the determination of L, which directly influences the calculated complex permittivity of MUT. The sensor is well suited for the characterization of dielectric samples at frequencies higher than 0.5 GHz. Nevertheless, the above is not a fixed limit as aspects such as the sensor‘s strip length and spacing, hence a total mutual capacitance that affects the measured transmission between single-ended ports influences the frequency range of operation.

The end-shorted coupled-line sensor S^{II} requires the coupled strips to be ground shorted with negligibly low inductance. Such a connection can be realized as a metallization pad shorted to the ground using a metal ribbon or foil. Therefore, in such a circuit there are no errors related to the determination of extra calibration constants, and the complex permittivity of MUT is calculated directly based on measurements. The calibration procedure is simple since only the C_0 value needs to be determined. However, the required metallization pad increases the entire sensor's size. Similarly, as in the case of S^{I}, the sensor S^{II} is well-suited for the characterization of dielectric samples at frequencies higher than 0.5 GHz.

The open-ended coupled-line sensor S^{III} features similarly to the S^{I} sensor a very simple construction, which occupies a smaller area than that of S^{II}. However, in the physical realization the sensor fringing capacitance at the end of the lines occurs and its value must be determined during the calibration process. Similarly, as in the case of the end-shorted sensor, in the open-ended one, the error related to the determination of C occurs, which directly influences the calculated complex permittivity of MUT. Moreover, this capacitance value varies with the permittivity of MUT [31], which additionally decreases its accuracy and limits its utilization to the detection of dielectric sample permittivity change. However, the utilization of a more complex calibration process, as shown in Sect. 1.5 allows for the sensor's accuracy improvement.

A summary of features for each coupled-line sensor is presented in Table 5. As can be seen, the coupled-line sensor needs to be chosen as a trade-off between the occupied area, the bandwidth of operation, the ease of calibration process, and the

Table 5 Comparison between proposed coupled-line sensors

	S^{I}	S^{II}	S^{III}
Size	++	+	++
Bandwidth of operation	++	++	++
Simple calibration process	+	++	+
Accuracy of MUT permittivity	++	++	+

accuracy of MUT permittivity determination. Moreover, manufacturing technology is an important aspect and needs to be considered depending on the application. One can use laminate technology for component scale integration and relatively narrower band applications or e.g., Metal–Insulator-Metal (MIM) technology for better integration and increased bandwidth of operation.

2 Two-Port Differential Broadside Coupled-Line Section Sensors

To provide an insight into the properties of the proposed sensing element in a broadside coupled configuration, a theoretical study was conducted to describe its behavior qualitatively and quantitatively. A cross-sectional view of the proposed microwave coupled-line sensor in a symmetrical stripline dielectric arrangement is presented in Fig. 24 where metal strips m_1 and m_2 are broadside coupled. The liquid-under-test sample features a relative complex permittivity ε_{MUT}^{comp}

$$\varepsilon_{MUT}^{comp} = \varepsilon_{MUT} - j\varepsilon_{MUT}'' = \varepsilon_{MUT}(1 - j\tan(\delta)_{MUT}) \qquad (43)$$

where ε_{MUT} is the real part, ε_{MUT}'' is the imaginary part, and $tan(\delta)_{MUT}$ is the loss tangent. Such liquid sample is placed in the microfluidic channel in between the coupled strips. The sensor can be described using the per-unit-length capacitive network as drawn in Fig. 25a. Two types of capacitances can be distinguished, i.e.,

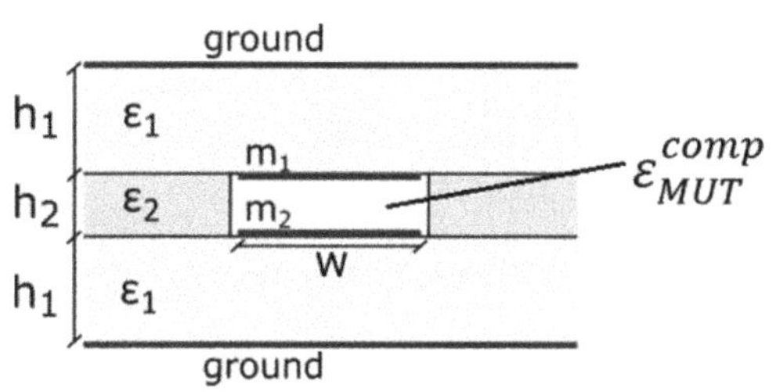

Fig. 24 A cross-sectional view of the proposed microwave broadside-coupled-line sensor. The microfluidic channel is realized in between the two aligned coupled strips marked as m_1 and m_2. Reprinted with permission from Sorocki et al. [18]; copyright 2020 IEEE

the self-capacitance between each strip and the ground plane described by C_1, and a mutual capacitance between the two strips described by C_m, representing the capacitance within the liquid sample, and C_{mp} representing the residual capacitance. Such a three-conductor system composed of two signal conductors placed over a common ground plane can be further decomposed and analyzed under even- and odd-mode excitation, as shown in Fig. 25b and c, respectively [22]. For the even-mode conductors, m_1 and m_2 are excited with equal amplitude and in-phase voltages, thus a magnetic wall is created, and the circuit simplifies to the one in Fig. 25b. For the odd-mode excitation, both conductors are excited with equal amplitude and out-of-phase voltages, thus an electric wall is created, and the circuit simplifies to the one in Fig. 25c. Therefore, the total capacitance for even- and odd-mode networks can be written as:

$$C_e = C_1 \tag{44}$$

$$C_o = C_1 + 2(C_m + C_{mp}) \tag{45}$$

The most important features of the proposed coupled-line sensing element are that the capacitance C_m is solely and directly affected by the material-under-test present

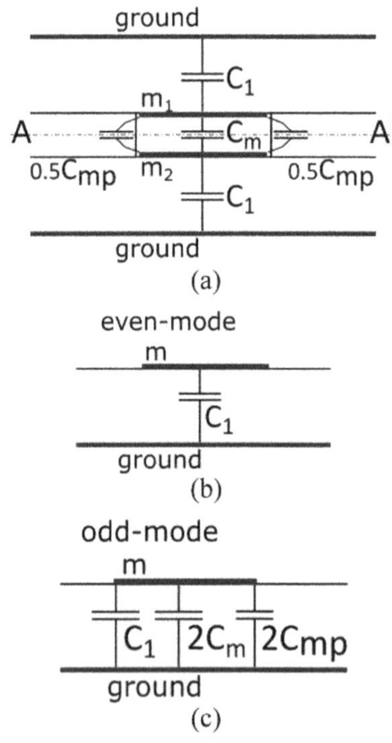

Fig. 25 The per-unit-length capacitive network of the proposed sensor (**a**) and its decomposition under even (**b**) and odd-mode (**c**) excitation along AA‘. Reprinted with permission from Sorocki et al. [18]; copyright 2020 IEEE

in the channel (see Fig. 25a), the microfluidic channel has a fixed and well-defined geometry and the tested liquid flows through its entire cross-section. Excluding the fringing field effect (if the channel width is equal to strip widths, this field is not enclosed within the sample and thus is associated with C_{mp}) the relation between C_m and complex permittivity ε_{MUT}^{comp} can be written from the parallel-plate capacitor formula as:

$$C_m = \frac{\varepsilon_0 \varepsilon_{MUT}^{comp} w}{h_2} \tag{46}$$

where ε_0 is the permittivity of free space, while w is metal m_1 and m_2 strips width and h_2 is the channel thickness (both w and h_2 are the design parameters). On the other hand, the capacitance C_1 is constant and relates (excluding fringe field effects) to the design parameters:

$$C_1 \approx \frac{\varepsilon_0 \varepsilon_1 w}{h_1} \tag{47}$$

The following conclusions can be drawn based on (44)–(47):

- Capacitance for even-mode excitation exhibits no sensitivity on the MUT.
- Capacitance for odd-mode excitation is affected by the MUT.
- The capacitance C_m is proportional to ε_{MUT}^{comp} with the coefficient of proportionality being w/h_2 (design parameters).
- The sensitivity of the proposed sensor on the dielectric properties of MUT can be maximized by:
  - decreasing ε_1 or increasing thickness h_1 to reduce C_1 and thus its contribution to C_o capacitance,
  - increasing w or decreasing h_2 to increase the C_m coefficient of proportionality,
  - make the channel wide and use thin metal to minimize C_{mp}.

Based on the above-stated analysis the proposed broadside coupled-line sensor can be utilized for liquid dielectric materials characterization. Depending on the application, there are two possibilities for performing the measurements of material properties:

- MUT properties change monitoring—in such a case complex permittivity change can be monitored being an indicator of MUTs properties variability based on the measurement of differential S-parameters change (both amplitude and phase) and no additional de-embedding or calibration, as well as knowledge about the sensor‘s geometry, is required.
- MUT dielectric properties determination—in such a case the complex permittivity is established based on measured differential S-parameters after de-embedding with no setup calibration required. However, when additional input signal lines are utilized for measurement purposes, a cal-kit needs to be used to de-embed the

lines' influence on the measurements. In such a case the geometry of the sensor needs to be known.

2.1 Material-Under-Test Dielectric Properties Monitoring

The use of the proposed sensor for monitoring dielectric properties change of the MUT within microwave range can be carried out through monitoring of its scattering parameters (S-parameters) change. In such a case, an indirect quantity being proportional to the MUT variation can be measured.

A schematic diagram of the proposed sensor with marked four single-ended as well as two differential ports and transmission line parameters (Z_{oe}, Z_{oo} - even- and odd-mode characteristic impedances, θ_e, θ_o-even- and odd-mode electrical lengths) is presented in Fig. 26a. Since the sensor exhibits high sensitivity only for odd-mode capacitance, such mode of operation is to be excited and resulting differential-mode S-parameters are of interest. Under differential excitation, the coupled-line section simplifies to a single transmission line (see Fig. 26b) having an electrical length equal to θ_o and characteristic impedance Z_{diff} equal to:

$$Z_{diff} = 2Z_{oo} \tag{48}$$

$$Z_{oo} = \frac{1}{v_{po}C_o} \tag{49}$$

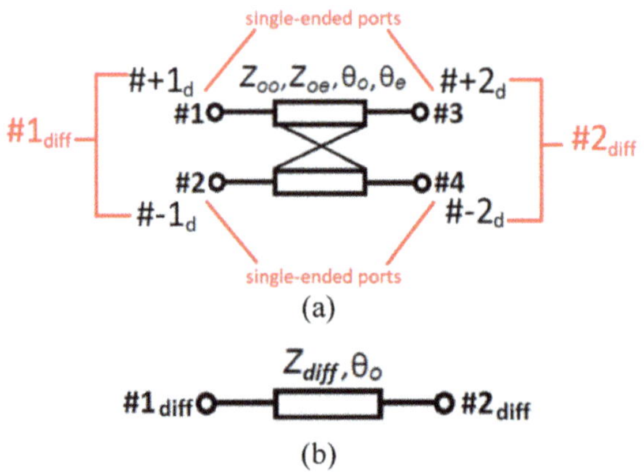

Fig. 26 General view of the sensor with marked four single-ended as well as two differential ports (**a**) and its simplification under differential excitation (**b**). Z_{oe} and Z_{oo} represent even- and odd-mode characteristic impedances, and θ_e, θ_o are the even- and odd-mode electrical lengths of the coupled-line sensor, $Z_{diff} = 2Z_{oo}$. Reprinted with permission from Sorocki et al. [18]; copyright 2020 IEEE

where v_{po} is the odd-mode phase velocity.

A similar analysis can be performed for the circuit from Fig. 26a for a common-mode excitation (signals at single-ended ports #1, #2 as well as #3, and #4 provided with the same phase and amplitude). In such a case, the simplified circuit also reduces to a single transmission line featuring electrical length equal to θ_e and characteristic impedance Z_{comm} equal to:

$$Z_{comm} = 0.5Z_{oe} \tag{50}$$

$$Z_{0e} = \frac{1}{v_{pe}C_e} \tag{51}$$

where v_{pe} is the even-mode phase velocity.

Using (45)–(49) and formulas for the scattering parameters of a lossy section of transmission line [44], the resulting differential S-parameters can be derived:

$$S_{11diff} = S_{22diff} = \frac{(z - y)\sinh(\gamma l)}{2\cosh(\gamma l) + (z + y)\sinh(\gamma l)} \tag{52}$$

$$S_{12diff} = S_{21diff} = \frac{2}{2\cosh(\gamma l) + (z + y)\sinh(\gamma l)} \tag{53}$$

where z and y are characteristic impedance Z_{diff} and admittance Y_{diff} of a simplified transmission line under differential excitation normalized to measurement system impedance and admittance, γ is the complex propagation constant (related to the complex effective permittivity of the dielectric stack-up, thus affected by the MUT as well) and l is the physical length of the sensor. Therefore, an indirect relationship between ε_{MUT}^{comp} and S-parameters can be obtained. Based on the performed analysis it can be seen that measurement of differential scattering parameters change allows for monitoring of dielectric material permittivity change.

2.2 *Material-Under-Test Dielectric Properties Measurement*

The proposed broadside coupled-line sensor features well-defined geometry, as can be seen in Fig. 24 with a microfluidic channel placed in-between metal strips. Due to this fact, the presented sensing setup can be used for the determination of the complex permittivity of MUT. For such an application, it is required to measure the 4-port single-ended S-parameters (see Fig. 26a), recalculate it into Z-parameters and determine the impedances Z_{oe}, Z_{oo} as well as complex propagation constants γ_e, γ_o using the following formulas [10]:

$$Z_{oe} = \frac{Z_{11} + Z_{12}}{\coth(l\gamma_e)} \tag{54}$$

$$Z_{oo} = \frac{Z_{11} - Z_{12}}{\coth(l\gamma_o)} \tag{55}$$

$$l\gamma_e = \cosh^{-1}\left(\frac{Z_{11} + Z_{12}}{Z_{13} + Z_{14}}\right) \tag{56}$$

$$l\gamma_o = \cosh^{-1}\left(\frac{Z_{11} - Z_{12}}{Z_{13} - Z_{14}}\right) \tag{57}$$

Having known Z_{oe}, Z_{oo} the complex modal electrical lengths θ_e and θ_o as well as, the C_e, C_o can be calculated:

$$\theta_e = -j\gamma_e l \tag{58}$$

$$\theta_o = -j\gamma_o l \tag{59}$$

$$C_e = \frac{1}{v_{pe} Z_{0e}} = \frac{\theta_e}{\omega l} \cdot \frac{1}{Z_{0e}} \tag{60}$$

$$C_o = \frac{1}{v_{po} Z_{0o}} = \frac{\theta_o}{\omega l} \cdot \frac{1}{Z_{0o}} \tag{61}$$

where ω is the angular frequency in radians and l is the physical length of the coupled-line sensor in meters. Rewriting (44) and (45) as:

$$C_m + C_{mp} = 0.5(C_o - C_e) \tag{62}$$

and plugging (54)–(61) enables the formula for complex ε_{MUT}^{comp} to be derived:

$$\varepsilon_{MUT}^{comp} = \frac{h_2}{\varepsilon_0 w}\left\{0.5(C_o - C_e) - C_{mp}\right\} \tag{63}$$

where h_2/w is the sensitivity (proportionality) coefficient. Finally, to complete (63), C_{mp} needs to be found. This is a residual capacitance related to the fringing field. Depending upon channel design, the fringing field is enclosed either (I) within ε_2 region (channel and metal strips of the same width) or (II) within ε_{MUT}^{comp} (channel wider than metal strips). Moreover, a wider strip w and a thinner metal t result in smaller fringing field effects. Either way, the formula for fringe field capacitance [45] can be adopted (under the assumption that the field encloses only within the h_2 layer) as a reference point to determine the per-unit-length value of C_{mp}:

$$C_{mp} \approx \frac{\varepsilon_r \varepsilon_0}{\pi}\left[1 + ln\left(\frac{2\pi w}{h_2}\right) + ln\left(1 + \frac{2t}{h_2} + 2\sqrt{\frac{t}{h_2} + \frac{t^2}{h_2^2}}\right)\right] \tag{64}$$

For a channel having a width of w, the C_{mp} can be calculated in a straightforward manner [ε_r in (64) equals ε_2]. Such configuration is desired when channel volume needs to be minimized. Otherwise for channels wider than strips, thus confining the majority fringe field, the Eq. (63) would need to be rearranged as ε_r in (64) equals ε_{MUT}^{comp}.

2.3 Sensor's Physical Length Versus Range of Operation

The proposed sensor is realized as a coupled-line section; thus, it exhibits a respective periodic behavior in frequency. The sensor has a maximal coupling at $\pi/2 + k\pi$ of its electrical length θ and transmission zeros at $0 + k\pi$. Moreover, the equivalent transmission line behavior under common/differential excitation is bounded with hyperbolical relation as seen in Eqs. (52) and (53) what manifest themselves through periodic discontinuities in e.g. (54) and (55) at $0.5\pi + k\pi$ and nonlinear behavior when the argument of hyperbolic function approaches discontinuity. Thus, the above-presented phenomena must be taken into account while designing the sensor to properly select its length based on the targeted application as well as permittivity and frequency range of operation. For monitoring applications, the physical length l can be selected to either yield an electrical length of $\pi/2$ at the center frequency of operation for an empty channel or the length should be selected as:

$$l \leq \frac{\pi}{\beta_{max}} = \frac{c}{2 f_{max} \sqrt{\varepsilon_{effmax}}} \tag{65}$$

where ε_{effmax} is the expected maximum effective permittivity of the sensor-MUT system (for lossless case γ reduces to β and $\theta = \beta l$). On the other hand, when permittivity measurement is concerned the physical length should be:

$$l \ll \frac{0.5\pi}{\beta_{max}} \tag{66}$$

the more the argument γl differs from $0.5\pi + k\pi$, the lower error is introduced being proportional to $(\tan(x) - x)$.

2.4 Theoretical Study of the Sensor Performance

The performance of the proposed sensor was investigated and compared against other transmission line-type sensors. The study was conducted for three types of transmission line-based sensors: the proposed broadside-coupled-line sensor A, an edge-coupled-line sensor B, and a single-strip transmission line sensor (see Fig. 27). Full-Wave EM simulations were carried out using *NI AWR Microwave Office* with the

ANALYST 3D EM solver. The dielectric stack-up used is shown in Fig. 24 where the h_2 layer is entirely filed with MUT for simplicity. Strip widths and lengths assumed in calculations were $w = 1.45$ mm and $l = 8$ mm, respectively. For edge-coupled configuration, strips were arranged $s = 0.1$ mm apart while for broadside-coupled strips were placed with no offset. Grid sizes of 0.1 mm in x- and 0.05 mm in y-direction were used. For sensors B and C the bottom boundary was set as a perfect conductor, whereas, the top boundary was set as approximately open. For sensor A, the top and bottom boundaries were set as a perfect conductor to simulate stripline configuration. The calculated electric field distribution overlaid on the stack-up cross-section is provided in Fig. 27. Moreover, sensors A and B are considered under differential excitation.

The electric field (magnitude) distribution for sensors A, B, and C calculated in *NI AWR Design Environment* software using a full-wave EM solver is shown in Fig. 28 allowing for a qualitative comparison. It can be observed for sensor A that most of

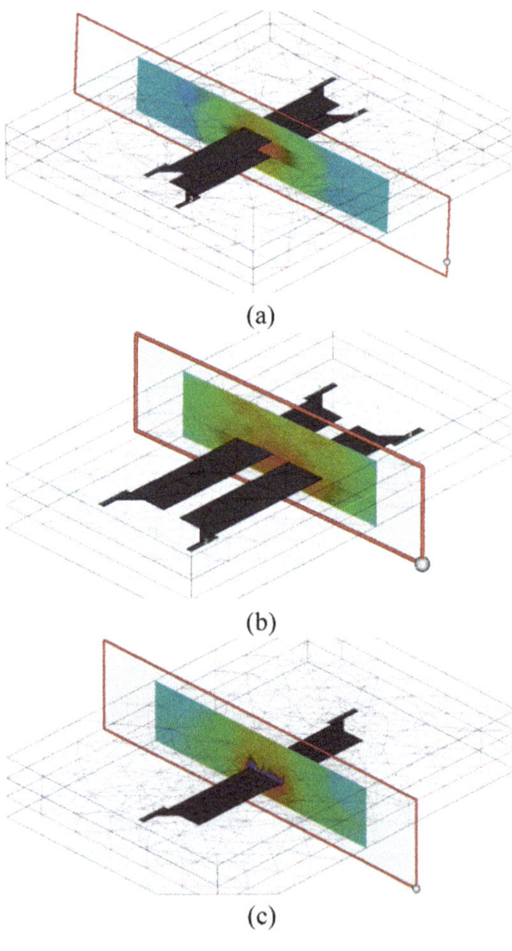

Fig. 27 Electric field cut-planes at 1.5 GHz with marked mesh utilized for EM simulations: within the proposed broadside-coupled-line sensor (**a**), edge-coupled-line sensor (**b**), both under a differential excitation and within a single transmission line (**c**). Maximum field concentration marked with red color, the magnitude in log scale. Reprinted with permission from Sorocki et al. [18]; copyright 2020 IEEE

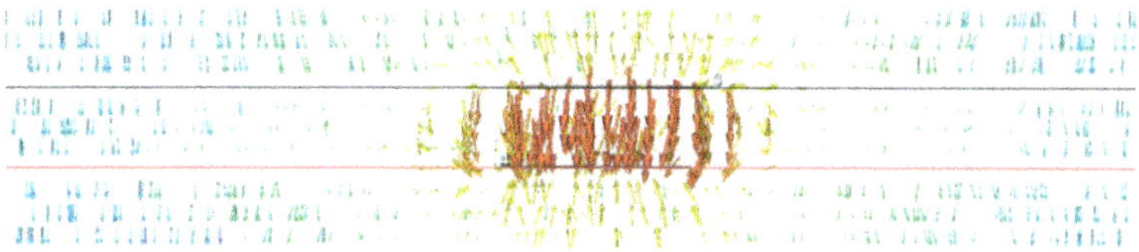

(a)

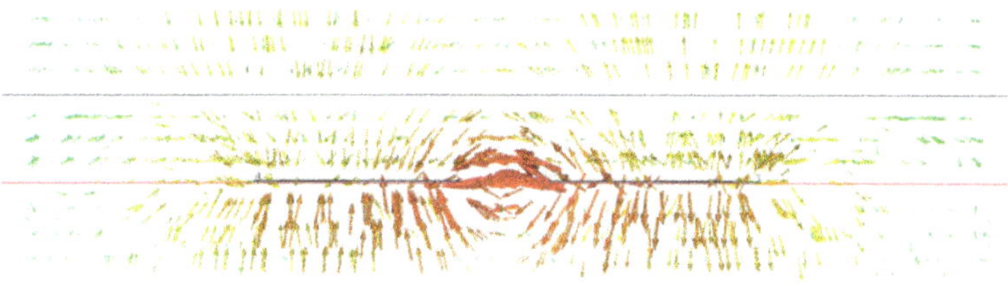

(b)

(c)

Fig. 28 Electric field distribution expressed through E-field vectors within the cross-section of proposed broadside-coupled-line sensor A (**a**), edge-coupled-line sensor B (**b**), both under a differential excitation and within a single transmission line sensor C (**c**). The maximum field strength is marked red, the colored scale is referenced to the same max value and the color gradient reflects a logarithmic scale. Reprinted with permission from Sorocki et al. [18]; copyright 2020 IEEE

the field is confined within the MUT, and the highest field strength (red region) is in the region in-between strips (microfluidic channel) resulting in high sensitivity. Following, for sensor B the highest field strength is observed as well in-between strips, however, due to the lateral (thus not vertical as for A) location, only part of that field is enclosed within the MUT leading to reduced sensitivity. Finally, for sensor C the highest field strength is observed at the strip edges while the field is more scattered in the volume of dielectrics, thus only a small fraction of the fringing field is enclosed within the MUT resulting in the lowest sensitivity in the set. Therefore, as it was investigated in Sect. 1 the edge coupled-line sensor—sensor B features higher sensitivity on the measured MUT than the sensor C.

The sensitivity of sensors A and B was also studied to derive a quantitative description. The influence of MUT complex permittivity ε_{MUT}^{comp} on the sensors' scattering parameters was determined through full-wave EM calculations in the same simulation setup, as shown in Fig. 27. A comparison between sensors of their differential-mode S_{21diff} (both amplitude and phase) as a function of ε_{MUT}^{comp} (ε_{MUT} in the range of 1–40; $tan(\delta)_{MUT} = 0.001$ and 0.1) at a given frequency is provided in Fig. 29. Sensitivity for both sensors expressed as a slope of linear regression for the ε_{MUT} in the range of 1–40 is summarized in Table 6 at low frequencies where close to a linear relationship is observed. As expected, much greater parameter change is obtained for the proposed sensor A than for B, thus certifying its superiority in these metrics.

It can also be noted that for higher frequencies and higher permittivity, the response of A and B deviates away from being close to the linear trend. This is because both

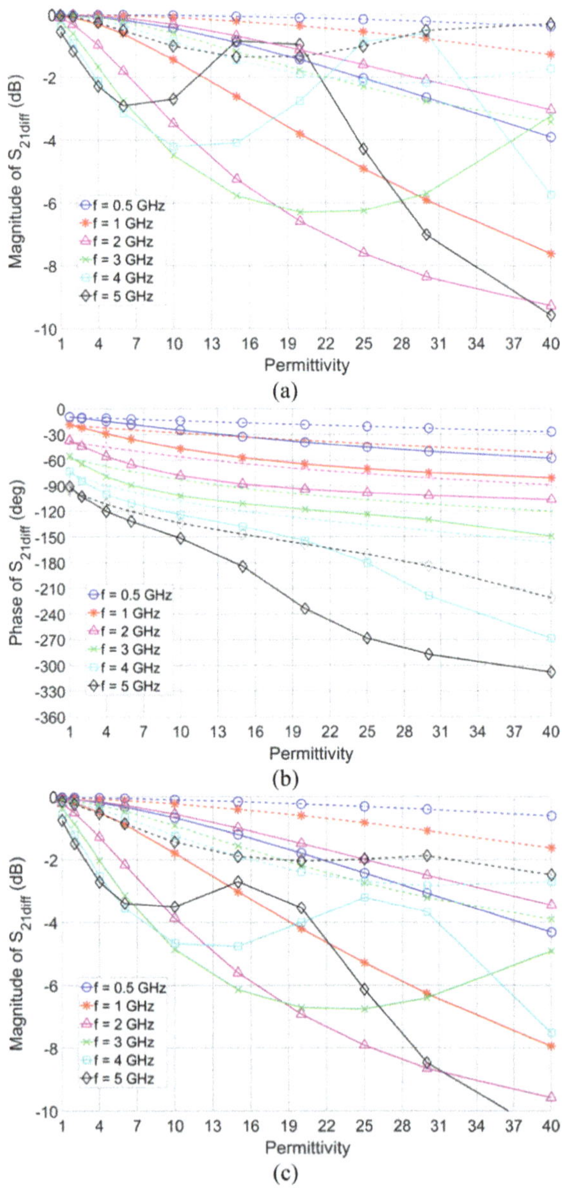

Fig. 29 Full-wave EM simulation-based sensitivity of the proposed broadside-coupled-line sensor (solid lines) and edge-coupled-line sensor (dashed lines) under a differential excitation expressed as the change of magnitude and phase of S_{21diff} as a function of ε_{MUT} for $tan\delta = 0.001$ (**a**–**b**) and $tan\delta = 0.1$ (**c**–**d**). Reprinted with permission from Sorocki et al. [18]; copyright 2020 IEEE

Fig. 29 (continued)

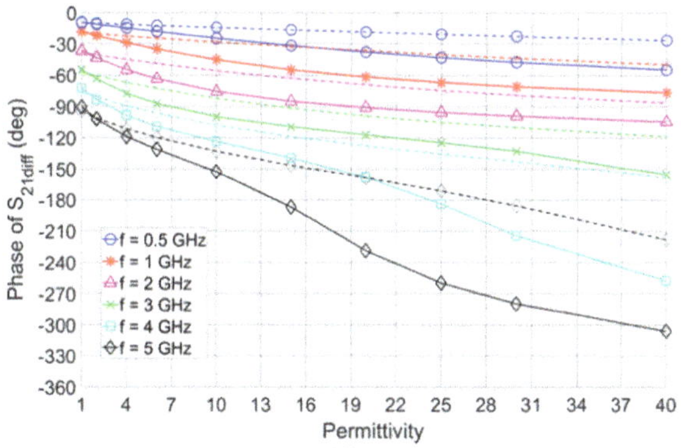

(d)

Table 6 Sensitivity comparison between the proposed broadside-coupled Sensor A and edge-coupled Sensor B calculated for ε_{MUT} 1–40

Freq. (GHz)	tanδ	$\lvert S_{21diff}\rvert$ slope (dB/ε_{MUT})		$\angle S_{21diff}$ slope (deg/ε_{MUT})	
		Sensor A	Sensor B	Sensor A	Sensor B
0.5	0.001	−0.0854	−0.0074	−1.6957	−0.8161
	0.1	−0.0993	−0.0135	−1.6217	−0.8074
1	0.001	−0.1890	−0.0252	−2.6062	−1.5804
	0.1	−0.2026	−0.0357	−2.4904	−1.5528
2	0.001	−0.2744	−0.0666	−3.6916	−2.9096
	0.1	−0.2866	−0.0806	−3.6053	−2.8534

the impedance and propagation constant of the equivalent line are affected by the MUT as well as the coupled-line section itself exhibits a periodic characteristic.

2.5 Development of a Microwave-Microfluidic Sensing Setup Employing 3D Printing Technology

To verify the performance of the proposed sensor and its applicability towards liquid properties monitoring and complex permittivity measurement, an exemplary general-purpose setup was developed and manufactured. A combination of laminate and 3D printing technologies was used since additive manufacturing allows for the reduction of manufacturing cost and prototyping time and therefore is an ideal candidate for industrial applications. A four-port broadside-coupled-line sensor was realized in a dielectric structure shown in Fig. 24 where upper and lower layers were realized using a microwave laminate while the middle layer was realized as a 3D printed insert hosting the microfluidic channel. The physical stack-up selection was optimized with the main focus being on manufacturability, mechanical strength, and ease of assembly.

2.5.1 Assembly Part: Microwave Laminate Layers

For the sensor realization, the *Arlon 25N* material ($\varepsilon_{r1} = 3.38$, $\tan\delta_1 = 0.0035$) with 17 μm thick copper cladding was selected to realize the two outer dielectric layers hosting the broadside coupled strips. To ensure enough rigidity of the combined laminate-3Dprint-laminate mechanical structure on one hand and to minimize the C_1 capacitance while keeping connecting lines relatively narrow on the other, a $h_1 = 0.81$ mm thick laminate was used. Since the strips' width plays an important role in the per-unit-length capacitance distribution, it needs to be selected carefully. In this study, $w = 1.45$ mm wide strips were realized. This results in a coefficient of proportionality being $h_2/w = 1.45/0.7 \approx 2$ and an empty channel odd-mode impedance is roughly equal to $Z_{0oempty} \approx 40\ \Omega$ ($Z_{diff} \approx 80$). The length of the coupled-line section was set to be as short as $l = 8$ mm (yielding quarter-wavelength at $f \approx 6$ GHz for the empty channel, see Sect. 2.3 for reference) as the longer the sensor the narrower the usable band of operation (especially when high permittivity materials are to be measured). Finally, sections of 1.25 mm wide transmission lines having roughly 50 Ω (assuming the insert is made of pure VeroWhite) were added to the sensor to allow for measurements using a Vector Network Analyzer.

2.5.2 Assembly Part: 3D Printed Insert

To manufacture the microfluidic insert, PolyJet 3D printing technology was used. A commercially available *Objet Connex350 Multi-material* 3D Printing System was employed as it allows printing fine-featured parts composed of various materials with build resolution of X-axis: 600 dpi; Y-axis: 600 dpi; Z-axis: 1600 dpi as well as a high-quality surface finish with fast and clean prototyping at the same time. A rigid VeroWhite polymer (assumed $\varepsilon_r \approx 2.8$, $\tan\delta \approx 0.01$) was used along with rubber-like TangoBlack material. Such a combination was used to provide structural strength and realize liquid terminals on one hand and proper channel sealing on the other. The downside is that electrical properties, i.e., complex permittivity ε_2 is hard to define and need to be determined during calibration. The total thickness of the insert was selected to be $h_2 = 0.7$ mm where a 0.1 mm thick TangoBlack is on both sides of a 0.5 mm thick VeroWhite core to maximize the capacitance C_m while keeping structural integrity. The width of the channel in the sensor area was set to be 1.5 mm to be slightly oversized compared to w to account for manufacturing and assembly tolerances. Total channel volume equals 29.4 μL (plus an extra 16.7 μL per terminal). A 3D view of the developed insert is shown in Fig. 30. Note the pin-hole system for precision alignment of each component of the assembly. Photographs of the manufactured sensor setup assembly parts are presented in Fig. 31. To ensure leak-proofing, components were bonded together using an epoxy adhesive.

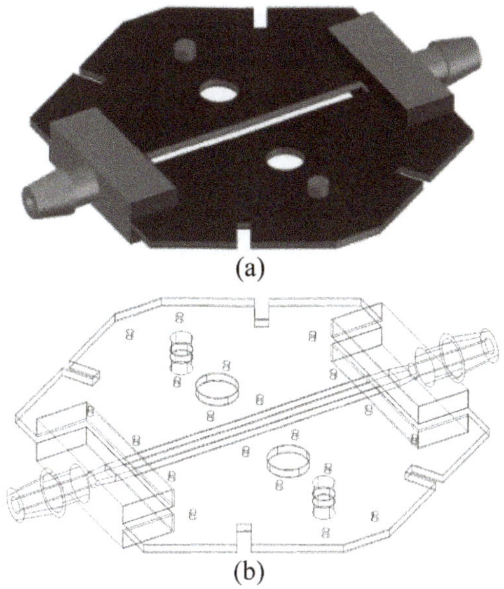

Fig. 30 3D **a** and wireframe **b** view of the developed insert hosting a microfluidic channel and liquid terminals. Rigid VeroWhite (light gray) and rubber-like TangoBlack (dark gray) materials were used to provide structural strength and channel leakproofing. Reprinted with permission from Sorocki et al. [18]; copyright 2020 IEEE

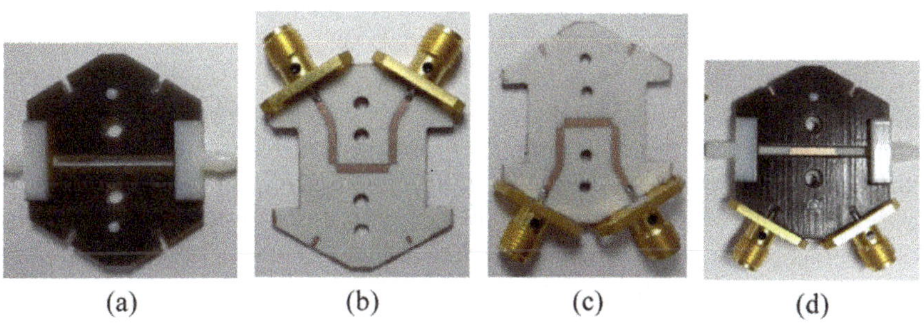

Fig. 31 Photograph of the sensor's components and assembly: 3D printed middle insert with dedicated liquid ports (**a**), top (**b**), and bottom (**c**) laminate with etched stripes. Partially assembled liquid properties monitoring setup with the microfluidic channel (**d**). Reprinted with permission from Sorocki et al. [18]; copyright 2020 IEEE

2.5.3 TRL de-Embedding Kit for Permittivity Measurements

Due to the sensor's construction, additional connecting lines are necessary, however, they need to be de-embedded to allow for measurement of the sensing element S-parameters. Therefore, in addition to the developed sensor, a calibration board was manufactured and the photograph of the physically realized setup is shown in Fig. 32.

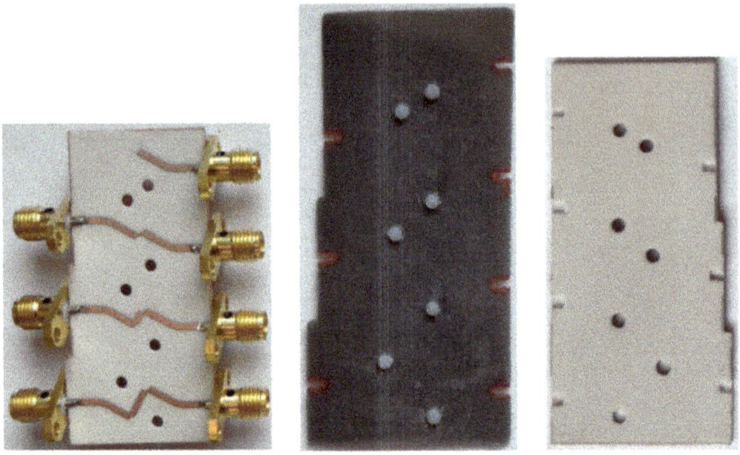

(a) (b) (c)

Fig. 32 Photograph of the manufactured TRL calibration standards. An open-ended one-port, back-to-back shorted, 2 and 5 mm long line standards were realized. Bottom (**a**), middle (**b**), and top (**c**) parts of the assembly. Reprinted with permission from Sorocki et al. [18]; copyright 2020 IEEE

For the de-embedding purpose, a TRL (Thru, Reflect, Line) calibration technique [46] was selected. The *Line* standard width was chosen as $w_{TLR} = 1.25$ mm to yield 50 Ω impedance while the length was chosen as $l_{TRL} = 5$ mm as a compromise between one-line calibration and to cover the intended experimental study bandwidth being up to 15 GHz. Such lengths yield 20°–160° electrical length for the assumed dielectric stack-up roughly between 1.8 to 15 GHz, however, the 2nd Tier calibration will be valid within a slightly wider band which is especially crucial for lower frequencies. It must be underlined that the calibration board was realized in the same dielectric stack-up as the proposed sensor as well as it maintains the geometry of the sensor's connecting lines. Moreover, both circuits i.e., the calibration setup and broadside coupled-line sensor were manufactured in the same manufacturing run to maintain the same material properties and manufacturing tolerances.

2.6 Experimental Verification of the Exemplary General-Purpose Sensor

The developed microwave-microfluidic sensor was experimentally investigated to verify its performance as being suitable for industrial applications. Two scenarios were considered: (I) measurement of quantities proportional to the MUT permittivity such as solvent concentration for monitoring applications as discussed in Sect. 2.1 and (II) measurement of MUT permittivity itself for measurement applications as discussed in Sect. 2.2.

Within scenario I the differential-mode S-parameters of the sensor filled with a solution of ethanol and water having varied concentrations were measured. Those liquids feature a significant difference in their relative permittivity (1:10) and therefore enable us to verify the performance of the setup using stimuli with high dynamic range. Following, within scenario II, various liquids being compounds of the alcohol group featuring frequency-dependent complex permittivity of the high dynamic range were measured, and the complex permittivity of ε_{MUT}^{comp} was derived based on the differential-mode S-parameters. The obtained results were compared against tabularized data [37] to determine the accuracy of the proposed method.

All S-parameter measurements were done using an *Agilent PNA N5224A* Vector Network Analyzer. Within scenario I, only One-Tier VNA calibration was performed up to SMA connectors, thus measurements are for a cascade of sensing (proposed sensor) and non-sensing (signal lines) parts of the setup. In the case of scenario II, a Two-Tier calibration was used, namely 1st Tier VNA calibration and 2nd Tier de-embedding of the manufactured setup. The custom-manufactured TRL cal-kit was used to enable the measurement of *S*-parameters of the coupled-line sensor only.

All the liquids were kept at room temperature beforehand in the same room as the measurement equipment, thus such is assumed. A syringe-based dosing system shown in Fig. 33 was used for liquid manipulation. To avoid the creation of air pockets within the channel and reduce the potential influence of uneven or discontinuous liquid flow through the channel, the non-flow conditions, i.e., pressurized liquid standing still inside the sensor during measurement, were ensured every time.

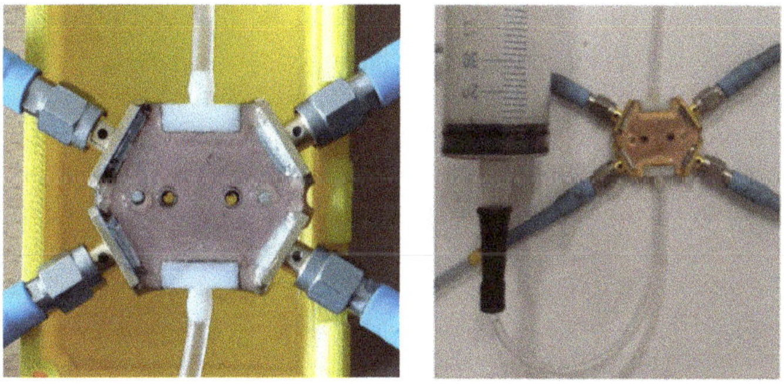

(a) (b)

Fig. 33 Photograph of the developed microwave-microfluidic setup during measurements. Close-up (**a**) and the dosing system using a syringe (**b**). Reprinted with permission from Sorocki et al. [18]; copyright 2020 IEEE

2.7 Material-Under-Test Dielectric Properties Monitoring by Means of Differential-Mode S-parameters Measurement

As indicated, Scenario I of the experimental study is focused on the verification of the dielectric properties and monitoring the performance of the developed sensor setup. For such use, there is no need to de-embed the system as proportional quantities are measured and a calibration procedure can take into account additional non-MUT-affected parts of the system.

For the testing purpose, a solution of water and ethanol having varied water concentrations was chosen. There were two reasons for choosing the ethanol–water solution:

1. Ethanol and water are both excellent solvents that use hydrogen bonds to dissolve the solute and therefore form homogeneous solutions in all proportions.
2. As it was previously mentioned, water and ethanol feature significant differences in their relative permittivity in a wide frequency range at room temperature.

Samples of the water–ethanol solution having water solvent concentration in the range of 0–50% with 10% steps (with an extra 5% concentration) were prepared outside the measurement setup and then introduced into the microfluidic channel. A total of 62.8 μL of MUT is required to fill the entire channel volume from one liquid terminal to another. After each filling of the channel, a set of differential *S*-parameter measurements was gathered and then the channel was flushed and cleaned by running compressed air through it. Important to note that, in general, the tested solution's permittivity is a complex number so as such requires at least another complex number for unambiguous projection. The proposed sensor provides complex S_{21diff} and S_{11diff}.

The performance of the sensing setup is assessed based on the data normalized to 0% of water concentration at selected frequencies provided in Fig. 34. It can be observed, that in each case, the ethanol concentration can be easily differentiated (either from S_{21diff} or S_{11diff}) and S_{21diff} exhibits higher sensitivity to the stimulus change. Moreover, at selected frequencies, the relation between the measured phase and magnitude of S_{21diff} and water content is close to linear, which drastically reduces the number of test samples for the calibration procedure during which the concentration estimation polynomial coefficients are to be found. Such flexibility is possible due to the inherent broadband nature of the proposed sensor.

Sensitivity defined as the slope of linear regression and R^2 goodness-of-fit metrics is provided in Table 7 at 6 GHz (being the intended empty-channel center frequency). Having that, the function for water content estimation $water\% = f(\boldsymbol{S_{diff}})$ can easily be derived (as an inverse to the relation $\boldsymbol{S_{diff}} = f(water\%)$) and would require only two different water concentrations in the solution for the coefficients to be determined. Matching the sensor's peak sensitivity and linear relation for a specified stimuli range to the measurement capabilities (in terms of bandwidth and resolution) of the equipment in the targeted application, is done through proper selection of the sensor's

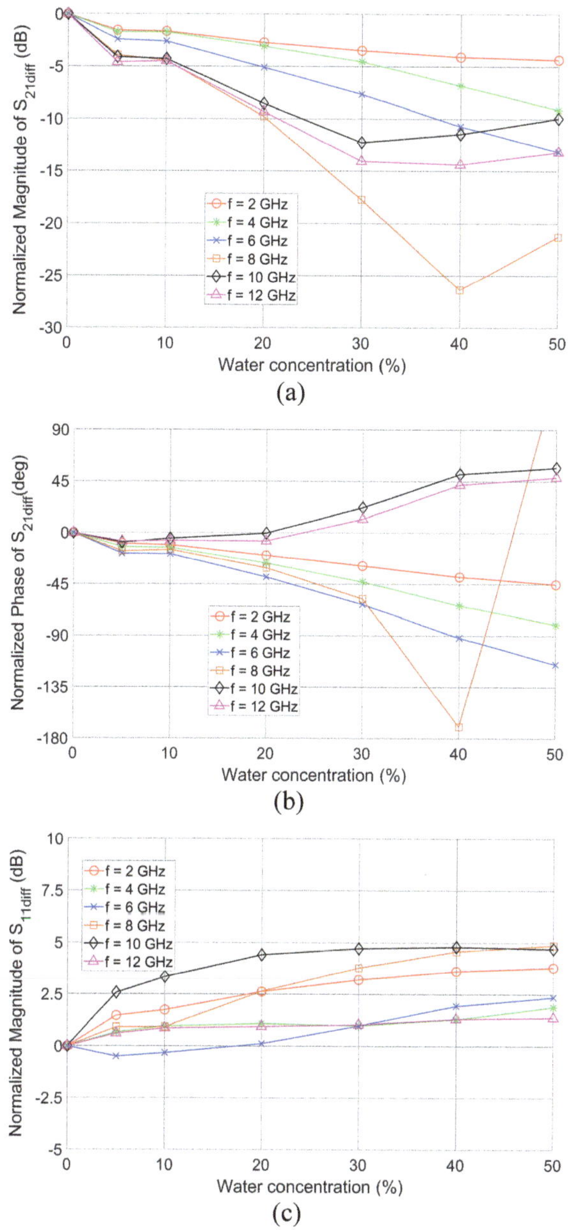

Fig. 34 Normalized amplitude and phase of the measured differential transmission coefficient (**a**) and (**b**), respectively as well as reflection coefficient (**c**) and (**d**), respectively as a function of water concentration in water–ethanol solution at different frequencies. Reprinted with permission from Sorocki et al. [18]; copyright 2020 IEEE

Fig. 34 (continued)

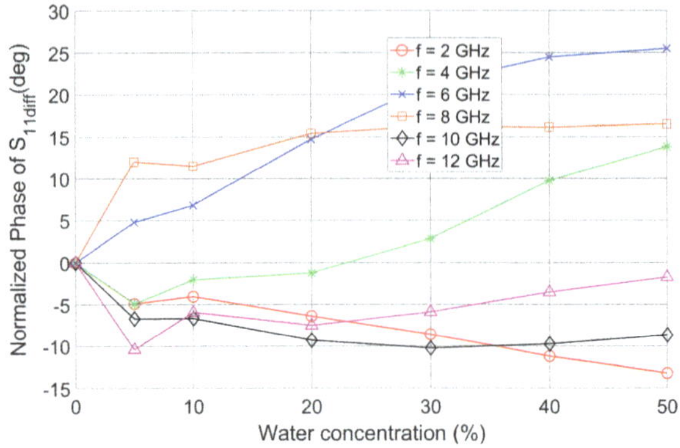

(d)

Table 7 The sensitivity of the developed sensing setup for measurement of water–ethanol solution for water content 0–50% at 6 GHz

		S_{21diff}	S_{11diff}
Slope	*abs* (dB/water%)	−0.2631	0.0401
	arg (deg/water%)	−0.0401	0.5939
R^2	*abs* (%)	0.9903	0.7711
	arg (%)	0.7711	0.9322

geometry (see Sect. 2.3). This is especially important for applications where VNA cannot be used.

It must be underlined that the information about both the amplitude and phase of S_{21diff} can be used not only for estimation of solute content in the solution, e.g., here water content in ethanol, but also allows for fast detection if the measured sample contains any additional substances, thus its deviation from the general trend [47]. Moreover, measurements performed with the proposed sensor generate an overdetermined system for the determination of solute content in the solution, i.e., a 2 × 2 [S_{diff}(freq)] matrix is used to produce a one-dimensional result. Such features (N-dimensional measurement data) of the sensor can be highly desirable for industrial applications such as the food industry since it enables monitoring and content determination (when properly calibrated) of complex solutions, i.e., more than two-ingredient solutions. The above-presented results certify the superior performance of the developed setup for dielectric properties change monitoring.

2.8 *Material-Under-Test Dielectric Properties Determination Based on Four-Port Single-Ended Measurements*

As indicated, scenario II of the experimental study is focused on the verification of the dielectric permittivity measurement performance of the developed sensor setup.

For such a use, there is a need to de-embed the input signal transmission lines from the measurement system beforehand so that the reference plane is set at the edges of the coupled-line section. As presented in Sect. 2.5.2, the developed TRL cal-kit was used for that purpose.

In the next step, liquids having frequency-dependent complex permittivity were tested and the resulting S-parameters were measured and taken as input data for the developed algorithm for ε_{MUT}^{comp} determination provided in Sect. 2.2. For the sake of this study, a set of alcohols were used as MUTs, namely: methanol, ethanol, propan-1-ol, isopropanol, and butanol. Similarly, as in the previous cases, each of the tested liquids was introduced into the microfluidic channel. After each filling of the channel, a set of measurements was performed and then the channel was flushed and cleaned by running compressed air through it. The obtained results being measured real and imaginary parts of ε_{MUT}^{comp} as a function of frequency are shown in Figs. 35 and 36, respectively. The obtained data was referenced against alcohols' permittivity provided in Gregory and Clarke [37] in tabularized form for up to 5 GHz (up to 3 GHz in the case of isopropanol). Due to that and considering the information described in Sect. 2.3, the extracted and displayed data was limited to 5 GHz. Nominal design geometry was assumed for ε_{MUT}^{comp} calculation and the C_{mp} was determined with the fringe field formula approach as a starting point.

It is seen that a good agreement was obtained between the measured data and the one provided in Gregory and Clarke [37]. Increased noise for data below roughly 2 GHz is, among others, an effect of de-embedding. The data deviation behavior matches sensor performance observations stated in Sect. 2.4. Since the exemplary sensor was developed as a general-purpose one rather than application-specific, a higher error is observed for alcohols having a complex permittivity of higher value and/or at higher frequencies with a good example being Methanol. A percentage error of the determined MUT complex permittivity for each tested liquid was calculated as related to the reference data from Gregory and Clarke [37] at 25 °C and is provided in Fig. 37. Note, that neither additional calibration procedure (apart from de-embedding) was applied, nor reference liquids were used.

A comparison between the range of operation of the proposed sensor vs the two-wire sensor presented in Piekarz et al. [48] is provided in Table 8 assuming a $\pm 10\%$ inaccuracy and the reference data from Gregory and Clarke [37] at 25 °C. As seen, the proposed broadside coupled-line sensor features a similar or greater frequency range of operation in comparison to the one presented in Piekarz et al. [48], where two-sample calibration was required. A more detailed error analysis is provided in the following section.

2.9 Permittivity Measurement Error Analysis

In Sect. 1 and Piekarz et al. [48] an extensive study was conducted to identify and discuss the main measurement uncertainty and error sources for the coupled-line section-based class of sensing elements. Based on the performed

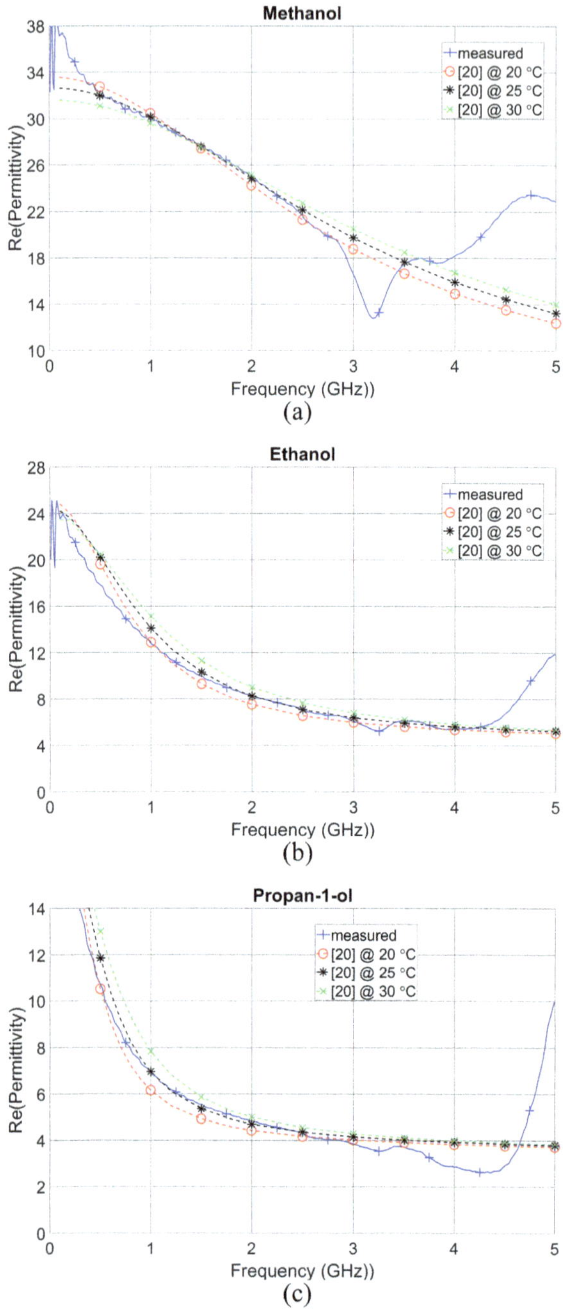

Fig. 35 Measured real part of complex permittivity (solid lines) of methanol (**a**), ethanol (**b**), propan-1-ol (**c**), isopropanol (**d**) and butanol (**e**) along with the reference data (dashed lines) as provided in Gregory and Clarke [37]. Reprinted with permission from Sorocki et al. [18]; copyright 2020 IEEE

Fig. 35 (continued)

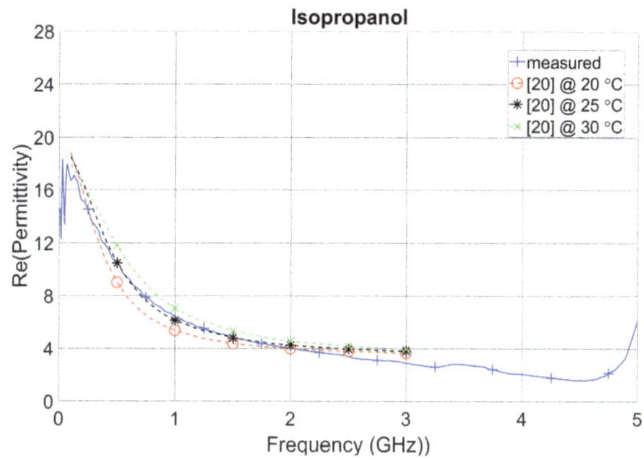

(d)

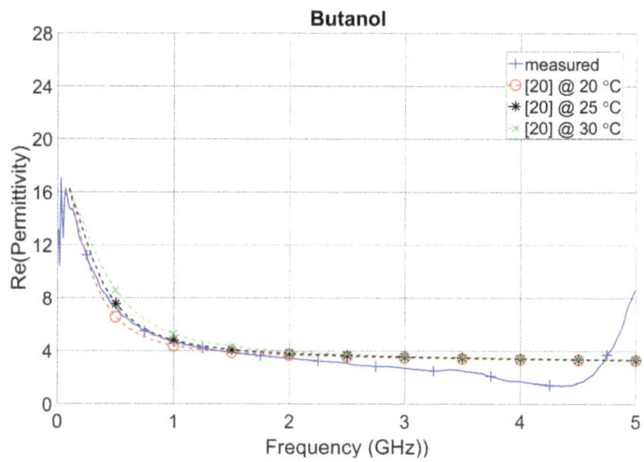

(e)

study, it can be pointed out, that the accuracy of measurements using the proposed broadside-coupled-line sensor is influenced by the following uncertainty and error sources:

1. Uncertainty of *S*-parameters' measurements using Vector Network Analyzer or any other.
2. Uncertainty of manufactured sensor's coupled-line section length, width, and thickness determination.
3. Systematic error due to the quality and accuracy of sensors manufacturing tolerances and assembly alignment.
4. Systematic error due to asymmetry in the sensor's stack-up.
5. Systematic error due to the quality of the manufactured custom TRL de-embedding standards and the resulting reference plane position.
6. An error due to leak-proofing limitations of the microfluidic channel and possible seepage on the connecting transmission line.

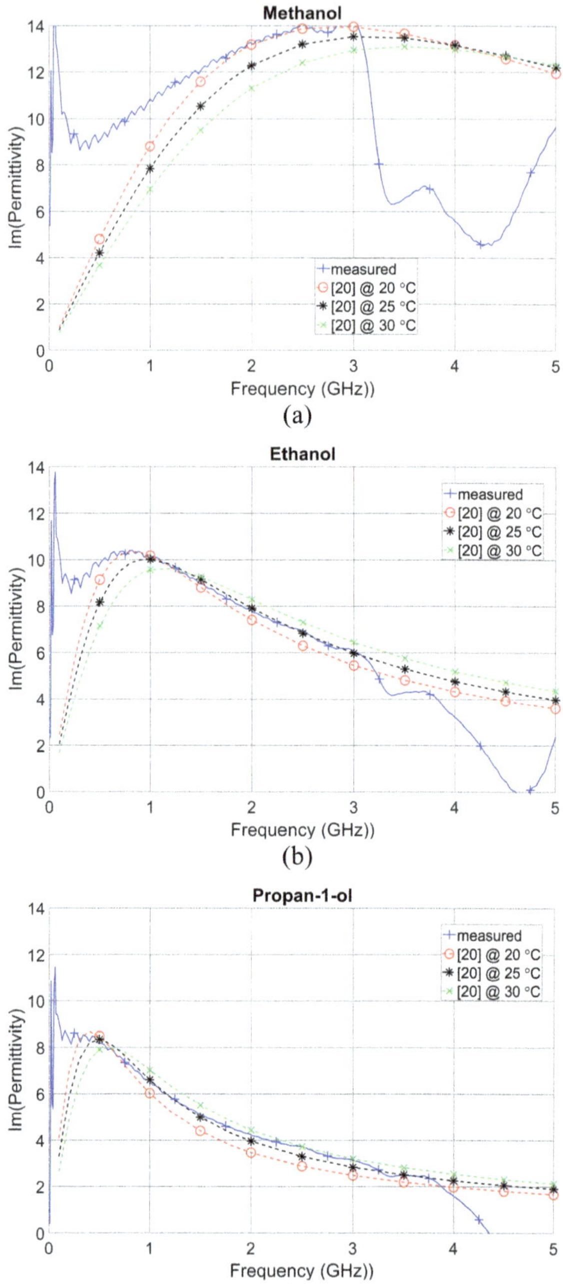

Fig. 36 The measured imaginary part of complex permittivity (solid lines) of methanol (**a**), ethanol (**b**), propan-1-ol (**c**), isopropanol (**d**), and butanol (**e**) along with the reference data (dashed lines) as provided in Gregory and Clarke [37]. Reprinted with permission from Sorocki et al. [18]; copyright 2020 IEEE

Fig. 36 (continued)

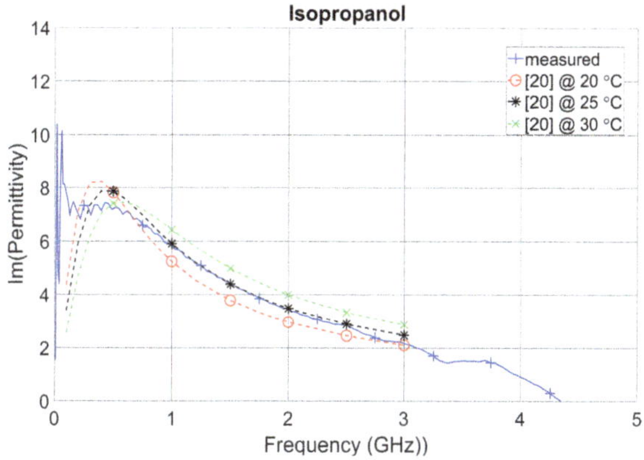

(d)

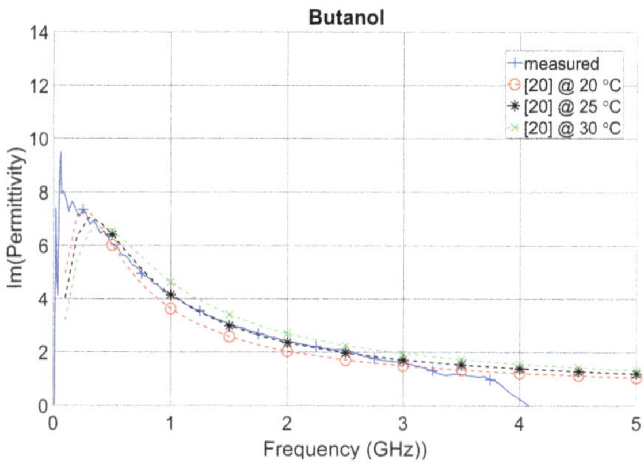

(e)

7. An error related to either air bubble(s) trapped in the channel when MUT is present or equivalently, a disruptive liquid flow.
8. An error related to the determination of C_{mp} as if the related field is enclosing either fully in the channel or fully outside the channel or in both.
9. Systematic error due to nonlinearities in the model used to derive ε_{MUT}^{comp}, as discussed in Sect. 2.3.

The above-given error sources list provides insight on what are the key contributors and indicates the direction of how to minimize those errors. Apart from manufacturing-related issues, an additional calibration procedure might be introduced to further reduce measurement error, however at the expense of the necessity of having one or more well-defined calibration liquids.

Since the manufacturing and assembly tolerances are of high importance, a worst-case study was conducted for the presented exemplary sensor design. This is due to the analytical formulas used, which rely on the design-specified channel and stack-up geometry dimensions as well as on the assumption of horizontal circuit

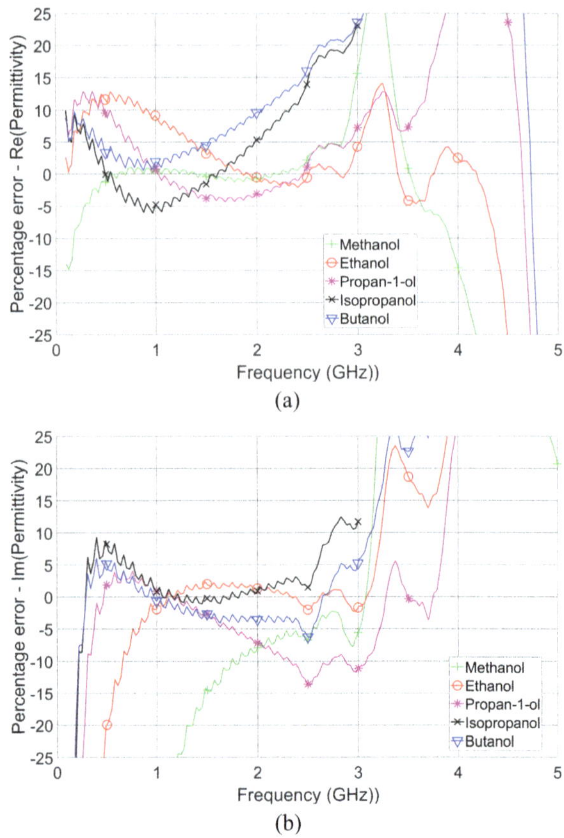

Fig. 37 Percentage error of determined ε_{MUT}^{comp}: real part (**a**) and imaginary part (**b**) for each of the measured liquids as related to the reference data provided in Gregory and Clarke [37] for 25 °C. Reprinted with permission from Sorocki et al. [18]; copyright 2020 IEEE

Table 8 The comparison between the range of operation of the proposed sensor versus the two-wire sensor presented in Piekarz et al. [48] assuming a ±10% inaccuracy

Liquid sample	ε_{MUT}^{comp}	Range (GHz) ± 10% proposed sensor	Range (GHz) ± 10% [48]
Methanol	*Re*	0.25–2.9	No data
	Im	1.8–3.1	No data
Ethanol	*Re*	1.0–4.4	No data
	Im	0.6–3.2	No data
Propan-1-ol	*Re*	0.5–3.6	0.1–1.2
	Im	0.4–2.2	0.15–5
Isopropanol	*Re*	0.01–2.3	1.5–3
	Im	0.4–2.8	0.1–2
Butanol	*Re*	0.01–2	1.1–5
	Im	0.4–3.1	0.1–2.4

symmetry. Each time nominal dimensions of the channel-sensor setup were assumed; MUT having $\varepsilon_{MUT} = 5$, $\tan\delta_{\mathrm{MUT}} = 0.1$ present in the channel; calculated complex permittivity is then related to the nominal value for nominal geometry. The following scenarios were considered where only one dimension is altered:

1. Channel narrower by 100 μm.
2. Channel wider by 100 μm.
3. The channel shifted by 100 μm with respect to coupled strips.
4. Channel thicker by 50 μm.
5. Channel thinner by 50 μm.
6. Air bubble in the middle of the channel filling its cross-section and being 100 μm long.
7. One laminate thicker by 50 μm (broken symmetry).
8. One laminate thinner by 50 μm (broken symmetry).

The resulting error is provided in Fig. 38. It is observed that the imaginary part of the calculated ε_{MUT}^{comp} is more sensitive to the assumed versus real dimensions' deviation. Moreover, the most impact is observed from wider and thinner channels. This could be expected as w/h_2 is the proportionality coefficient of the relation (64). If tight manufacturing tolerances cannot be achieved, then the assembly needs to be well characterized or sensor calibrated. Next, a small air bubble is affecting the calculations. The broken cross-section symmetry has the least impact.

Based on the above-presented measurement error analysis, the following can be stated:

- Accurate determination of the residual capacitance C_{mp} related to the fringing field is important for the measurement error minimization perspective. For extracting the permittivity of MUT, it was assumed that C_{mp} is frequency invariant and encloses within ε_2. Since the measurement error varies with frequency as well as with the type of liquid, it can be attributed to the fact that the fringe field encloses partially within the MUT and partially within ε_2 and both mediums can feature frequency-depended behavior.
- Manufacturing and assembly tolerances contribute to the error as in the presented sensor the final thickness of h_2 was a result of 3D printed part thickness, adhesive layer thickness, and clamping force during glue-up. Moreover, the glue-up process needs to be controlled as an excess of adhesive may partially cover the coupled strips. Layers' misalignment can lead to the channel not being at the edges or extending beyond the strips on both sides. The mean error value offset from zero can be attributed to the discussed issues.
- Consistency in the process and resulting tolerances for manufacturing both the sensing setup and the cal-kit is of importance. Offset in the reference planes leading to them not being at coupled-line section edges contributes to an error in S-parameters measurement thus impacting the permittivity measurement.
- A test and calibration procedure allowing to identify the above-presented issues and delivering proper corrections would reduce the overall error as if such is required for the application. Such could be provided by using data collected for

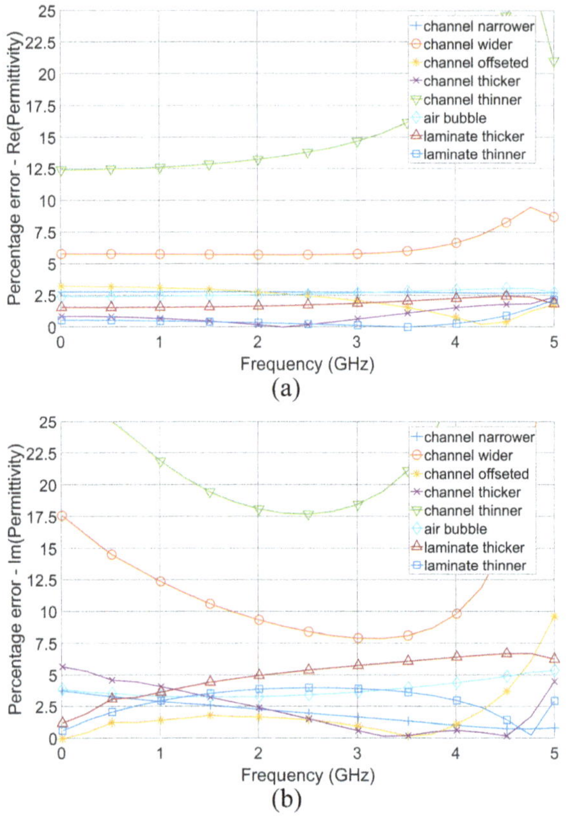

Fig. 38 The error of MUT complex permittivity determination normalized to the nominal calculated value for cases 1–8; the real part of ε_{MUT}^{comp} (**a**) and imaginary part of ε_{MUT}^{comp} (**b**). Reprinted with permission from Sorocki et al. [18]; copyright 2020 IEEE

an empty channel sensor to determine the fringe field (under the assumption that it is fully enclosed outside the channel) and at least one test liquid with known permittivity for the testing procedure.

It can be concluded that MUT complex permittivity measurements performed using a general-purpose sensor could satisfy certain applications while when higher performance is required an application-specific design that can easily be carried out following the guidelines provided in this section needs to be performed.

2.10 Summary on the Broadside Coupled-Line Sensors

A presented microwave-microfluidic coupled-line sensor setup allows for monitoring and determination of complex permittivity of liquid dielectric samples. The application of broadside coupling configuration enabled the sample to be placed in-between coupled strips, thus a high sensitivity can be obtained as the electric field for differential excitation is confined within the sample‘s volume. Moreover, the use of 3D printing technology allowed for the realization of an insert with a microfluidic channel and dedicated liquid ports. Two application scenarios are possible, i.e., material properties monitoring and complex permittivity measurement. For material properties monitoring no pre-processing is required and the relation between measured S-parameters and the signal proportional to the MUT permittivity can be established. For measurement of complex permittivity, a straightforward de-embedding technique using a custom TRL cal-kit can be used, and parameters of the unknown sample can be determined based on direct analytical relation between S-parameters and MUT permittivity due to the well-defined geometry of the sensor-sample setup. The conducted error analysis has shown that accuracy can be further improved if necessary, when additional calibration procedures would be applied. Moreover, it is possible to reduce the number of measurement ports by leaving one of the differential ports open, however at the expense of a more sophisticated calibration procedure requiring two known liquid samples to be measured [49].

3 Measurement Setup Composed of Marchand Balun and an Open-Ended Edge Coupled-Line Sensor

Baluns are well-known components utilized for various applications, where there is a need for the conversion of unbalanced signals to balanced ones, and vice versa. Among them, the Marchand balun is widely used due to its simplicity and wideband performance [50–53]. In this Section, a measurement technique is presented, which utilizes an open-ended, out-of-phase excited edge coupled-line sensor and a Marchand balun providing an appropriate excitation and balanced to unbalanced conversion to characterize the dielectric sample covering the lines. The proposed method requires only two measurements to include the influence of the Marchand balun on the measurement setup, which simplifies and accelerates the calibration process.

3.1 Measurement Setup Analysis

The proposed measurement setup is shown in Fig. 39a which consists of an open-ended coupled-line sensor connected to a Marchand balun which provides out-of-phase excitation and differential to single-ended conversion to allow one-port measurements. When such a balun is utilized, the proposed measurement setup can be analyzed considering its axis of symmetry [54]. Then, the structure decomposes into two simple one-port circuits (see Fig. 39b, c), which are easier to analyze. The coupled-line sensor excited with out-of-phase signals can be represented by two uncoupled open-ended transmission lines having impedance (assuming lossless case):

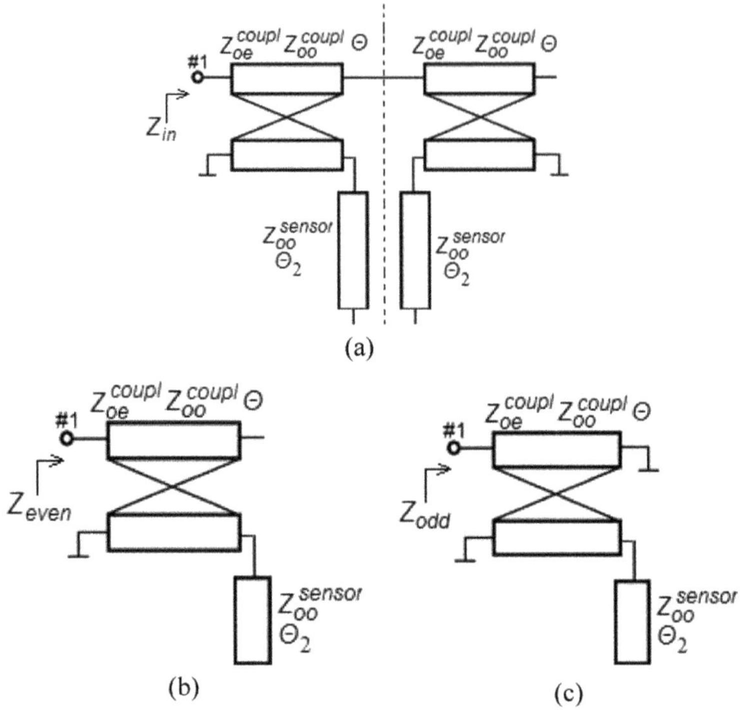

Fig. 39 The measurement setup is composed of a Marchand balun and an open-ended coupled-line sensor (**a**) analyzed as two uncoupled open-ended transmission lines having impedances equal to $Z_L = -jZ_{oo}^{sensor}\ cot(\theta_2)$ (**b**), (**c**). The Marchand balun has been realized as two identical coupled-line sections having electrical lengths θ and even- and odd-mode impedances equal to $Z_{oe}{}^{coupl}$, and $Z_{oo}{}^{coupl}$, respectively. Reprinted with permission from Piekarz et al. [16]; copyright 2016 IEEE

$$Z_L = -\mathrm{j}Z_{oo}^{sensor}\cot(\theta_2) \tag{67}$$

since the balun on its differential port sees as a load only the odd-mode impedance of the sensor.

Considering the symmetry axis of the circuit, the input impedance can be considered as a mean of two impedances (see Fig. 39b–c):

$$Z_{in} = \frac{(Z_{even} + Z_{odd})}{2} \tag{68}$$

The impedance Z_{even} (see Fig. 39b) is calculated based on the Marchand balun coupled lines' parameters i.e., their even- and odd-mode impedances ($Z_{oe}{}^{coupl}$, $Z_{oo}{}^{coupl}$) and electrical length θ [11] as:

$$Z_{even} = -j\frac{2Z_{oo}^{coupl}Z_{oe}^{coupl}}{Z_{oo}^{coupl} + Z_{oe}^{coupl}}\cot(\theta) \tag{69}$$

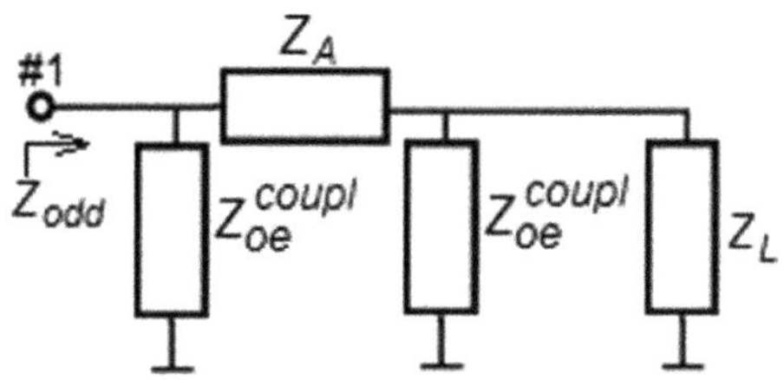

Fig. 40 The Equivalent circuit of the structure shown in Fig. 39c utilized for the calculation of Z_{odd}. Reprinted with permission from Piekarz et al. [16]; copyright 2016 IEEE

The impedance Z_{odd} (see Fig. 39c) depends on the load impedance Z_L at the balanced port (in the general case) and based on the equivalent circuit presented in Fig. 40 [11] it can be calculated as:

$$Z_{odd} = \frac{\frac{-Z_L Z_{oe}^{coupl^2} Z_A tan^2(\theta)}{Z_L + j Z_{oe}^{coupl} \tan(\theta)} - Y}{X + j\frac{Z_L Z_{oe}^{coupl} \tan(\theta)\left[Z_A - Z_{oe}^{coupl} tan^2(\theta)\right]}{Z_L + j Z_{oe}^{coupl} tan(\theta)}} \tag{70}$$

where

$$Y = Z_{oe}^{coupl} Z_A^2 \tan^2(\theta) \tag{70a}$$

$$X = j\left[Z_{oe}^{coupl} Z_A \tan(\theta) + Z_A^2 \tan(\theta)\right] \tag{70b}$$

$$Z_A = \frac{2Z_{oe}^{coupl} Z_{oo}^{coupl}}{Z_{oe}^{coupl} - Z_{oo}^{coupl}} \tag{70c}$$

Inserting Eqs. (69) and (70) into Eq. (68) one obtains an expression describing balanced Z_L to unbalanced Z_{in} impedance transformation of the Marchand balun.

3.2 Sample Effective Permittivity Determination Procedure

The sample-under-test effective permittivity $\varepsilon_{effo}^{sensor}$ can be determined following the described procedure:

1. First, the reflection coefficient S_{11} at port #1 (see Fig. 39a) is measured and the impedance Z_{in} seen at port #1 is calculated as:

$$Z_{in} = Z_T \frac{1 + S_{11}}{1 - S_{11}} \tag{71}$$

where Z_T is the reference impedance of the measurement system (typically equal to 50 Ω).

2. Second, the impedance Z_{even} is calculated based on the Marchand balun coupled lines' parameters obtained during the calibration process ($Z_{oe}{}^{coupl}$, $Z_{oo}{}^{coupl}$, θ) using Eq. (69).
3. Next, based on the obtained values of Z_{in} and Z_{even} the impedance Z_{odd} is calculated as:

$$Z_{odd} = 2Z_{in} - Z_{even} \tag{72}$$

4. In the next step, by comparing Eq. (72) to Eq. (70), the formula to determine load impedance Z_L is given as:

$$Z_L = \frac{-YB - Z_{odd}XB}{Z_{odd}X + CZ_{odd} + A + Y} \tag{73}$$

where

$$A = Z_{oe}^{coupl^2} Z_A \tan^2(\theta) \tag{74}$$

$$B = jZ_{oe}^{coupl} Z_A tan(\theta) \tag{75}$$

$$C = jZ_{oe}^{coupl} tan(\theta)\left[Z_A - Z_{oe}^{coupl} \tan^2(\theta)\right] \tag{76}$$

5. Having found Eq. (73) describing Z_L, it is compared with Eq. (67). Based on (2b) and (2d) $Z_{oo}{}^{sensor}$ can be expressed as:

$$Z_{oo}^{sensor} = \frac{1}{cC_{oo}^{sensor}\sqrt{\varepsilon_{effo}^{sensor}}} \tag{77}$$

where C_{oo}^{sensor} is the odd-mode per-unit-length capacitance of the coupled-line sensor in the presence of a uniform dielectric medium having a relative dielectric constant equal to 1 (defined in Sect. 1) determined during the calibration process. Moreover, an approximation of the cot(x) function with its first two elements of Maclaurin series expansion ($\cot(x) \approx \frac{1}{x} + \frac{x}{3}$) is used.
6. Finally, after some mathematical manipulations, the effective permittivity equals:

$$\varepsilon_{effo}^{sensor} \approx \frac{H}{M} \tag{78}$$

where

$$H = \frac{-j}{2\pi C_{oo}^{sensor} l f} \tag{79}$$

$$M = Z_L - \frac{j2\pi l f}{3c^2 C_{oo}^{sensor}} \tag{80}$$

and l is the length of the utilized open-ended coupled lines (in meters), and f is the frequency (in Hz) at which the effective permittivity is calculated.

It must be noted that due to the utilization of Maclaurin series expansion of cotangent function and cotangent periodicity, the shorter the coupled-line sensor the higher permittivity of the covering sample can be detected. Moreover, the above-described analysis only holds within the operating frequency range of the Marchand balun. However, the coupled-line sensor does not have to be matched to the impedance of Marchand balun as long as the value C_{oo}^{sensor} of the coupled-line sensor can be determined. Therefore, the entire measurement system, especially balun, can be designed to achieve the widest operating frequency range in a given dielectric structure.

3.3 Calibration Process

To calibrate the sensor, the following procedure is applied which is done only once for a given measurement setup. There are four constant parameters, which need to be determined to allow for MUT effective permittivity calculation, i.e. parameters of the coupled-line sections composing balun Z_{oe}^{coupl}, Z_{oo}^{coupl}, θ along with C_{oo}^{sensor} (odd-mode per-unit-length capacitance of the coupled-line sensor in the presence of uniform dielectric medium having a relative dielectric constant equal to 1).

In order to determine the values of Z_{oe}^{coupl}, Z_{oo}^{coupl} and θ of the measurement setup, an extra single coupled-line section (see Fig. 41) constituting Marchand balun needs to be manufactured in the same process and fabrication run as the sensor circuit for system calibration purposes. From the measured Z-parameters of the calibration coupled lines (Marchan balun component), the Z_{oe}^{coupl}, Z_{oo}^{coupl} and θ of can are calculated as:

$$Z_{oe}^{coupl} = j\frac{Z_{11} + Z_{12}}{\cot(\theta_e)} \tag{81}$$

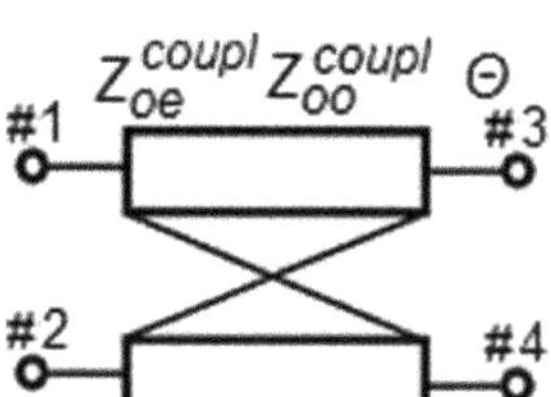

Fig. 41 A coupled-line section used for calibration of the measurement system. Reprinted with permission from Piekarz et al. [16]; copyright 2016 IEEE

$$Z_{oo}^{coupl} = j\frac{Z_{11} - Z_{12}}{\cot(\theta_o)} \tag{82}$$

$$\theta = \frac{\theta_e + \theta_o}{2} \tag{83}$$

where Z_{11}, Z_{12}, Z_{13}, and Z_{14} are elements of the coupled-line section's Z-matrix and θ_e, θ_o are the even- and odd-mode electrical lengths of the coupled-line section (given in radians) computed from:

$$\theta_e = \cos^{-1}\left(\frac{Z_{11} + Z_{12}}{Z_{13} + Z_{14}}\right) \tag{84}$$

$$\theta_o = \cos^{-1}\left(\frac{Z_{11} - Z_{12}}{Z_{14} - Z_{13}}\right) \tag{85}$$

The ${C_{oo}}^{sensor}$ value can be obtained utilizing the procedure presented in Sect. 3.2, Eq. (77), for the sample having known effective permittivity $\varepsilon_{effo}^{sensor}$ (for example air cover) and the calculated impedance Z_{oo}^{sensor} based on the performed S_{11} measurements.

The proposed calibration process requires only measurements of the coupled-line section composing the Marchand balun, and the determination of the C_{oo}^{sensor} which significantly reduces calibration and measurement time. This is contrary to the calibration shown e.g. in [55] where the Marchand balun circuit is entirely de-embedded from the measurements requiring ten calibration standards. At the expense of measurement accuracy reduction, a time and resource-efficient method of effective permittivity determination is obtained.

3.4 Measurement Setup Design

To verify the provided theoretical analysis, an exemplary measurement setup composed of a coupled-line sensor and a Marchand balun, operating at the center frequency $f_0 = 1$ GHz was designed, manufactured, and measured. The coupled-line sensor was designed in a dielectric structure presented in Fig. 1 (assuming air cover, dielectric permittivity, and thickness are equal $\varepsilon_{r1} = 3.38$ and $h_1 = 1.524$ mm). Sensor strips' width w and spacing between the lines s have been set to: $w = 1$ mm, $s = 0.08$ mm (minimum spacing is limited by the available manufacturing technology). The resulting odd-mode characteristic impedance of the coupled lines has been calculated using *Linpar* software [20] and equals $Z_{oo}^{sensor} = 40\ \Omega$. The length of the sensor has been set to $l = 7.8$ mm which is much less than 90° at f_0.

The Marchand type balun was designed in a dielectric structure presented in Fig. 42 to be matched at the input to $Z_{in} = 50\ \Omega$ and at the output to $Z_{out} = 40\ \Omega$. The symmetric homogeneous dielectric structure has been chosen to ensure the conditions

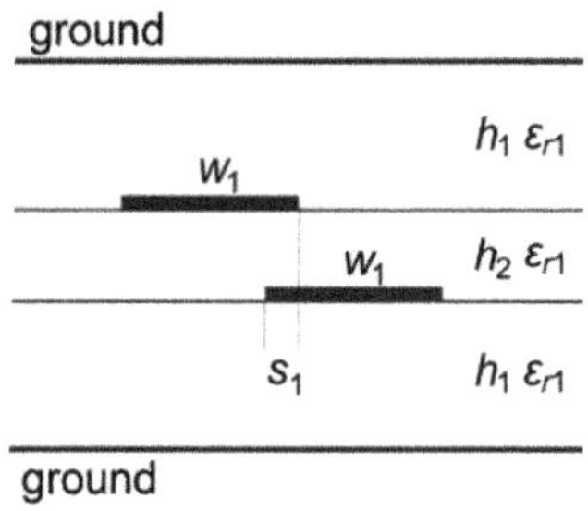

Fig. 42 Cross-sectional view of the coupled lines composing Marchand baluns: $w_1 = 0.87$ mm, $s_1 = 0.03$ mm, $h_1 = 1.524$ mm, $h_2 = 0.508$ mm, $\varepsilon_{r1} = 3.38$. Reprinted with permission from Piekarz et al. [16]; copyright 2016 IEEE

of an ideal coupled-line section's realization. The design approach described in Ahn and Nam [52] was used to calculate the modal impedances of coupled-line sections composing the balun. For design simplicity, the characteristic impedance of the lines has been assumed as $Z_0 = 50\ \Omega$ and resulting modal impedances are equal: $Z_{oe}^{coupled} = 103.3\ \Omega$, $Z_{oe}^{coupled} = 24.13\ \Omega$ ($k = 0.621$). The geometry of the coupled lines was determined using *Linpar* [20] software and is given in Fig. 42. The designed Marchand balun loaded at the outputs with 40 Ω was electromagnetically (EM) analyzed with the use of *AWR Microwave Office* software. The obtained frequency characteristics are presented in Fig. 43. It is seen that the center frequency has slightly shifted due to asymmetry in coupled-strips length (in the area of connection between coupled-line sections). The final layout and cross-sectional view of the designed measurement setup are presented in Fig. 44. As can be seen, the coupled sections composing the balun have been connected perpendicularly to each other to minimize the influence of the connecting line length on the circuit's parameters [52, 53] and to minimize the influence of the balun on the coupled-line sensor.

3.5 *Measurement Results*

The manufactured circuit is shown in Fig. 45 where the coupled-line sensor and the Marchand balun are visible. A photograph of the measurement setup in use, i.e., the manufactured balun and the coupled-line sensor covered by sample-under-test, connected to the vector network analyzer (*Agilent 5224A*) using a coaxial cable is presented in Fig. 46. First of all, the aforementioned calibration process was performed and parameters of the manufactured coupled-line sections composing the Marchand balun have been measured and equal $Z_{oe} = 100.1\ \Omega$, $Z_{oo} = 29.14\ \Omega$ ($Z_0 = 54\ \Omega$, $k = 0.56$). Additionally, the input signal line has been de-embedded from the measured data to obtain S-parameters of the measurement setup. The measured center frequency of the setup has shifted to $f_0 = 1.03$ GHz. The difference from the assumed values is related to finite manufacturing accuracy. Then, the proposed

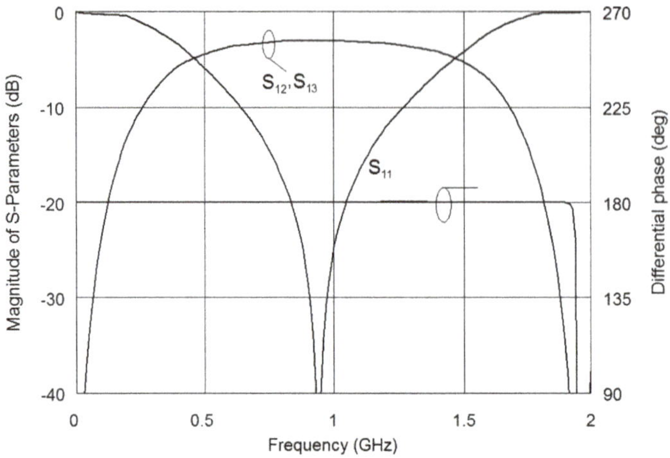

Fig. 43 Amplitude and phase characteristics of the designed Marchand balun terminated at the input port #1 with 50 Ω and at the output ports #2, and #3 with 40 Ω. Reprinted with permission from Piekarz et al. [16]; copyright 2016 IEEE

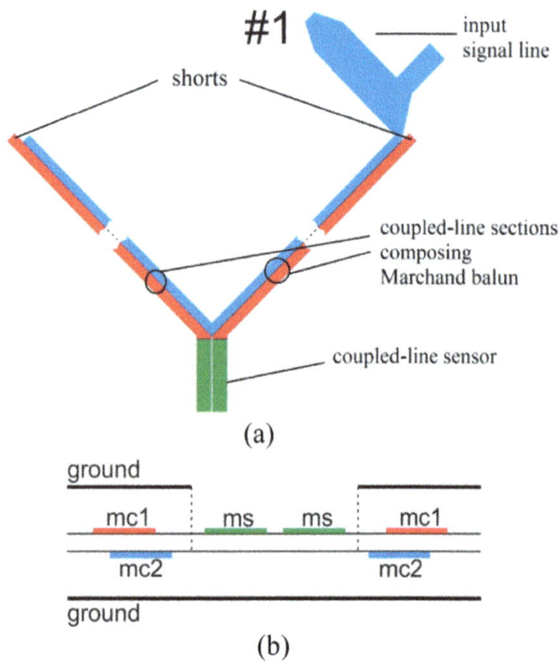

Fig. 44 View of the layout (**a**) and cross-sectional view (**b**) of the designed measurement setup composed of an open-ended coupled-line sensor (metal strips *ms*) and a Marchand balun (metal strips *mc*1 and *mc*2). Reprinted with permission from Piekarz et al. [16]; copyright 2016 IEEE

method of sample's effective permittivity determination was experimentally verified by covering the coupled-line sensor (see Fig. 46) with various dielectric materials and calculating $\varepsilon_{effo}^{sensor}$ in a wide frequency range. Pieces of microwave laminates having different thicknesses h_2 and permittivity ε_{r2} served as sample-under-tests. Based on the measured complex reflection coefficient, the effective permittivity has been calculated following the proposed method. The resulting value versus frequency is shown in Fig. 47 together with the values calculated based on the EM simulated setup for comparison purposes. As can be seen, a good agreement between electromagnetically calculated and measured $\varepsilon_{effo}^{sensor}$ is obtained. A summary of effective permittivity values calculated using numerical analysis with *Linpar* software [20] and calculated with the proposed approach at f_0 using the EM analyzed and manufactured measurement setup is presented in Table 9. As can be seen, there is a satisfactory agreement between different approaches. It is worth mentioning that the quality of contact between the sensor and the measured sample is an important factor that has a direct influence on the measured effective permittivity, i.e. when there is even a small air gap between these two, the effective permittivity changes. The differences between predicted (obtained based on EM simulations) and measured effective permittivity values can be caused by manufacturing precision e.g., parameters of the coupled-line sections (Z_{oe}^{coupl}, Z_{oo}^{coupl}), physical realization of shorts as well as input signal line's length.

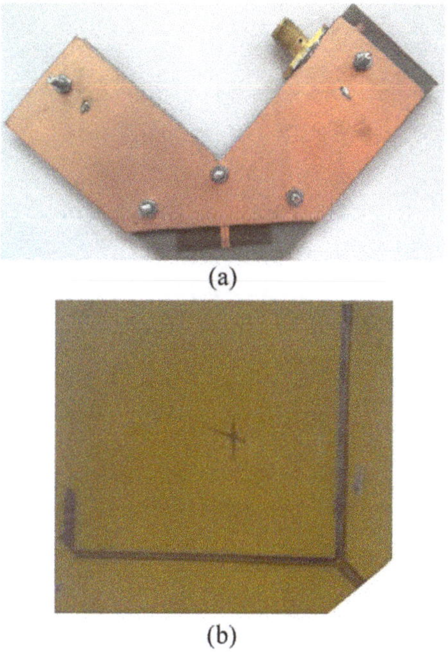

Fig. 45 Photograph of the manufactured measurement setup (**a**) and view of thin center laminate where the Marchand balun and the coupled-line sensor are visible (**b**). Reprinted with permission from Piekarz et al. [16]; copyright 2016 IEEE

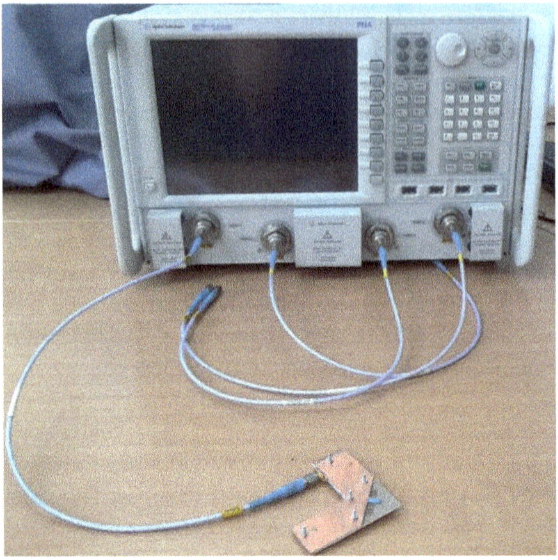

Fig. 46 Photograph of the measurement setup with the manufactured sensor circuit. Reprinted with permission from Piekarz et al. [16]; copyright 2016 IEEE

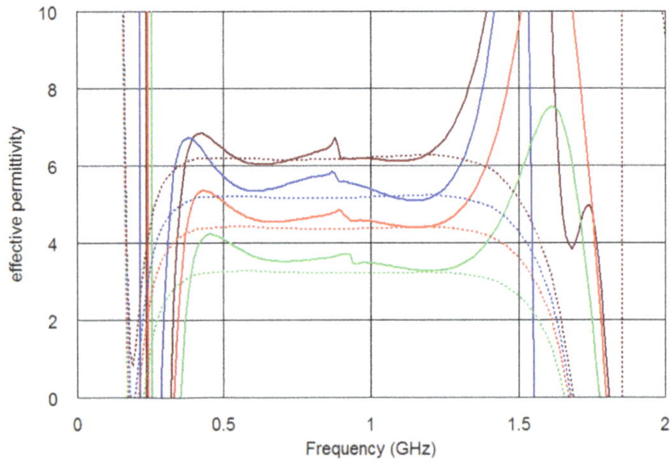

Fig. 47 Calculated (dotted lines) and measured (solid lines) effective permittivity $\varepsilon_{effo}^{sensor}$ for various dielectric covers in a wide frequency range: $\varepsilon_{r2} = 3.38$, $h_2 = 1.524$ mm—green line, $\varepsilon_{r2} = 7.7$, $h_2 = 1$ mm—blue line, $\varepsilon_{r2} = 6.15$, $h_2 = 0.79$ mm—red line, $\varepsilon_{r2} = 10.35$, $h_2 = 0.76$ mm—brown line. Reprinted with permission from Piekarz et al. [16]; copyright 2016 IEEE

Table 9 The comparison between $\varepsilon_{effo}^{sensor}$ obtained with *Linpar*, and using the proposed method based on EM simulations and based on measurements

Dielectric sample coverage	ε_{MUT} and h_u of the dielectric sample (microwave laminate)	$\varepsilon_{effo}^{sensor}$ calculated with Linpar software [21]	$\varepsilon_{effo}^{sensor}$ at f_0 calculated with the proposed formulas based on EM calculations	$\varepsilon_{effo}^{sensor}$ at f_0 calculated with the proposed formulas based on measurements
Entire sensor	$\varepsilon_{r2} = 3.38$ $h_2 = 1.52$ mm	3.36	3.21	3.44
Entire sensor	$\varepsilon_{r2} = 6.15$ $h_2 = 0.79$ mm	4.52	4.38	4.50
Entire sensor	$\varepsilon_{r2} = 7.7$ $h_2 = 1$ mm	5.3	5.25	5.5
Entire sensor	$\varepsilon_{r2} = 10.35$ $h_2 = 0.76$ mm	6.32	6.22	6.14

3.6 Error Analysis

The measurement precision relates to the calibration process, which can be performed in a fast and simple way, however, at the expense of measurement accuracy reduction, as mentioned in Sect. 3.3. The proposed time-saving and easy calibration method allows correcting parameters of the single coupled-line section composing Marchand balun (Z_{oo}^{sensor}, Z_{oe}^{sensor}, θ) and C_{oo}^{sensor} value, which are required to calculate effective permittivity of the sample-under-test. Due to the reduced number of calibration standards, the influence of manufacturing precision, especially Marchand balun's physical realization of shorted coupled strips was not included and can cause a discrepancy between the predicted and measured MUT effective permittivity. Physical realization of shorts requires the connection of particular strips of coupled-line sections (see Fig. 44a) with top and bottom ground planes (see Fig. 44b) with the use of a thin wire, which introduces additional inductance to the circuit. Moreover, the ideal strip-to-wire connection is also not possible and requires in many cases extra electrically short transmission lines (a few degrees), which deteriorate the circuit's parameters.

A sensitivity analysis of the circuit (see Fig. 48) was performed to assess the influence of physically realized shorts in Marchand balun on the effective permittivity values. The EM-based test circuit was composed of electromagnetically calculated Marchand balun's coupled-line sections with the coupled-line sensor being covered with various dielectric materials. The calculated effective permittivity in the case of ideal shorts ($L_1 = 0$ nH, $\theta_1 = 0°$) presented in Fig. 47 (dotted lines) is shown as a reference while the influence of physically realized shorts can be observed in Fig. 49. As seen the added inductance and electrically short transmission lines deteriorate the effective permittivity characteristics. Even a small value of inductance L_1 (below 0.5 nH) results in characteristic peaks near the center frequency of the

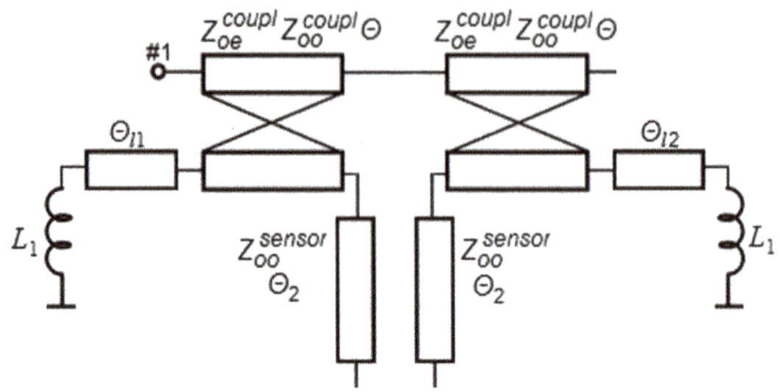

Fig. 48 Schematic diagram of the circuit used for sensitivity analysis. Reprinted with permission from Piekarz et al. [16]; copyright 2016 IEEE

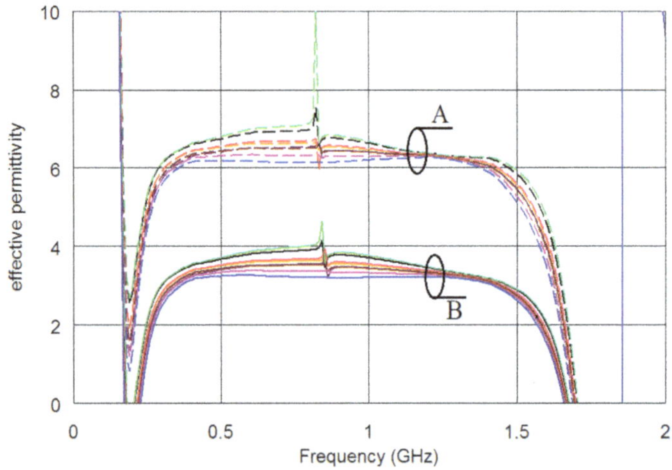

Fig. 49 Calculated influence of physically realized shorts (different values of L_1, θ_{l1}, θ_{l2}—see Fig. 50) for two dielectric covers: A—$\varepsilon_{r2} = 10.35$, $h_2 = 0.76$ mm, B—$\varepsilon_{r2} = 3.38$, $h_2 = 1.524$ mm. Values of parasitics from Fig. 50: $\theta_{l1} = \theta_{l2} = 0°$, $L_1 = 0$ nH—blue lines, $\theta_{l1} = \theta_{l2} = 0°$, $L_1 = 0.1$ nH—pink lines, $\theta_{l1} = \theta_{l2} = 0°$, $L_1 = 0.2$ nH—brown lines, $\theta_{l1} = \theta_{l2} = 0°$, $L_1 = 0.3$ nH—red lines, $\theta_{l1} = \theta_{l2} = 0°$, $L_1 = 0.5$ nH—green lines, $\theta_{l1} = 0.2°$, $\theta_{l2} = 0.5°$, $L_1 = 0.4$ nH—black lines, $\theta_{l1} = 0.8°$, $\theta_{l2} = 0.5°$, $L_1 = 0.2$ nH—orange lines, $\theta_{l1} = 0.1°$, $\theta_{l2} = 0.9°$, $L_1 = 0.1$ nH—violet lines. Reprinted with permission from Piekarz et al. [16]; copyright 2016 IEEE

designed measurement system, as observed in effective permittivity calculated based on measurements. Moreover, a very short transmission line section connected to ends of coupled strips (below 1°) can result in a slightly higher effective permittivity in comparison to the reference value.

Additionally, it is important to ensure proper quality of contact between the coupled-line sensor and measured sample as mentioned in Sect. 3.5. Even a thin layer of air between the sensor and sample can result in lower effective permittivity

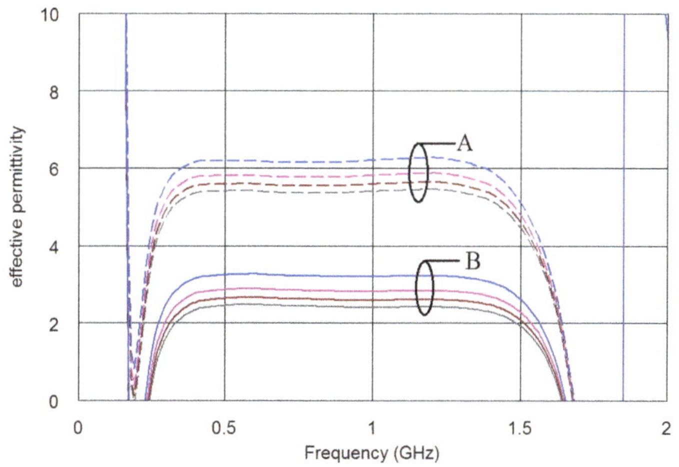

Fig. 50 Calculated influence of air layer between the coupled-line sensor and covering dielectric material. Analysis performed for two samples: A—$\varepsilon_{r2} = 10.35$, $h_2 = 0.76$ mm, B—$\varepsilon_{r2} = 3.38$, $h_2 = 1.524$ mm. Thickness of air layer: no air—blue lines, 0.02 mm—pink lines, 0.05 mm—brown lines, 0.1 mm—grey lines. Reprinted with permission from Piekarz et al. [16]; copyright 2016 IEEE

compared to the reference value. Analysis of the influence of air layer thickness vs resulting effective permittivity is shown in Fig. 50. As seen, the thicker the air layer, the greater the reduction of effective permittivity.

The presented experimental study was carried out using pieces of microwave laminates (see Fig. 46) as a set u MUT samples. Therefore, the theoretical effective permittivity of the MUT-sensor stack-up was calculated with the use of *Linpar* software [20], which is a numerical tool for multiconductor systems evaluation. Moreover, the measured effective permittivity values were compared with the ones calculated based on EM calculations in *AWR Microwave Office* software to verify the presented theoretical analysis (see Fig. 47). To assess the theoretically calculated (*Linpar* software, *AWR Microwave Office* software) and measured effective permittivity values within the operation range of the measurement setup the percent error was calculated and is provided in Fig. 51 As seen, a satisfactory agreement between measured and predicted effective permittivity values were obtained since the difference between EM calculated and measured parameters do not exceed $\pm 7\%$ and the difference between calculated in *Linpar* and measured parameters does not exceed $\pm 11\%$ (in the frequency range for which the impedance match is better than -12 dB, see Fig. 43).

Based on the obtained measurement results and results of error analysis it can be asserted that the proposed system is well suitable for fast and easy dielectric sample detection.

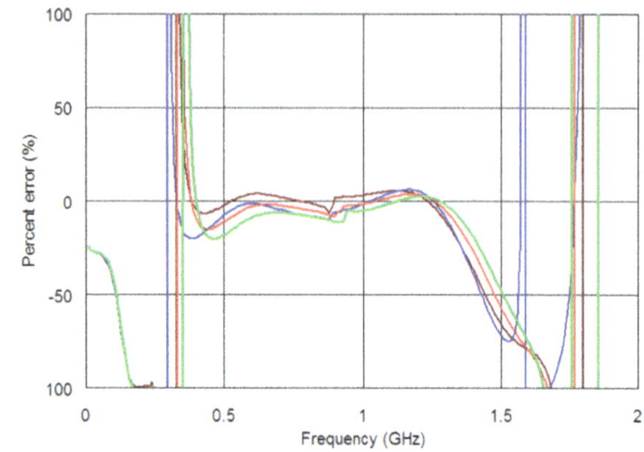

(a)

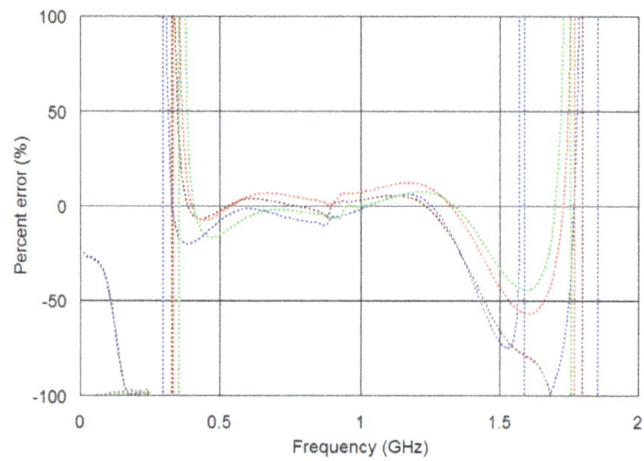

(b)

Fig. 51 Percent error calculated for the measured sample-under-test effective permittivity in comparison to the values obtained with the use of *AWR Microwave Office* software (**a**) and *Linpar* software (**b**) for various dielectric covers: $\varepsilon_{r2} = 3.38$, $h_2 = 1.524$ mm—green line, $\varepsilon_{r2} = 7.7$, $h_2 = 1$ mm —blue line, $\varepsilon_{r2} = 6.15$, $h_2 = 0.79$ mm—red line, $\varepsilon_{r2} = 10.35$, $h_2 = 0.76$ mm—brown line. Reprinted with permission from Piekarz et al. [16]; copyright 2016 IEEE

3.7 Summary

The presented measurement setup composed of Marchand balun and an open-ended edge coupled-line sensor requires a reduced number of ports, namely a single unbalanced one while ensuring proper excitation to the sensing element. Thanks to the use of a wideband balun circuit, it is possible to characterize the dielectric samples in a wide frequency range. Derivation of measured samples' effective permittivity is based on the measured one-port S-parameters and analytical formulas describing Marchand balun's input impedance in a wide frequency range. It must be noted that the presented procedure is preferable for the characterization of low-loss solid samples due to the modeling approach.

References

1. Levy R (1963) General synthesis of asymmetric multi-element coupled- transmission-line directional couplers. IEEE Trans Microw Theory Techn MTT-11(4):226–237, July 1963
2. Cristal EG, Young L (1965) Theory and tables of optimum symmetrical TEM-mode coupled-transmission-line directional couplers. IEEE Trans Microw Theory Techn MTT-13(5):544–558, September 1965
3. Shelton JP, Mosko JA (1966) Synthesis and design of wide-band equal-ripple TEM directional couplers and fixed phase shifters. IEEE Trans Microw Theory Techn MTT-14(10):462–473, October 1966
4. Sachse K (1990) The scattering parameters and directional coupler analysis of characteristically terminated asymmetric coupled transmission lines in an inhomogeneous medium. IEEE Trans Microw Theory Techn 38(4):417–425
5. Gruszczynski S, Wincza K, Sachse K (2006) Design of compensated coupled-stripline 3-dB directional couplers, phase shifters, and magic-T's—part I: Single-section coupled-line circuits. IEEE Trans Microw Theory Techn 54(11):3986–3994
6. Gruszczynski S, Wincza K, Sachse K (2006) Design of compensated coupled-stripline 3-dB directional couplers, phase shifters, and magic-T's—part II: Broadband coupled-line circuits. IEEE Trans Microw Theory Techn 54(9):3501–3507
7. Wincza K, Gruszczynski S (2010) Miniaturised quasi-lumped coupled-line single-section and multisection directional couplers. IEEE Trans Microw Theory Techn 58(11):2924–2931
8. Piekarz I, Sorocki J, Wincza K, Gruszczynski S, Muller J, Welker T (2013) Miniaturised quasi-lumped coupled-line single-section directional coupler designed in multilayer LTCC technology. Microw Opt Technol Lett 55(6):1401–1405
9. Piekarz I, Sorocki J, Wincza K, Gruszczynski S (2015) Modelling and performance improvement of folded coupled lines in miniaturised quasi-lumped directional couplers. Int J RF Microw Comput-Aided Eng 25(1):1–9, January 2015
10. Jonesand EMT, Bolljahn JT (1956) Coupled-strip-transmission-line filters and directional couplers. IRE Trans Microw Theory Techn MTT-4(2):75–81, April 1956
11. Matthaei GL, Young L, Jones EMT (1964) Microwave filters, impedance-matching networks, and coupling structures. McGraw Hill, New York
12. Malherbe JAG (1979) Microwave transmission line filters. Artech House, Norwood, MA
13. Sorocki J, Piekarz I, Wincza K, Gruszczynski S (2015) Right/left-handed transmission lines based on coupled transmission line sections and their application towards bandpass filters. IEEE Trans Microw Theory Techn 63(2):384–396
14. Sorocki J, Piekarz I, Wincza K, Gruszczynski S (2015) Semi-distributed approach to dual-composite right/left-handed transmission lines and their application to bandstop filters. IEEE Microw Wirel Compon Lett 25(12):784–786
15. Schiffman BM (1958) A new class of broadband microwave 90-degree phase shifters. IRE Trans Microw Theory Technol 6(2):232–237
16. Piekarz I, Sorocki J, Wincza K, Gruszczynski S (2016) Coupled-line sensor with Marchand balun as RF system for dielectric sample detection. IEEE Sens J 16(1):88–96, January 1, 2016
17. Piekarz I, Sorocki J, Wincza K, Gruszczynski S (2017) Microwave sensors for dielectric sample measurement based on coupled-line section. IEEE Trans Microw Theory Technol 65(5):1615–1631
18. Sorocki J, Piekarz I, Wincza K, Gruszczynski S, Papapolymerou J (2020) Broadband microwave microfluidic coupled-line sensor with 3-D-printed channel for industrial applications. IEEE Trans Microw Theory Technol 68(7):2808–2822
19. Pozar D (2005) Microwave engineering. Wiley, New York
20. Djordievic A et al (1990) Matrix parameters for multiconductor transmission lines, software and user's manual. Artech House, Norwood, MA
21. Bockelman DE, Eisenstadt WR (1995) Combined differential and common-mode scattering parameters: theory and simulation. IEEE Trans Microw Theory Technol 43(7):1530–1539

22. Bockelman DE, Eisenstadt WR (1997) Pure-mode network analyzer for on-wafer measurements of mixed-mode S-parameters of differential circuits. IEEE Trans Microw Theory Technol 45(7):1071–1077
23. Fan W, Lu A, Wai LL, Lok BK (2003) Mixed-mode S-parameter characterization of differential structures. In: Proceedings of 5th conference on electronics package technology, pp 533–537
24. Halim MA, Hamid MAK (1970) End effect in parallel-coupled transmission-line filters. Electron Lett 6(5):130–131
25. Gajda GB, Stuchly SS (1983) Numerical analysis of open-ended coaxial lines. IEEE Trans Microw Theory Technol 31(5):380–384
26. Anderson LS, Gajda GB, Stuchly SS (1986) Analysis of an open-ended coaxial line sensor in layered dielectrics. IEEE Trans Microw Theory Techn IM-35(1):13–18, March 1986
27. Rong AS, Tripathi VK, Sun ZL (1994) Effects of thin-film nonuniformity on thin-film open end microstrip line for miniature multilayer MMICs. Electron Lett 30(19):1606–1608
28. Papió Toda A, De Flaviis F (2015) 60-GHz substrate materials characterization using the covered transmission-line method. IEEE Trans Microw Theory Technol 63(3):1063–1075, March 2015
29. Vaid JK, Prakash A, Mansingh A (1979) Measurement of dielectric parameters at microwave frequencies by cavity perturbation technique. IEEE Trans Microw Theory Technol 27:791–795
30. Krupka J (2006) Frequency domain complex permittivity measurements at microwave frequencies. Meas Sci Technol 17(6):R55–R77
31. Stuchly MA, Stuchly SS (1980) Coaxial line reflection methods for measuring dielectric properties of biological substances at radio and microwave frequencies—a review. IEEE Trans Instrum Meas IM-29(3):176–183, September 1980
32. Grenier K, Dubuc D, Poleni P-E, Kumemura M, Toshiyoshi H, Fujii T, Fujita H (2009) Integrated broadband microwave and microfluidic sensor dedicated to bioengineering. IEEE Trans Microw Theory Technol 57(12):3246–3253
33. Abduljabar AA, Rowe DJ, Porch A, Barrow DA (2014) Novel microwave microfluidic sensor using a microstrip spkit-ring resonator. IEEE Trans Microw Theory Technol 62(3):679–688
34. Abduljabar AA, Choi H, Barrow DA, Porch A (2015) Adaptive coupling of resonators for efficient microwave heating of microfluidic system. IEEE Trans Microw Theory Technol 63(11):3681–3690
35. Guarin G, Hofmann M, Nehring J, Weigel R, Fischer G, Kissinger D (2015) Miniature microwave biosensors: noninvasive applications. IEEE Microw Mag 16(4):71–86
36. Stuchly SS, Bassey CE (1998) Microwave coplanar sensor for dielectric measurements. Meas Sci Technol 9:1324–1329
37. Gregory AP, Clarke RN (2012) Tables of the complex permittivity of dielectric reference liquids at frequencies up to 5 GHz, NPL Report, MAT 23
38. Sandoval R, MacLachlan RA, Oh MY, Riviere CN (2007) Positioning accuracy of neurosurgeons. In: Proceedings of 29th annual international IEEE EMBS conference, Lyon, France, pp 206–209
39. Riviere CN, Khosla PK (1996) Accuracy in positioning of handheld instruments. In: Proceedings of 18th international IEEE engineering medicine and biology society conference, Amsterdam, The Netherlands, pp 212–213
40. Ocera A, Dionigi M, Fratticcioli E, Sorrentino R (2006) A novel technique for complex permittivity measurement based on a planar four-port device. IEEE Trans Microw Theory Technol 54(6):2568–2575
41. Narayanan PM (2014) Microstrip transmission line method for broadband permittivity measurement of dielectric substrates. IEEE Trans Microw Theory Technol 62(11):2784–2790
42. Shete M, Shaji M, Akhtar MJ (2013) Design of a coplanar sensor for RF characterization of thin dielectric samples. IEEE Sens J 13(12):4706–4715
43. Ghodgaonkar DK, Varadan VV, Varadan VK (1990) Free-space measurement of complex permittivity and complex permeability of magnetic materials at microwave frequencies. IEEE Trans Instrum Meas 39(2):387–394
44. Morgan MA (2019) Principles of RF and microwave design. Artech House, Norwood, USA

45. Leus, Elata D (2004) Fringing field effect in electrostatic actuators. Technical Report ETR-2004-2, May 2004
46. Engen GF, Hoer CA (1979) Thru-Reflect-Line: an improved technique for calibration the dual six-port automatic network analyzer. IEEE Trans Microw Theory Technol 27(12):987–998
47. Staszek K, Piekarz I, Sorocki J, Koryciak S, Wincza K, Gruszczynski S (2018) Low-cost microwave vector system for liquid properties monitoring. IEEE Trans Ind Electron 65(2):1665–1674
48. Piekarz I, Sorocki J, Wincza K, Gruszczynski S (2018) Liquids permittivity measurement using two-wire transmission line sensor. IEEE Sens J 18(18):7458–7466
49. Piekarz I, Sorocki J, Delmonte N, Silvestri L, Marconi S, Alaimo G, Auricchio F, Bozzi M (2021) Microwave-microfluidic sensor in hybrid 3-D printing and laminate technology for chemicals monitoring from differential reflection. In: International microwave symposium (IMS'21), Atlanta, GA, USA, 6–11 June 2021
50. Marchand N (1944) Transmission line conversion transformers. Electronics 17(12):142–145
51. Ang KS, Robertson ID (2001) Analysis and design of impedance-transforming planar Marchand baluns. IEEE Trans Microw Theory Technol 49(2):402–406
52. Ahn HR, Nam S (2011) New design formulas for impedance-transforming 3-dB Marchand baluns. IEEE Trans Microw Theory Technol 59(11):2816–2823
53. Piekarz I, Sorocki J, Gruszczynski S, Wincza K (2014) Input match and output balance improvement of Marchand balun with connecting line. IEEE Microw Wirel Compon Lett 24(10):683–685
54. Leong YC, Ang KS, Lee CH (2009) A derivation of a class of 3-port baluns from symmetrical 4-port networks. In: Proceedings of IEEE MTT-S international microwave symposium digest, June 2009
55. Kooho J, Hayden LA, Crisalle OD, Eisenstadt WR, Fox RM, Hanaway P, Campbell RL, McCuen C, Lewis M (2008) A new characterization and calibration method for 3-dB-coupled on-wafer measurements. IEEE Trans Microw Theory Technol 56(5):1193–1200

Coupled-Line Directional Coupler Permittivity Sensors

Vahid Nayyeri and Omar M. Ramahi

Abstract Coupled-line directional couplers have played a significant role in microwave systems since their introduction by Oliver [1]. This chapter presents a new application of this component: dielectric constant sensing. To this end, the material under test is placed on the top of a microstrip coupled-line section, and by analyzing the coupler's response, the dielectric constant of the sample is detected. The response can be either the coupling or isolation levels of the coupler. This sensing approach benefits from high sensitivity, particularly when the isolation level is used. In the following, we will first review the theory of coupled lines and the basics of directional couplers, then present the idea of using a coupler for dielectric constant sensing.

1 Introduction

A microstrip coupled-line directional coupler (MLDC) is a passive microwave device used in radio frequency and microwave communication systems. It is commonly used to separate or combine signals in a transmission line. The coupler consists of two parallel transmission lines, known as coupled lines, which are placed close to each other on a dielectric substrate. These lines are coupled through mutual capacitance and inductance between them, enabling the transfer of energy between the main transmission line and the coupled line. The main function of a directional coupler is to send a portion of the input power to the coupled line while allowing the remaining power to continue along the main transmission line. The coupling between the lines

V. Nayyeri (✉)
School of Advanced Technologies, Iran University of Science and Technology, Tehran 16846-13114, Iran
e-mail: nayyeri@iust.ac.ir

O. M. Ramahi
Department of Electrical and Computer Engineering, University of Waterloo, Waterloo, ON N2L3G1, Canada
e-mail: oramahi@uwaterloo.ca

F. Martın and E. Bronchalo (eds.), *Coupled Structures for Microwave Sensing*, Lecture Notes in Electrical Engineering 1150,
https://doi.org/10.1007/978-3-031-53861-2_4

can be adjusted to achieve a desired power split ratio. This allows the directional coupler to separate the input signal into two output signals: one is the coupled output, which carries a fraction of the input power, and the other is the through output, which carries the majority of the input power.

Directional couplers find extensive applications in various RF and microwave systems, such as power dividers, signal samplers, impedance-matching networks, and in measurement setups for signal monitoring. They can be designed for different coupling ratios and operating frequencies, making them versatile components in wireless communication systems.

This chapter presents a new application of MCLDC, which is a dielectric permittivity sensor. Two methods using a MCLDC for dielectric constant measurement are introduced. In both methods, the material under test (MUT) is put on the coupled-line section of the coupler. The first method, which is a resonant method, relies on the change in the coupling coefficient of the coupler, while in the second method, which is a non-resonant method, the sensing relies on the isolation coefficient of the coupler. We will show that for a unique estimation of the dielectric constant of the MUT from the isolation level, a conventional MCLDC, which has a low level of isolation, does not work properly. And the coupler needs modifications for a high level of isolation, (i.e., a high directivity). Notably, the introduced methods based on MCLDC need only the amplitude of one transmission coefficient (not its phase). Therefore, there is no need for a vector network analyzer (VNA), but instead, either a scalar network analyzer (SNA) or a frequency synthesizer and a power detector, which makes the measurements simpler and lower cost.

2 Dielectric Loading Effect on Microstrip Coupled-Line Directional Coupler

Figure 1 depicts a MCLDC configuration, where the MUT is placed on top of the coupled-line section. To simplify the analysis, the coupled microstrip lines are often assumed to operate in the transverse electromagnetic (TEM) mode, which is approximately valid but not rigorously so. By making this assumption and employing an even-odd mode analysis, the magnitude of parameter S_{31}, denoted as $|S_{31}|$, can be determined as (see Chap. 2):

$$|S_{31}| = \frac{C|\tan\theta|}{|\sqrt{1 - C^2} + j\tan\theta|} \tag{1}$$

Here, $\theta = \beta_{\text{eff}} L = \beta_0 \sqrt{\epsilon_r^{\text{eff}}} L$ represents the electrical length of the coupled lines, L its physical length, β_{eff} the effective propagation constant in the microstrip medium, β_0 the propagation constant in free space, ϵ_r^{eff} the effective dielectric constant of the microstrip medium, and C the maximum coupling coefficient, which is equal to $|S_{31}|$ when $\theta = \frac{\pi}{2}$. A plot of $|S_{31}|$ for a coupler (with $C = -10$ dB) as a function of θ is

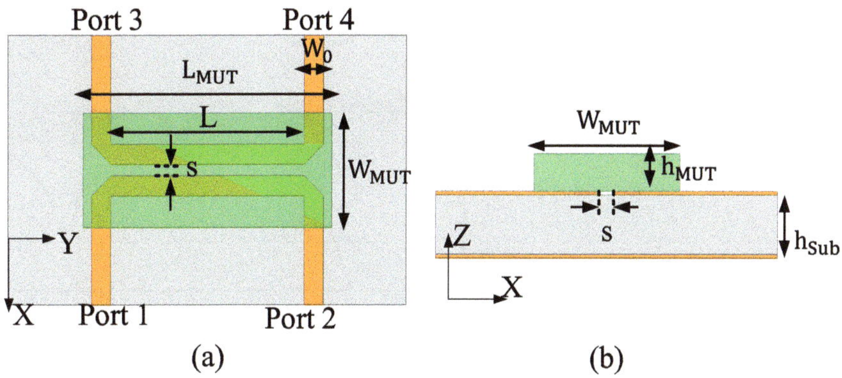

Fig. 1 A MCLDC when an MUT is placed on the coupled-line section. **a** Top view. **b** Cross-sectional view. $W_{\text{MUT}} = 18$ mm, $L_{\text{MUT}} = 43$ mm, $S = 2$ mm, $L = 35$mm, $W_0 = 3.5$ mm, $h_{\text{MUT}} = 1$ mm, $h_{\text{Sub}} = 1.6$ mm. Reprinted with permission from [19]; copyright 2022 IEEE

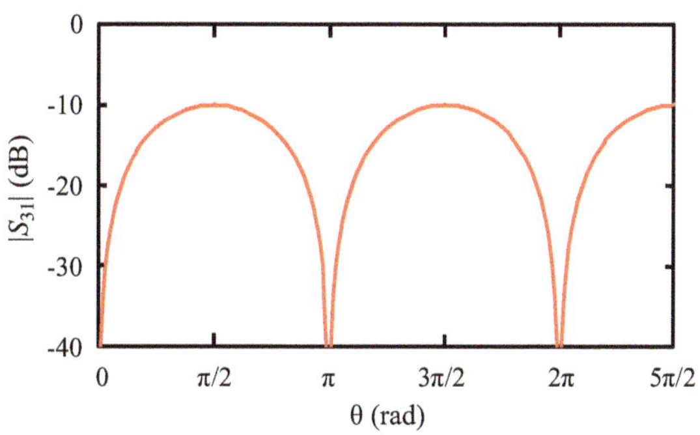

Fig. 2 Theoretical coupling ($|S_{31}|$) of a MCLDC versus the electric length (θ) of the coupled-line when $C = -10$ dB. Reprinted with permission from [19]; copyright 2022 IEEE

presented in Fig. 2, illustrating that $S_{31} = 0$ for $\theta = \pi$ and for all integers that are multiples of π. It should be noted that the derivation of Eq. (1) assumes that the even and odd modes of the coupled line structure possess the same propagation velocities (i.e., the same effective propagation constant), ensuring that the line exhibits the same electrical length for both modes. However, for coupled microstrip lines that do not support a pure TEM mode, this condition is not exactly satisfied, resulting in slightly different propagation velocities (and consequently propagation constants and electrical lengths) for the even and odd modes. In Eq. (1), θ (and β_{eff}) represents the average of the electrical lengths (and effective propagation constants) of the two modes.

In microstrip circuits, the electric field exists in both the substrate and the medium above the signal strip, referred to as the superstrate. The effective dielectric constant ϵ_r^{eff} depends on both the dielectric constants of the substrate and superstrate, with its value lying between them. Consequently, by placing different MUTs on the coupled-line section, ϵ_r^{eff} and consequently θ vary. As a result, the frequency at which $S_{31} = 0$, denoted as the zero-coupling frequency (f_0), changes. This frequency can be utilized

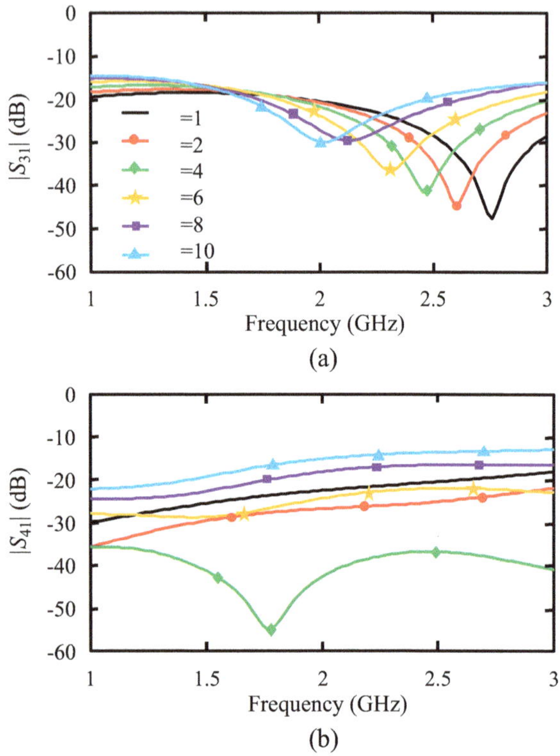

Fig. 3 Simulated $|S_{31}|$ (**a**) and $|S_{41}|$ (**b**) of the MCLDC shown in Fig. 1 when loaded with different MUTs. Reprinted with permission from [19]; copyright 2022 IEEE

to determine the dielectric constant of the MUT. It is worth noting that the microstrip line behaves as a dispersive transmission line above a certain threshold, typically within the 3–10 GHz range for common substrates. For frequencies higher than this threshold, the relative effective permittivity gradually approaches the substrate's permittivity in such a way that the medium above the strip has a less significant impact on the effective dielectric constant.

Let's consider a coupler with the dimensions indicated in the caption of Fig. 1. Employing Rogers RO4003C as the substrate, which possesses a dielectric constant of $\epsilon_r^{\text{Sub}} = 3.55$, a loss tangent of 0.0027, and a thickness of $h_{\text{Sub}} = 1.6$ mm, the width of the microstrip lines is determined to achieve a characteristic impedance of $Z_0 = 50\Omega$. Additionally, the length of the coupled lines and the separation between them are adjusted to ensure that the maximum value of $|S_{31}|$ is approximately $C \approx -18$ dB at around 1.4 GHz.

Figure 3a displays the simulated $|S_{31}|$ for both an unloaded coupler and a coupler loaded with MUTs possessing different dielectric constants (ϵ_r^{MUT}). All the simulations were performed using the frequency-domain solver of CST Microwave Studio. As illustrated in this figure, the zero-coupling frequency of the unloaded coupler is around $f_0 = 2.76$ GHz. With an increase in ϵ_r^{MUT} from 1 to 10, ϵ_r^{eff} rises, subsequently leading to a decrease in f_0 from 2.76 GHz to 2 GHz. Thus, ϵ_r^{MUT} can be uniquely determined by analyzing f_0. Furthermore, as ϵ_r^{MUT} increases from 1 to 10, the effective electric length of the coupled-line section, and thus f_0, decreases.

In Fig. 3b, it can be observed that the change in ϵ_r^{MUT}, which affects the phase velocity in the medium above the coupled microstrip lines, significantly impacts the isolation level ($|S_{41}|$) of the coupler. It's important to note that when assuming a TEM mode and employing an even-odd mode analysis, the analyzed $|S_{41}|$ for a coupled-line coupler is zero for all values of θ. However, in a MCLDC operating in a quasi-TEM mode (not pure TEM), the difference in propagation velocity (and consequently propagation constant and electrical length) between the even and odd modes leads to a non-zero $|S_{41}|$. The imperfect isolation of MCLDC allows utilizing $|S_{41}|$ as a second approach to measure ϵ_r^{MUT}. Nevertheless, as demonstrated in Fig. 4, $|S_{41}|$ plotted against ϵ_r^{MUT} at 2 GHz is not a monotonic function. Consequently, it becomes impossible to uniquely extract ϵ_r^{MUT} since each value of $|S_{41}|$ within the range of -43 dB to -22 dB corresponds to two different values of ϵ_r^{MUT}. To address this issue, modifications are required in the coupler's design to ensure that $S_{41} \approx 0$ occurs for $\epsilon_r^{MUT} = 1$ instead of $\epsilon_r^{MUT} = \epsilon_r^{Sub}$. In other words, a coupler with high isolation ($S_{41} \approx 0$) and, consequently, high directivity in the unloaded state must be utilized.

It should be noted that the change in the dielectric constant of the MUT results in a variation in the isolation level ($|S_{41}|$) of the coupler significantly greater than in the transmission coefficient ($|S_{21}|$) of a microstrip line loaded with the same MUT. Figure 5 provides the simulated transmission coefficients ($|S_{21}|$) of a 50 Ω microstrip line on the same substrate for various ϵ_r^{MUT} values. The sample size of the MUT is assumed to be identical to the sample size in the coupler. By comparing Figs. 3b and 5, it is evident that increasing ϵ_r^{MUT} from 1 to 10 leads to a change in the coupler's isolation level ($|S_{41}|$) of several tens of decibels, while the maximum change in the microstrip line's transmission level ($|S_{21}|$) is approximately 0.2 dB. This comparison underscores the high sensitivity of using the coupler's isolation for sensing ϵ_r^{MUT}.

In previous simulations, it was assumed that the MUT had no losses. However, it is important to investigate how the losses in the MUTs affect the sensor's responses, specifically the magnitudes of $|S_{31}|$ and $|S_{41}|$. By examining the impact of loss tangent variations (which represent conductivity) in the MUT, while keeping its permittivity constant, we can determine whether significant changes in $|S_{31}|$ or $|S_{41}|$ occur. Such changes would indicate undesired cross-sensitivity in the sensor's response, potentially leading to inaccurate determination of ε_r^{MUT}, the relative permittivity of the MUT.

Figure 6 illustrates the simulated values of $|S_{31}|$ and $|S_{41}|$ for different MUTs, each possessing varying conductivity values ranging from $\sigma = 0$ (lossless material) to $\sigma = 6.7 \times 10^{-2}$ S/m (equivalent to a loss tangent of 0.1 at 2 GHz). The dielectric constant of the MUTs was kept constant at $\varepsilon_r^{MUT} = 6$. The results reveal that the dielectric losses in the MUTs do not significantly affect the zero-coupling frequency (f_0) or $|S_{41}|$. It is important to emphasize that similar results were obtained for other values of ε_r^{MUT}. Hence, the proposed sensor shows minimal cross-sensitivity to losses in the MUTs. However, this insensitivity to loss hinders the precise characterization of the MUT's complex permittivity.

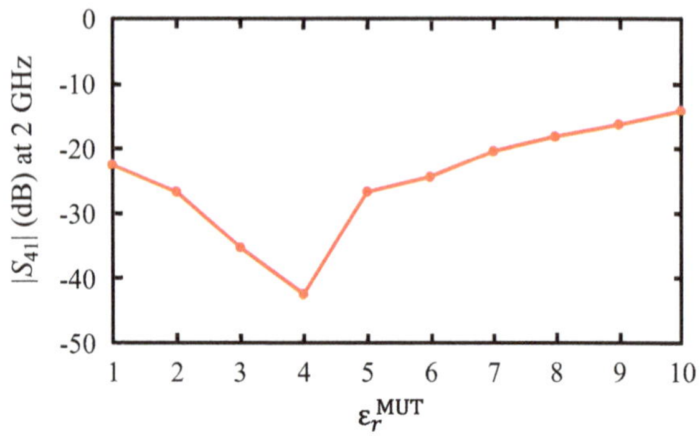

Fig. 4 The isolation level ($|S_{41}|$) of the coupler shown in Fig. 1 at 2 GHz as a function of $\varepsilon_r^{\mathrm{MUT}}$. Reprinted with permission from [19]; copyright 2022 IEEE

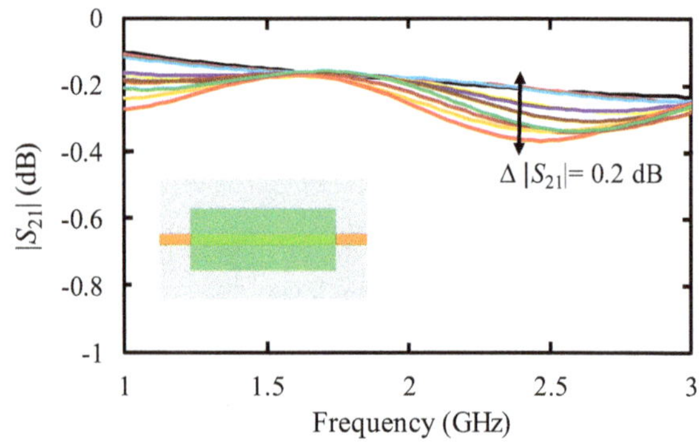

Fig. 5 Simulated transmission coefficients ($|S_{21}|$) of a 50Ω microstrip TL loaded with an MUT (as the inset of the figure shows) for different $\varepsilon_r^{\mathrm{MUT}}$ between 1 and 10 (by step 1). Reprinted with permission from [19]; copyright 2022 IEEE

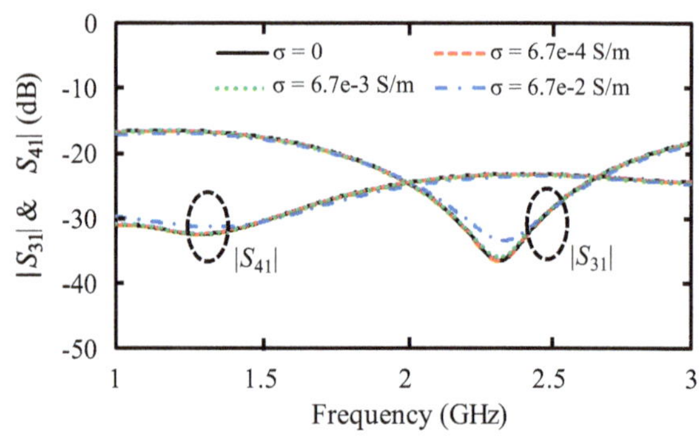

Fig. 6 Simulated $|S_{31}|$ and $|S_{41}|$ of the MCLDC shown in Fig. 1 when loaded with lossy MUTs having $\varepsilon_r^{\mathrm{MUT}} = 6$ and different values of conductivity (loss tangent). Reprinted with permission from [19]; copyright 2022 IEEE

3 Directivity Enhancement of Microstrip Coupled-Line Directional Coupler

Poor directivity (and isolation) is unavoidable in microstrip coupled-line couplers due to their inhomogeneous structure, which stems from the inherent issue of unequal phase velocities between the even and odd modes of microstrip parallel coupled lines. Various techniques have been successfully employed to improve directivity in microstrip couplers. Among these techniques, the reactive-loading method [2–7] stands out as the most attractive and popular option because of its relatively simple design procedure.

Figure 7 illustrates popular reactive methods used to enhance directivity in microstrip directional couplers. By carefully designing the structure, it is possible to balance the phase velocities of the even and odd modes, resulting in enhanced directivity (isolation). In the realm of microwave frequencies, the lumped reactive elements can be implemented with a distributed circuit design. For example, the two stubs at the ends of the coupled-line section, as depicted in Fig. 8, can play the role of the capacitors between the strips shown in Fig. 7a. Assuming a 1.6 mm thick Rogers RO4003C laminate as the substrate, the structure depicted in Fig. 8 with $W_0 = 3.5$ mm, $W_s = 3$ mm, $L_c = 20$ mm, $L_s = 3.5$ mm, $S = 2$ mm, and $S_g = 0.75$ mm yields a coupling coefficient ($|S_{31}|$) of approximately -18 dB around 1.4 GHz, and an isolation level ($|S_{41}|$) better than -40 dB. Moreover, the ports are well matched to 50 Ω.

4 Sensing Permittivity with High Directivity Microstrip Coupled-Line Directional Coupler

As shown in Fig. 8, a microstrip coupled line coupler with enhanced directivity using capacitive loading technique was fabricated and measured using a SNA. Then, by putting different standard laminates on the coupled-line section (sensing area in Fig. 9a), such as Rogers RT/duroid 5880, Rogers RO4003C, FR4, and Rogers AD1000 with dielectric constants of 2.2, 3.55, 4.4, and 10.2, respectively, the coupler's $|S_{31}|$ and $|S_{41}|$ were tested (see Fig. 9b). The results are compared with the full-wave simulation results in Fig. 10, and a good match is observed between them. The measured and simulated $|S_{41}|$ shows that using capacitive stubs achieved a high level of isolation (around -40 dB) in the unloaded state.

An air gap between the coupler and solid MUTs is inevitable. To minimize this gap, four teflon screws were used, as shown in Fig. 9b. However, the air gap cannot be completely removed, so that the simulation and measurement results can be further aligned by assuming a suitable amount of a gap in the simulations. The dotted lines in Fig. 10 show the simulation results by considering a 15 μm air gap between the coupler and the MUT. The slight difference between the simulated and measured results, even when assuming a gap in the simulations, is due to the assumption of the same gap in the simulations of different MUTs. But in practical implementation, we cannot achieve the same gap when changing the samples because of the surface roughness of the MUT and the main board in the measurement setup. These discrepancies can also be attributed to the reflection from the non-ideal ports, fabrication inaccuracies, and measurement errors. In addition, simulations and measurements show that increasing the MUT thickness about approximately 1 mm does not affect the responses due to negligible strength of the fields far away from the coupler.

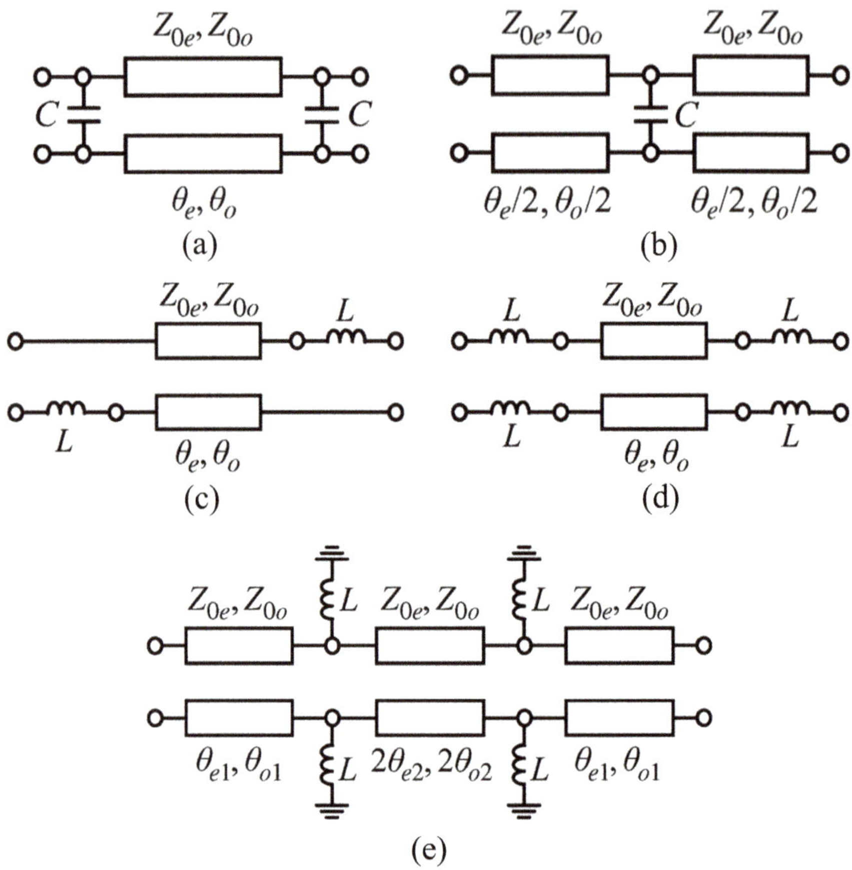

Fig. 7 Various types of reactive-loading technique for directivity enhancement of microstrip coupled-line couplers employed in **a** [2, 3], **b** [4], **c** [5], **d** [6], and **e** [7]

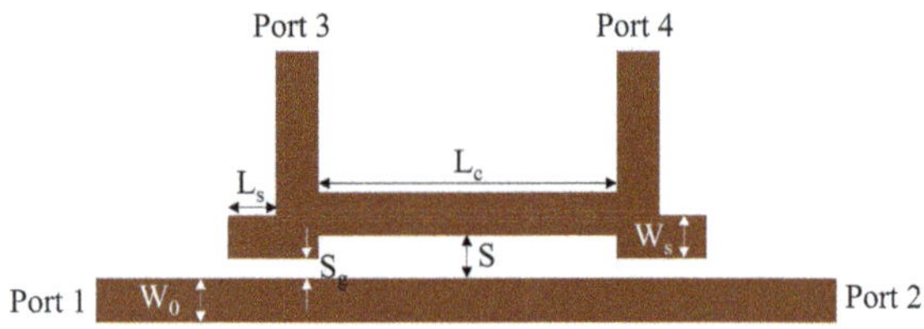

Fig. 8 A distributed implementation of a capacitive-loading technique for directivity enhancement of microstrip coupled-line couplers shown in Fig. 7a

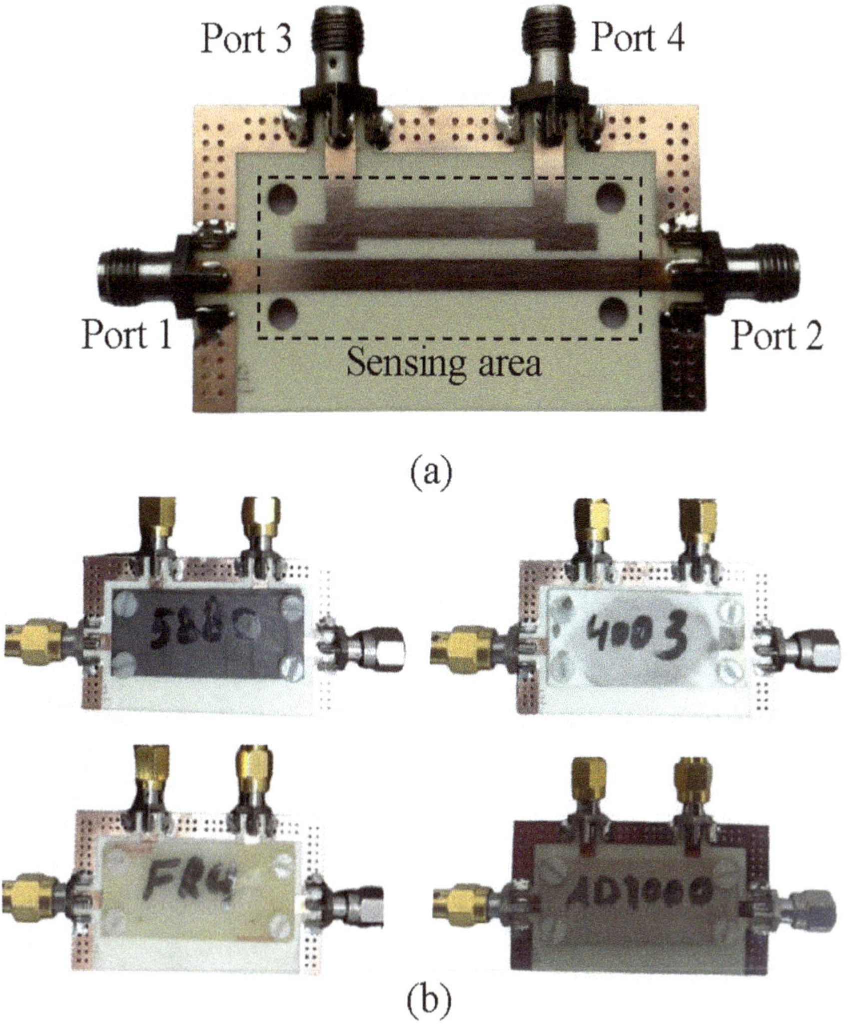

Fig. 9 **a** Fabricated high-directivity microstrip coupled-line coupler; **b** different test setups. Reprinted with permission from [19]; copyright 2022 IEEE

As shown in Fig. 10a, by increasing $\varepsilon_r^{\mathrm{MUT}}$, the zero-coupling frequency (f_0 at which $|S_{31}|$ is minimum) decreases. Moreover and importantly, while in a conventional MCLDC, as presented in Section 2, the $|S_{41}|$ dependence on $\varepsilon_r^{\mathrm{MUT}}$ is not monotonic, Fig. 10b demonstrates that using the high-directivity coupler shown in Fig. 9, $|S_{41}|$ increases monotonically by increasing $\varepsilon_r^{\mathrm{MUT}}$, making the unique retrieval of $\varepsilon_r^{\mathrm{MUT}}$ possible. The measurement results of f_0 and $|S_{41}|$ at 2 GHz, for different MUTs, are listed in Table 1. It is seen that by increasing $\varepsilon_r^{\mathrm{MUT}}$ from 1 (unloaded) to 10.2 (AD1000), f_0 decrease and $|S_{41}|$ increase by about 42% and 30 dB, respectively.

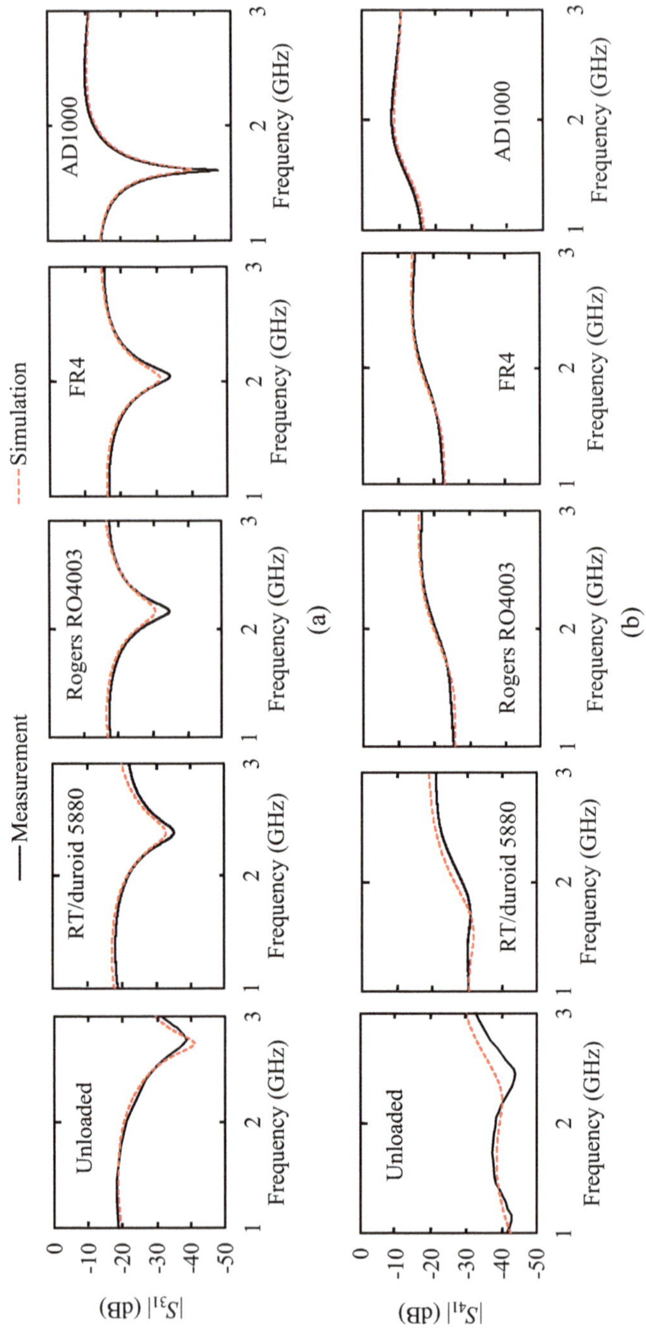

Fig. 10 Simulated and measured results of the proposed sensor (high isolation MCLDC shown in Fig. 9) for different values of $\varepsilon_r^{\mathrm{MUT}}$: **a** coupling coefficient ($|S_{31}|$) and **b** isolation coefficient ($|S_{41}|$). Reprinted with permission from [19]; copyright 2022 IEEE

Table 1 Measured f_0 and $|S_{41}|$ for different MUTs

MUT	ε_r^{MUT}	f_0 (GHz)	$|S_{41}|$(dB) at 2GHz
Air/(Unloaded)	1	2.77	−37.9
RT/duroid 5880	2.2	2.39	−28.7
RO4003C	3.55	2.17	−20.4
FR4	4.6	2.04	−16.9
AD1000	10.2	1.62	−8.0

Therefore, each of these two parameters can be used to retrieve ε_r^{MUT} with a high sensitivity.

To obtain closed-form expressions of ε_r^{MUT} in terms of f_0 and $|S_{41}|$, by setting aside the data of RO4003C, which is considered as test data, a curve fitting method was applied to the data of the other four MUTs (i.e., air, RT5880, FR4, and AD1000) given in Table 1. The curve fitting resulted in:

$$\varepsilon_r^{MUT} = a_3(f_0)^3 + a_2(f_0)^2 + a_1(f_0) + a_0, \tag{2}$$

$$\varepsilon_r^{MUT} = b_3(S_{41})^3 + b_2(S_{41})^2 + b_1(S_{41}) + b_0. \tag{3}$$

where, $a_3 = -2.91$, $a_2 = 2.60 \times 10^1$, $a_1 = -7.92 \times 10^1$, $a_0 = 8.26 \times 10^1$, $b_3 = 5.72 \times 10^{-4}$, $b_2 = 5.12 \times 10^{-2}$, $b_1 = 1.63$, $b_0 = 2.02 \times 10^1$, and f_0 and $|S_{41}|$ are in GHz and dB units. Figure 11a and b show ε_r^{MUT} as functions of the zero-coupling frequency (f_0) and the isolation level ($|S_{41}|$), respectively. In this figure, a good agreement between the measured results and the fitted curves is observed. It should be noted that either (2) (i.e. Fig. 11a) or (3) (i.e. Fig. 11b) can be used for the retrieval of ε_r^{MUT} alternatively, in other words, measuring one of $|S_{31}|$ and $|S_{41}|$ is sufficient. It is noticeable that although for retrieving ε_r^{MUT} using the isolation level, we employed $|S_{41}|$ at 2 GHz, its value at other frequencies can be utilized as well. However, the change in $|S_{41}|$ with ε_r^{MUT} varying from 1 to 10.2 is higher for 2 GHz than for other frequencies; thus, the highest possible sensitivity is achieved.

To assess the accuracy of the proposed retrieval methods, the measured data of RO4003C, which was omitted from the curve fitting process, was used. The retrieved dielectric constant from (2) and (3) using zero-coupling frequency and isolation level, respectively, are given in Table 2. The table also shows the relative error of each method, calculated as

$$\text{Relative Error} = \frac{\left|\varepsilon_r^{\text{retrieved}} - \varepsilon_r^{\text{reference}}\right|}{\varepsilon_r^{\text{reference}}} \tag{4}$$

where $\varepsilon_r^{\text{retrieved}}$ and $\varepsilon_r^{\text{reference}}$ are the retrieved and reference (provided by the manufacturer) values of the dielectric constant, respectively. It is seen that the relative retrieval error in both methods is very small, around 1%.

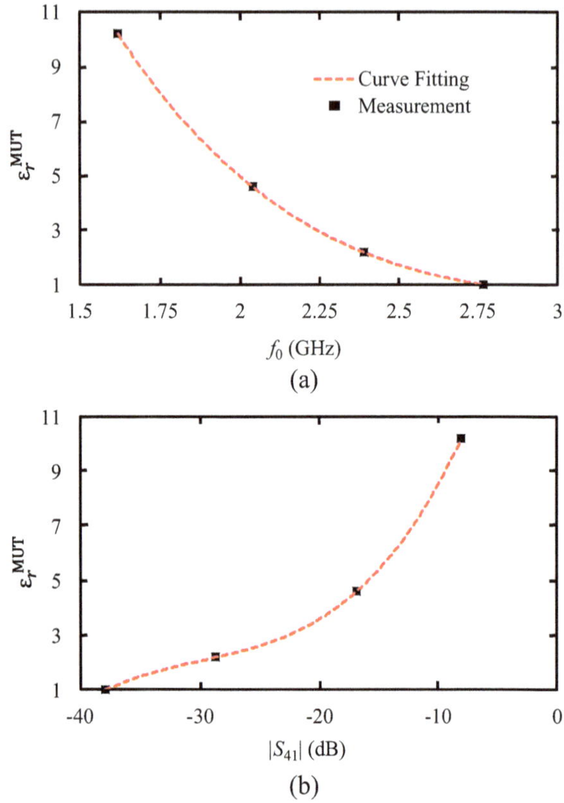

Fig. 11 The dielectric constant of the MUT (ε_r^{MUT}) as a function of the **a** zero-coupling frequency (f_0) and **b** amplitude of isolation coefficient ($|S_{41}|$); Markers: data obtained by measurement. Dash line: fitted curve. Reprinted with permission from [19]; copyright 2022 IEEE

Table 2 Reference and retrieved dielectric constant of RO4003C

MUT	Reference value	Retrieved value using f_0	Retrieved value using $\|S_{41}\|$
RO4003C	3.55	3.51 (1.1 %)[a]	3.50 (1.4 %)[a]

[a]The numbers in the parentheses show the relative errors

It is worth mentioning that we do not rely on simulation data in the retrieval process. But we first calibrate the sensor with some known samples and extract the governing formula (i.e. (2) or (3)). Then, the sensor can be used to retrieve the dielectric constant of unknown MUTs.

As shown in Fig. 12a, a set of walls made of Rogers RO4003C with a height of about 4 mm were added to the structure to form a small pool around the sensing area, so that the sensor can measure the dielectric constant of liquids. Before testing liquids, the measured $|S_{31}|$ and $|S_{41}|$ of the coupler loaded with the Rogers RT/duroid 5880 with and without walls are compared, and the results are shown in the Fig. 12b. It can be seen that the dielectric walls have a negligible effect on the sensor's responses.

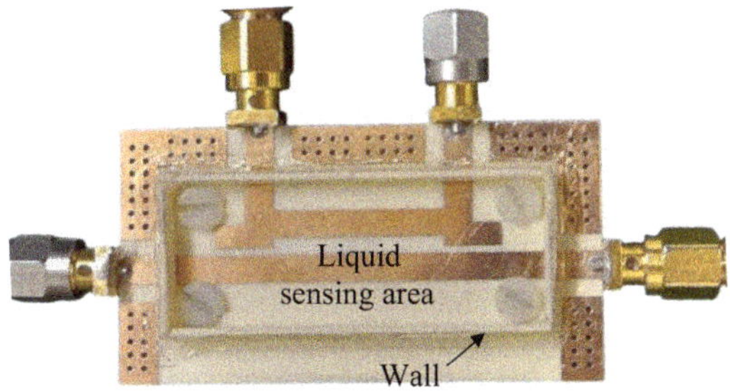

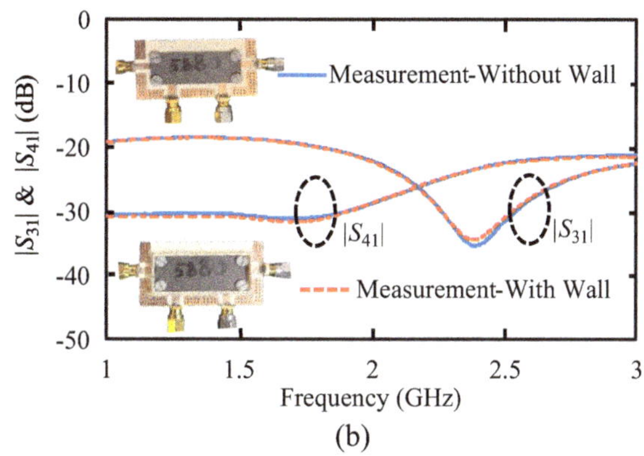

Fig. 12 **a** Fabricated coupler with walls around sensing area for liquid measurement. **b** Negligible effect of the walls on the measured $|S_{31}|$ and $|S_{41}|$ of the coupler. Reprinted with permission from [19]; copyright 2022 IEEE

To demonstrate the applicability of the proposed methods to liquids, diesel fuel and chloroform were used as the MUT. The simulated and measured $|S_{31}|$ and $|S_{41}|$ of the coupler loaded with diesel fuel and chloroform are shown in Fig. 13a and b, respectively. In the simulations, the dielectric constants of diesel fuel and chloroform were set to 2.05 and 4.75, respectively, as reference values, obtained by performing a set of waveguide measurements using a VNA and applying the well-known Nicolson-Ross-Weir (NRW) method [8, 9]. A good match is observed between the measured and simulated results demonstrating that both the proposed methods (which use coupler's zero-coupling frequency and isolation level) are applicable to retrieve the dielectric constants of liquids. However, it is essential to note that the formulas obtained earlier (i.e., (2) and (3)) are valid only for solids and should not be relied upon for liquids since a liquid MUT does not leave a gap between the sensor and itself. In other words, to retrieve the dielectric constant of unknown liquids, a set of new calibrating measurements with some known liquid samples whose reference dielectric constants are at hand are necessary. Then, equations similar to (2) and (3) can be derived for retrieving the dielectric constant of liquids. It should be also noted that due to the direct contact between the liquid samples and the coupler, a cleaning procedure is required after each measurement.

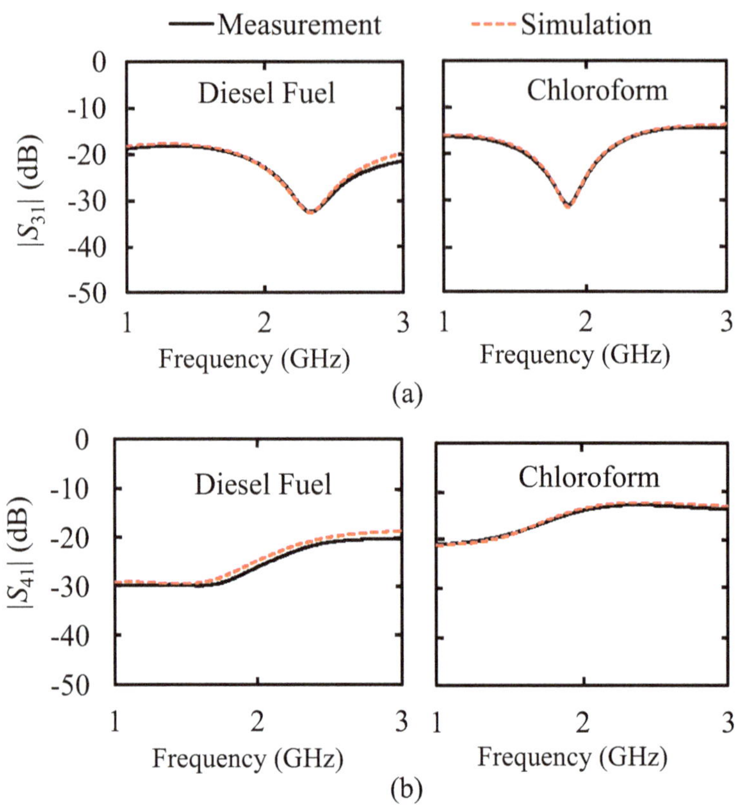

Fig. 13 Simulated and measured results of the proposed sensor when loaded with two liquids (diesel fuel and chloroform): **a** $|S_{31}|$ and **b** $|S_{41}|$. Reprinted with permission from [19]; copyright 2022 IEEE

5 Comparison to Other Methods

A comparison between the characteristics and performance of the methods based on MCLDC and other prior published works is given in Tables 3 and 4. It should be noted that two different methods, both based on a MCLDC, were presented here. The method that uses zero-coupling frequency is a resonant method, while the one using the isolation level is a non-resonant method. So, the former is compared with other resonant methods, whereas the latter is compared with other similar non-resonant methods that use only the amplitude of a transmission coefficient, in a similar way to the method exposed here, which uses only the S_{41} amplitude.

Table 3 provides a comparison between the resonant sensors based on a microstrip resonator. While the other works use the zero-transmission frequency (i.e., the frequency at which the transmission coefficient of a two-port circuit is almost zero), in the method presented here, zero-coupling frequency is used. In such

Table 3 Comparison between the characterization technique based on MCLDC when zero-coupling frequency is used and some other works based on microstrip resonators

Resonator	Response	f_0^{unloaded} (GH)	f_0^{FR4} (GH)	Δf_0 (%)
CSRR [11]	Zero-Trans. Freq.	2.45	2.25	2.3
SRR [12]	Zero-Trans. Freq.	2.2	2	2.5
CSRR[13]	Zero-Trans. Freq.	1.40	1.19	4.2
CSRR[14]	Zero-Trans. Freq.	1.16	0.96	4.8
CLDC	Zero-Coupl. Freq.	2.77	2.05	7.2

resonant methods, since the change in the dielectric constant of the MUT is sensed through the change in the resonance frequency, when loaded with a specific sample, the higher the resonance frequency shift the loaded sensor exhibits with respect to the unloaded one, the higher the sensitivity of the sensor [10]. Therefore, the normalized resonance frequency shift of the sensor when loaded with a specific sample (with respect to the unloaded state) can be a measure of the sensitivity. Since FR4 has been commonly used as MUT in the literature, Table 3, lists the results of different works for this material to perform a comparison between the shift in the resonant frequency when the sensor is loaded with this material with respect to the unloaded sensor. The measured unloaded (f_0^{unloaded}) and loaded (f_0^{FR4}) resonant frequencies are given in the table, and the normalized resonant frequency shift is calculated as,

$$\Delta f_0 = 100 \, \frac{f_0^{\text{unloaded}} - f_0^{\text{FR4}}}{f_0^{\text{unloaded}}(\varepsilon_r^{\text{FR4}} - 1)} \tag{5}$$

It is observed from Table 3 that the normalized resonant frequency shift, equivalently the sensitivity, in the presented method is considerably higher than previously reported works.

The presented method based on the isolation level is a kind of (non-resonant) amplitude-only transmission method since it only requires the amplitude of S_{41}, the transmission coefficient between the input and isolated ports. In Table 4, a qualitative comparison between different transmission media that can be applied in the amplitude-only transmission methods is provided. Among them, microstrip TL, as well as MCLDC used in our work, with a planar structure, have low-cost fabrication advantage. Also, when using these microstrip-based structures, preparing the MUT is quite easy, while in the case of metallic waveguides or coaxial lines, cutting and drilling samples precisely in a specific manner is required.

In amplitude-only transmission methods, since measuring change in dielectric constant of MUT by change in amplitude of a transmission coefficient is done, a higher change in this parameter means higher sensitivity. Therefore, to quantify comparison between sensitivity of different approaches, change in transmission/isolation level when sensor is loaded with two specific MUTs (with dielectric constants of 3 and 4) is considered as measure of sensitivity as

Table 4 Comparison between characterization method based on MCLDC when isolation level is used and other amplitude-only transmission methods

Transm. medium	Response	Structure	Sample preparation	Sensitivity
Microstrip line[a]	Trans. level	Planar	Easy	Low
Coax. line [15–17]	Trans. level	Non-planar	Difficult	Moderate
Waveguide [18]	Trans. level	Non-planar	Difficult	Moderate
MCLDC	Isola. level	Planar	Easy	High

[a]As can be seen in Figs. 8 and 12, the transmission level of a microstrip line is so insensitive to the change in the MUT's dielectric constant that the authors could not find any work based on an amplitude-only transmission method in which such a TL is applied

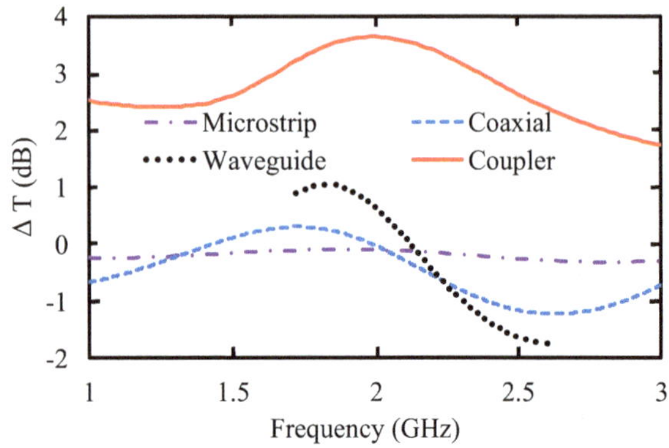

Fig. 14 Change in the simulated transmission/isolation level for different transmission media when $\varepsilon_r^{\mathrm{MUT}}$ changes from 3 to 4. Reprinted with permission from [19]; copyright 2022 IEEE

$$\Delta T(\mathrm{dB}) = S_{ij}(\varepsilon_r^{\mathrm{MUT}} = 4) - S_{ij}(\varepsilon_r^{\mathrm{MUT}} = 3), \tag{6}$$

where S_{ij} is transmission/isolation level in dB scale. By modeling a 50Ω microstrip line on Rogers RO4003C substrate with thickness of 1.6 mm, coaxial line with inner and outer conductor diameters of 2.5 mm and 5.8 mm respectively (having characteristic impedance of 50Ω in unloaded state), and WR430 rectangular waveguide with operating frequency range from 1.7 to 2.6 GHz in full-wave simulator ΔT was obtained for different structures indicated in Table 4 which is shown in Fig. 14. It should be noted that a gap of 15 μm was considered between MUT and substrate when simulating microstrip TL and MCLDC. As can be seen in Fig. 14, ΔT (isolation) with MCLDC is noticeably higher than ΔT (transmission) with microstrip line, coaxial line, or waveguide, indicating a higher sensitivity of the proposed retrieval method based on MCLDC.

6 Conclusions

This chapter explains how a MCLDC can be used to measure the dielectric constant of a MUT, using either the zero-coupling frequency or the isolation level at a specific frequency.

The presented methods depend on the change in the effective permittivity of the microstrip line and the phase velocity in the medium above the microstrip line when different MUTs are placed on the coupled-line section, which influences the coupling and isolation levels of the coupler.

It has been shown that for measuring unambiguously the MUT permittivity, the coupler should have a high directivity in the unloaded state, which is not achievable with a conventional microstrip coupled-line coupler. By attaching two capacitive stubs to the coupled lines, a high-isolation coupler was designed and built. The $|S_{31}|$ and $|S_{41}|$ parameters were measured for the coupler loaded with different solid and liquid samples. The measurement results showed that the presented methods are highly sensitive to the change in the dielectric constant of the MUT, such that by increasing ε_r^{MUT} from 1 to 10.2, f_0 and $|S_{41}|$ change by about 42% and 30 dB, respectively. Comparison with other similar works shows that the presented methods have a higher sensitivity.

References

1. Oliver BM (1954) Directional electromagnetic couplers. Proc IRE 42(11):1686–1692
2. Dydyk M (1990) Accurate design of microstrip directional couplers with capacitive compensation. In: 1990 IEEE MTT-S international microwave symposium (IMS), vol 1, pp 581–584
3. March S (1982) Phase velocity compensation in parallel-coupled microstrip. In: IEEE MTT-S international microwave symposium (IMS), pp 410–412
4. Dydyk M (1999) Microstrip directional couplers with ideal performance via single-element compensation. IEEE Trans Microw Theory Technol 47(6):956–964
5. Phromloungsri R, Chongcheawchamnan M, Robertson I (2006) Inductively compensated parallel coupled microstrip lines and their applications. IEEETrans Microwave Theory Technol 54(9):3571–3582
6. Phromloungsri R, Chamnanphrai V, Chongcheawchamnan M (2006) Design high-directivity parallel-coupled lines using quadrupled inductive-compensated technique. In: Asia-Pacific microwave conference (APMC), pp 1380–1383
7. Lee S, Lee Y (2010) A design method for microstrip directional couplers loaded with shunt inductors for directivity enhancement. IEEE Trans Microwave Theory Technol 58(4):994–1002
8. Nicolson AM, Ross GF (1970) Measurement of the intrinsic properties of materials by time-domain techniques. IEEE Trans Instrum Meas 19(4):377–382
9. Weir WB (1974) Automatic measurement of complex dielectric constant and permeability at microwave frequencies. Proc IEEE 62(1):33–36
10. Saadat-Safa M, Nayyeri V, Ghadimi A, Soleimani M, Ramahi OM (2019) A pixelated microwave near-field sensor for precise characterization of dielectric materials. Sci Rep 9(1):1–12
11. Saadat-Safa M, Nayyeri V, Khanjarian M, Soleimani M, Ramahi OM (2019) A csrr-based sensor for full characterization of magneto-dielectric materials. IEEE Trans Microwave Theory Technol 67(2):806–814

12. Rammah Ali A et al (2017) High-q sensor based on symmetrical split ring resonator with spurlines for solids material detection. IEEE Sens J 17(9):2766–2775
13. Paris V et al (2021) Single-frequency amplitude-modulation sensor for dielectric characterization of solids and microfluidics. IEEE Sens J 21(10):12-189–12-201
14. Boybay MS, Ramahi OM (2012) Material characterization using complementary split-ring resonators. IEEE Trans Instrum Meas 61(11):3039–3046
15. Mosavirik T, Soleimani M, Nayyeri V, Mirjahanmardi SH, Ramahi OM (2021) Permittivity characterization of dispersive materials using power measurements. IEEE Trans Instrum Meas 70:1–8
16. Mosavirik T, Hashemi M, Soleimani M, Nayyeri V, Ramahi OM (2021) Accuracy-improved and low-cost material characterization using power measurement and artificial neural network. IEEE Trans Instrum Meas 70:1–9
17. Mosavirik T, Nayyeri V, Hashemi M, Soleimani M, Ramahi OM (2023) Direct permittivity reconstruction from power measurements using a machine learning aided method. IEEE Trans Microwave Theory Technol, Early access:1–12
18. Hasar U (2009) Elimination of the multiple-solutions ambiguity in permittivity extraction from transmission-only measurements of lossy materials. Microw Opt Technol Lett 51(2):337–341
19. Omam ZR, Nayyeri V, Javid-Hosseini S-H, Ramahi OM (2022) Simple and high-sensitivity dielectric constant measurement using a high-directivity microstrip coupled-line directional coupler. IEEE Trans Microwave Theory Technol 70(8):3933–3942

Reflective-Mode Permittivity Sensors Based on Distributed and Semi-Lumped Coupled Resonators

Pau Casacuberta, Paris Vélez, Jonathan Muñoz-Enano, Lijuan Su, and Ferran Martín

Abstract One-port reflective-mode phase-variation permittivity sensors based on weakly coupled resonators (either distributed or semi-lumped) are presented in this chapter. It is demonstrated that by weakly coupling the resonators, the split resonances that appear as consequence of coupling are very close one to each other, and the phase of the reflection coefficient experiences a significant variation (360° if losses are negligible or small) between such frequencies. The consequence is an enhancement in the sensitivity when the device is tuned in between the two resonance frequencies, as demonstrated in the chapter. It is shown that the sensitivity increases as the inverse square of the coupling coefficient, and unprecedented sensitivities are reported in the various examples included in the chapter, where inter-resonator coupling is very weak. The effects of losses in the material under test (MUT), either small or significant, are also studied in the chapter.

As indicated in Chap. 1, there is a class of planar microwave sensors based on the variation of the phase of the reflection coefficient in one-port structures caused by changes in the dielectric constant of the material under test (MUT) [1–6].[1] In these sensors, the sensing element is either a distributed [1–3] or a semi-lumped [4–6] resonator, and for sensitivity enhancement, the quality factor of the sensing resonator must be high. Moreover, as it was shown in Chap. 1, the sensitivity can be boosted up by cascading impedance inverters, with alternating high and low impedance, between the sensing resonator and the access line. Very high sensitivities in the limit of small perturbations of the dielectric constant of the MUT in the vicinity of the reference (REF) dielectric constant have been achieved with these sensors (e.g., a maximum sensitivity of 528.7° has been demonstrated in one of the sensors reported in [1]).

[1] The phase of the reflection coefficient is also used as output variable in reflective-mode phase-variation sensors devoted to the measurement of linear [7] and angular [8-10] displacements.

P. Casacuberta · P. Vélez · J. Muñoz-Enano · L. Su · F. Martín (✉)
Departament d'Enginyeria Electrònica, CIMITEC, Universitat Autònoma de Barcelona, 08193 Bellaterra, Spain
e-mail: Ferran.Martin@uab.cat

F. Martín and E. Bronchalo (eds.), *Coupled Structures for Microwave Sensing*, Lecture Notes in Electrical Engineering 1150,
https://doi.org/10.1007/978-3-031-53861-2_5

Such technique of sensitivity optimization for small perturbations is very efficient, but at the expense of increasing the overall sensor size, since the high/low impedance inverters needed for sensitivity enhancement should be implemented by means of quarter-wavelength transmission line sections.[2] The stratagem behind this sensitivity optimization approach (based on a cascade of impedance inverters) is to achieve a high phase slope in the phase response of the reflection coefficient at the operating frequency, which should be set to the resonance frequency of the sensing resonator when it is loaded with the REF material.

In the present chapter, an alternative, and very efficient, approach to significantly increasing the phase slope at the operating frequency of the sensor, without the need to cascade quarter-wavelength transmission line sections, is proposed [11–13]. It consists in coupling an identical resonator to the sensing resonator terminating the host line of the sensor. The result is resonance frequency splitting, where the separation between the new generated resonances is dictated by the coupling level between the resonant elements (see Chap. 2 and [14]). Thus, by weakly coupling the resonant elements, the frequency separation is small. Since the phase excursion between the split resonances should be 360° (provided losses are negligible or small), it follows that the phase slope in the frequency region between such resonances should be high. Consequently, a high sensitivity is expected, provided the operating frequency of the sensor is set to a value within the frequency range comprised between the split resonances. The present chapter reports the working principle of these reflective-mode phase-variation sensors based on weakly coupled resonators. A detailed sensitivity analysis that provides analytical expressions for the sensitivity in the limit of small perturbations is carried out for both considered sensor structures, one based on weakly coupled distributed resonators [11], and one based on weakly coupled semi-lumped resonators [12, 13]. Some illustrative examples, representative of both sensor types (distributed and semi-lumped coupled resonators), are reported. Finally, the effects of losses of the material under test (MUT) on the sensitivity are studied.

1 Working Principle

Let us consider a one-port reflective-mode sensor consisting of a transmission line terminated with a planar resonator (the sensing element) and let us assume that the output variable is the phase of the reflection coefficient at a certain (operating) frequency, dependent on the dielectric constant of the MUT (the input variable). The effect of a variation in the dielectric constant of the MUT is an overall shift in the phase response (reflection coefficient). This shift generates a variation in the phase of the reflection coefficient at the operating frequency, and such variation increases as the phase slope increases, as illustrated in Fig. 1. Thus, it follows that for sensitivity optimization in reflective-mode phase-variation sensors, it is necessary to achieve a

[2] Note, however, that the size of the sensing region, dictated by the dimensions of the sensing resonator, is not affected.

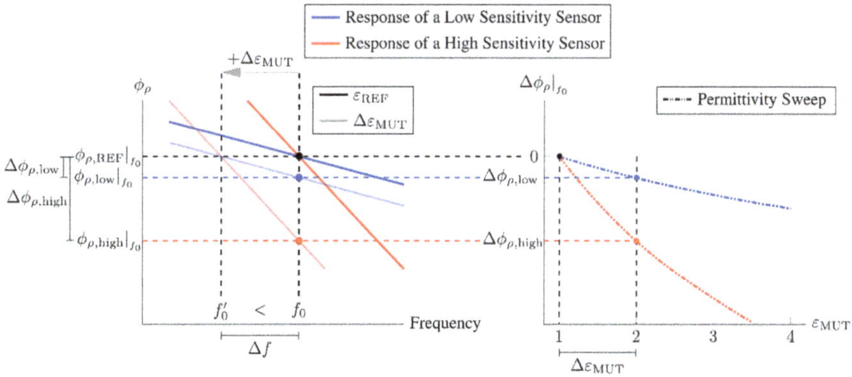

Fig. 1 Sketch showing that the variation of the phase of the reflection coefficient at a certain frequency, f_0, caused by an overall shift in the phase response, increases with the phase slope. When a change in the dielectric constant of the MUT, $\Delta\varepsilon_{MUT}$, occurs, the variation experienced by the phase of the reflection coefficient at f_0 for the high-sensitivity sensor (higher phase slope), $\Delta\phi_{\rho,high}$, is larger than for the low-sensitivity sensor, $\Delta\phi_{\rho,low}$

high phase slope at the operating frequency. Such frequency should be set to the resonance frequency of the sensing resonator when it is loaded with the REF material, since the phase slope is maximum at this frequency, provided the resonator is conveniently designed, i.e., with a high-quality factor (an aspect discussed in Chap. 1). By this means, the sensitivity of the phase of the reflection coefficient with the dielectric constant of the MUT in the vicinity of the REF dielectric constant is optimized. It was also shown in Chap. 1 that the sensitivity can be further optimized by cascading high and low impedance inverters (alternating) to the sensing resonator (either distributed or semi-lumped). The reason is that such cascaded inverters increase the phase slope at resonance, as far as the high and low impedance inverters are properly cascaded to the sensing resonator (see further details in Chap. 1).

An alternative, and efficient, approach for sensitivity optimization consists in using a pair of weakly coupled resonators for sensing. It is well known that a pair of identical coupled resonators exhibits frequency splitting [14]. This means that the fundamental resonance of the uncoupled resonator, f_0, splits into two frequencies, one located above, f_+, and the other one located below, f_-, the resonance frequency of the uncoupled resonator, f_0. The frequency difference between these split frequencies, $f_+ - f_-$, depends on the coupling level between the resonators. If the coupling is weak, such frequencies are closely spaced. By considering that the pair of coupled resonators is the one-port structure under study, the excursion experienced by the phase of the reflection coefficient between these two frequencies is 360° (provided losses are negligible or small). Consequently, the phase slope between f_- and f_+ can be made arbitrarily large by weakly coupling the sensing resonators. This aspect is illustrated in Fig. 2, where the phase of the reflection coefficient of an isolated grounded series resonator and the phase of the reflection coefficient of a grounded series resonator electrically coupled to an identical resonator are depicted.[3] Such phase slope enhancement technique, based on weakly coupled resonators, is by far more efficient than simply increasing the quality factor of the (single) resonator. The reason is that the quality factor necessary to achieve a certain phase slope at resonance might require very extreme reactive element values (i.e., very high inductance and very small capacitance for a series resonator). It is apparent that with a high phase slope at the operating frequency of the sensor, the sensitivity of the phase of the reflection coefficient to variations of the dielectric constant in the medium surrounding the resonator, or pair of coupled resonators, can be made large (an aspect that will be demonstrated later).[4]

For sensor implementation, the pair of weakly coupled sensing resonators can be either distributed (e.g., quarter-wavelength resonators [11]) or semi-lumped, i.e., electrically small planar resonators (e.g., step impedance resonators—SIRs [12, 13]). Let us next analyze in detail both types of sensors. In Sect. 2, the focus is on highly sensitive reflective-mode phase-variation sensors based on coupled quarter-wavelength resonators implemented in microstrip technology. Section 3 analyzes a sensor based on a pair of coupled SIRs implemented in coplanar waveguide (CPW) technology. In both cases, a detailed sensitivity analysis that provides the sensitivity in the limit of small perturbations is carried out, and prototype examples are reported.

2 Reflective-Mode Phase-Variation Sensors Based on Pairs of Coupled Quarter-Wavelength Resonators

A quarter wavelength transmission line resonator can be terminated by either an open circuit or a short circuit [15]. In the first case, the impedance seen from the input port at resonance is null, and the resonator resembles a grounded series resonator. By contrast, if the quarter-wavelength transmission line is short-circuited to ground, the behavior is like that of a parallel resonant tank (exhibiting an open circuit at resonance). According to these two types of isolated quarter-wavelength resonators, there are two possible configurations of coupled quarter-wavelength resonators [11].

---

[3] Note that, in Fig. 2, the frequency of the isolated resonator $f_{0,\text{is}}$ is different than the frequency of the uncoupled resonator, f_0. The former is the frequency of the resonator when it is isolated (and hence C_{is} is the capacitance of the isolated resonator). The frequency of the uncoupled resonator, f_0, is the one given by the resonator's self-inductance, L, and self-capacitance, C, see Fig. 2, where C is the capacitance of the resonator when the other resonator is present but short-circuited to ground. C and C_{is} are very similar, but not identical, unless the separation between the resonators is very high (a case that lacks interest). For that reason, strictly speaking, $f_0 \neq f_{0,\text{is}}$, although these frequency are very close one to each other in any practical situation.

[4] It should be mentioned that if losses are significant, the excursion experienced by the phase of the reflection coefficient between the two resonance frequencies might be smaller than 360°. However, it does not mean that losses degrade the sensitivity. On the contrary, losses may aid sensitivity, if the sensor is adequately designed, an aspect that will be discussed in Sect. 4.

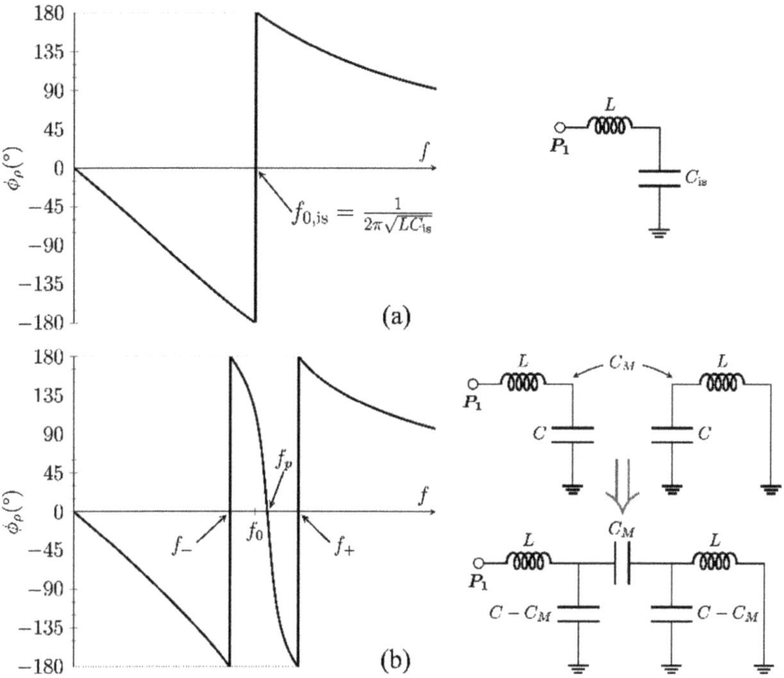

Fig. 2 Sketch showing the phase response of an isolated series resonator (**a**) and a pair of weakly coupled resonators (**b**). The phase variation between f_- and f_+ is 360°. C_{is} is the capacitance of the isolated resonator, C is the self-capacitance of the uncoupled resonator (i.e, with the presence of the other resonator short-circuited to ground), C_M is the mutual capacitance, and L is the inductance of the resonator

Such configurations are depicted in Fig. 3 and are designated as sensor-type A and sensor-type B. Let us next calculate the resonance conditions for both configurations.

2.1 Resonance Frequencies

For sensor-type A, the impedance seen from the input port can be expressed in terms of the elements of the order-four impedance matrix of the pair of coupled lines, **Z**, as

$$Z_{in,A} = Z_{11} - \frac{Z_{13}^2}{Z_{11}} \tag{1}$$

whereas for sensor-type B, the input impedance is given by

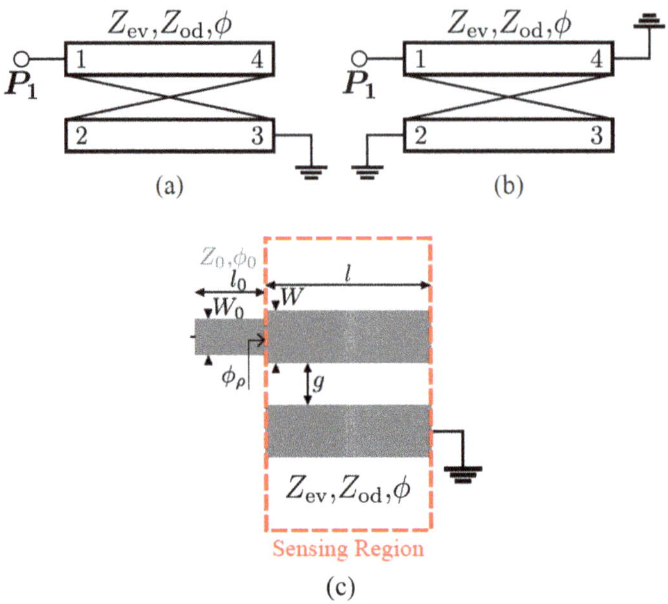

Fig. 3 Schematic of the reflective-mode phase-variation sensor based on a pair of coupled quarter-wavelength resonators. **a** Sensor-type A, with an open-ended resonator connected to the input port; **b** sensor-type B, with a short-circuited resonator connected to the input port; **c** typical layout (for sensor A) and relevant dimensions. The width W_0 and length l_0 corresponds to the access line, with characteristic impedance equal to the reference impedance of the ports Z_0

$$Z_{in,B} = \frac{Z_{11}(Z_{11}^2 - Z_{12}^2 - Z_{13}^2 - Z_{14}^2) + 2Z_{12}Z_{13}Z_{14}}{Z_{11}^2 - Z_{13}^2} \tag{2}$$

where the matrix elements, by excluding losses, are [15, 16]

$$Z_{11} = Z_{22} = Z_{33} = Z_{44} = -\frac{j}{2}\{Z_{ev}\cot\phi_{ev} + Z_{od}\cot\phi_{od}\} \tag{3a}$$

$$Z_{12} = Z_{21} = Z_{34} = Z_{43} = -\frac{j}{2}\{Z_{ev}\cot\phi_{ev} - Z_{od}\cot\phi_{od}\} \tag{3b}$$

$$Z_{13} = Z_{31} = Z_{24} = Z_{42} = -\frac{j}{2}\left\{Z_{ev}\frac{1}{\sin\phi_{ev}} - Z_{od}\frac{1}{\sin\phi_{od}}\right\} \tag{3c}$$

$$Z_{14} = Z_{41} = Z_{23} = Z_{32} = -\frac{j}{2}\left\{Z_{ev}\frac{1}{\sin\phi_{ev}} + Z_{od}\frac{1}{\sin\phi_{od}}\right\} \tag{3d}$$

In (3a, 3b, 3c, 3d), Z_{ev} and Z_{od} are the even- and odd-mode characteristic impedances, respectively, of the coupled lines, whereas ϕ_{ev} and ϕ_{od} are the even- and odd-mode electrical lengths.

Introduction of (3a, 3b, 3c, 3d) in (1) provides the lossless input impedance for sensor-type A[5], i.e.,

$$Z_{in,A} = -\frac{j}{2}\frac{2Z_{ev}Z_{od}(\cos\phi_{ev}\cos\phi_{od}+1)-(Z_{ev}^2+Z_{od}^2)\sin\phi_{ev}\sin\phi_{od}}{Z_{ev}\cos\phi_{ev}\sin\phi_{od}+Z_{od}\cos\phi_{od}\sin\phi_{ev}} \tag{4}$$

Since for sensitivity optimization, weak coupling between the resonant elements is necessary, it is reasonable to assume that the phase constants for the even and odd modes are similar. Under this weak coupling approximation, $\phi_{ev} \approx \phi_{od} \equiv \phi$, and expression (4) simplifies to

$$Z_{in,A} = -j\frac{2Z_{ev}Z_{od}(\cos^2\phi+1)-(Z_{ev}^2+Z_{od}^2)\sin^2\phi}{2(Z_{ev}+Z_{od})\cos\phi\sin\phi} \equiv j\chi_{in,A} \tag{5}$$

where $\chi_{in,A}$ is the input reactance. Expression (5) is also valid under moderate coupling provided the REF material exhibits a dielectric constant identical to that of the substrate, since in that case the even and odd modes exhibit the same phase velocity. The zeros of (5) provide the resonance frequencies, given by the solutions of

$$\cos\phi = \pm\frac{Z_{ev}-Z_{od}}{Z_{ev}+Z_{od}} \equiv \pm C \tag{6}$$

where C is the coupling factor.[6] Note that in the limit of negligible coupling, $Z_{ev} = Z_{od}$ (or $C = 0$), it follows that $\phi = 90°$, corresponding to the resonance condition of a quarter-wavelength resonator. According to (6), the proximity of the zeros, or resonance frequencies, is controlled by the coupling factor, and it is apparent that if the coupling factor is small, the resonance frequencies are closely spaced. Namely, the two zeros, f_+ and f_- (present at both sides of the resonance frequency of the isolated quarter-wavelength resonator), can be as closely spaced as desired, by simply weakly coupling the lines. It is interesting to mention that for the frequency satisfying $\phi = 90°$, the denominator of (5) vanishes, whereas the numerator is finite (provided $Z_{ev} \neq Z_{od}$). Consequently, the input impedance, $Z_{in,A}$, exhibits a pole at that frequency, f_p.

According to the previous paragraph, a significant variation of the phase of the reflection coefficient is achievable in the vicinity of the pole frequency by weakly coupling the lines. The reason is that the reflection coefficient is $\rho = -1$ and ρ

[5] The input impedance by including losses will be introduced in Sect. 4, where the effects of losses are discussed in detail.

[6] Note that the coupling factor of a pair of coupled lines has been named C, the most usual designation. It should not be confused with the self-capacitance of a pair of coupled resonators (also designated as C), introduced in Fig. 2 and used in Sect. 3. Nevertheless, the use of C either as coupling factor or as self-capacitance occurs in different contexts, and, for that reason, the authors prefer to keep the usual designations of such terms.

= +1 at the zero and pole frequencies, respectively. Thus, the phase of the reflection coefficient should experience an excursion of 360° between two closely spaced frequencies (f_+ and f_-), provided the phase of the reflection coefficient is $\phi_\rho = \pm 180°$ at f_+ and f_- (or ϕ_+ and ϕ_-), and it is $\phi_\rho = 0°$ at f_p (or $\phi = 90°$). Note that the phase of the reflection coefficient is a decreasing function of frequency.[7]

For sensor B, the lossless input impedance [that results by introducing (3a, 3b, 3c, 3d) in (2)] is found to be

$$Z_{in,B} = j\frac{2Z_{ev}Z_{od}(Z_{ev}\cot\phi_{od} + Z_{od}\cot\phi_{ev})\sin\phi_{ev}\sin\phi_{od}}{2Z_{ev}Z_{od}(\cos\phi_{ev}\cos\phi_{od} + 1) - (Z_{ev}^2 + Z_{od}^2)\sin\phi_{ev}\sin\phi_{od}} \tag{7}$$

and for very weakly coupled lines, or for a REF material identical to the one of the substrate, with $\phi_{ev} \approx \phi_{od} \equiv \phi$, expression (7) simplifies to

$$Z_{in,B} = j\frac{2Z_{ev}Z_{od}(Z_{ev} + Z_{od})\cos\phi\sin\phi}{2Z_{ev}Z_{od}(\cos^2\phi + 1) - (Z_{ev}^2 + Z_{od}^2)\sin^2\phi} \equiv j\chi_{in,B} \tag{8}$$

Note that the numerator of (8) nulls for $\phi = 90°$, whereas the denominator is finite for such phase condition. This means that there is a zero at that frequency, f_z. At both sides of such frequency, two poles arise as consequence of coupling. Such poles are given by the two frequencies, f_+ and f_-, that null the denominator of (8), i.e., the resonance frequencies corresponding to the phases ϕ_+ and ϕ_- that are the solutions of equation (6). The difference between sensor-type B and sensor-type A is that, at the frequencies f_+ and f_-, an open-circuit, rather than a short-circuit, is seen from the input port of the coupled lines for sensor B. For sensor B, the phase of the reflection coefficient is $\phi_\rho = 0°$ at f_+ and f_- (or ϕ_+ and ϕ_-), and it is $\phi_\rho = \pm 180°$ at f_z (or $\phi = 90°$). Thus, an identical excursion of the unwrapped phase (360°) is expected for frequencies varying between f_+ and f_-.

2.2 Sensitivity Analysis

Under the weak coupling approximation, or for a REF material with dielectric constant identical to the one of the substrate, the phases satisfy $\phi_{ev} \approx \phi_{od} \equiv \phi$, and the sensitivity, or derivative of the phase of the reflection coefficient, ϕ_ρ, with the dielectric constant of the MUT, ε_{MUT}, can be written as

$$S = \frac{d\phi_\rho}{d\varepsilon_{\text{MUT}}} = \frac{d\phi_\rho}{d\phi}\frac{d\phi}{d\varepsilon_{\text{MUT}}} + \frac{d\phi_\rho}{dZ_{ev}}\frac{dZ_{ev}}{d\varepsilon_{\text{MUT}}} + \frac{d\phi_\rho}{dZ_{od}}\frac{dZ_{od}}{d\varepsilon_{\text{MUT}}} \tag{9}$$

an expression valid for both sensor types, A and B. The reflection coefficient is given by

[7] This is true provided losses are negligible or small, the assumption made in the present analysis.

$$\rho = \frac{Z_{in} - Z_0}{Z_{in} + Z_0} = \frac{j\chi_{in} - Z_0}{j\chi_{in} + Z_0} \tag{10}$$

and the phase of the reflection coefficient is

$$\phi_\rho = 2\arctan\left(-\frac{\chi_{in}}{Z_0}\right) \tag{11}$$

where $\chi_{in} = \chi_{in,A}$ for sensor A [see (5)], $\chi_{in} = \chi_{in,B}$ for sensor B [see (8)], and Z_0 is the reference impedance of the ports (typically, $Z_0 = 50\ \Omega$).

Let us now calculate the sensitivity for $\phi = 90°$, corresponding to f_p for sensor A and to f_z for sensor B, i.e., the intermediate frequency in the range between f_+ and f_- for both sensor types. Indeed, the last two terms of the right-hand side member of (9) evaluated at $\phi = 90°$ are null for both sensors, since $d\phi_\rho/dZ_{ev} = d\phi_\rho/dZ_{od} = 0$ for such phase condition. Concerning the first term, the key derivative is the first one. For arbitrary values of ϕ, such derivative is given by

$$\frac{d\phi_\rho}{d\phi} = -4Z_0\frac{(Z_{ev} + Z_{od})^3\cos^2\phi + (Z_{ev} + Z_{od})(Z_{ev} - Z_{od})^2(\sin^2\phi - \cos^2\phi)}{4Z_0^2(Z_{ev} + Z_{od})^2\sin^2\phi\cos^2\phi + \left[(Z_{ev} + Z_{od})^2\cos^2\phi - (Z_{ev} - Z_{od})^2\right]^2} \tag{12}$$

for sensor A, and by

$$\frac{d\phi_\rho}{d\phi} = -4Z_0Z_{ev}Z_{od}\frac{(Z_{ev} + Z_{od})^3\cos^2\phi - (Z_{ev} + Z_{od})(Z_{ev} - Z_{od})^2(\sin^2\phi - \cos^2\phi)}{4Z_{ev}^2Z_{od}^2(Z_{ev} + Z_{od})^2\sin^2\phi\cos^2\phi + Z_0^2\left[(Z_{ev} + Z_{od})^2\cos^2\phi - (Z_{ev} - Z_{od})^2\right]^2} \tag{13}$$

for sensor B. Evaluation of (12) and (13) for $\phi = 90°$ gives

$$\left.\frac{d\phi_\rho}{d\phi}\right|_{A,\phi=90°} = -4Z_0\frac{Z_{ev} + Z_{od}}{(Z_{ev} - Z_{od})^2} = \frac{-4Z_0}{C^2(Z_{ev} + Z_{od})} \tag{14}$$

for sensor A, and

$$\left.\frac{d\phi_\rho}{d\phi}\right|_{B,\phi=90°} = -4\frac{Z_{ev}Z_{od}}{Z_0}\frac{Z_{ev} + Z_{od}}{(Z_{ev} - Z_{od})^2} = \frac{-4}{Z_0C^2(Y_{ev} + Y_{od})} \tag{15}$$

for sensor B, with $Y_{ev} = 1/Z_{ev}$ and $Y_{od} = 1/Z_{od}$.

It is apparent from (14) and (15) that, for sensitivity optimization in the vicinity of the dielectric constant of the REF material (where $\phi = 90°$), the coupling factor C must but as low as possible. Note that the dependence of (14) and (15) with such term is quadratic, which means that it is possible to achieve very high sensitivity by weakly coupling the quarter-wavelength resonators. On the other hand, low even-

and odd-mode impedances for sensor A, and high even- and odd-mode impedances for sensor B, contribute to enhance the terms (14) and (15) of the sensitivity.

Note that, according to (9), the whole sensitivity depends also on the derivative of the electrical length with the dielectric constant of the MUT, $d\phi/d\varepsilon_{\text{MUT}}$. Within the weak coupling approximation, it is reasonable to consider that the dependence of ϕ with ε_{MUT} is that of an isolated line with identical width, i.e.,

$$\phi = \frac{\omega_0 l}{c}\sqrt{\varepsilon_{eff}} \tag{16}$$

where c is the speed of light in vacuum, l is the length of the line, and the effective dielectric constant is [15][8]

$$\varepsilon_{eff} = \frac{\varepsilon_r + \varepsilon_{\text{MUT}}}{2} + \frac{\varepsilon_r - \varepsilon_{\text{MUT}}}{2}F \tag{17}$$

where ε_r is the dielectric constant of the substrate and F is a geometry factor given by

$$F = \left(1 + 12\frac{h}{W_s}\right)^{-\frac{1}{2}} \tag{18a}$$

for $W_s/h \geq 1$, or by

$$F = \left(1 + 12\frac{h}{W_s}\right)^{-\frac{1}{2}} + 0.04\left(1 - \frac{W_s}{h}\right)^2 \tag{18b}$$

for $W_s/h < 1$. In (18a, 18b), h and W_s are the substrate thickness and the width of the coupled lines, respectively, and it is assumed that $t << h$, where t is the thickness of the metallic layer. It should be mentioned that the validity of (17) is subjected to the fact that the MUT is thick enough, so that the electromagnetic field generated by the coupled lines does not reach the MUT/air interface (i.e., this means that the MUT can be considered to be semi-infinite in the vertical direction).[9] Using (16) and (17),

[8] Such effective dielectric constant corresponds to microstrip technology. For coplanar waveguide technology, the effective dielectric constant is given by (17) with $F = 0$.

[9] If the thickness of the MUT is not enough to consider it semi-infinite in the vertical direction, the effective dielectric constant, ε_{eff}, as a function of the dielectric constant of the MUT, ε_{MUT}, should be calculated numerically or by means of specific software tools. In that case, the term $d\phi/d\varepsilon_{\text{MUT}}$ is not analytical, and, consequently, the overall sensitivity must be inferred numerically. Nevertheless, the main conclusions of the analysis (i.e., the significant role of the coupling coefficient and even- and odd-mode impedances) are valid even for narrow MUT samples. For CPW technology, the validity of (17), with $F = 0$, requires that both the substrate and the MUT are semi-infinite in the vertical direction. Nevertheless, for narrow substrates, but thick enough MUTs, expression (17), with $F = 0$, is valid provided ε_r is replaced with $\varepsilon_{r,\,eq}$, the so-called equivalent dielectric constant of the substrate, defined in [17] as the dielectric constant of an hypothetical semi-infinite substrate providing the same contribution to the effective dielectric constant (see [17,18] for further details).

the derivative of the phase of the coupled lines with the dielectric constant of the MUT is found to be

$$\frac{d\phi}{d\varepsilon_{\text{MUT}}} = \frac{\omega_0 l(1-F)}{4c\sqrt{\varepsilon_{eff}}} = \frac{\phi(1-F)}{4\varepsilon_{eff}} \tag{19}$$

Thus, the overall sensitivity, as given by (9), for ε_{MUT} corresponding to the dielectric constant of the REF material, ε_{REF}, and the sensor designed to exhibit a phase of $\phi = 90°$ when the line pair is covered with such material, is

$$S = \left.\frac{d\phi_\rho}{d\varepsilon_{\text{MUT}}}\right|_{A,\varepsilon_{\text{MUT}}=\varepsilon_{\text{REF}}} = \frac{-Z_0\pi(1-F)}{2C^2(Z_{ev}+Z_{od})\varepsilon_{eff,\text{REF}}} \tag{20}$$

for sensor A, and

$$S = \left.\frac{d\phi_\rho}{d\varepsilon_{\text{MUT}}}\right|_{B,\varepsilon_{\text{MUT}}=\varepsilon_{\text{REF}}} = \frac{\pi(1-F)}{2Z_0C^2(Y_{ev}+Y_{od})\varepsilon_{eff,\text{REF}}} \tag{21}$$

for sensor B. To infer (20) and (21), expressions (14) and (15) have been used, and $\varepsilon_{eff,\text{REF}}$ is the effective dielectric constant, given by (17), when the line pair is covered with the REF material (i.e., $\varepsilon_{\text{MUT}} = \varepsilon_{\text{REF}}$). Expressions (20) and (21) can be used for sensor design, as far as such equations provide a link between the sensitivity for the REF material, the main relevant parameter, and the electrical and geometrical parameters of the coupled lines. Note that if the dielectric constant of the REF material is identical to the one of the substrate, the weak coupling approximating is not necessary, and expressions (20) and (21) are valid by replacing $\varepsilon_{eff,\text{REF}}$ with ε_r.

2.3 *Prototype Examples*

As indicated, the weak coupling approximation is also valid when the REF material exhibits the same dielectric constant than the substrate ($\varepsilon_{\text{REF}} = \varepsilon_r$), regardless of the coupling level, since, under these circumstances, $\phi_{ev} \approx \phi_{od} \equiv \phi$ in the limit of small perturbations. The purpose of this section is to validate the previous sensitivity analysis, by considering two different cases, reported in [11]. In one case, the prototype sensor is of type A, the coupling is moderate, and $\varepsilon_{\text{REF}} = \varepsilon_r$. In the second prototype example, a sensor-type B with weak coupling and $\varepsilon_{\text{REF}} = 1$ (corresponding to air) is reported. In both cases, the operating frequency is set to $f_0 = 2$ GHz, and the considered substrate is *Rogers RO4003C* with dielectric constant $\varepsilon_r = 3.55$, thickness $h = 1.524$ mm, and loss tangent $\tan\delta = 0.0022$ (the sensors are implemented in microstrip technology).

2.3.1 Sensor-Type A with Moderate Coupling and $\varepsilon_{\mathrm{REF}} = \varepsilon_r = 3.55$

In this subsection, a sensor of type A with moderate coupling and $\varepsilon_{\mathrm{REF}} = \varepsilon_r = 3.55$ is reported (i.e., the dielectric constant of the REF material, semi-infinite, covering the coupled lines is $\varepsilon_{\mathrm{REF}} = \varepsilon_r = 3.55$). The transverse geometry of the coupled lines is the one providing an even- and odd-mode impedances of $Z_{ev} = 42.7\ \Omega$ and $Z_{od} = 22.8\ \Omega$, respectively. With such even and odd mode impedances, the coupling coefficient is found to be $C = 0.30$, corresponding to a moderate (rather than weak) coupling. Nevertheless, since the dielectric constant of the REF material is identical to the one of the substrate, it can be assumed that for small variations of the dielectric constant of the MUT in the vicinity of $\varepsilon_{\mathrm{REF}}$, the phases satisfy $\phi_{ev} \approx \phi_{od} \equiv \phi$, and hence the analysis of the previous section can be considered to be valid. The photograph of the fabricated sensor is depicted in Fig. 4a. Grounding the required termination of the coupled lines was carried out by means of vias, and a 50-Ω access line was cascaded to the input port of the pair of coupled lines for connector soldering.

The first validation of the sensor was done at simulation level by obtaining the phase of the reflection coefficient at f_0 as a function of the dielectric constant of the MUT (the *CST Microwave Studio* commercial software was used for that purpose). The considered thickness of the MUT was set to 18.3 mm, corresponding to a stacking of 12 samples of 1.524 mm, available in our laboratory for several substrates of

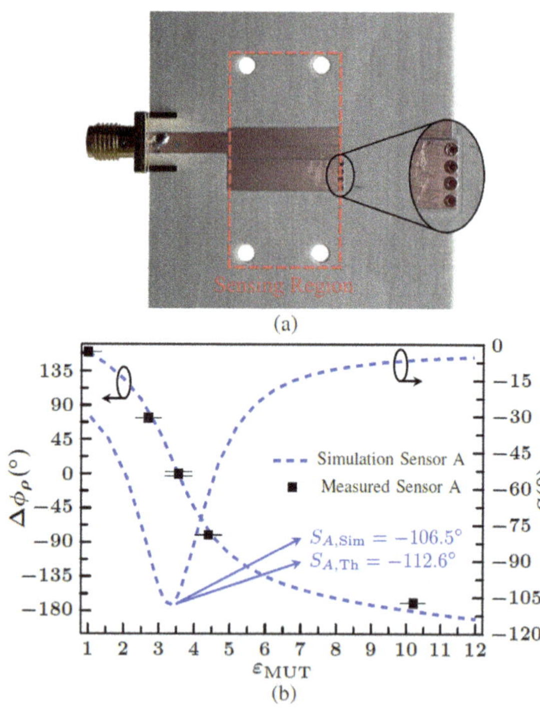

Fig. 4 Photograph of the fabricated sensor A (**a**) and phase of the reflection coefficient at f_0 by varying the dielectric constant of the MUT and sensitivity (**b**). The sensing region is indicated with a dashed rectangle. Dimensions (in mm) in reference to Fig. 3c are: $W = 5$, $g = 0.2$, $l = 18.32$. The dimensions of the 50-Ω access line (in mm) are: $W_0 = 3.33$ and $l_0 = 12.8$. Reprinted with permission from [11]; copyright 2023 IEEE

different dielectric constants. With such thickness, the semi-infinite MUT approximation can be adopted. The considered input dynamic range is the interval between $\varepsilon_{MUT} = 1$ and $\varepsilon_{MUT} = 12$, and the dielectric constant step between adjacent simulations was set to 0.5, except between $\varepsilon_{MUT} = 3$ and $\varepsilon_{MUT} = 4$, which was set to 0.05. With this increase in the density of data points in the vicinity of ε_{REF}, the sensitivity can be calculated more accurately. The phase variation, $\Delta\phi_\rho$, with ε_{MUT}, and the sensitivity are depicted in Fig. 4b, where $\Delta\phi_\rho = \phi_\rho - \phi_{\rho,\varepsilon REF}$ ($\phi_{\rho,\varepsilon REF}$ is the phase of the reflection coefficient when the MUT is the REF material). The sensitivity for $\varepsilon_{MUT} = \varepsilon_{REF}$, inferred from the simulated data points, $S_{A,Sim}$, is indicated in the figure. The value is in reasonable good agreement with the predictions of expression (20), indicated in the figure as $S_{A,Th}$.

Experimental validation was carried out by considering samples of different dielectric constant, namely, uncladded commercially available substrates and fused deposition modelled (FDM) 3D-printed PLA (polylactic acid). The specific dielectric constants are $\varepsilon_{MUT} = 2.7$ (for PLA), $\varepsilon_{MUT} = 3.55$ (for *Rogers RO4003C*), $\varepsilon_{MUT} = 4.4$ (for *FR4*), and $\varepsilon_{MUT} = 10.2$ (for Rogers *RO3010*). As indicated before, 12 pieces of roughly 1.5 mm were stacked in each case, or directly printed (case of PLA). The area of the samples is identical to the one indicated by the dashed rectangle in Fig. 4a. The phase of the reflection coefficient at f_0 was measured for each sample. To avoid the well-known effects of the air gap, or air layer potentially present between the MUT and the sensor substrate, the MUTs were pressured against the sensor by means of Nylon screws. The measured data points are also included in Fig. 4b and show good agreement with the simulated values. It should be mentioned that, in order to ensure repetitiveness of the results, three independent measurements were performed for each sample (thus, the mean value plus the error bars are included in Fig. 4).

2.3.2 Sensor-Type B with Weak Coupling and $\varepsilon_{REF} = 1$

The second prototype is a type B sensor with weak coupling, and the REF material is air ($\varepsilon_{REF} = 1$). The geometry of the device (see Fig. 5a), provides even- and odd-mode characteristic impedances of $Z_{ev} = 111.6\ \Omega$ and $Z_{od} = 76.4\ \Omega$, respectively, corresponding to a coupling coefficient of $C = 0.19$. The simulated phase of the reflection coefficient, $\Delta\phi_\rho$, at f_0 as a function of the dielectric constant of the MUT is shown in Fig. 5b. The thickness of the MUT was also set to 18.3 mm, and the input dynamic range is identical to the one of the previous sensor. The sensitivity inferred from the simulated phase points is also included in Fig. 5b. The value of the sensitivity for $\varepsilon_{MUT} = \varepsilon_{REF} = 1$ was found to be $-736.0°$, i.e., in reasonably good agreement with the prediction of the theory, despite the fact that the REF material is air, with dielectric constant ($\varepsilon_{REF} = 1$) significantly different to the one of the substrate ($\varepsilon_r = 3.55$).

Due to the lack of available MUT materials exhibiting a dielectric constant close to the one of air, validation of the capacity of the sensor to discriminate very small variations in the dielectric constant of the MUT in the vicinity of $\varepsilon_{REF} = 1$ was carried out in [11] by a different approach. In particular, it was considered a variable

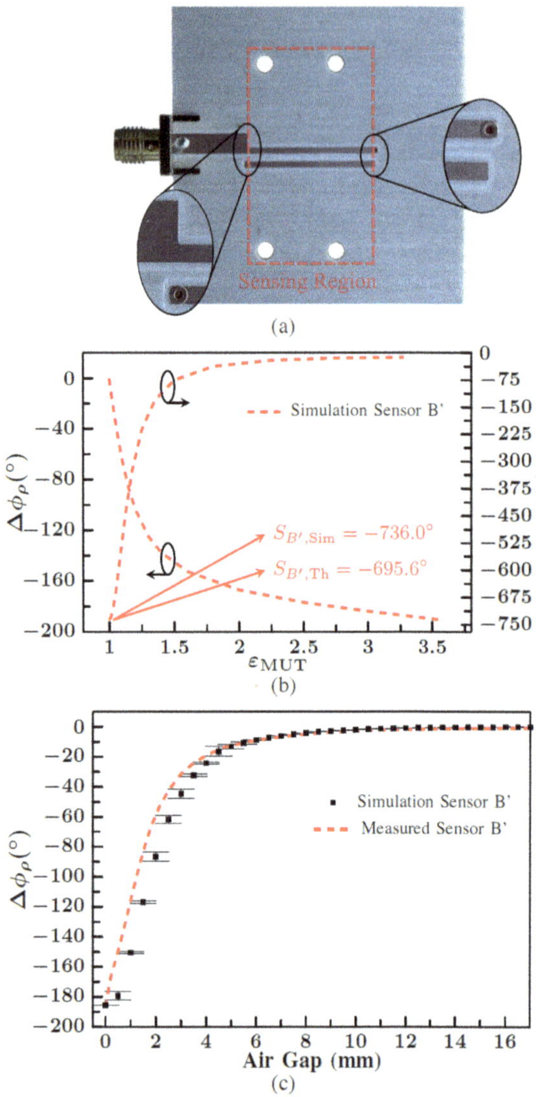

Fig. 5 Photograph of the fabricated sensor B with $\varepsilon_{REF} = 1$ (**a**), phase of the reflection coefficient at f_0 by varying the dielectric constant of the MUT and sensitivity (**b**), and phase of the reflection coefficient at f_0 as a function of the distance between the sensor and the movable sample (**c**). The sensing region is indicated with a dashed rectangle. Dimensions (in mm) in reference to Fig. 3c are: $W = 1$, $g = 1.25$, $l = 21.84$. The dimensions of the 50-Ω access line (in mm) are: $W_0 = 3.33$ and $l_0 = 12.8$. Reprinted with permission from [11]; copyright 2023 IEEE

vertical distance between the sensor and a stack of uncladded *Rogers RO4003C* substrates with a total thickness 18.3 mm and dielectric constant 3.55, identical to the dielectric constant of the substrate. When the sample is sufficiently separated from the sensor, it is equivalent to consider a bare sensor (i.e., with $\varepsilon_{MUT} = \varepsilon_{REF} = 1$). However, as the sample progressively approaches the sensor, the equivalent dielectric constant of the MUT progressively increases, and it takes the maximum value when the dielectric sample is in contact with the sensor. The distance between the sensor and the dielectric (movable) slab was achieved by means of the *Thorlabs*

MVS010/M—25 MM Travel Vertical Translation Stage. The simulated and measured phases of the reflection coefficient as a function of the distance between the sensor and the movable sample are depicted in Fig. 5c, and the agreement is very good. The quasi-saturation effect for high distances can be perfectly appreciated. For low separations between the sensor and the movable sample, the variation of the phase of the reflection coefficient with such separation is very significant, a consequence of the high sensitivity of such sensor.

An additional prototype sensor of type B with weak coupling and $\varepsilon_{REF} = \varepsilon_r = 3.55$ was reported in [11]. The agreement between the sensitivity in the limit of small perturbations inferred from simulations ($S_{sim} = -659.6°$) and the theoretical prediction ($S_{Th} = -688.3°$) is also good for that sensor.

The results presented in this section validate the sensitivity analysis of Sect. 3.2 and point out that extremely high sensitivities can be achieved by weakly coupling the quarter-wavelength sensing resonators. Nevertheless, it should be mentioned that sensitivity optimization in limit of small perturbations is done at the expense of sacrificing linearity, as it can be appreciated in Figs. 4b, 5b. Thus, these sensors are of special interest in applications where the input dynamic range is restricted to a small span, and to detect small variations of the dielectric properties of the MUT as compared to the REF sample. For example, these sensors are useful as defect detectors, based on the dielectric constant changes experienced by the samples under study, caused by such defects.

The main figure of merit (FoM) in phase-variation sensors is the ratio between the maximum sensitivity and the area of the sensing region expressed in terms of the square wavelength. For the sensor of Fig. 5, the figure of merit is FoM $= 12690°/\lambda^2$, a very competitive value. Such FoM is inferior to the one of the sensor based on a quarter-wavelength sensing resonator cascaded to a pair of high/low impedance inverters, and reported in [1], with FoM $= 21148°/\lambda^2$ and a maximum sensitivity of $-528.7°$. Such very high FoM is consequence of the fact that the sensing region in the sensor of [1] is restricted to a small region that contains the (single) quarter-wavelength sensing resonator (rather than a pair of weakly coupled resonators). Nevertheless, the overall sensor size in the device reported in [1] is $0.131\ \lambda^2$, whereas the total size of the sensor of Fig. 5 coincides with the size of the sensing region ($0.058\ \lambda^2$), provided no additional microwave circuitry is required in the sensor of Fig. 5 to boost up the sensitivity. Thus, a relevant advantage of the reflective-mode phase-variation sensors subject of this section, and based on coupled lines, is the overall small sensor size combined with a very high (maximum) sensitivity, easily controllable with the coupling factor. In the next section, a similar approach, where the weakly coupled sensing resonators are semi-lumped (rather than distributed) elements, is studied in detail.

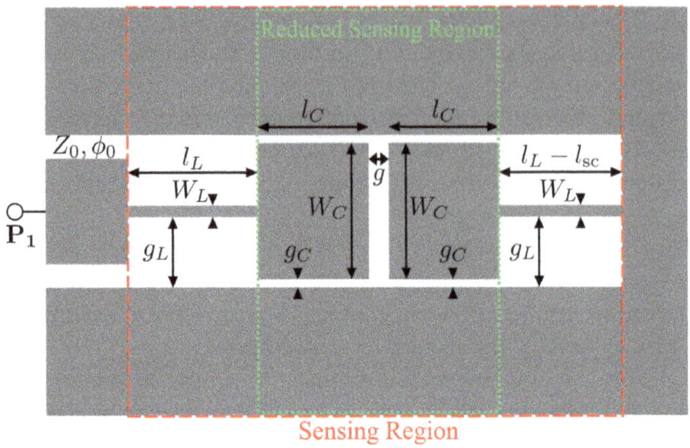

Fig. 6 Typical layout of a reflective-mode phase-variation sensor based on a pair of coupled SIRs and implemented in CPW technology, and relevant dimensions

3 Reflective-Mode Phase-Variation Sensors Based on Pairs of Coupled Step-Impedance Resonators (SIRs)

Reflective-mode phase-variation sensors based on pairs of coupled SIRs were first reported in [12], where two prototype device sensors implemented in microstrip technology, sensitive to variations of the dielectric constant of the MUT, were presented (the maximum sensitivity in one of such sensors was found to be -468°, a competitive value). Later, in [13], the focus was the implementation of this type of sensors in CPW technology (the typical topology, with relevant dimensions indicated, is depicted in Fig. 6). The advantage of CPW over microstrip technology is the potentially higher achievable sensitivities, by virtue of the fact that the field lines in CPW technology are significantly distributed in the MUT region (or air, if the MUT is not present), as compared to microstrip technology. In this section, reflective-mode phase-variation sensors based on coupled SIRs are studied in detail. After analytically obtaining the relevant frequencies of the pair of coupled SIRs, a detailed sensitivity analysis (valid for sensor implementation in CPW technology) is reported. The section concludes with a validation subsection, where some prototype examples are presented.

3.1 Resonance Frequencies

The equivalent circuit model of the pair of coupled SIRs of Fig. 6, by excluding losses, is the one depicted in Fig. 2b (the reactive elements are indicated in the caption).[10] The inductance of the SIRs, *L*, is dictated by the length and width of the narrow strip

[10] The circuit model with losses will be presented in Sect. 4.

(or line section), whereas the self-capacitance, C, is determined by the width of the wide line section (patch) of the SIRs and its distance to the ground planes.[11] Finally, the mutual capacitance, C_M, is controlled by the separation between the SIR patches. This model is valid for sensor implementation not only in CPW technology, but also in microstrip technology, as it was demonstrated in [12, 13]. Let us mention that the validity of the model requires that the electrical lengths of the line sections of the SIR are small and that the characteristic impedances are extreme, i.e., high for the narrow line section and low for the wide line section (under these conditions, the narrow line behaves inductively and the wide line capacitively, as required).

The input impedance of the circuit of Fig. 2b is

$$Z_{in} = j\frac{\left(1-k^2\right)\frac{\omega^4}{\omega_0^4} - 2\frac{\omega^2}{\omega_0^2} + 1}{C\omega\left\{\frac{\omega^2}{\omega_0^2}\left(1-k^2\right) - 1\right\}} \equiv j\chi_{in} \tag{22}$$

where ω is the angular frequency, $\omega_0 = 2\pi f_0 = 1/\sqrt{LC}$ is the angular resonance frequency of the uncoupled resonator, and $k = C_M/C$ is the electric coupling coefficient. The resonances of the coupled SIRs pair are given by those angular frequencies that null the reactance, namely,

$$\omega_{\pm} = 2\pi f_{\pm} = \frac{\omega_0}{\sqrt{1 \mp k}} \tag{23}$$

According to (23), if the coupling is weak, then $k << 1$, and consequently, f_- and f_+ are very close one to each other, as discussed before. Under these conditions, the phase of the reflection coefficient experiences a strong variation between f_- and f_+, as anticipated. Thus, by tuning the operating frequency of the sensor between f_- and f_+, the sensitivity of the phase of the reflection coefficient, ϕ_ρ, with the dielectric constant of the MUT, ε_{MUT}, is expected to be high.[12] The circuit of Fig. 2b exhibits a pole at the angular frequency where the denominator of (22) is null, and hence $\chi_{in} = \infty$ and $\rho = 1$ (or $\phi_\rho = 0°$). Such pole is found to be

$$\omega_p = 2\pi f_p = \frac{\omega_0}{\sqrt{1-k^2}} \tag{24}$$

and the corresponding frequency, $f_p = \omega_p/2\pi$, is located slightly to the right of f_0 and below f_+, provided the coupling is weak. All these relevant frequencies (f_0, f_p, f_+ and f_-) are indicated in Fig. 2b.

[11] Note that, according to Fig. 6, the length of the narrow strip of the SIR on the right, $l_L - l_{sc}$, is slightly shorter than that of the SIR on the left, l_L. This is required to achieve an identical SIR inductance, L, in both elements [19].

[12] The optimum frequency of operation is an aspect that will be discussed in the Sect. 3.2.

3.2 Sensitivity Analysis

The objective of this subsection is to obtain an expression providing the sensitivity of the CPW sensor based on a pair of mirrored (coupled) SIRs as a function of the circuit and substrate parameters. The main hypothesis of the analysis is that the electric coupling coefficient, $k = C_M/C$, does not appreciably change by varying the dielectric constant of the material under test, ε_{MUT}. This hypothesis is justified since the involved capacitances C and C_M are edge capacitances. Thus, it is expected that such capacitances experience the same percentage of variation with ε_{MUT}. Let us consider that the reference material is an arbitrary material with dielectric constant ε_{REF}. Following the nomenclature in [13], let us designate by C_{REF} and $C_{\text{M,REF}}$ the self-capacitance and mutual capacitance, respectively, of the SIRs pair when the REF material is on top of the coupled resonators (C and C_M are the capacitances corresponding to the coupled SIRs covered with air). If the REF material is semi-infinite in the vertical direction, C_{REF} and $C_{M,\text{REF}}$ can be expressed in terms of C and C_M, respectively, according to [4, 5, 17]

$$C_{\text{REF}} = C\frac{\varepsilon_{r,\text{eq}} + \varepsilon_{\text{REF}}}{\varepsilon_{r,\text{eq}} + 1} \tag{25a}$$

$$C_{M,\text{REF}} = C_M\frac{\varepsilon_{r,\text{eq}} + \varepsilon_{\text{REF}}}{\varepsilon_{r,\text{eq}} + 1} \tag{25b}$$

where $\varepsilon_{r,\text{eq}}$ is the equivalent dielectric constant of the substrate, defined before (see Chap. 1, Sect. 2.1.1). The equivalent dielectric constant of the substrate depends slightly on the ratio between the gap space of the considered capacitance and the substrate thickness, and it approaches the substrate dielectric constant, ε_r, if such ratio is very small (since under these conditions the semi-infinite substrate approximation holds, and $\varepsilon_{r,\text{eq}} = \varepsilon_r$). Not necessarily the equivalent dielectric constant of the substrate must be identical for both capacitances, C_{REF} and $C_{M,\text{REF}}$, but the value is expected to be very similar if the gap space between both SIRs and the width of the slots between the SIR patches and the ground planes of the CPW do not differ so much. Let us designate by C' the self-capacitance of the SIRs covered with the MUT and by C'_M the mutual capacitance of the coupled SIRs also covered with the MUT. It is obvious that such capacitances can be expressed as

$$C' = C_{\text{REF}}\frac{\varepsilon_{r,\text{eq}} + \varepsilon_{\text{MUT}}}{\varepsilon_{r,\text{eq}} + \varepsilon_{\text{REF}}} = C\frac{\varepsilon_{r,\text{eq}} + \varepsilon_{\text{MUT}}}{\varepsilon_{r,\text{eq}} + 1} \tag{26a}$$

$$C'_M = C_{M,\text{REF}}\frac{\varepsilon_{r,\text{eq}} + \varepsilon_{\text{MUT}}}{\varepsilon_{r,\text{eq}} + \varepsilon_{\text{REF}}} = C_M\frac{\varepsilon_{r,\text{eq}} + \varepsilon_{\text{MUT}}}{\varepsilon_{r,\text{eq}} + 1} \tag{26b}$$

Thus, if we do assume that $\varepsilon_{r,\text{eq}}$ is identical for both capacitances, as indeed it is explicitly considered in expressions (25a, 25b) and (26a, 26b), then it follows that k does not depend on the dielectric constant of the MUT (note that, under the considered hypothesis, $k = C_M/C = C_{M,\text{REF}}/C_{\text{REF}} = C'_M/C'$). The invariance of k with the dielectric constant of the MUT was verified in [13] by simulating the responses of a specific designed sensor by considering the coupled SIRs covered with MUTs of different dielectric constants. Extraction of the circuit parameters revealed that the inductance L did not change, whereas the capacitances C' and C'_M experienced a proportional variation with ε_{MUT}, and hence k was found to be invariable with the MUT (see further details in [13]). Thus, k is a constant in the sensitivity analysis, to be developed next.

The sensitivity can be expressed as

$$S = \frac{d\phi_\rho}{d\varepsilon_{\text{MUT}}} = \frac{d\phi_\rho}{dC'}\frac{dC'}{d\varepsilon_{\text{MUT}}} \tag{27}$$

namely, a variation in the dielectric constant of the MUT, ε_{MUT}, modifies the capacitance of the SIR, C', and it in turn perturbs the phase of the reflection coefficient, ϕ_ρ. Obviously, a change in ε_{MUT} has also an effect on the mutual capacitance C'_M, but the reflection coefficient, given by (10) does not explicitly depend on C'_M, since $Z_{in} = j\chi_{in}$ does not, see expression (22). The dependence of the reflection coefficient on C'_M is through k, constant as justified before. Consequently, in coherence with the considered hypothesis, expression (27) does not include a term of the form $\frac{d\phi_\rho}{dC'_M}\frac{dC'_M}{d\varepsilon_{\text{MUT}}}$, as it would be necessary if an explicit dependence of χ_{in} on C'_M was present in (22).

The starting point for the sensitivity analysis is expression (22), where ω_0, the angular resonance frequency of the uncoupled SIR surrounded by air must be replaced with the angular resonance frequency of the uncoupled SIR covered with the MUT, ω'_0. Additionally, the self-capacitance C of the uncoupled SIR covered by air in (22) must be replaced with C'. Note that ω'_0 is simply

$$\omega'_0 = \frac{1}{\sqrt{LC'}} \tag{28}$$

and it can be expressed in terms of ω_0, or in terms of the angular resonance frequency of the SIR loaded with the REF material, $\omega_{0,\text{REF}}$, in this latter case, as

$$\omega'_0 = \frac{1}{\sqrt{L(C_{\text{REF}} + \Delta C)}} = \frac{\omega_{0,\text{REF}}}{\sqrt{1 + \frac{\Delta C}{C_{\text{REF}}}}} \tag{29}$$

where ΔC is the variation in the self-capacitance, with regard to the REF self-capacitance, when it is loaded with the MUT (i.e., $C' = C_{\text{REF}} + \Delta C$). For the general case of the coupled SIRs loaded with an arbitrary MUT, expression (22) is

$$\chi_{in} = \frac{(1-k^2)\frac{\omega^4}{\omega_{0,\text{REF}}^4}\left(1 + \frac{\Delta C}{C_{\text{REF}}}\right)^2 - 2\frac{\omega^2}{\omega_{0,\text{REF}}^2}\left(1 + \frac{\Delta C}{C_{\text{REF}}}\right) + 1}{(C_{\text{REF}} + \Delta C)\omega\left\{\frac{\omega^2}{\omega_{0,\text{REF}}^2}\left(1 + \frac{\Delta C}{C_{\text{REF}}}\right)(1-k^2) - 1\right\}} \tag{30}$$

where ω_0 and C have been replaced with ω'_0 and C', respectively, and expression (29) has been used.

For sensitivity optimization, the operating frequency must be set within the range delimited by f_- and f_+, as indicated before. Within that range, there are two canonical frequencies, i.e., f_0, the resonance frequency of the uncoupled SIR, and f_p, the pole frequency. Let us perform the sensitivity analysis for both cases separately, i.e., by considering operation at both frequencies. Nevertheless, for the first case, the operating frequency should actually be set to the resonance frequency of the uncoupled SIR when it is loaded with the REF material, $f_{0,\text{REF}}$, rather than air. Similarly, for the second case, the pole frequency should be the one when the SIRs pair is loaded with the REF material, $f_{p,\text{REF}}$.

Case A: Operation at $f_{0,REF}$

Evaluation of the input reactance (30) at $\omega_{0,\text{REF}} = 2\pi f_{0,\text{REF}}$ gives

$$\chi_{in}\left(\omega_{0,\text{REF}}\right) = \frac{\left(1-k^2\right)\left(1+\frac{\Delta C}{C_\text{REF}}\right)^2 - 2\left(1+\frac{\Delta C}{C_\text{REF}}\right)+1}{(C_\text{REF}+\Delta C)\omega_{0,\text{REF}}\left\{\left(1+\frac{\Delta C}{C_\text{REF}}\right)\left(1-k^2\right)-1\right\}} \tag{31}$$

an expression that can be simplified to

$$\chi_{in}\left(\omega_{0,\text{REF}}\right) = \frac{\left(1-k^2\right)\frac{\Delta C^2}{C_\text{REF}^2} - 2k^2\frac{\Delta C}{C_\text{REF}} - k^2}{(C_\text{REF}+\Delta C)\omega_{0,\text{REF}}\left\{\frac{\Delta C}{C_\text{REF}} - k^2 - \frac{\Delta C}{C_\text{REF}}k^2\right\}} \tag{32}$$

Introducing (32) in (10), the reflection coefficient can be expressed as

$$\rho = \frac{jN - Z_0 D}{jN + Z_0 D}, \tag{33}$$

N and D being the numerator and the denominator, respectively, of (32). The phase of the reflection coefficient, the relevant (output) variable, is thus

$$\phi_\rho = 2\arctan\left\{-\frac{\left(1-k^2\right)\frac{\Delta C^2}{C_\text{REF}^2} - 2k^2\frac{\Delta C}{C_\text{REF}} - k^2}{Z_0(C_\text{REF}+\Delta C)\omega_{0,\text{REF}}\left\{\frac{\Delta C}{C_\text{REF}} - k^2 - \frac{\Delta C}{C_\text{REF}}k^2\right\}}\right\} \tag{34}$$

For the calculation of the sensitivity, let us first obtain the first derivative of the right-hand side member of (27). After some straightforward (but tedious) calculation, the following result is obtained

$$\frac{d\phi_\rho}{d\Delta C} = -\frac{2Z_0\omega_{0,\text{REF}}\left\{k^2 + \left(1-k^2\right)k^2\frac{\Delta C^2}{C_\text{REF}^2} - \left(1-k^2\right)^2\frac{\Delta C^3}{C_\text{REF}^3}\right\}}{Z_0^2 D^2 + N^2} \tag{35}$$

where it has been assumed that $d\phi_\rho/dC' = d\phi_\rho/d\Delta C$. In the limit of small perturbations of the dielectric constant with regard to the one of the REF material, i.e., for $C' \to C_{\text{REF}}$, or $\Delta C \to 0$, expression (35) gives

$$\left.\frac{d\phi_\rho}{d\Delta C}\right|_{\Delta C=0} = -\frac{2Z_0\omega_{0,\text{REF}}}{k^2\left(1 + Z_0^2\omega_{0,\text{REF}}^2 C_{\text{REF}}^2\right)} \tag{36}$$

Concerning the second term of the right-hand side member of (27), according to 26(a), it is simply

$$\frac{dC'}{d\varepsilon_{\text{MUT}}} = \frac{C_{\text{REF}}}{\varepsilon_{r,\text{eq}} + \varepsilon_{\text{REF}}} \tag{37}$$

Thus, introducing (36) and (37) in (27), the sensitivity in the limit of small perturbations for a semi-infinite MUT is found to be

$$S|_{\varepsilon_{\text{MUT}}=\varepsilon_{\text{REF}}} = -\frac{2Z_0\omega_{0,\text{REF}}C_{\text{REF}}}{k^2\left(\varepsilon_{r,\text{eq}} + \varepsilon_{\text{REF}}\right)\left(1 + Z_0^2\omega_{0,\text{REF}}^2 C_{\text{REF}}^2\right)} \tag{38}$$

Case B: Operation at $f_{p,REF}$

Evaluation of the input reactance (30) at $\omega_{p,\text{REF}} = 2\pi f_{p,\text{REF}}$ gives

$$\chi_{in}\left(\omega_{p,\text{REF}}\right) = \frac{\frac{1}{(1-k^2)}\left(1 + \frac{\Delta C}{C_{\text{REF}}}\right)^2 - 2\frac{1}{(1-k^2)}\left(1 + \frac{\Delta C}{C_{\text{REF}}}\right) + 1}{(C_{\text{REF}} + \Delta C)\frac{\omega_{0,\text{REF}}}{\sqrt{1-k^2}}\frac{\Delta C}{C_{\text{REF}}}} \tag{39}$$

an expression that can be simplified to

$$\chi_{in}\left(\omega_{p,\text{REF}}\right) = \frac{\frac{\Delta C^2}{C_{\text{REF}}^2} - k^2}{(C_{\text{REF}} + \Delta C)\sqrt{1 - k^2}\,\omega_{0,\text{REF}}\frac{\Delta C}{C_{\text{REF}}}} \tag{40}$$

The reflection coefficient is given by (33), N and D being the numerator and the denominator, respectively, of (40). The phase of the reflection coefficient, the relevant (output) variable, is thus

$$\phi_\rho = 2\arctan\left\{-\frac{\frac{\Delta C^2}{C_{\text{REF}}^2} - k^2}{Z_0(C_{\text{REF}} + \Delta C)\sqrt{1 - k^2}\,\omega_{0,\text{REF}}\frac{\Delta C}{C_{\text{REF}}}}\right\} \tag{41}$$

The derivative of the right-hand side member of (27) is found to be

$$\frac{d\phi_\rho}{d\Delta C} = -\frac{2Z_0\sqrt{1-k^2}\omega_{0,\mathrm{REF}}\left\{\frac{\Delta C^2}{C_{\mathrm{REF}}^2} + k^2\frac{(C_{\mathrm{REF}}+2\Delta C)}{C_{\mathrm{REF}}}\right\}}{Z_0^2 D^2 + N^2} \tag{42}$$

where, similar to case A, it has been assumed that $d\phi_\rho/dC' = d\phi_\rho/d\Delta C$. In the limit of small variations of the dielectric constant in the vicinity of the one corresponding to the REF material, i.e., for $C' \to C_{\mathrm{REF}}$, or $\Delta C \to 0$, expression (42) gives

$$\left.\frac{d\phi_\rho}{d\Delta C}\right|_{\Delta C=0} = -\frac{2Z_0\omega_{0,\mathrm{REF}}\sqrt{1-k^2}}{k^2} \equiv -\frac{2Z_0\omega_{p,\mathrm{REF}}\left(1-k^2\right)}{k^2} \approx -\frac{2Z_0\omega_{p,\mathrm{REF}}}{k^2} \tag{43}$$

where the last approximation is legitimated as far as weak coupling is considered (i.e., $k << 1$). Using (37) and (43), the sensitivity (27) in the limit of small perturbations for a semi-infinite MUT is found to be

$$S|_{\varepsilon_{\mathrm{MUT}}=\varepsilon_{\mathrm{REF}}} = -\frac{2Z_0\omega_{p,\mathrm{REF}}C_{\mathrm{REF}}}{k^2\left(\varepsilon_{r,\mathrm{eq}} + \varepsilon_{\mathrm{REF}}\right)} \tag{44}$$

According to (38) and (44), operation at $f_{p,\mathrm{REF}}$ provides higher sensitivity than operation at $f_{0,\mathrm{REF}}$. Indeed, it was demonstrated in [13] that the optimum frequency for sensitivity optimization (for $\varepsilon_{\mathrm{MUT}} = \varepsilon_{\mathrm{REF}}$) is very close to $f_{p,\mathrm{REF}}$ (but it is not exactly $f_{p,\mathrm{REF}}$). According to (44), a high value of C_{REF} and a low value of the equivalent dielectric constant (and hence dielectric constant) of the substrate favor the sensitivity. Nevertheless, the important parameter for sensitivity optimization is the electric coupling coefficient, k, since it appears squared in the denominator of (44) [and also in the denominator of (38)]. By weakly coupling the resonators ($k << 1$), the sensitivity in the limit of small perturbations can be significantly enhanced, and sensitivities of hundreds of degrees per unit of dielectric constant variation can easily be achieved (an aspect that will be later demonstrated). Indeed, the sensitivity can be boosted up to any desired value by simply reducing the electric coupling coefficient to the necessary value given by (44). In practice, a small coupling coefficient is always achievable by separating enough the SIRs. However, it must be considered that losses have been excluded in the previous analysis, and that their effects are expected to be more significant for high sensitivities or small coupling coefficients.[13] Moreover, by weakly coupling the resonators, tuning the operating frequency to $f_{p,\mathrm{REF}}$ is expected to be subjected to certain difficulty, due to the proximity between f_+ and f_-. Thus, in practice, a tradeoff is necessary, and excessively weak coupling is not convenient. Nevertheless, as it will be shown, very high sensitivities, combined with small sensing regions, are achievable with this sensing strategy, representing a very competitive approach in terms of the main Figure of Merit (FoM), or ratio between the maximum sensitivity and the area of the sensing region expressed in guided square wavelengths.

[13] Nevertheless, it will be shown in Sect. 4 that losses tend to improve the sensitivity.

3.3 Prototype Example

Let us report in this section a prototype example of a reflective-mode phase-variation sensor implemented by means of a pair of weakly coupled SIRs in CPW technology [13], with the purpose of validating the sensitivity analysis of Sect. 3.2. The frequency of operation was tuned to $f_{p,\mathrm{REF}} = 2.003$ GHz, corresponding to case B of that section (the resonance frequency of the uncoupled resonator being $f_{0,\mathrm{REF}} = 1.986$ GHz). The reference material was considered to be a stack of dielectric slabs (uncladded *Rogers RO4003C* low-loss microwave substrates) with dielectric constant $\varepsilon_{\mathrm{REF}} = 3.55$. The total thickness of the REF sample (made of 4 slabs) is roughly 6 mm, enough to consider the REF material semi-infinite. The sensor was implemented in the *Rogers RO4003C* microwave substrate with dielectric constant $\varepsilon_{\mathrm{r}} = 3.55$, thickness $h = 1.524$ mm and loss tangent $\tan\delta = 0.0022$. The photograph of the fabricated sensor is depicted in Fig. 7a, where dimensions are indicated. With such dimensions and substrate parameters, the reactive elements of the circuit model of the sensor (Fig. 2b) loaded with the REF material were found to be $L = 5.035$ nH, $C_{M,\mathrm{REF}} = 0.166$ pF and $C_{\mathrm{REF}} = 1.276$ pF (the parameter extraction procedure is detailed in [13]). With such capacitances, the coupling coefficient is $k = 0.130$. Figure 7b depicts the phase response of the sensor inferred from full-wave electromagnetic simulation (using the *CST Microwave Studio* commercial software), compared to that obtained by circuit simulation with the extracted reactive elements. The results are almost undistinguishable, indicative of the validity of the model and the parameter extraction procedure.

Figure 7c depicts the response of the sensor, that is, the phase of the reflection coefficient at $f_{\mathrm{p,REF}}$ as a function of the dielectric constant of the MUT, $\varepsilon_{\mathrm{MUT}}$, as inferred by electromagnetic simulation. The figure also includes the measured phase of the reflection coefficient corresponding to specific MUTs, with dielectric constants $\varepsilon_{\mathrm{MUT}} = 1$ (air), $\varepsilon_{\mathrm{MUT}} = 2.8$ (a piece of 3D-printed polylactic acid –PLA), $\varepsilon_{\mathrm{MUT}} = \varepsilon_{\mathrm{REF}} = 3.55$ (uncladded *Rogers 4003C*), $\varepsilon_{\mathrm{MUT}} = 4.4$ (FR4), and $\varepsilon_{\mathrm{MUT}} = 10.2$ (uncladded *Rogers RO3010*). Due to the lack of a significant number of experimental data points, the sensitivity was obtained in [13] from the derivative of the simulated data points. The sensitivity is also included in Fig. 7c. The sensitivity in the limit of small perturbations was found to be $S_{\mathrm{sim}} = -774°$, in excellent agreement with the theoretical value, given by expression (44), i.e., $S_{\mathrm{Th}} = -775°$. It should be mentioned that the simulated phases of Fig. 7c were inferred by excluding losses, in coherence with the sensitivity analysis of the previous section (where losses were excluded). The good agreement between the simulated sensitivity and the one predicted by expression (44) validates the previous sensitivity analysis. As it was demonstrated in [13], the results by including the effects of substrate and MUT losses (as well as conductive losses) are very similar to those corresponding to the lossless case, as expected if losses are small. Nevertheless, the simulated sensitivity in the limit of small perturbations by including the effects of losses was found to be slightly better than the one inferred by excluding losses. Particularly, the inferred "lossy" sensitivity in the limit of small perturbation was $S_{\mathrm{Lossy}} = -818°$. These "lossy" simulations were

Fig. 7 Photograph of the fabricated sensor based on a pair of weakly coupled SIRs in CPW technology (**a**), phase response corresponding to the pair of SIRs covered with the REF material (**b**), and phase (at $f_{p,\mathrm{REF}}$) dependence of the sensor with $\varepsilon_{\mathrm{MUT}}$, and sensitivity (**c**). In (**c**), the phase is the one in reference to the phase when the sensor is loaded with the REF sample, i.e., $\Delta\phi_\rho = \phi_\rho - \phi_{\rho,\mathrm{REF}}$. Dimensions, in reference to Fig. 6, in mm, are: $l_0 = 6.0$ mm, $w_0 = 3.39$ mm, $g_0 = 0.305$ mm, $l_L = 6.40$ mm, $w_L = 0.3$ mm, $g_L = 1.85$ mm, $l_C = 4.86$ mm, $w_C = 3.8$ mm, $g_C = 0.2$ mm, $g = 0.62$ mm, and $l_{\mathrm{sc}} = 0.29$ mm. Reprinted with permission from [13]; copyright 2023 IEEE

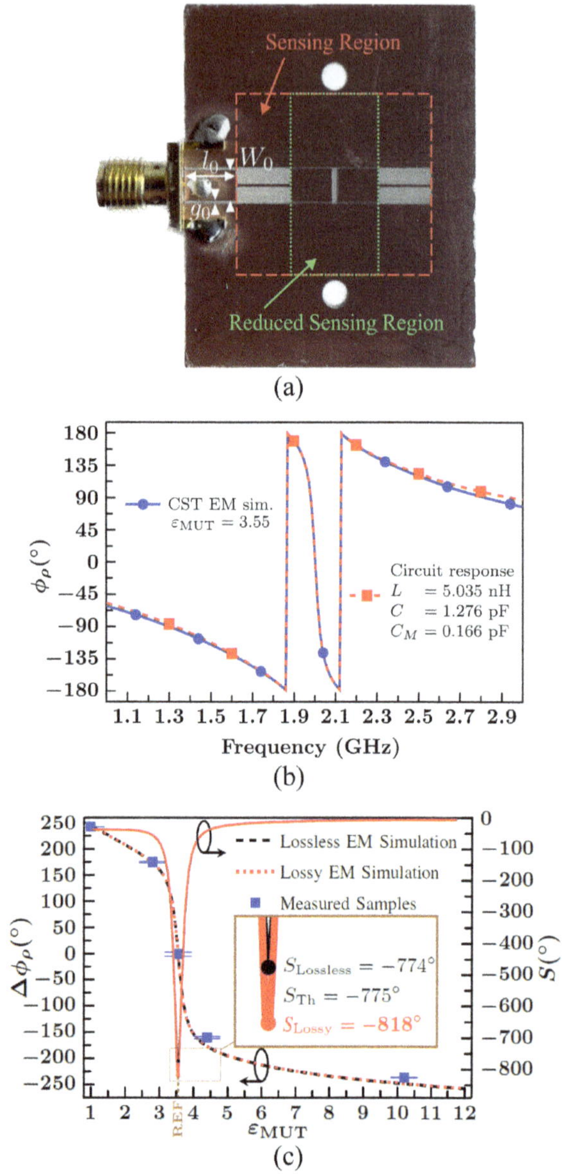

carried out in [13] by considering that the loss tangent of the MUT does not depend on the dielectric constant of the MUT (specifically, $\tan\delta_{\mathrm{MUT}} = \tan\delta_{\mathrm{REF}} = 0.0022$). This does not represent a loss of generality and points out that losses can help to boost up (rather than degrading) the sensitivity in the limit of small perturbations, an aspect that will be discussed and justified in the next section.

It should be clarified that the above simulated sensitivity in the limit of small perturbations was achieved by considering the whole resonator pair (not only the capacitive patches) coated with the MUT (this also applies to the phase response of Fig. 7c). The reason is that the capacitance of the narrow strips, although very small, contributes somehow to the self-capacitance of the SIR. Thus, the agreement with the model and sensitivity analysis is better by considering the extended sensing region. By considering MUTs circumscribed to the sensing region designated as reduced sensing region in Fig. 7a, the ("lossy") sensitivity is somehow degraded ($S_{\text{Lossy}} = -620°$ in the limit of small perturbations), but the sensing region is substantially reduced, and this benefits the FoM.[14] Thus, the FoM by considering the reduced sensing region is FoM $= 26551°/\lambda^2$, whereas the one corresponding to the extended sensing region is found to be FoM $= 15469°/\lambda^2$. The former is very competitive, thanks to the combination of the high sensitivity and small area of the sensing region. Note also that the whole sensor is very small by virtue of the electrically small size of the SIRs pair.

Table 1 compares various phase-variation permittivity sensors, not necessarily operating in reflection mode, where it can be appreciated that the FoM of the sensor reported in [13], the one of Fig. 7, is very competitive, and, at the same time, the overall sensor size is very small. The sensor reported in [1], consisting in an open-ended quarter-wavelength sensing resonator terminating a pair of high/low impedance inverters, exhibits an excellent FoM ($= 21148°//\lambda^2$), but at the expense of adding further elements (the pair of high/low impedance inverters) to the sensing element, and hence increasing the overall sensor size. In [12], the reported sensor is also based on a pair of weakly coupled SIRs but implemented in microstrip technology. The FoM and size of that sensor are also excellent, though that sensor was not optimized. According to Table 1, weakly coupled resonators, especially based on semi-lumped elements (e.g., SIRs), offer an interesting solution for microwave sensing in applications where high sensitivity and reduced sensor size are required.

4 Effects of Losses

Let us analyze in this section the effects of losses. To a first-order approximation, substrate losses can be neglected provided low-loss microwave substrates are used for sensing. The considered MUTs used for validation purposes of the sensors presented in the previous sections are also low-loss samples, or moderately low-loss samples (the case of FR4). However, these reflective-mode phase-variation sensors based on weakly coupled resonators can also be applied to liquid sensing (or to sensing other lossy materials). The typical loss tangents of liquids are relatively high. Thus, neglecting losses (as it has been done in the analysis carried out so far in this chapter) is not necessarily a valid approximation, in general. In Sect. 4.1, the effects of losses are

[14] Note that this decrease in the sensitivity when the sensing region is circumscribed to the SIRs patches (rather than to the whole SIRs area) is not related to the effect of losses.

Table 1. Comparison of various phase-variation permittivity sensors

References	Mode	Sensing size* (λ^2)	Overall size (λ^2)	Max. Sensitivity	FoM (°/λ^2)
[1]	RX	0.025	0.131	528.7°	21148
[2]	RX	0.100	0.297	45.5°	455
[3]	RX	0.065	0.126	101.3°	1558
[4]	RX	0.015	0.073	83.35°	5643
[5]	RX	0.018	0.076	66.5°	3643
[12]	RX	0.017	0.102	468°	27419
[20]	TX	12.90	12.90	415.6°	32.2
[21]	TX	0.030	0.090	7.7°	257
[22]	TX	0.040	0.054	20.0°	500
[23]	TX	0.038	0.053	12.3°	332
Fig. 4	RX	0.069	0.069	106.5°	1535
Fig. 5	RX	0.058	0.058	736°	12690
Fig. 7 (ext.)**	RX	0.053	0.053***	818°	15469
Fig. 7 (red.)**	RX	0.023	0.042***	620°	26551

* This size corresponds to the sensing region, not to the size of the whole sensing structure, referred to as overall size and indicated in the adjacent column to the right.
** Ext. and red. stand for extended and reduced sensing region, respectively.
*** The entries vary because the guided wavelength differs in cases where the MUT fully covers the SIR pair (extended sensing region) and when it only covers the capacitive patches (reduced sensing region). This yields a different guided wavelength for the SIRs' inductive section, leading to distinct values of the whole sensing region when expressed as a function of the guided wavelength. However, the total sensor area in cm^2 is unique and does not depend on the size of the sensing region and MUT.

studied in reference to the sensors based on coupled distributed quarter-wavelength resonators, presented in Sect. 2. Section 4.2 deals with the influence of losses in the sensors reported in Sect. 3, based on weakly coupled SIRs. Though the typical considered substrates for sensor implementation exhibit low or moderate loss tangents, in the analysis of the next subsection substrate losses are also considered.

4.1 *Effects of Losses in Sensors Based on Weakly Coupled Quarter-Wavelength Resonators*

In [11], it was demonstrated that if losses (including substrate and MUT losses) are relatively small, the phase response of the sensor does not vary appreciably. The effects of losses in the sensing structures of [11] were included by considering that the real part of the complex propagation constant, i.e., the attenuation constant, α, is not negligible. Specifically, it was proved in [11] that the low-loss approximation requires that

$$\alpha l \ll \frac{Z_{ev} - Z_{od}}{2Z_0} C \tag{45}$$

where C is the coupling factor. Thus, if (45) is satisfied to a good approximation, then the effects of losses can be neglected. However, the sensitivity in the limit of small perturbations by including (moderate or significant) losses was not calculated in [11], and this is the main objective of this subsection (the sensitivity analysis by including losses was reported in [24]). Without loss of generality, the specific implementation of Sensor-type A, see Fig. 3, is the considered subject of study.

Let us assume that the weak coupling approximation with the presence of losses can be applied. Under this approximation, it is reasonable to consider that the complex propagation constant is identical for both modes (i.e., $\gamma_{ev} \equiv \alpha_{ev} + j\beta_{ev} = \gamma_{od} \equiv \alpha_{od} + j\beta_{od} = \gamma \equiv \alpha + j\beta$, where α and β are the attenuation and phase constants, respectively). If l is the length of the coupled lines, the elements of the impedance matrix with the presence of losses, and under the weak coupling approximation, are found to be

$$Z_{11} = Z_{22} = Z_{33} = Z_{44} = \frac{1}{2}\{Z_{ev} + Z_{od}\} \coth \gamma l \tag{46a}$$

$$Z_{12} = Z_{21} = Z_{34} = Z_{43} = \frac{1}{2}\{Z_{ev} - Z_{od}\} \coth \gamma l \tag{46b}$$

$$Z_{13} = Z_{31} = Z_{24} = Z_{42} = \frac{1}{2}\{Z_{ev} - Z_{od}\} \frac{1}{\sinh \gamma l} \tag{46c}$$

$$Z_{14} = Z_{41} = Z_{23} = Z_{32} = \frac{1}{2}\{Z_{ev} + Z_{od}\} \frac{1}{\sinh \gamma l} \tag{46d}$$

From these elements, the input impedance can be obtained, and, from it, the phase of the reflection coefficient, the output variable, can be inferred. Using (1), the input impedance for sensor type A, by considering losses, is found to be

$$Z_{in,A} = \frac{(Z_{ev} + Z_{od})^2 \cosh^2 \gamma l - (Z_{ev} - Z_{od})^2}{2(Z_{ev} + Z_{od}) \cosh \gamma l \sinh \gamma l} \tag{47}$$

and after some straightforward mathematical calculation, it can be expressed as

$$Z_{in,A} = \frac{(Z_{ev} + Z_{od})^2 (\cos 2\phi \cosh 2\alpha l + j \sin 2\phi \sinh 2\alpha l) - Z_{ev}^2 - Z_{od}^2 + 6 Z_{ev} Z_{od}}{2(Z_{ev} + Z_{od})(\cos 2\phi \sinh 2\alpha l + j \sin 2\phi \cosh 2\alpha l)} \tag{48}$$

where $\phi = \beta l$. The reflection coefficient is

$$\rho = \frac{Z_{in,A} - Z_0}{Z_{in,A} + Z_0} = \frac{N - Z_0 D}{N + Z_0 D} \tag{49}$$

where N and D are the numerator and the denominator, respectively, of the right-hand side member of (48).

The phase of the reflection coefficient, the output variable of the sensor, can be expressed as

$$\phi_\rho = \arctan\left(\frac{I_N}{R_N}\right) - \arctan\left(\frac{I_D}{R_D}\right) \tag{50}$$

where I_N and R_N are the imaginary and the real part, respectively, of the numerator in (49), and I_D and R_D are the imaginary and the real part, respectively, of the denominator in (49), i.e.,

$$I_N = (Z_{ev} + Z_{od})^2 \sin 2\phi \sinh 2\alpha l - 2Z_0(Z_{ev} + Z_{od}) \sin 2\phi \cosh 2\alpha l \tag{51a}$$

$$I_D = (Z_{ev} + Z_{od})^2 \sin 2\phi \sinh 2\alpha l + 2Z_0(Z_{ev} + Z_{od}) \sin 2\phi \cosh 2\alpha l \tag{51b}$$

$$R_N = (Z_{ev} + Z_{od})^2 \cos 2\phi \cosh 2\alpha l - Z_{ev}^2 - Z_{od}^2 + 6Z_{ev}Z_{od} - 2Z_0(Z_{ev} + Z_{od}) \cos 2\phi \sinh 2\alpha l \tag{51c}$$

$$R_D = (Z_{ev} + Z_{od})^2 \cos 2\phi \cosh 2\alpha l - Z_{ev}^2 - Z_{od}^2 + 6Z_{ev}Z_{od} + 2Z_0(Z_{ev} + Z_{od}) \cos 2\phi \sinh 2\alpha l \tag{51d}$$

The derivative (12) with losses is

$$\frac{d\phi_\rho}{d\phi} = \frac{1}{1 + \left(\frac{I_N}{R_N}\right)^2} \cdot \frac{R_N I_N' - I_N R_N'}{R_N^2} - \frac{1}{1 + \left(\frac{I_D}{R_D}\right)^2} \cdot \frac{R_D I_D' - I_D R_D'}{R_D^2} \tag{52}$$

with

$$I_N' = 2(Z_{ev} + Z_{od})^2 \cos 2\phi \sinh 2\alpha l - 4Z_0(Z_{ev} + Z_{od}) \cos 2\phi \cosh 2\alpha l \tag{53a}$$

$$R_N' = -2(Z_{ev} + Z_{od})^2 \sin 2\phi \cosh 2\alpha l + 4Z_0(Z_{ev} + Z_{od}) \sin 2\phi \sinh 2\alpha l \tag{53b}$$

$$I_D' = 2(Z_{ev} + Z_{od})^2 \cos 2\phi \sinh 2\alpha l + 4Z_0(Z_{ev} + Z_{od}) \cos 2\phi \cosh 2\alpha l \tag{53c}$$

$$R_D' = -2(Z_{ev} + Z_{od})^2 \sin 2\phi \cosh 2\alpha l - 4Z_0(Z_{ev} + Z_{od}) \sin 2\phi \sinh 2\alpha l \tag{53d}$$

and evaluation at $\phi = 90^\circ$ gives

$$\left.\frac{d\phi_\rho}{d\phi}\right|_{\phi=90^\circ} = \frac{A_1 + A_2}{A_3 + A_4} - \frac{A_1 - A_2}{A_3 - A_4} = \frac{2(A_2A_3 - A_1A_4)}{A_3^2 - A_4^2} \tag{54}$$

with

$$A_1 = -2(Z_{ev} + Z_{od})^2 \sinh 2\alpha l \tag{55a}$$

$$A_2 = 4Z_0(Z_{ev} + Z_{od}) \cosh 2\alpha l \tag{55b}$$

$$A_3 = -(Z_{ev} + Z_{od})^2 \cosh 2\alpha l - Z_{ev}^2 - Z_{od}^2 + 6Z_{ev}Z_{od} \tag{55c}$$

$$A_4 = 2Z_0(Z_{ev} + Z_{od}) \sinh 2\alpha l \tag{55d}$$

After some simple (but tedious) calculation, it is found that (54) can be expressed as

$$\left.\frac{d\phi_\rho}{d\phi}\right|_{\phi=90^\circ} = \frac{-4Z_0}{C^2(Z_{ev} + Z_{od})} \frac{2(Z_{ev} - Z_{od})^2}{\frac{\left[(\cosh 2\alpha l+1)(Z_{ev}+Z_{od})^2-8Z_{ev}Z_{od}\right]^2-4Z_0^2(Z_{ev}+Z_{od})^2\sinh^2 2\alpha l}{(\cosh 2\alpha l+1)(Z_{ev}+Z_{od})^2-8Z_{ev}Z_{od}\cosh 2\alpha l}} \tag{56}$$

where the first term of the product in the right-hand side coincides with (14), the lossless case. Indeed, if losses are neglected ($\alpha l = 0$), the second term of the product of the right-hand side of (56) is 1, and the derivative corresponding to the losses case, i.e., (14), results, as expected. By using some hyperbolic trigonometric identities, expression (56) can be further developed and expressed as

$$\left.\frac{d\phi_\rho}{d\phi}\right|_{\phi=90^\circ} = \frac{-4Z_0}{C^2(Z_{ev} + Z_{od})} \frac{2(Z_{ev} - Z_{od})^2}{\frac{\left[2\cosh^2\alpha l(Z_{ev}-Z_{od})^2+8Z_{ev}Z_{od}\sinh^2\alpha l\right]^2-16Z_0^2(Z_{ev}+Z_{od})^2\sinh^2\alpha l\cosh^2\alpha l}{2\cosh^2\alpha l(Z_{ev}-Z_{od})^2-8Z_{ev}Z_{od}\sinh^2\alpha l}} \tag{57}$$

Thus, the sensitivity in the limit of small perturbations should be calculated by multiplying (57) and (19).[15]

An important aspect to analyze is whether the sensitivity can be enhanced by the effect of losses, as it occurs in reflective-mode phase-variation sensors based on sensing resonators (either distributed or semi-lumped) terminating a cascade of high/low impedance inverters [6, 25]. Indeed, it was demonstrated in [6, 25] that the sensitivity in those sensors can be made (ideally) infinite if losses (modelled by the attenuation constant or by a conductance, depending on the specific type of sensing element) and the design parameters of the sensor (impedance of the inverters, and electrical characteristics of the sensing element) satisfy a certain condition.

[15] Note that, similar to the lossless case, the last two terms of the right-hand side member of (9) evaluated at $\phi = 90°$ are null when losses are included. The reason is that $d\phi_\rho/dZ_{ev} = d\phi_\rho/dZ_{od} = 0$ for such phase condition when losses are not neglected, i.e., when ϕ_ρ is given by (50).

Furthermore, the sensitivity, of negative sign when losses are absent, can experience a change in its sign if losses exceed a certain limit (that depends on the design parameters of the sensor). Thus, two operational regimes were considered in [6, 25], the low-loss regime, with negative sensitivity, and the high-loss regime, with positive sensitivity. In the boundary between both regimes, the sensitivity takes the maximum (ideally infinite) value. Note that the sensitivity for sensor-type A by excluding losses, as given by (20), cannot be infinite, except for a null coupling coefficient (i.e., $C = 0$), a non-realistic situation. However, inspection of (57) reveals that the sensitivity when losses are present can be infinite without the need that $C = 0$. This occurs if the numerator of the denominator in the last term in (57) is null, whilst the denominator is not null. Thus, let us force that term to be null, in order to obtain the attenuation constant that leads to an infinite sensitivity, i.e.,

$$\left[2\cosh^2\alpha l(Z_{ev} - Z_{od})^2 + 8Z_{ev}Z_{od}\sinh^2\alpha l\right]^2 - 16Z_0^2(Z_{ev} + Z_{od})^2\sinh^2\alpha l\cosh^2\alpha l = 0 \tag{58}$$

The previous equation can be simplified to the following quadratic equation in the variable $\tanh\alpha l$

$$4Z_{ev}Z_{od}\tanh^2\alpha l - 2Z_0(Z_{ev} + Z_{od})\tanh\alpha l + (Z_{ev} - Z_{od})^2 = 0 \tag{59}$$

and the solutions are

$$\tanh\alpha l = \frac{Z_0(Z_{ev} + Z_{od})}{4Z_{ev}Z_{od}} \cdot \left\{1 \pm \sqrt{1 - 4\frac{Z_{ev}Z_{od}}{Z_0^2}C^2}\right\} \tag{60}$$

where (6) has been used. Under the weak-coupling approximation (of interest for sensitivity optimization), the coupling factor C should be small. Moreover, for sensor-type A, Z_{ev} and Z_{od} should also be small (as compared to Z_0) for enhancing the sensitivity. Thus, the square root in (60) can be approximated by the Taylor series expansion up to the first term, and the two solutions of (60) can be expressed as

$$\tanh\alpha l = \frac{Z_0(Z_{ev} + Z_{od})}{4Z_{ev}Z_{od}} \cdot \left\{1 \pm \left(1 - 2\frac{Z_{ev}Z_{od}}{Z_0^2}C^2\right)\right\} \tag{61}$$

namely,

$$\tanh\alpha l_+ = \frac{Z_0(Z_{ev} + Z_{od})}{2Z_{ev}Z_{od}} \cdot \left\{1 - \frac{Z_{ev}Z_{od}}{Z_0^2}C^2\right\} \tag{62a}$$

$$\tanh\alpha l_- = \frac{(Z_{ev} + Z_{od})}{2Z_0}C^2 = \frac{(Z_{ev} - Z_{od})}{2Z_0}C \tag{62b}$$

The two solutions given by (62a, 62b) null the denominator of (57). Nevertheless, the one of interest is $\tanh \alpha l_-$, the lower one, since this solution provides the boundary between the loss-loss and the high-loss regime. Let us further elaborate on this aspect. For the lossless case ($\alpha l = 0$), the sensitivity, given by (14), is negative. As losses increase, the sign of the sensitivity will not change up to $\tanh \alpha l_-$, where the denominator of the last term in (57) is null and therefore the magnitude of the sensitivity goes to infinity. Note that this is true since the denominator of the denominator of (57) is positive up to $\tanh \alpha l_-$. This is obvious, since the following equation

$$2\cosh^2 \alpha l (Z_{ev} - Z_{od})^2 - 8Z_{ev}Z_{od}\sinh^2 \alpha l = 0 \tag{63}$$

has a solution given by

$$\tanh \alpha l_0 = \frac{(Z_{ev} + Z_{od})}{2\sqrt{Z_{ev}Z_{od}}}C > \tanh \alpha l_- \tag{64}$$

and (63) is positive up to $\tanh \alpha l_0$.

From the previous analysis, it can be concluded that losses can benefit sensitivity, especially if the attenuation constant and the even and odd mode characteristic impedances of the coupled lines (which also dictate the coupling factor C) satisfy (62b). Nevertheless, it should be mentioned that it is not convenient to design the sensor at the exact boundary between the two cited operational regimes (high- and low-loss), since a change of regime as the properties of the material under study vary must be avoided. It should also be mentioned that in the limit when the sensitivity goes to infinity, i.e., when (59) is satisfied, the reflection coefficient is null, and, therefore, it is not possible to retrieve the phase of the reflection coefficient in that limit.[16] Nevertheless, this is a limiting case, which should be avoided, as indicated above. The main conclusion of this analysis is that the effects of losses in these reflective-mode phase-variation sensors based on pairs of weakly coupled quarter-wavelength resonators are similar to those in reflective-mode phase-variation sensors based on semi-lumped or distributed sensing resonators (eventually combined with a cascade of high/low impedance inverters).

Interestingly, (62b) coincides with the right-hand side member of (45). Thus, (62b) provides the limit of the low-loss regime, and losses, as given by the attenuation constant, should be much smaller than (62b) in order to obtain a phase response and sensor sensitivity undistinguishable to the one corresponding to the lossless case (note that $\tanh \alpha l \approx \alpha l$ if losses are very small).

The validation of the effects of losses on sensitivity, as derived from the previous analysis, has been done from electromagnetic simulations, due to the lack of a significant number of available materials with representative (and different) values of the

---

[16] Indeed, it can be demonstrated that equation (59) is equivalent to $N - Z_0 D = 0$ [the numerator of (49)] evaluated at $\phi = 90°$, where N and D are the numerator and the denominator, respectively, of the right-hand side member of (48). Therefore, it is clear that the solution of (59), i.e., (60), nulls the reflection coefficient given by (49).

dielectric constants and loss tangents. For such validation, the sensor of Fig. 8a (of type A) is considered (dimensions and substrate parameters are given in the caption) [24]. For such sensor, the even and odd mode characteristic impedances are $Z_{ev} = 44.71\ \Omega$ and $Z_{od} = 31.63\ \Omega$, respectively, and the coupling factor is $C = 0.17$ (corresponding to weak coupling). For such sensor, the phase response and sensitivity as the dielectric constant of the MUT varies between $\varepsilon_{\text{MUT}} = 1$ and $\varepsilon_{\text{MUT}} = 12$ has been obtained by considering different values of the loss tangent of the MUT (corresponding to different levels of losses). The REF material has been considered to exhibit a dielectric constant of $\varepsilon_{\text{REF}} = 4.42$ (identical to the one of the substrate), so that the pair of weakly coupled quarter-wavelength resonators exhibit an electrical length (for the even and odd modes) of 90° at the operating frequency, $f_0 = 2$ GHz, when the lines are loaded with a material with such dielectric constant. Figure 8b depicts the phase responses and sensitivities for the different values of the loss tangent of the MUT. As it can be appreciated, the magnitude of the sensitivity in the limit of small perturbations in the vicinity of the REF dielectric constant increases as the loss tangent of the MUT increases, in agreement with the previous analysis. Figure 8b also includes the phase responses inferred experimentally by coating the sensor with different materials (indicated). In all cases, the MUTs, achieved by stacking up several slabs, can be considered to be semi-infinite. The agreement between the simulated and the measured data points is very reasonable in all cases.

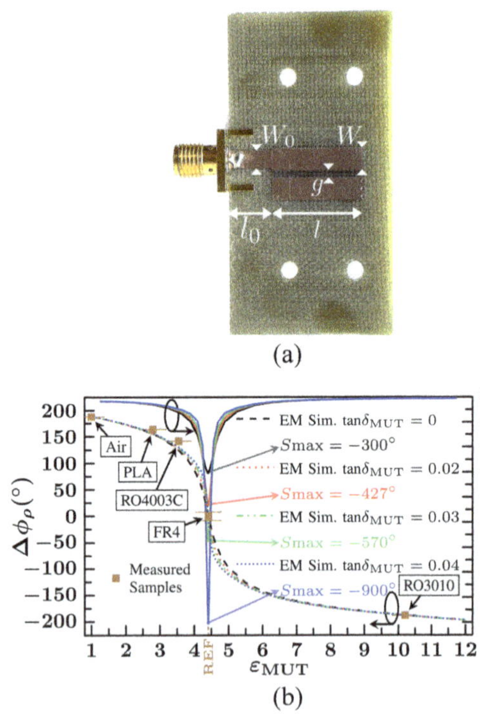

Fig. 8 Layout of the sensor based on a pair of weakly coupled quarter-wavelength resonators of type A (**a**), and phase response and sensitivity for various values of the loss tangent of the MUT (**b**). Substrate parameters are those of the *FR4* with dielectric constant $\varepsilon_r = 4.42$, thickness $h = 1.524$ mm, and loss tangent $\tan\delta = 0.024$. Sensor dimensions (in mm) are: $W = 3.4$, $g = 1.2$, $l = 16.47$, $W_0 = 2.78$, $l_0 = 8$. Reprinted with permission from [24]; copyright 2023 IEEE

4.2 Effects of Losses in Sensors Based on Weakly Coupled SIRs

It was shown in Sect. 3.3 that losses help to boost up the sensitivity in sensors based on pairs of weakly coupled SIRs. The difference between the maximum sensitivity inferred from the lossless and the lossy simulated data points (phase of the reflection coefficient as a function of the dielectric constant of the MUT) is not very relevant (see Sect. 3.3) since losses (including conductive, substrate, and MUT losses) are small in the considered case example. However, in situations where losses are significant (e.g., liquid samples) sensitivity enhancement can be substantial. The main objective of this section is to demonstrate this important aspect. The circuit model by including losses is depicted in Fig. 9a. Such model includes the dielectric losses of both the MUT and substrate, modeled as conductances parallel-connected to the different capacitive elements of the model, i.e., the self-capacitance, C', and the coupling capacitance, C'_M, see Fig. 9a. Ohmic losses, small when the metallic layer is made of a high conductive material such as Cu (the typical metal in low-loss microwave substrates), and radiation losses, are not explicitly considered to avoid an excessively complex lossy model. Nevertheless, their potential contributions to overall losses are accounted for by the conductances that appear in Fig. 9a to a good approximation, as reveals the good agreement between the lossy model response and the lossy electromagnetic simulations, to be demonstrated later.

As it can be appreciated in the lossy model, the conductances parallel-connected to the capacitances are not independent. That is, if the conductance in parallel with $C' - C'_M$ is designated by G, then the conductance in parallel with C'_M, designated by G_M, must satisfy [13]

$$G_M = \frac{k}{1-k} G \tag{65}$$

as depicted in Fig. 9a, where k is the electric coupling coefficient, defined before. To demonstrate (65), let us assume that there is a quasi-magnetic wall in the slot between the SIR patches (parallel to the sensor surface), and also in the slots between the SIR patches and the ground plane of the CPW (a reasonable hypothesis, according to [17]). Under this hypothesis, the electric field lines in such slots are roughly tangential (or parallel) to the interface between the substrate and the MUT, and both the self-capacitance C' and the mutual capacitance C'_M can be considered to be constituted by the contribution of the substrate plus the contribution of the MUT, see Fig. 10. Thus, the self-capacitance can be expressed as $C' = C'_{MUT} + C'_{SUB}$, and the mutual capacitance as $C'_M = C'_{M,MUT} + C'_{M,SUB}$. Similarly, the conductances can be expressed as the contribution of the substrate and MUT conductances, i.e., $G = G_{MUT} + G_{SUB}$, and $G_M = G_{M,MUT} + G_{M,SUB}$.

The loss tangent of the MUT can be alternatively expressed by

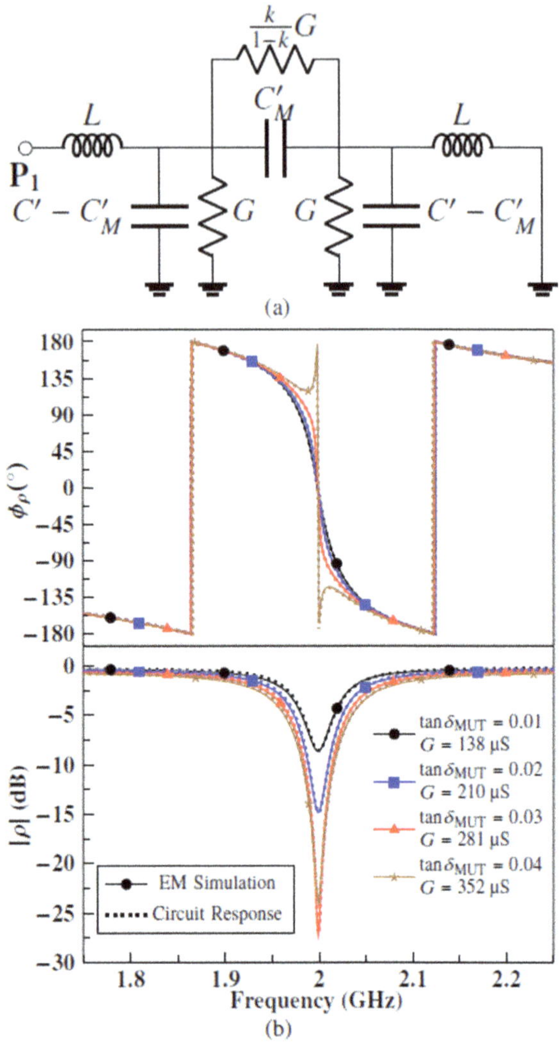

Fig. 9 Circuit model of the pair of coupled SIRs taking into account losses modeled by means of three conductances dependent on only one value of G (**a**) and phase and magnitude response of the reflection coefficient for different cases of loss of the MUT, for the REF case (**b**). The extracted reactive element values are: $C' = 1.276$ pF, $C'_M = 0.166$ pF, $L = 5.035$ nH. Reprinted with permission from [13]; copyright 2023 IEEE

$$\tan\delta_{\mathrm{MUT}} = \frac{G_{\mathrm{MUT}}}{\omega\left(C'_{\mathrm{MUT}} - C'_{M,\mathrm{MUT}}\right)} \tag{66}$$

or by

$$\tan\delta_{\mathrm{MUT}} = \frac{G_{M,\mathrm{MUT}}}{\omega C'_{M,\mathrm{MUT}}}, \tag{67}$$

whereas the loss tangent of the substrate is given by

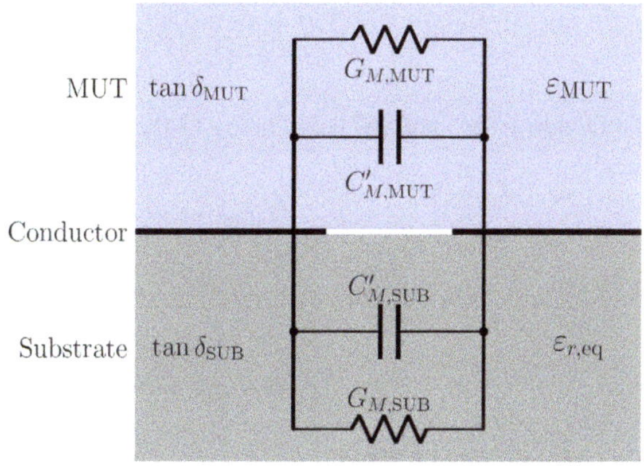

Fig. 10 Contributions of the MUT and substrate to the mutual capacitance, C'_M, and to the associated conductance, G_M. The indicated slot corresponds to the separation, g, between the SIR patches. For the parallel capacitance, $C' - C'_M$, and associated conductance, G, a similar figure applies

$$\tan\delta_{\mathrm{SUB}} = \frac{G_{\mathrm{SUB}}}{\omega\left(C'_{\mathrm{SUB}} - C'_{M,\mathrm{SUB}}\right)} \tag{68}$$

or by

$$\tan\delta_{\mathrm{SUB}} = \frac{G_{M,\mathrm{SUB}}}{\omega C'_{M,\mathrm{SUB}}} \tag{69}$$

Under the semi-infinite MUT approximation, it follows that the contribution of the MUT and substrate capacitances are related by

$$C'_{\mathrm{MUT}}\varepsilon_{r,\mathrm{eq}} = C'_{\mathrm{SUB}}\varepsilon_{\mathrm{MUT}} \tag{70}$$

for the self-capacitance, and by

$$C'_{M,\mathrm{MUT}}\varepsilon_{r,\mathrm{eq}} = C'_{M,\mathrm{SUB}}\varepsilon_{\mathrm{MUT}} \tag{71}$$

for the mutual capacitance. Thus, from (70) it follows that

$$C' = C'_{\mathrm{SUB}}\left(1 + \frac{\varepsilon_{\mathrm{MUT}}}{\varepsilon_{r,\mathrm{eq}}}\right) \tag{72}$$

and from (71),

$$C'_M = C'_{M,\mathrm{SUB}}\left(1 + \frac{\varepsilon_{\mathrm{MUT}}}{\varepsilon_{r,\mathrm{eq}}}\right) \tag{73}$$

Hence, it is apparent that

$$\frac{C'_M}{C'} = \frac{C'_{M,\text{SUB}}}{C'_{\text{SUB}}} = \frac{C'_{M,\text{MUT}}}{C'_{\text{MUT}}} = k \tag{74}$$

Dividing (66) and (67), and using (74), it follows that

$$G_{M,\text{MUT}} = G_{\text{MUT}} \frac{k}{1-k} \tag{75}$$

Similarly, dividing (68) and (69), one obtains

$$G_{M,\text{SUB}} = G_{\text{SUB}} \frac{k}{1-k} \tag{76}$$

and therefore,

$$G_M = G_{M,\text{MUT}} + G_{M,\text{SUB}} = G_{\text{MUT}} \frac{k}{1-k} + G_{\text{SUB}} \frac{k}{1-k} = G \frac{k}{1-k} \tag{77}$$

in agreement with (65).

Let us next compare the phase and the magnitude response of the lossy electromagnetic simulation with the circuit response by including losses. For that purpose, a set of simulations changing the value of the loss tangent of the MUT ($\tan\delta_{\text{MUT}}$) was carried out in [13]. After extracting the reactive parameter values from the phase response of the reflection coefficient for a base value of losses ($\tan\delta_{\text{MUT}} = 0.01$), see caption of Fig. 9, the value of G was tuned to match the same level of the magnitude of the reflection coefficient for each of the $\tan\delta_{\text{MUT}}$ values reported in Fig. 9. It can be seen from Fig. 9b that the circuit model responses are in very good agreement with the lossy EM simulations. This validates the lossy circuit model, and points out that all loss mechanisms, including also ohmic and radiation losses, are adequately accounted for by the model.

An analytical expression for the sensitivity in the limit of small perturbations when losses are considered cannot be easily obtained from the model of Fig. 9a (the involved algebra is extremely cumbersome). However, the effects of losses on sensitivity can be studied by combining (and comparing) electromagnetic simulations and circuit simulations [13]. An approximation to the value of the sensitivity in the limit of small perturbations can be derived by obtaining two responses for close values of the dielectric constant of the MUT, i.e., $\varepsilon_{\text{MUT}} = \varepsilon_{\text{REF}}$ and $\varepsilon_{\text{MUT}} = \varepsilon_{\text{REF}} - 0.01$, and then calculating the difference of the phase response at $f_{p,\text{REF}}$. By dividing such phase difference by the variation in the permittivity (-0.01 in the considered case), the resulting value should be very close to the sensitivity, provided the considered dielectric constant step is very small. Using this procedure, it is possible to extract a set of values of sensitivity for a hypothetical reference MUT with a different value of the loss tangent, while maintaining the real part of the dielectric constant invariable. With the premise that the lossy circuit model accurately describes the sensor response, the sensitivity that results from the circuit model for the values of G that make the value of the magnitude of the reflection coefficient at $f_{p,\text{REF}}$ to coincide with the

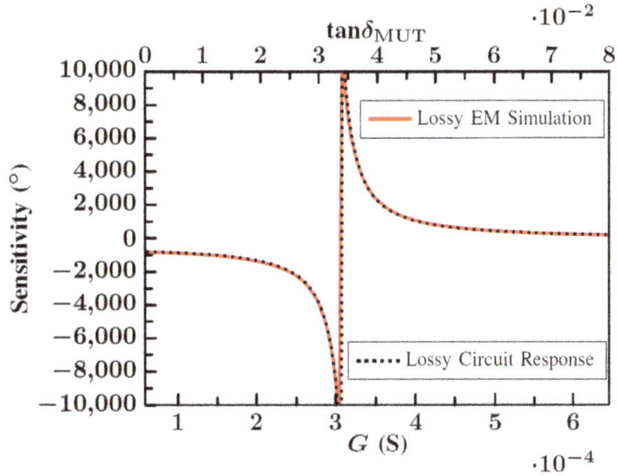

Fig. 11 Sensitivity extracted from lossy EM simulations for different values of dielectric losses in the MUT, and sensitivity extracted from the circuit model for different values of the conductance *G*. Reprinted with permission from [13]; copyright 2023 IEEE

electromagnetic simulation taking into account all loss contributions, has also been obtained. The sensitivities inferred from electromagnetic and circuit simulation are depicted in Fig. 11. It can be clearly observed that the magnitude of the sensitivity first increases, it exhibits an asymptotic behavior at a particular value of $\tan\delta_{MUT}$ (and at the corresponding value of *G*), and then it decreases. If we consider the correspondence between the extracted *G* values and the values of $\tan\delta_{MUT}$, see Fig. 9, it follows that the resulting sensitivities in the limit of small perturbations are very similar (see Fig. 11, where the two curves are roughly undistinguishable).

The value of $\tan\delta_{MUT}$, or *G*, where the magnitude of the sensitivity asymptotically increases delimits the so-called low-loss regime from the high-loss regime of the sensor [13]. Under the low-loss regime, the sensitivity is negative, and the phase slope at the operating frequency is also negative, see Fig. 9b. By contrast, both the sensitivity and the phase slope at the operating frequency are positive under the high loss regime. As the loss level increases (by increasing the value of $\tan\delta_{MUT}$, or *G*), the magnitude of the (negative) phase slope at the operating frequency progressive increases, it exhibits a maximum (infinite) value for $\tan\delta_{MUT}$, or *G*, corresponding to the change of regime, and then the phase slope reverses its sign and decreases. There is thus a correlation between the phase slope and the maximum sensitivity, as anticipated before in this chapter. Note that in the considered case study (Fig. 9), the change of regime takes place for a loss factor comprised between $\tan\delta_{MUT} = 0.03$ and $\tan\delta_{MUT} = 0.04$ (or for a conductance comprised between $G = 281$ μS and $G = 352$ μS), as both Figs. 9 and 11 illustrate. It should be mentioned that the point at which the sensitivity exhibits the asymptotic behavior (change of regime) correlates with the resonators' critical coupling. At this point, the impedance seen from the input port matches the reference impedance, leading to complete absorption of the signal by the resonators. Thus, operation at this critical point should be avoided, as no signal is reflected back to the source[17] (moreover, as it was argued in Sect. 4.1, it is not convenient to design the sensor at the exact boundary between the two operational regimes, since a change of regime as the properties of the material under study vary must be avoided).

In summary, it has been shown that losses help to boost up the sensitivity in reflective-mode phase-variation sensors based on weakly coupled resonators. If the substrate and the considered MUT samples exhibit moderate or low loss tangents, the impact on sensitivity is moderate (i.e., the lossy sensitivity is slightly enhanced as compared to the lossless sensitivity), the case of the prototypes reported in the present chapter. However, a significant improvement in the sensitivity in the limit of small perturbations is expected if losses are significant and are close to the loss level where the change of regime (from the low-loss to the high-loss) occurs.

References

1. Muñoz-Enano J, Vélez P, Su L, Gil M, Casacuberta P, Martín F (2021) On the sensitivity of reflective-mode phase variation sensors based on open-ended stepped-impedance transmission lines: theoretical analysis and experimental validation. IEEE Trans Microw Theory Techn 69(1):308–324
2. Su L, Muñoz-Enano J, Velez P, Casacuberta P, Gil M, Martin F (2021) Highly sensitive phase variation sensors based on step-impedance coplanar waveguide (CPW) transmission lines. IEEE Sensors J 21(3):2864–2872
3. Casacuberta P, Muñoz-Enano J, Vélez P, Su L, Gil M, Martín F (2020) Highly sensitive reflective-mode detectors and dielectric constant sensors based on open-ended stepped-impedance transmission lines. Sensors 20, paper 6236
4. Su L, Muñoz-Enano J, Vélez P, Gil M, Casacuberta P, Martín F (2021) Highly sensitive reflective-mode phase-variation permittivity sensor based on a coplanar waveguide (CPW) terminated with an open complementary split ring resonator (OCSRR). IEEE Access 9:27928–27944
5. Casacuberta P, Vélez P, Muñoz-Enano J, Su L, Gil M, Ebrahimi A, Martín F (2022) Circuit analysis of a coplanar waveguide (CPW) terminated with a step-impedance resonator (SIR) for highly sensitive one-port permittivity sensing. IEEE Access 10:62597–62612
6. Muñoz-Enano J, Vélez P, Casacuberta P, Su L, Martín F, Reflective-mode phase-variation microwave sensor based on a step-impedance microstrip line terminated with a slot resonator for solid and liquid characterization. IEEE Trans Microw Theory Techn. https://doi.org/10.1109/TMTT.2023.3315832
7. Muñoz-Enano J, Vélez P, Su L, Gil M, Martín F (2020) A reflective-mode phase-variation displacement sensor. IEEE Access 8:189565–189575
8. Jha AK, Lamecki A, Mrozowski M, Bozzi M (2020) A highly sensitive planar microwave sensor for detecting direction and angle of rotation. IEEE Trans Microw Theory Techn 68(4):1598–1609
9. Horestani AK, Shaterian Z, Martín F (2020) Rotation sensor based on the cross-polarized excitation of split ring resonators (SRRs). IEEE Sensors J 20:9706–9714

[17] Indeed, in the boundary between the two regimes, where the sensitivity goes asymptotically to infinity, the input signal is completely absorbed by the sensor, regardless on the sensor type, as demonstrated in [24].

10. Mehrjoo Z, Ebrahimi A, Beziuk G, Martín F, Ghorbani K (2023) Microwave rotation sensor based on reflection phase in transmission lines terminated with lumped resonators. IEEE Sensors J 23(7):6571–6580
11. Casacuberta P, Vélez P, Muñoz-Enano J, Su L, Martín F (2023) Highly sensitive reflective-mode phase-variation permittivity sensors using coupled line sections. IEEE Trans Microw Theory Techn 71(7):2970–2984
12. Casacuberta P, Vélez P, Muñoz-Enano J, Su L, Gil-Barba M, Martín F (2022) Reflective-mode phase-variation permittivity sensors based on coupled resonators. IEEE Sensors 2022, Dallas, Texas, USA
13. Casacuberta P, Vélez P, Muñoz-Enano J, Su L, Martín F (2024) Highly sensitive coplanar waveguide (CPW) reflective-mode phase-variation permittivity sensors based on weakly coupled step-impedance resonators (SIRs). IEEE Trans Microw Theory Techn 72(3):1739–1753
14. Hong J, Lancaster MJ (2001) Microstrip filters for RF and microwave applications. Wiley, Hoboken, NJ, USA
15. Pozar DM (2011) Microwave engineering, 4th ed. Wiley, Hoboken, NJ, USA
16. Zysman GI, Johnson AK (1969) Coupled transmission line networks in an inhomogeneous dielectric medium. IEEE Trans Microw Theory Techn 17(10):753–759
17. Muñoz-Enano J, Martel J, Vélez P, Medina F, Su L, Martín F (2021) Parametric analysis of the edge capacitance of uniform slots and application to frequency-variation permittivity sensors. Appl Sci 11, paper 7000
18. Martín F, Vélez P, Muñoz-Enano J, Su L (2022) Planar microwave sensors. Wiley/IEEE Press, Hoboken, NJ, USA
19. Simons RN (2001) Coplanar waveguide circuits, components, and systems. In: Chapter 9: coplanar waveguide discontinuities and circuit elements. Wiley, Hoboken, NJ, USA
20. Muñoz-Enano J, Vélez P, Gil M, Martín F (2020) An analytical method to implement high sensitivity transmission line differential sensors for dielectric constant measurements. IEEE Sensors J 20:178–184
21. Coromina J, Muñoz-Enano J, Vélez P, Ebrahimi A, Scott J, Ghorbani K, Martín F (2020) Capacitively-loaded slow-wave transmission lines for sensitivity improvement in phase-variation permittivity sensors. 50th European microwave conference, Utrecht, The Netherlands
22. Ebrahimi A, Coromina J, Muñoz-Enano J, Vélez P, Scott J, Ghorbani K, Martín F (2021) Highly sensitive phase-variation dielectric constant sensor based on a capacitively-loaded slow-wave transmission line. IEEE Trans Circ Syst I: Reg Papers 68(7):2787–2799
23. Su L, Muñoz-Enano J, Vélez P, Casacuberta P, Gil M, Martín F (2021) Phase-variation microwave sensor for permittivity measurements based on a high-impedance half-wavelength transmission line. IEEE Sensors J 21(9):10647–10656
24. Casacuberta P, Vélez P, Muñoz-Enano J, Su L, Martín F (2023) Losses-assisted sensitivity enhancement in reflective mode-phase-variation permittivity sensors based on weakly coupled distributed resonators. IEEE Sensors Lett 7(8):1–4, Art no. 1500704
25. Vélez P, Canalies X, Muñoz-Enano J, Casacuberta P, Su L, Martín F (2024) Effects of losses on the sensitivity of reflective-mode phase-variation liquid sensors. IEEE Trans Microw Theory Techn 72(2):903–918

Transmission-Mode Permittivity Sensors Based on Coupled Resonators

Zhixia Xu, Yi Wang, and Shaojun Fang

Abstract Loading resonators on a two-port transmission line creates transmission zeros, where each transmission zero corresponds to a resonant mode. By placing an additional resonator with a higher quality (Q) factor close to the existing lower-Q resonator and ensuring that their resonant frequencies overlap, the strong coupling between them makes the original single transmission zero transform into a high-Q transmission peak with a pair of transmission zeros on both sides. The strongly coupled resonators induce this transmission mode. The high Q factor and sensitivity make the transmission mode suitable for two-port transmission-type sensing measurements. In this chapter, we explain the basic theory, design several transmission-mode sensors based on coupled resonators, and use these sensors to detect the permittivity of various samples.

Permittivity is a crucial parameter that characterizes the electrical response of materials to external electromagnetic fields. It plays a vital role in the design of various electric components and devices that operate in the microwave regime, such as transmission lines [1], antennas [2], microwave integrated circuits [3], water-based antennas [4], and metamaterials [5, 6]. Therefore, accurate measurement of permittivity is essential for many applications. One standard method to measure permittivity is based on resonators, which are structures that can resonate at specific frequencies and interact with the materials under test (MUTs) [7]. Resonant frequency shift (Δf) and quality factor (Q) change depending on the permittivity of the MUTs. Different types of resonators can be used for permittivity sensing, such as waveguide cavities [8, 9], and planar resonators [10]. The latter has the advantage of being compact and easy to fabricate. Especially with the emergence of metamaterials, the split

Z. Xu (✉) · Y. Wang · S. Fang
The School of Information Science and Technology, Dalian Maritime University, Dalian 116026, China
e-mail: zxxucn@163.com

Z. Xu
State Key Laboratory of Millimeter Waves, Southeast University, Nanjing 210096, China

F. Martín and E. Bronchalo (eds.), *Coupled Structures for Microwave Sensing*, Lecture Notes in Electrical Engineering 1150,
https://doi.org/10.1007/978-3-031-53861-2_6

ring resonator (SRR) [11] and their complementary counterpart (CSRR) [12] have been widely used in resonant sensors. Moreover, various SRR or CSRR arrays are proposed for advanced sensor designs [13, 14]. At the early stage, planar resonant sensors were based on a single mode, and they can detect permittivity MUTs around the resonant frequency precisely [15]. Now, they can support multiple resonant modes that enable broadband or multi-frequency sensing [16, 17]. Moreover, SRR sensors without substrate have been proposed to eliminate the intrinsic dielectric loss of the sensor itself to increase the sensitivity [18].

In traditional resonant sensors above, loading resonators on a two-port transmission line creates transmission zeros (TZs), where each TZ corresponds to a resonant mode. In this chapter, we present a novel type of planar resonant sensor that exploits the phenomenon of transmission mode based on strong coupling between two resonators [19]. The two resonators have the same resonant frequency but different Q. One is a bright-mode resonator that can be directly excited by the source, while the other is a dark-mode resonator that can only be excited by coupling with the bright-mode resonator [20, 21]. Their strong coupling transforms a TZ into a high-quality transmission peak (TP) with two TZs on both sides. This TP is sensitive to the permittivity of the MUTs and can be used for various sensing applications. Note that a similar transmission-mode phenomenon is also named electromagnetically induced transparency in the physical society [22, 23]. We use coupled-mode theory to explain the generation of the transmission mode and design three prototypes of transmission-mode sensors based on SRRs and spoof localized surface plasmon (spoof-LSP) resonators. We first design the single transmission-mode sensor based on stub-SRR configuration [24], then replace the SRR with a spoof-LSP resonator [25–27] to design a multipole transmission-mode sensor [28], and finally cascade two stub-SRR structures to realize a dual-band sensor. We demonstrate the performance of these sensors by measuring the complex permittivity of various MUTs based on curve-fitting methods.

1 Origin of Transmission Mode Based on Bright-Dark-Mode Coupling

The bright-dark-mode coupling theory explains the origin of the resonant transmission mode discussed in this chapter. The dark mode cannot be directly excited by the excitation source but by strong coupling with the bright mode, and the dark mode shows a very narrow-resonant band with a very high Q factor. The excitation source can directly excite the bright mode: the main transmission line in this chapter. It has the characteristics of a broader resonant band with a lower Q factor. In addition, the bright and dark modes have overlapping center frequencies. Under this condition, a sharp TP appears in the transmission spectrum, which can be utilized to design sensors. In this section, we discuss a series of mechanical oscillator models and deduce the mathematical formulations to describe the spectrum of the high-Q resonant transmission mode in universal wave systems. This section is based on our previous review article [19].

1.1 View of Coupling Harmonic Oscillators

It is usually challenging to obtain the analytical solutions of microwave coupling structures. However, the behavior of mechanical harmonic oscillators can be well deduced through kinetic differential equations. Therefore, we can deduce the universal mathematical model to describe the spectrum of the strong coupling transmission-mode microwave structures: (1) We solve the double-oscillators equations, which describe a primary transmission-mode spectrum corresponding to the stub-SRR structure in Sect. 2. (2) We extend the mechanical model to the triple-harmonic oscillator model. (3) We provide solutions to higher-order models. These higher-order models describe the multipole transmission-mode sensors based on the stub-LSP structure in Sect. 3.

1.2 Double-Oscillators Model

Figure 1 shows the classical double-oscillators model. Oscillator 1 can be considered the bright mode, while Oscillator 2 can be considered the dark mode because only Oscillator 1 is subjected to external harmonic forces. The whole system consists of two oscillators with specific masses and three springs with fixed stiffness coefficients, where the masses of the oscillators are m_1 and m_2 and the stiffness coefficients of the springs are k_1, K, and k_2, respectively. The springs with elastic coefficients k_1 and k_2 are fixed to the wall at one end and connected to the oscillators m_1 and m_2 at the other.

To simplify the model, we assume that $k_1 = k_2 = k$, $m_1 = m_2 = m$, and the self-resonant frequency of the oscillator $\omega_1 = \omega_2 = \omega$. The displacements are denoted by x_1(t) and x_2(t), the respective damping parameters are denoted by γ_1 and γ_2, and the coupling strength between the two oscillators is denoted as $\Omega_r^2 = K/m$. The

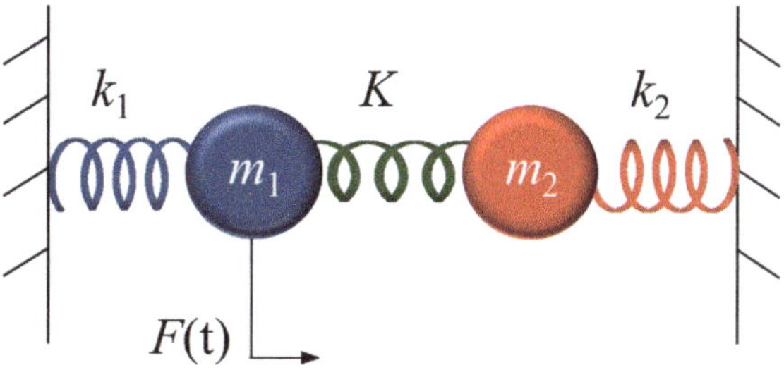

Fig. 1 Double-oscillators model. Reprinted with permission from [19]; copyright 2023 under a Creative Commons Attribution 4.0 License

differential equations to describe the displacements of the system can be expressed as

$$\begin{cases} \ddot{x}_1(t) + \gamma_1 \dot{x}_1(t) + \omega^2 x_1(t) - \Omega_r^2 x_2(t) = \frac{F}{m} e^{-i\omega_s t} \\ \ddot{x}_2(t) + \gamma_2 \dot{x}_2(t) + \omega^2 x_2(t) - \Omega_r^2 x_1(t) = 0 \end{cases} \quad (1)$$

The displacement of Oscillator 1 can be solved as

$$x_1(t) = \frac{(\omega^2 - \omega_s^2 - i\gamma_2\omega_s) F e^{-i\omega_s t}}{m\left[\left(\omega^2 - \omega_s^2 - i\gamma_1\omega_s\right)\left(\omega^2 - \omega_s^2 - i\gamma_2\omega_s\right) - \Omega_r^4\right]} \quad (2)$$

The absorbed power of Oscillator 1 in one period is calculated as

$$P_s(\omega_s) = -\frac{2\pi i F^2 \omega_s (\omega^2 - \omega_s^2 - i\gamma_2\omega_s)}{m\left[\left(\omega^2 - \omega_s^2 - i\gamma_1\omega_s\right)\left(\omega^2 - \omega_s^2 - i\gamma_2\omega_s\right) - \Omega_r^4\right]} \quad (3)$$

We further use Eq. (4) to express the normalized remaining power, corresponding to the transmission spectrum (S_{21}) in microwave bands.

$$T = 1 - P_s(\omega_s) \quad (4)$$

Based on (4), we further discuss the variation of the transmission spectrum of the system under different coupling strengths Ω_r, as shown in Fig. 2. The fixed parameters are $\omega = 2$ rad/s, $\gamma_1 = 4 \times 10^{-2}$ s^{-1}, $\gamma_2 = 1 \times 10^{-2}$ s^{-1}, $F = 0.1$ N, $m = 1$ kg. When two oscillators are coupled, the transmission power spectrum of the whole system will appear as a very sharp TP, and with the increase of the coupling strength Ω_r, the amplitude of the TP will be larger, and the two TZs will split farther.

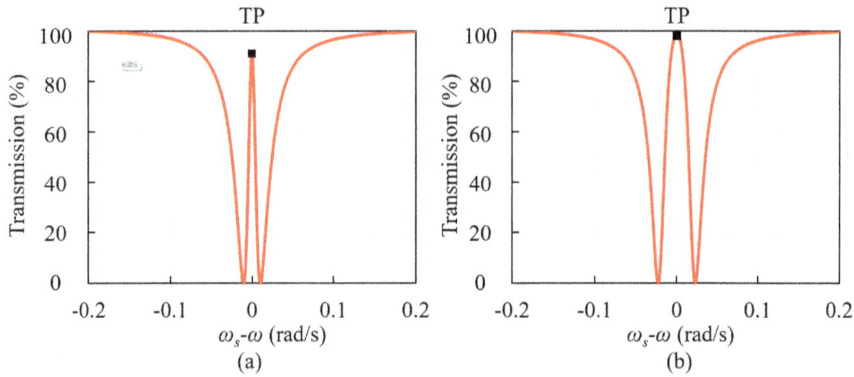

Fig. 2 The normalized transmission spectrum varies with different coupling coefficients **a** $\Omega_r = 0.2$ rad/s. **b** $\Omega_r = 0.3$ rad/s

1.3 Triple-Oscillators Model

By adding oscillators, we can obtain more transmission modes. We analyze the triple-oscillators model, as shown in Fig. 3. In this model, Oscillator 1, the bright mode, is subject to harmonic forces. At the same time, Oscillators 2 and 3 are dark modes, and the three oscillators are connected by springs with elastic coefficients k_1, k_2, k_3, k_{21}, and k_{31}. Three oscillators' displacements, damping losses, and masses are denoted as x_i(t), γ_i, $m_1 = m_2 = m_3 = m$. We assume the elastic coefficient $k_{21} = k_{31} = K$, the coupling coefficient $\Omega_r = (K/m)^{1/2}$, the self-resonant frequency $\omega_i = (k_i/m)^{1/2}$. The differential equations of the whole system can be expressed by

$$\begin{cases} \ddot{x}_1(t) + \gamma_1\dot{x}_1(t) + \omega_1^2 x_1(t) - \Omega_r^2 x_2(t) - \Omega_r^2 x_3(t) = \frac{F}{m}\mathrm{e}^{-\mathrm{i}\omega_s t} \\ \ddot{x}_2(t) + \gamma_2\dot{x}_2(t) + \omega_2^2 x_2(t) - \Omega_r^2 x_1(t) = 0 \\ \ddot{x}_3(t) + \gamma_3\dot{x}_3(t) + \omega_3^2 x_3(t) - \Omega_r^2 x_1(t) = 0 \end{cases} \tag{5}$$

The displacement of Oscillator 1 can be solved as

$$x_1(t) = \frac{F\mathrm{e}^{-\mathrm{i}\omega_s t}}{m\left[\left(\omega_1^2 - \omega_s^2 - i\gamma_1\omega_s\right) - \frac{\Omega_r^4}{\omega_2^2 - \omega_s^2 - i\gamma_2\omega_s} - \frac{\Omega_r^4}{\omega_3^2 - \omega_s^2 - i\gamma_3\omega_s}\right]} \tag{6}$$

Therefore, the absorbed power of Oscillator 1 in one period can be expressed as

$$P_s(\omega_s) = \frac{-2\pi\mathrm{i}F^2\omega_s}{m\left[\left(\omega_1^2 - \omega_s^2 - i\gamma_1\omega_s\right) - \frac{\Omega_r^4}{\omega_2^2 - \omega_s^2 - i\gamma_2\omega_s} - \frac{\Omega_r^4}{\omega_3^2 - \omega_s^2 - i\gamma_3\omega_s}\right]} \tag{7}$$

The normalized transmission spectrum of the triple-oscillators system defined in (4) is shown in Fig. 4. The fixed parameters are $\omega_1 = 2$ rad/s, $\omega_2 = 1.9$ rad/s, $\omega_3 =$ 2.0 rad/s, $\gamma_1 = 4 \times 10^{-2}$ s^{-1}, $\gamma_2 = \gamma_3 = 1 \times 10^{-4}$ s^{-1}, $F = 0.1$ N, $m = 1$ kg. It should be mentioned that the damping parameter of Oscillator 1 is set much larger than others to ensure that the bright mode has the lowest Q factor and its bandwidth is broad enough to couple with other dark modes. Two sharp TPs appear in the transmission

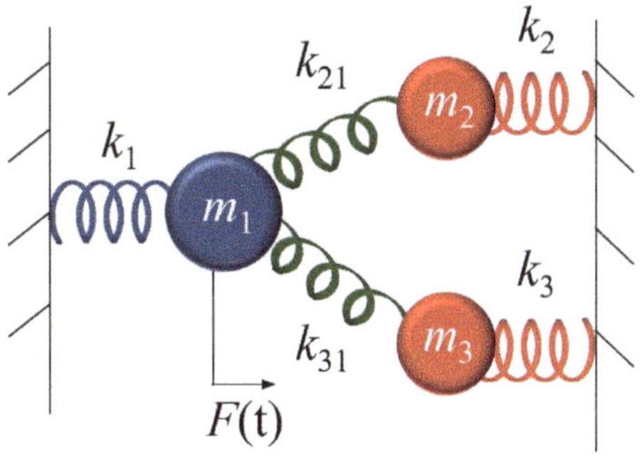

Fig. 3 Triple-oscillators model. Reprinted with permission from [19]; copyright 2023 under a Creative Commons Attribution 4.0 License

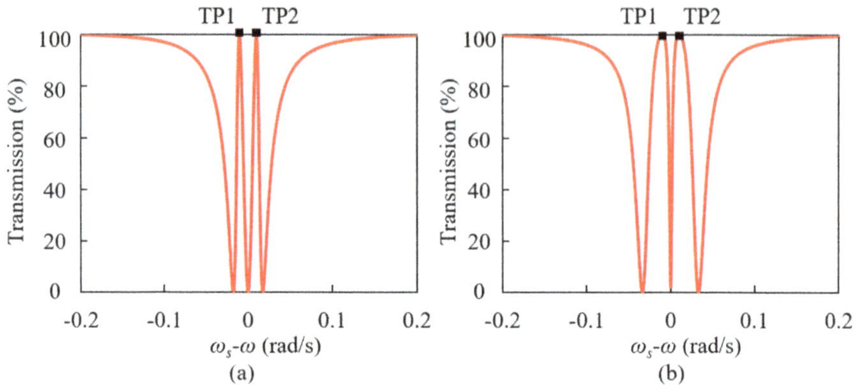

Fig. 4 The normalized transmission spectrum varies with different coupling coefficients **a** $\Omega_r = 0.2$ rad/s. **b** $\Omega_r = 0.3$ rad/s. The coupling coefficient affects the Q-factor of transmission modes, without changing the center frequency

spectrum, as shown in Fig. 4a. The center frequency of the two TPs depends on ω_2 and ω_3, respectively. With the increase of coupling strength, the magnitude of the transparent transmission peak becomes larger, and the TZs split farther, as shown in Fig. 4b. We further sweep the change of the self-resonant frequency in Fig. 5 to prove that the center frequency of TPs is related to the self-resonant frequency of individual dark modes ω_i.

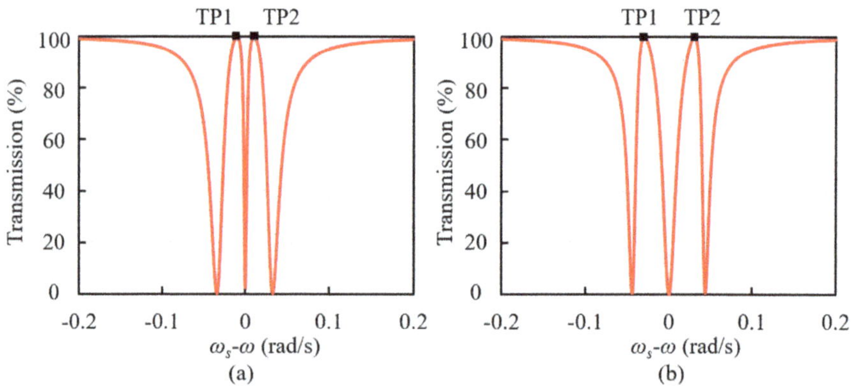

Fig. 5 The normalized transmission spectrum varies with different resonant frequencies. **a** $\omega_1 = 2$ rad/s, $\omega_2 = 1.99$ rad/s, $\omega_3 = 2.01$ rad/s **b** $\omega_1 = 2$ rad/s, $\omega_2 = 1.97$ rad/s, $\omega_3 = 2.03$ rad/s. The center frequency of transmission modes depends on the resonant frequency of dark modes

1.4 Higher-Order Models

We further extend our analysis to higher-order coupling-oscillators models and deduce the generalized solutions of the transmission spectrum, as shown in Fig. 6. Parameters of each oscillator include the mass m_i, the damping coefficient γ_i, the elastic coefficient k_i, the self-resonant frequency $\omega_i = (k_i/m_i)^{1/2}$ and the displacement x_i(t). The elastic coefficient connecting any two oscillators m and n is k_{mn}. Oscillator 1, the bright mode, is subject to the harmonic force $F(t) = F\mathrm{e}^{-\mathrm{i}\omega_s t}$. The mechanical displacement equations of the whole system can be expressed by

$$\begin{cases} \ddot{x}_1(t) + \gamma_1\dot{x}_1(t) + \omega_1^2 x_1(t) - \sum\limits_{i=2}^{n} \frac{k_{i1}}{m_1} x_i(t) = \frac{F}{m_1}\mathrm{e}^{-\mathrm{i}\omega_s t} \\ \ddot{x}_2(t) + \gamma_2\dot{x}_2(t) + \omega_2^2 x_2(t) - \dfrac{k_{21}}{m_2} x_1(t) = 0 \\ \dots \\ \ddot{x}_i(t) + \gamma_i\dot{x}_i(t) + \omega_i^2 x_i(t) - \frac{k_{i1}}{m_i} x_1(t) = 0 \end{cases} \tag{8}$$

Suppose the coupling coefficient between two oscillators are $\Omega_{ij}^4 = k_{ij}^2/m_i m_j$, we can deduce the displacement of Oscillator 1 as

$$x_1(t) = \frac{F\mathrm{e}^{-\mathrm{i}\omega_s t}}{m_1\left[\left(\omega_1^2 - \omega_s^2 - i\gamma_1\omega_s\right) - \sum\limits_{i=2}^{n} \frac{\Omega_{i1}^4}{\omega_i^2 - \omega_s^2 - i\gamma_i\omega_s}\right]} \tag{9}$$

The absorbed power of Oscillator 1 in one period can be written as

$$P_s(\omega_s) = -\frac{2\pi i F^2 \omega_s}{m_1\left[\left(\omega_1^2 - \omega_s^2 - i\gamma_1\omega_s\right) - \sum\limits_{i=2}^{n} \frac{\Omega_{i1}^4}{\omega_i^2 - \omega_s^2 - i\gamma_i\omega_s}\right]} \tag{10}$$

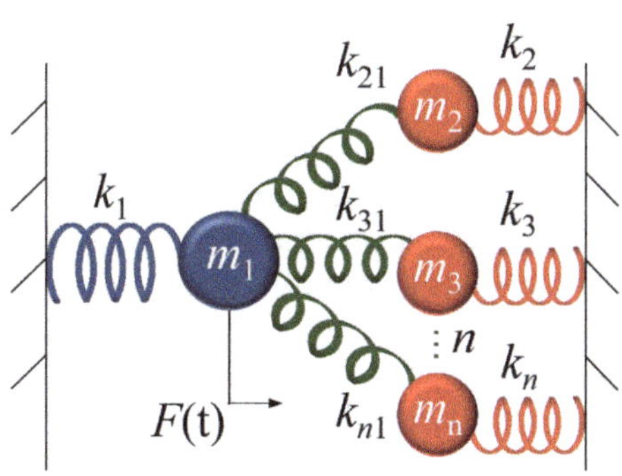

Fig. 6 High-order coupling-oscillators model. Reprinted with permission from [19]; copyright 2023 under a Creative Commons Attribution 4.0 License

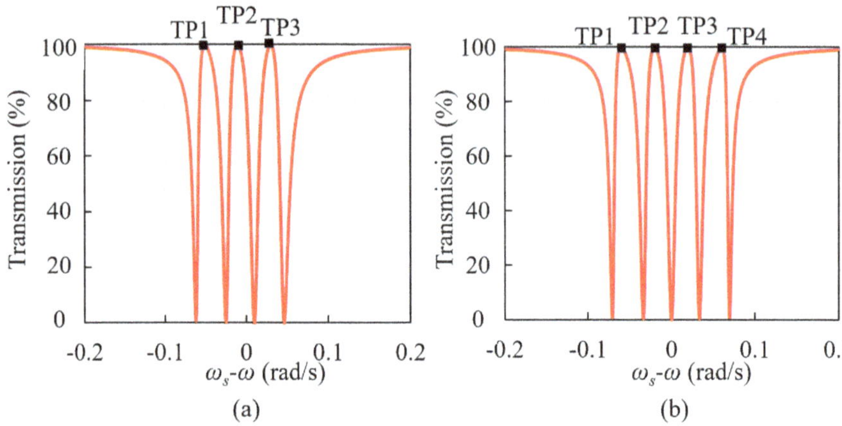

Fig. 7 The normalized transmission spectrum of high-order coupling-oscillators model. **a** The fourth-order-oscillators **b** the fifth-order-oscillators

Expression (10) is universal, and when n is set to 2 or 3, (10) corresponds to (3) and (7), respectively. The transmission spectrum of the four and five-oscillators are given in Fig. 7 as examples.

In the fourth-order mechanical model, the parameters are $\omega_1 = 2$ rad/s, $\omega_2 = 1.95$ rad/s, $\omega_3 = 1.99$ rad/s, $\omega_4 = 2.03$ rad/s, $\gamma_1 = 4 \times 10^{-2}\ \text{s}^{-1}$, $\gamma_2 = \gamma_3 = \gamma_3 = 1 \times 10^{-4}\ \text{s}^{-1}$, $F = 0.1$ N, $m = 1$ kg, $\Omega_r = 0.3$ rad/s. In the fifth-order mechanical model, the parameters are $\omega_1 = 2$ rad/s, $\omega_2 = 1.94$ rad/s, $\omega_3 = 1.98$ rad/s, $\omega_4 = 2.02$ rad/s, $\omega_5 = 2.06$ rad/s, $\gamma_1 = 4 \times 10^{-2}\ \text{s}^{-1}$, $\gamma_2 = \gamma_3 = \gamma_3 = \gamma_4 = 1 \times 10^{-4}\ \text{s}^{-1}$, $F = 0.1$ N, $m = 1$ kg, $\Omega_r = 0.3$ rad/s. The angular frequency ω_i corresponding to the center frequency of each TP is only decided by the self-resonant frequency of dark-mode oscillators, which can be calculated by the mass m_i and to the elastic coefficient k_i of the spring connected to the wall.

1.5 Summary

This section has introduced the bright-dark-mode theory behind the high-Q transmission mode induced by the strong coupling. We have used coupling-oscillators models to derive the mathematical model of the transmission spectrum and identify the system's key parameters. Some fundamental properties are summarized as follows: (1) The system requires that the dark modes have a higher Q factor than the bright mode. In other words, the bright mode's resonant curve should cover the dark mode's resonant curve. (2) The self-resonant frequency of dark modes decides the center frequency of TPs, and the number of TPs is the same as the number of dark modes. (3) The intrinsic damping of dark modes decides the Q factor of TPs. When the damping of dark modes increases, the Q factor decreases. (4) When the coupling coefficient between bright and dark modes increases, the transmission efficiency of TPs increases, and it directly decides the frequency gap between the TZs on both sides of the TP. Although all the analysis is based on simple mechanical models, we will show that the same phenomenon and conclusion are still applicable in the microwave regime in the following sections about transmission-mode microwave sensors.

2 Basic Transmission-Mode Permittivity Sensor

Corresponding to the double-oscillators model in Fig. 1, we designed a fundamental microwave transmission-mode sensor based on the stub and SRR [24]. This section discusses the structure design, equivalent circuit model, and experimental results.

2.1 *Structure Design*

As shown in Fig. 8, the transmission-mode structure comprises a stub (bright mode) and an SRR structure (dark mode), whose resonant frequency is the same. The electrical length of the stub is λ/4, and the open stub is bent at 90° to couple with the SRR. The main transmission line is connected directly with the stub, meaning the bright-mode stub can be excited by the source directly. The main transmission line cannot couple the dark-mode SRR but is coupled with a stub by the small gap. The geometric dimensions of the structures are given in Table 1. The entire sensor structure is etched on an FR-4 substrate with a complex permittivity of 4.5, a thickness of 0.8 mm, and a loss tangent of 0.025.

The simulation analysis of the entire structure is realized based on the frequency domain solver of *CST Microwave Studio* software. Figure 9 shows the transmission

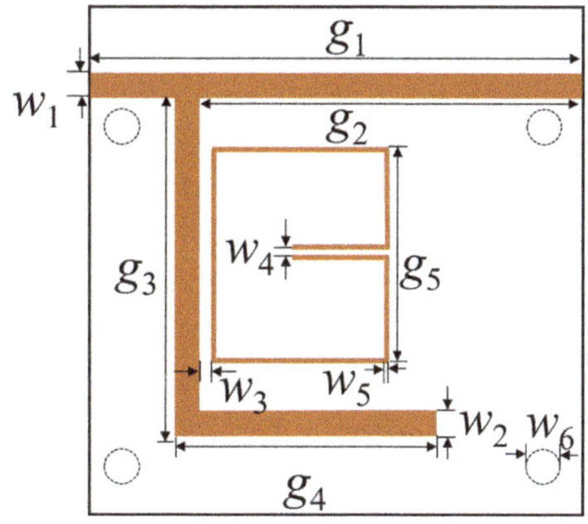

Fig. 8 Stub-SRR transmission-mode structure. The four circles marked by dotted lines are the holes for fixing the Teflon container in the following experiments, and the orange parts represent the microstrip structures. Reprinted with permission from [24]; copyright 2021 IEEE

Table 1 The geometric parameters of the stub-SRR structure

Parameters	w_1	w_2	w_3	w_4	w_5	w_6
Values	1.5 mm	1.5 mm	0.3 mm	0.2 mm	0.3 mm	2 mm
Parameters	g_1	g_2	g_3	g_4	g_5	g_6
Values	40 mm	30 mm	25 mm	18 mm	15 mm	10.2 mm

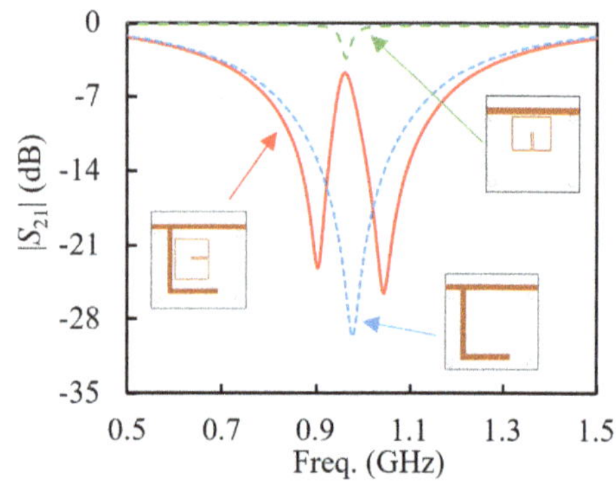

Fig. 9 S_{21} curves of the stub, SRR and strongly coupled stub-SRR structure. Reprinted with permission from [24]; copyright 2021 IEEE

curves of the open stub, the SRR, and the transmission-mode structure, respectively. When the stub and SRR are analyzed separately, the transmission has a TZ, as shown in the blue and green dashed curves. Note that the center frequency of these two TZs overlaps. When we combine the two resonators to construct a strong-coupling system, the spectrum splits, and we can find one TP and two TZs on both sides (red lines), which is the typical transmission spectrum of the transmission-mode structure. We can compare this result with Fig. 2.

In order to better demonstrate the physical procedure of the structure, we plot the field distributions in Fig. 10. The individual resonant modes are shown in Fig. 10a, b. When the main transmission line excites the two resonators separately, the TZ represents the resonant modes in the stub and SRR. Then, we move SRR farther away from the main transmission line, and it can only be coupled with the stub. The transmission mode, corresponding to the TP and two TZs on both sides of the stub-SRR structure, is shown in Fig. 10c–e. The field distributions of the SRR are anti-symmetrical in the left/right TZs.

2.2 *Equivalent Circuit*

The equivalent circuit of the structure is shown in Fig. 11. The stub can be regarded as a series resonant circuit, and the SRR is described by a resonant loop. The energy between the two resonant circuits is coupled through the mutual inductor, and the resistance is used to represent the damping coefficients brought by metal and other

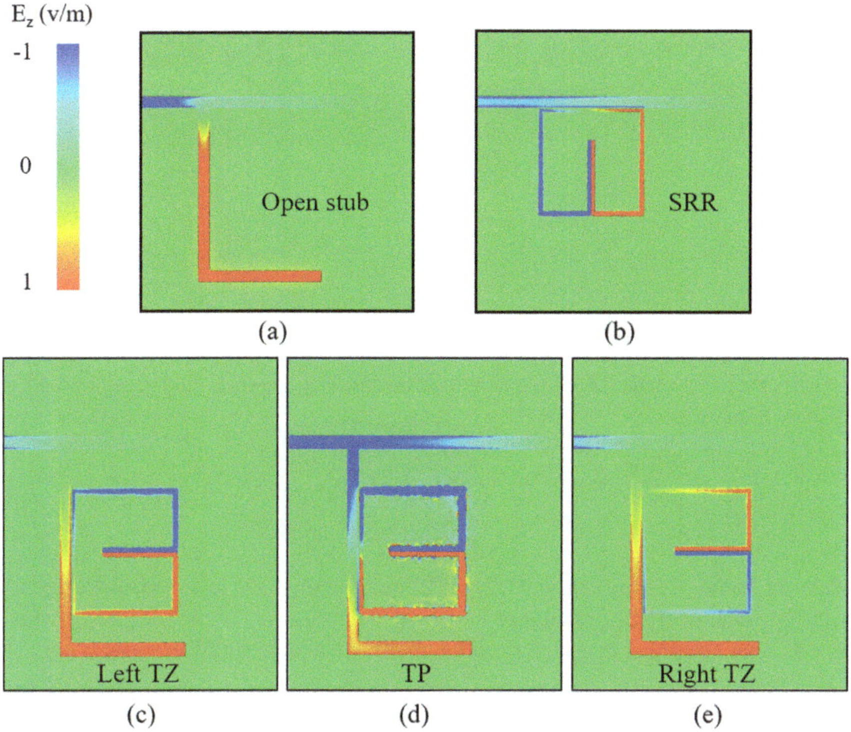

Fig. 10 Distributions of electric field of the stub, SRR, and strongly coupled combination at different frequencies. Reprinted with permission from [24]; copyright 2021 IEEE

losses. Let us analyze this microwave network. The input impedance at the parallel node of the stub to the main transmission line can be expressed as

$$Z = R_1 + \mathrm{i}\left(\omega L_1 - \frac{1}{\omega C_1}\right) + \frac{(\omega M)^2}{R_2 + \mathrm{i}\left(\omega L_2 - \frac{1}{\omega C_2}\right)} \tag{11}$$

where R_i, L_i, and C_i represent the resistance, inductance, and capacitance of the open stub ($i = 1$) and SRR ($i = 2$), respectively. M is the mutual inductance between L_1 and L_2, related to the coupling coefficient k as

$$k = \frac{M}{\sqrt{L_1 L_2}} \tag{12}$$

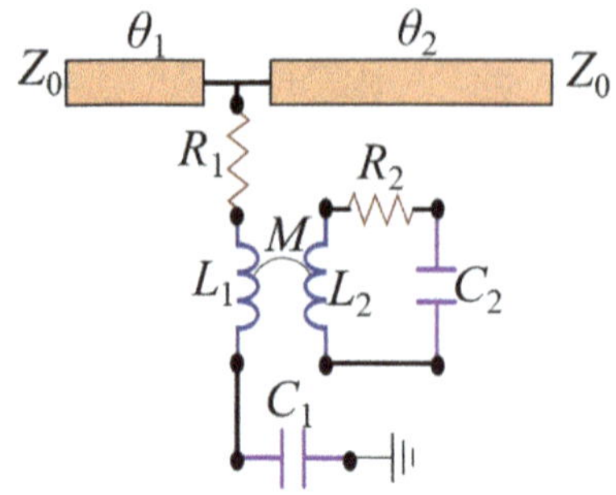

Fig. 11 The equivalent circuit of the stub-SRR structure. Reprinted with permission from [24]; copyright 2021 IEEE

The cascaded matrix T of the whole transmission network can be expressed as (13),

$$\mathrm{T} = \begin{bmatrix} A & B \\ C & D \end{bmatrix} = \begin{bmatrix} \cos\theta_1 & \mathrm{i}Z_0\sin\theta_1 \\ \mathrm{i}\frac{1}{Z_0}\sin\theta_1 & \cos\theta_1 \end{bmatrix} \begin{bmatrix} 1 & 0 \\ \frac{1}{Z} & 1 \end{bmatrix} \begin{bmatrix} \cos\theta_2 & \mathrm{i}Z_0\sin\theta_2 \\ \mathrm{i}\frac{1}{Z_0}\sin\theta_2 & \cos\theta_2 \end{bmatrix} \tag{13}$$

where θ_1 and θ_2 are the electrical lengths from the node to the left and right terminals with 50 Ω characteristic impedance Z_0. The transmission coefficient S_{21} of the whole network can be further deduced from the matrix T, which can be expressed as

$$S_{21} = \frac{2}{A+\frac{B}{Z_0}+CZ_0+D} \tag{14}$$

Table 2 shows the specific parameters of the equivalent circuit, which are obtained by the ADS curve fitting toolbox. We compare transmission curves obtained from the measurement, CST simulation, and equivalent circuit fitting, as shown in Fig. 12. It can be seen that all results match well and verify the correctness of our analysis. In order to verify the correctness of the mathematical model deduced from the coupling-oscillators models in Sect. 1, we use the curve-fitting method to match the simulated transmission with (3) and (4), as shown in Fig. 13. Mathematical models obtained in Sect. 1 is applicable to analyze the microwave structure.

Table 2 The equivalent circuit parameters of the single frequency sensor

Parameters	Z_0	L_1	L_2	C_1	C_2
Values	50 Ω	4.84 nH	4.85 nH	5.277 pF	5.634 pF
Parameters	R_1	R_2	θ_1	θ_2	M
Values	0.9 Ω	0.649 Ω	9.8°	34.74°	0.765 nH

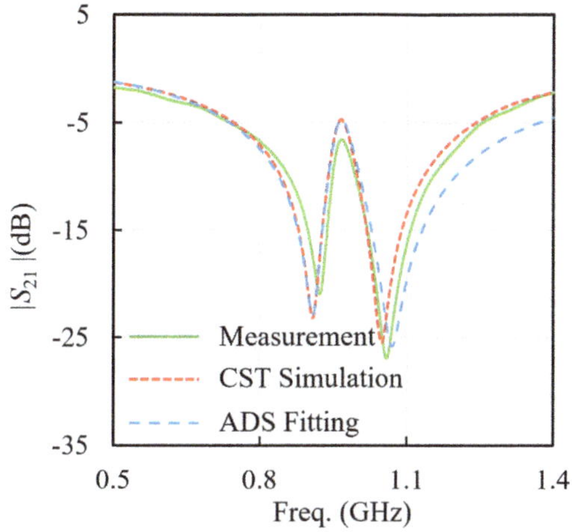

Fig. 12 S_{21} curves obtained by the measurement, CST simulation, and equivalent circuit fitting. Reprinted with permission from [24]; copyright 2021 IEEE

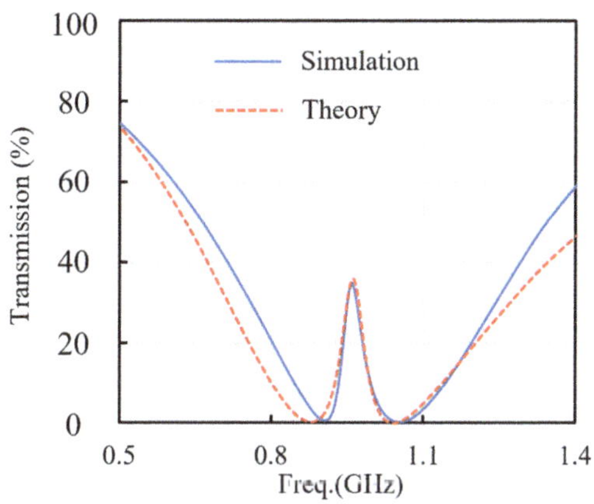

Fig. 13 S_{21} curves obtained by the CST simulation and coupling-oscillators model. Parameters of the theoretical model (3) and (4): $\gamma_1 = 4.77\ s^{-1}$, $\gamma_2 = 0.22\ s^{-1}$, $\Omega = 0.35 \times 2\pi \times 10^9$ rad/s, $\omega = 0.96 \times 2\pi \times 10^9$ rad/s, $F = 1$ N, $m = 1$ kg

2.3 *Experiments*

The stub-SRR structure is further used to measure the complex permittivity of liquids. The complex permittivity is dispersive. The real part ε' is related to the electron displacement polarization, and the imaginary part ε'' is related to the loss factor. The complex permittivity of a liquid mixture is expressed by the Debye model as

$$\varepsilon = \varepsilon_h + \frac{\Delta\varepsilon}{1+i\omega\tau} \tag{15}$$

where ε_h is the complex permittivity at infinite frequency, $\Delta\varepsilon$ is the dielectric attenuation, and τ is the relaxation time. These parameters depend on the different liquid mixtures and volume fractions, and the relevant values are available in [29]. The surface of the microstrip circuit is sensitive to the external environment, and the influence of MUTs on the surface will change the equivalent capacitance and affect

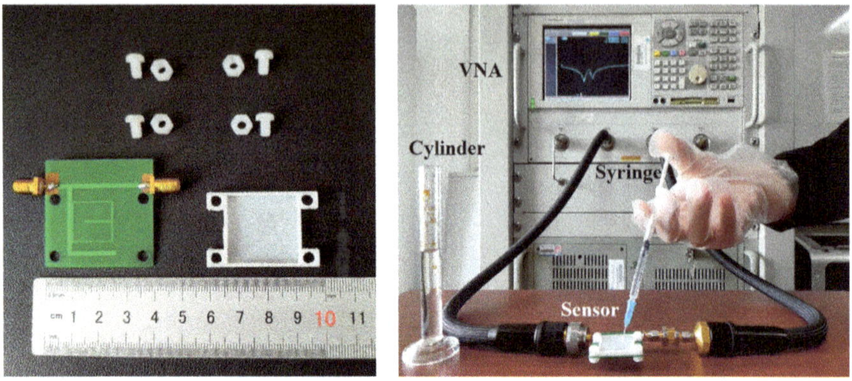

(a) (b)

Fig. 14 **a** Fabricated prototype. **b** Experimental setup. Reprinted with permission from [24]; copyright 2021 IEEE

the transmission spectrum. The MUTs in this experiment are binary liquid mixtures with different volume fractions (water–ethanol). The fabricated sensor with a Teflon container is shown in Fig. 14a, and the overall experimental setup is shown in Fig. 14b. The measurement instrument is a vector network analyzer (model *Agilent N5230A*), and the two ports of the sensor are connected to the vector network analyzer through coaxial cables to record the transmission spectrum.

A measuring cylinder is used to prepare different MUTs in the experiment, and a syringe is used to inject them. The center frequency of TP f_0 and Q factor varies with the change of different MUTs, and the real and imaginary parts of the complex permittivity can be further extracted. The water–ethanol solutions with different volume fractions were tested, as shown in Fig. 15. A total of 11 sets of S_{21} curves were recorded, where the volume fraction of the water varies from 0 to 100% in steps of 10%. As the water content increases from 0 to 100%, f_0 decreases from 890 to 861 MHz, and the Q value, defined as

$$Q = \frac{f_{\text{peak}}}{\text{BW}} \tag{16}$$

increases monotonically. In (16), f_{peak} is the frequency of the TP and BW denotes the 3 dB transmission bandwidth.

The values of f_0 and Q of each curve were extracted for quantitative analysis and fitted by a set of linear functions, which are presented in Eqs. (17)–(21).

$$\begin{bmatrix} \Delta f_0 \\ \Delta Q \end{bmatrix} = \begin{bmatrix} m_{11} & m_{12} \\ m_{21} & m_{22} \end{bmatrix} \begin{bmatrix} \Delta\varepsilon' \\ \Delta\varepsilon'' \end{bmatrix} \tag{17}$$

$$\Delta\varepsilon' = \varepsilon'_{sam} - \varepsilon'_{ref} \tag{18}$$

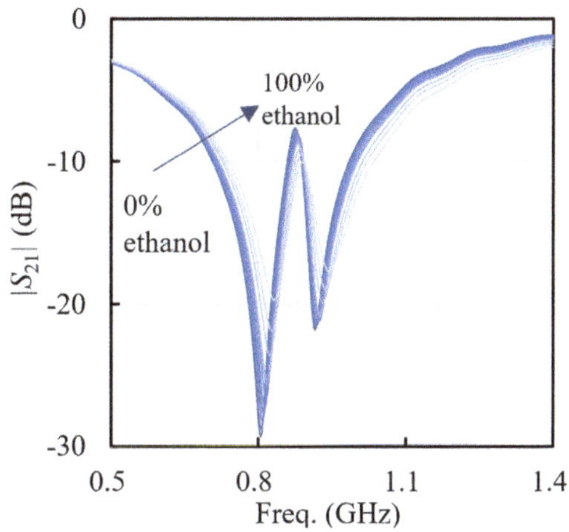

Fig. 15 Measured S_{21} curves of different water–ethanol mixtures. Reprinted with permission from [24]; copyright 2021 IEEE

$$\Delta\varepsilon'' = \varepsilon''_{sam} - \varepsilon''_{ref} \tag{19}$$

$$\Delta f_0 = f_{sam} - f_{ref} \tag{20}$$

$$\Delta Q = Q_{sam} - Q_{ref} \tag{21}$$

The subscript "*sam*" in the equations indicates the measured sample, and "*ref*" indicates the reference sample. The unknown coefficients of the matrix in Eq. (17) can be calculated from the measurement data based on the least squares method [30].

$$\begin{bmatrix} \Delta\varepsilon' \\ \Delta\varepsilon'' \end{bmatrix} = \begin{bmatrix} -3281.8 & -1.7 \\ -604.6 & -1.7 \end{bmatrix} \begin{bmatrix} \Delta f_0 \\ \Delta Q \end{bmatrix} \tag{22}$$

Equation (22) is the calculated empirical model based on the least squares method [30] for detect the complex permittivity of unknown MUTs. We compare our results with the Debye model and the results from [11] in Fig. 16. All results agree well with each other, verifying the working performance of the sensor.

The relative sensitivity of the sensor, defined as

$$S = \frac{\Delta f_0}{\Delta\varepsilon \cdot f_0} \times 100\% \tag{23}$$

was found to be 0.05% based on the experimental results shown in Fig. 16. However, the Teflon container affects the sensitivity. It is foreseeable that if we place MUTs directly on the surface of the circuit, the calculated sensitivity will increase dramatically. Figure 17 shows the simulated and measured transmission spectrum when detecting various solid MUTs, including 3-mm-think Teflon ($\varepsilon \approx 2.1$) and glass

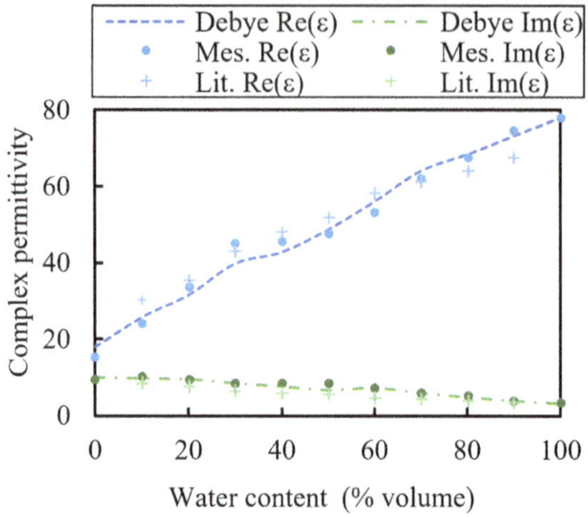

Fig. 16 Comparison of data of the measurement, the Debye model and the other literature [11]. Reprinted with permission from [24]; copyright 2023 IEEE

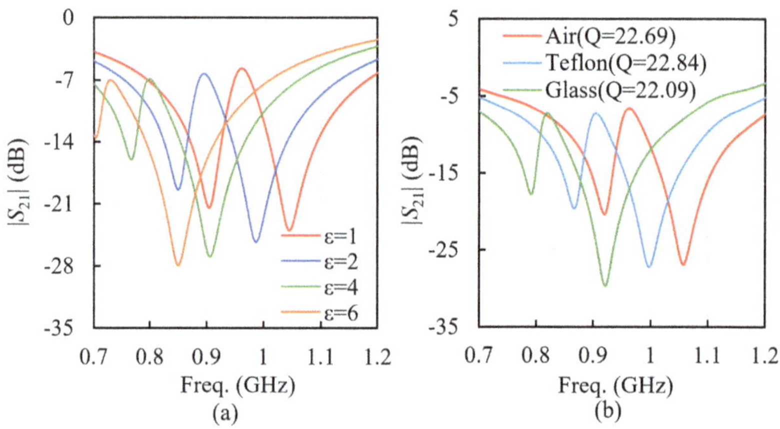

Fig. 17 Simulated and measured S_{21} curves with different MUTs placed on the surface. **a** Simulated S_{21} curves. **b** Measured S_{21} curves. Reprinted with permission from [24]; copyright 2021 IEEE

($\varepsilon \approx 4$) substrates. The calculated sensitivity increases from 0.05 to 5%, indicating that the proposed sensor can achieve a very high sensitivity when directly detecting MUTs placed on the surface.

2.4 Summary

As the microwave presentation of the double-oscillators model in Fig. 1, a fundamental transmission-mode sensor based on a stub-SRR structure is presented. The stub is the bright mode, and the SRR is the dark mode. The coupling-induced transmission mode exhibits high sensitivity to measure the permittivity of MUTs on the surface. The change in the complex permittivity can be derived from the change in the transmission curve. Water–ethanol mixtures and solid samples (Teflon and glass substrates) were tested.

3 Multipole Transmission-Mode Permittivity Sensor

This section discusses a higher-order transmission-mode structure based on a spoof-LSP resonator and a broadband fan-shaped stub [28]. The structure is a microwave realization of the mechanical higher-order coupling-oscillators model in Fig. 6, containing multipole transmission modes. It is a traditional method to realize higher-order transmission modes by adding a number of resonators [31]; however, this section only uses two resonators, and the multipole spoof-LSP resonances are the key to realizing the higher-order transmission modes. This section discusses the structure design, equivalent circuit model, and measurements.

3.1 Structure Design

Based on the prior stub-SRR structure, we designed a fan-shaped stub as a broadband resonator and replaced the SRR with a spoof-LSP resonator containing multipole resonances, as shown in Fig. 18. The fan-shaped stub provides the bright mode, and the spoof-LSP resonator provides the multipole dark modes because the fan-shaped stub is directly connected to the main transmission line, and the spoof-LSP resonator can only be excited by coupling with the fan-shaped stub. We first study the properties of the two resonators, respectively, then analyze the coupling transmission modes of the whole structure.

In order to realize multipole transmission modes, the bandwidth of the bright mode needs to be wide enough to cover all dark modes. The fan-shaped stub is a simple choice. By increasing the angle of the fan-shaped stub, we can broaden the frequency spectrum, as shown in Fig. 19. The relative bandwidth of the fan-shaped stub is much broader than the simple stub used in the prior stub-SRR structure. The center frequency is at 1.45 GHz, and the magnitude of S_{21} remains below -10 dB from 1.05 to 1.96 GHz.

The spoof-LSP resonator provides multipole resonant modes, as shown in Fig. 20a. The basic concept behind the traditional circular spoof-LSP resonator is like the

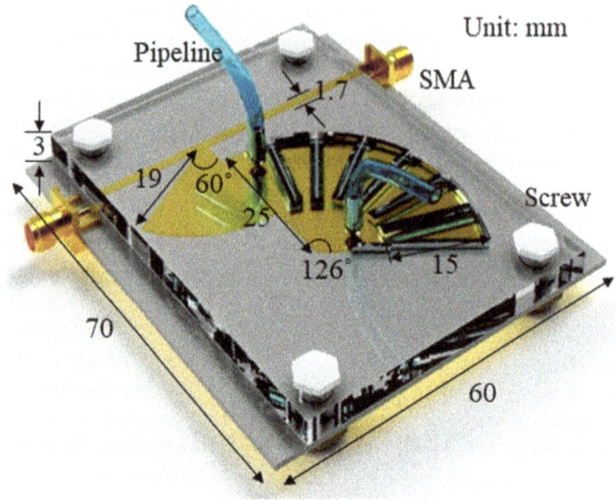

Fig. 18 Multipole transmission-mode sensor structure. Substrate is *Rogers 4350B* with permittivity of 3.48, loss tangent loss of 0.0037, and thickness of 0.762 mm. Microfluidic structure is made of polycarbonate, the permittivity of which is 3 and the loss tangent is 0.05. Pipe is 0.2 mm away from the spoof-LSP structure. Reprinted with permission from [28]; copyright 2022 American Physical Society

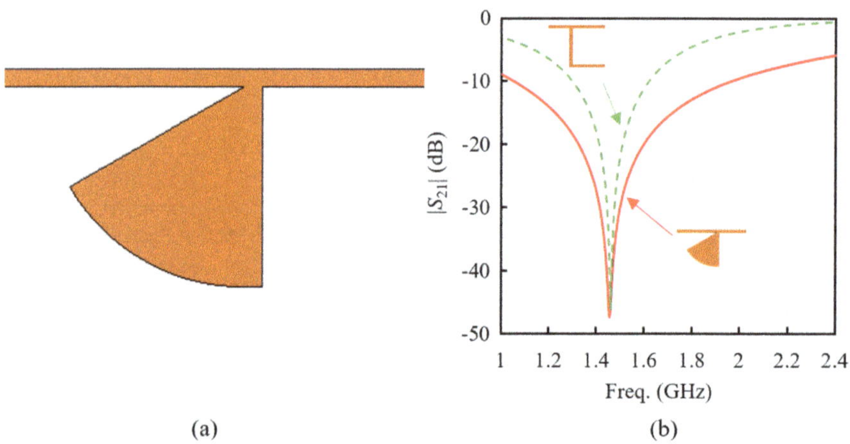

Fig. 19 **a** Fan-shaped stub. **b** S_{21} curves of the fan-shaped stub compared with the prior stub in Sect. 2. Reprinted with permission from [28]; copyright 2022 American Physical Society

whispering-gallery-mode (WGM) resonances [25]. WGM resonators support bidirectional transmission of clockwise and counterclockwise to establish resonances, whose resonant condition of the WGM resonator is written as below,

$$\beta \times C = 2\pi n \tag{24}$$

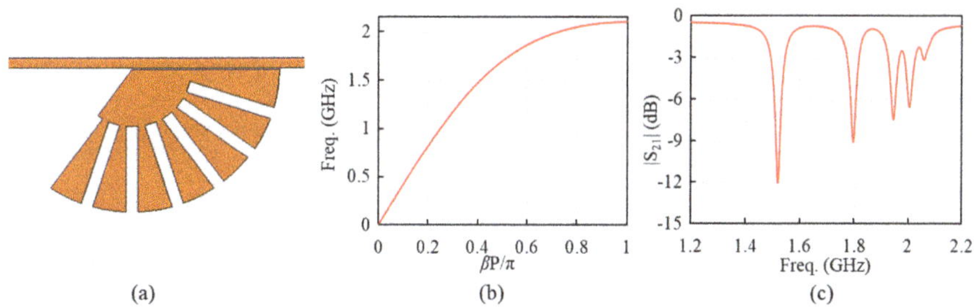

Fig. 20 **a** Spoof-LSP resonator. **b** Dispersion of the slow-wave along the periodic groove arc. **c** The simulated S_{21} curve. Reprinted with permission from [28]; copyright 2022 American Physical Society

where β is the wave vector propagation along the arc, C is the circumference of the WGM resonator, and n is the mode index. Different from the WGM resonators, fan-shaped spoof-LSP resonators utilized in this section consist of two open terminals, trapping waves in the periodic groove waveguide between two terminals to form standing waves.

$$\beta \times C \times \frac{\alpha}{2\pi} + \Delta\varphi_1 + \Delta\varphi_2 = n \tag{25}$$

where α is the angle of the fan-shaped periodic groove metal region, phase shifts at two reflecting terminals are denoted as $\Delta\varphi_1$ and $\Delta\varphi_2$. The wavevector of the slow wave mode, also named spoof surface plasmon, propagating along the periodic groove arc can be obtained by the eigenmode simulation, as shown in Fig. 20b. From the dispersion curves, we can confirm that multipole spoof-LSP resonances only exist below 2.2 GHz. More discussions about the mechanism of the spoof-LSP resonators can be found in [25, 32, 33]. From the transmission spectrum of the spoof-LSP resonator in Fig. 20c, we can obtain the center frequencies of these multipole resonant modes from 1.4 to 2.2 GHz. Note that resonances become denser as the frequency increases because the wave becomes slower and higher-order resonances appear according to (25).

We further combine the fan-shaped stub and the spoof-LSP resonator, as illustrated in Fig. 21a. The fan-shaped stub, the bright-mode resonator, can be directly excited by the main transmission line. The spoof-LSP resonator, the dark-mode resonator, cannot be excited by the main transmission line but can be coupled by the fan-shaped stub. We compare the S_{21} curves of the fan-shaped stub, the spoof-LSP structure, and the multipole transmission-mode structure, as shown in Fig. 21b. It is found that the Q value of the stub is much lower than that of the spoof-LSP resonator. When coupling occurs between two resonant structures, the multipole transmission modes appear as the TPs in the spectrum, whose center frequency is decided by the multipole dark modes, as discussed in Sect. 1. The transmission modes in Fig. 21b move towards a lower frequency band because the microfluidic chip is fixed on the top of the structure.

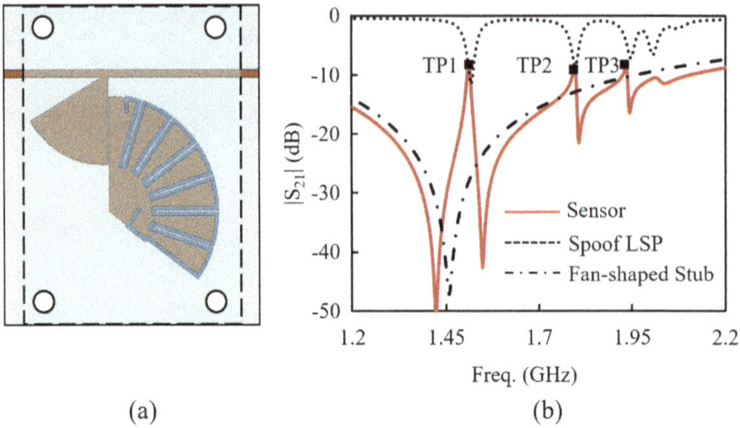

Fig. 21 **a** The layout of the stub-LSP transmission-mode structure, where the microfluidic chip is fixed on the top of the structure. **b** Simulated S_{21} curves of fan-shaped stub, spoof-LSP resonator, and multipole transmission-mode structure. Reprinted with permission from [28]; copyright 2022 American Physical Society

As marked in Fig. 21b, the first three TPs of multipole transmission modes are taken for study and named TP1, TP2, and TP3, respectively. The electric field distributions are shown in Fig. 22. There is a strong coupling between the spoof-LSP resonator and fan-shaped stub at these three TPs. Moreover, electric field distributions of the spoof-LSP resonator reveal the multipole resonances. With the increase of the mode order, the field is more confined in the periodic grooves, indicating a good sensitivity. Therefore, we set testing pipes along these grooves in the following experiments.

3.2 *Equivalent Circuit*

Figure 23 shows the equivalent circuit of the stub-LSP structure. The dark-mode spoof-LSP resonator is regarded as a periodic set of *RLC* series resonators, and neighboring sections are connected by the inductor L_2. The bright mode (stub) is still represented as a series *RLC* resonator, and the coupling between the bright and dark modes is represented as the mutual inductance between the inductance L_0 of the bright mode and the inductance L_1 of the first branch of the spoof-LSP resonator. In order to extract the parameters of the equivalent circuit, curve-fitting method is used. Firstly, the fan-shaped stub is analyzed, and the specific parameter values are determined by the curve fitting method based on the resonant frequencies and Q values. For the spoof-LSP structure, fitting parameters are obtained by using quasi-Newton optimization. Considering the difference of the coupling coefficient K in different frequencies, the equivalent circuits at different TPs are analyzed separately,

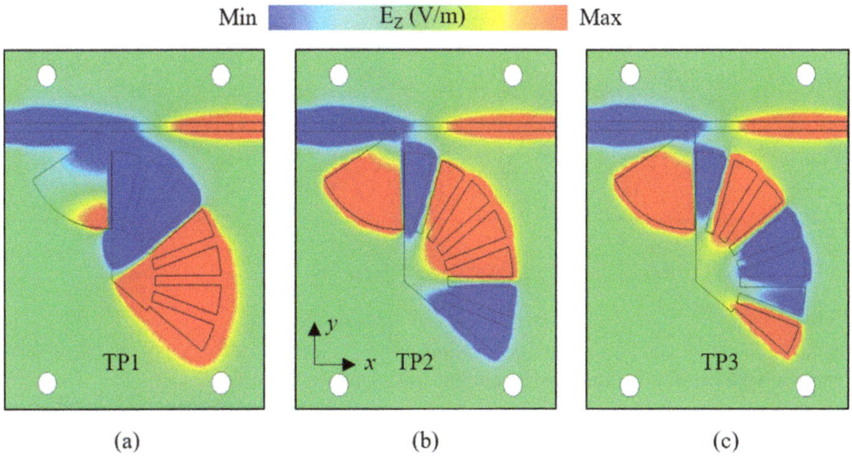

Fig. 22 Electric field distributions at three TPs. **a** TP1, **b** TP2, **c** TP3. Reprinted with permission from [28]; copyright 2022 American Physical Society

Fig. 23 Equivalent circuit model of the stub-LSP structure. Reprinted with permission from [28]; copyright 2022 American Physical Society

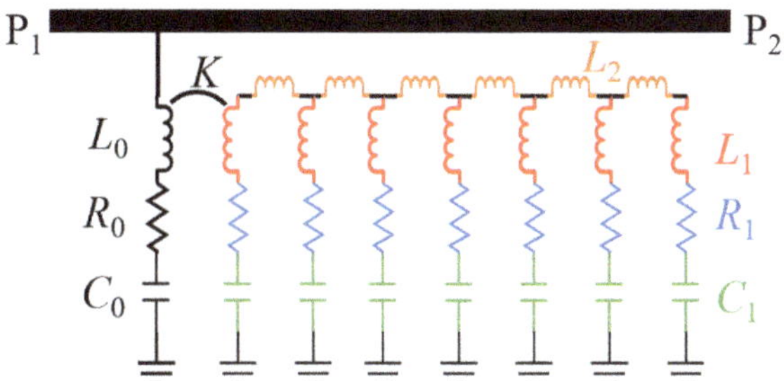

and the fitting curves are shown in Fig. 24. The calculated results of the equivalent circuit are in good agreement with the EM simulation results. The specific parameters are shown in Table 3.

We further verify the previously deduced theoretical model based on the coupling oscillators model in Fig. 6. We deduce the parameters of (10) based on the curve-fitting method and present these results in the caption of Fig. 25. The theoretical model matches well with the simulations at the first three transmission modes, and we can utilize higher-order theoretical modes to better match these higher-order transmission modes.

3.3 Measurements

A microfluidic chip made of polycarbonate was fixed on the surface of the spoof-LSP resonator for the experiment. The thickness of the microfluidic chip is 3 mm,

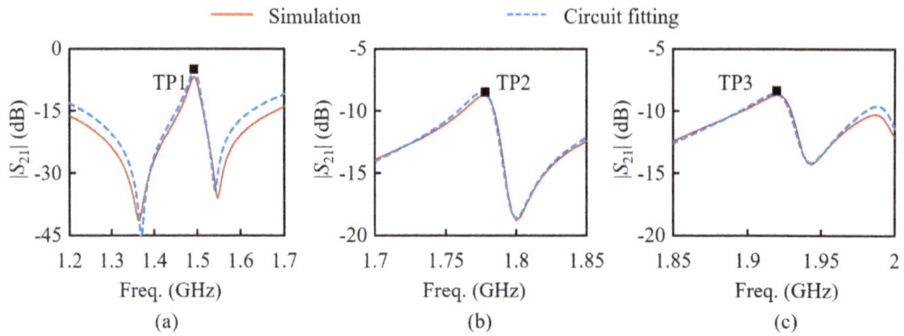

Fig. 24 Simulation and equivalent circuit fitting curves at different TPs: **a** TP1, **b** TP2, **c** TP3. Reprinted with permission from [28]; copyright 2022 American Physical Society

Table 3 Parameters of the equivalent-circuit model at different TPs

Parameters	TP1	TP2	TP3
L_0 (nH)	1.9	1.7	1.7
L_1 (nH)	19.6	18.8	18.5
L_2 (nH)	4	6.2	6.5
C_0 (pF)	6.36	6.36	6.36
C_1 (pF)	0.29	0.30	0.30
R_0 (Ω)	0.074	0.074	0.074
R_1 (Ω)	2.6	3.0	3.2
K	0.30	0.18	0.17

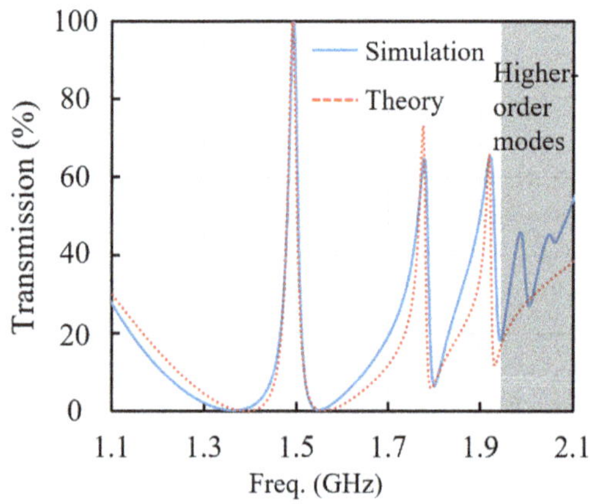

Fig. 25 The S_{21} curves obtained by the CST simulation and theoretical coupling-oscillators model. Parameters of the theoretical model (10): $\gamma_1 = 8.164$ s^{-1}, $\gamma_2 = 0.006$ s^{-1}, $\gamma_3 = 0.031$ s^{-1}, $\gamma_4 = 0.035$ s^{-1}, $\Omega_{21} = 0.5 \times 2\pi \times 10^9$ rad/s, $\Omega_{31} = 0.46 \times 2\pi \times 10^9$ rad/s, $\Omega_{41} = 0.44 \times 2\pi \times 10^9$ rad/s, $\omega_1 = 1.49 \times 2\pi \times 10^9$ rad/s, $\omega_2 = 1.49 \times 2\pi \times 10^9$ rad/s, $\omega_3 = 1.78 \times 2\pi \times 10^9$ rad/s, $\omega_4 = 1.92 \times 2\pi \times 10^9$ rad/s, $F = 1$ N, $m = 1$ kg. Reprinted with permission from [28]; copyright 2022 American Physical Society

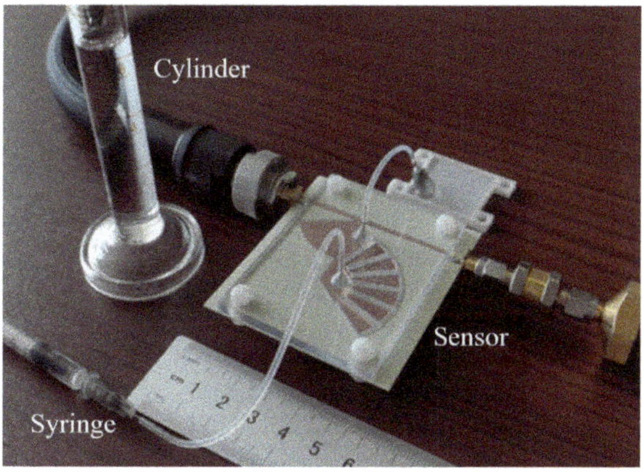

Fig. 26 Fabricated sensor and experimental setup. Reprinted with permission from [28]; copyright 2022 American Physical Society

where the lowest part of the pipe is 0.2 mm away from the sensor surface, and the liquid flows in the pipe along the metal edge of the spoof-LSP resonator. Four plastic screws at the corners fix the microfluidic pipe to the sensor. Figure 26 shows the final design of the physical model of the sensor and the experimental environment. Two plastic hoses are connected to the inlet and outlet, a measuring cylinder is used to dispense mixtures of different volume fractions, a syringe is connected to the inlet for injecting liquids, and the outlet is connected to the waste liquid bucket. The microstrip line has SMA connectors soldered to both ports and is connected to the vector network analyzer (*Agilent N5230A*) via a coaxial cable. In this experiment, the group delay (D_G) is used instead of the S_{21} to explore the sensing potential because the TPs revealed by the group delay spectrum are much more precise than those of the S_{21} spectrum, as aspect thoroughly discussed in [28]. The bidirectional energy transfer between bright and dark modes causes a large group delay accompanied by TPs, meaning the potential of the group delay spectrum as an option to design sensors. The measurement range is 1.3–1.9 GHz, and the number of sampling points is 16001. Water–ethanol and water–methanol mixtures with different volume fractions are selected as the MUTs. Figure 27 shows the measured results for different volume fractions of water–ethanol solutions. As the water content decreases, the real part of the complex permittivity decreases, and the loss tangent increases. As a result, the peak of the measured D_G shifts to higher frequencies, and the amplitude decreases.

The sensor was first calibrated based on water–ethanol solutions with a reference complex permittivity covering a wide range, according to [29]. Figure 28 shows the frequencies and D_G variation for the three TPs. Standard calibrations were conducted based on empirical models, usually in two steps: choosing the mathematical equations and obtaining the parameters based on curve fitting. A mathematical relationship between D_G and complex permittivity can be represented as

$$\varepsilon' = \frac{A}{(B \times f_0^2 + C \times f_0 + D)} + E \tag{26}$$

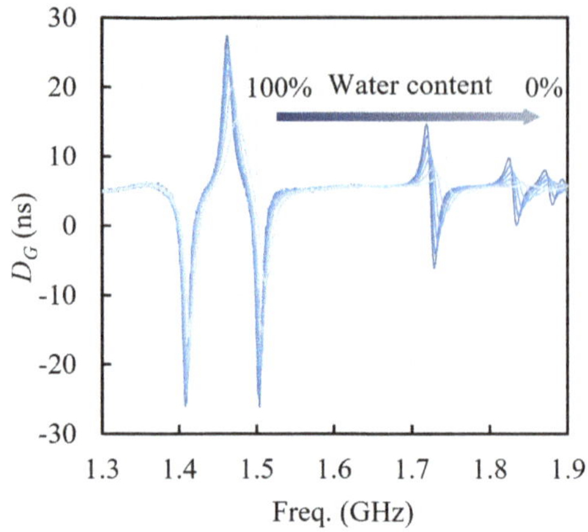

Fig. 27 Measured D_G spectrum of water–ethanol solutions. Reprinted with permission from [28]; copyright 2022 American Physical Society

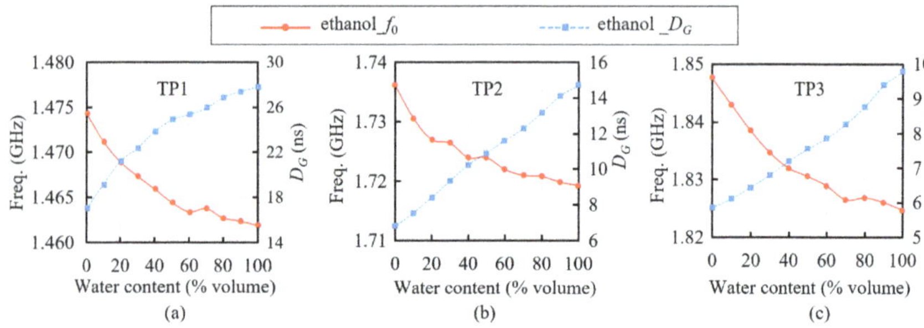

Fig. 28 Peak frequency and group delay of different water–ethanol solutions at three TPs: **a** TP1, **b** TP2, **c** TP3. Reprinted with permission from [28]; copyright 2022 American Physical Society

$$\varepsilon'' = F \times D_G + G \times f_0 + H \tag{27}$$

where f_0 and D_G represent the peak frequency and the corresponding group delay amplitude values. ε' and ε'' are the real and imaginary parts of the complex permittivity. The curve fitting parameters are represented from A to H. As ε' increases, f_0 decreases. D_G is related to the oscillation time of the strongly coupled system and is approximately linear with ε'. Larger ε'' corresponds to smaller D_G. Measured results at three TPs were matched to the mathematical model using the curve fitting toolbox in MATLAB. The specific parameter values and R^2 values obtained are listed in Table 4, where R^2 is a statistical parameter used to assess the closeness of the data to the fitted regression line. Figure 29 shows the performance of the curve fitting method.

We further compare the calculated and theoretical values of the complex permittivity of different water–ethanol mixture solutions at different TPs, as shown in

Table 4 Parameters of the curve fitting method

Parameters	TP1	TP2	TP3
A	65.3	69.1	95.1
B	42.1	30.0	24.5
C	−25.1	−26.6	−25.3
D	−52.6	−42.2	−34.6
E	−21.2	−24.6	−29.4
F	−5.2	−2.6	−4.9
G	−4165.8	−1102.6	−695.5
H	6241.2	1941.4	1323.9
$R^2(\varepsilon')$	0.97	0.98	0.99
$R^2(\varepsilon'')$	0.82	0.92	0.95

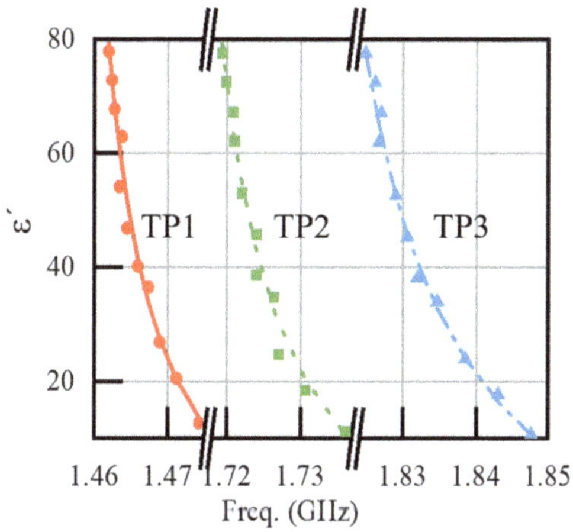

Fig. 29 Fitting performance of the real part of permittivity and the center frequency. Reprinted with permission from [28]; copyright 2022 American Physical Society

Fig. 30. The discrete points are the calculated complex permittivity from (26) and (27), and the solid and dashed lines are the theoretical complex permittivity for the real and imaginary parts [29]. The two results match well, indicating the correctness of the curve-fitting method.

We used the same parameters to detect water–methanol solutions to verify that the mathematical model derived from the water–ethanol solutions is reliable for measuring other solutions. The measured peak frequencies f_0 and D_G curves with different water–methanol solutions are plotted in Fig. 31. Expressions (26) and (27) obtained from the water–ethanol calibration were applied directly to the data from the water–methanol solutions. The theoretical results for the water–methanol solutions were compared with the actual calculated results and plotted in Fig. 32, where the discrete points are the calculated complex permittivity and the solid and dashed lines are the theoretical complex permittivity [29]. The results are consistent, verifying the accuracy of the mathematical model.

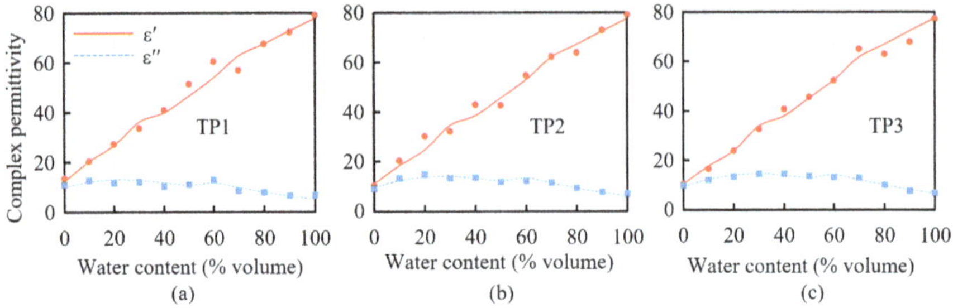

Fig. 30 Theoretical and measured values of water–ethanol solutions at different TPs: **a** TP1, **b** TP2, **c** TP3. Reprinted with permission from [28]; copyright 2022 American Physical Society

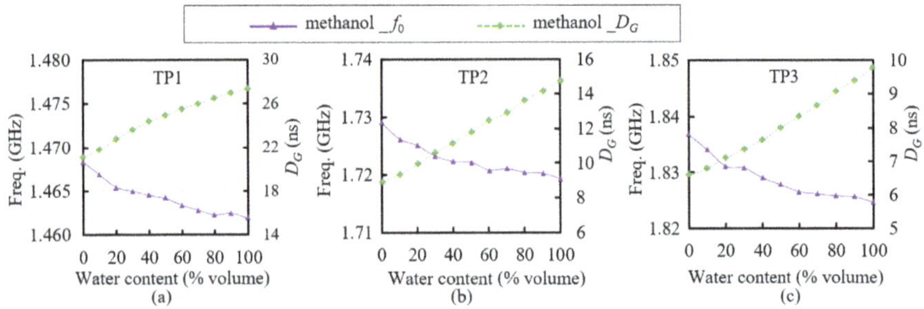

Fig. 31 Peak frequency and group delay of different water–methanol solutions at three TPs: **a** TP1, **b** TP2, **c** TP3. Reprinted with permission from [28]; copyright 2022 American Physical Society

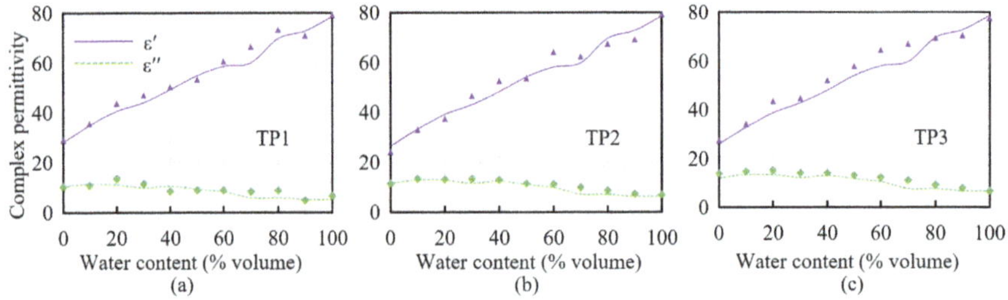

Fig. 32 Theoretical and measured values of water–methanol solutions at three TPs: **a** TP1, **b** TP2, **c** TP3. Reprinted with permission from [28]; copyright 2022 American Physical Society

The relative error is defined in (28) to better evaluate the sensor's accuracy.

$$\text{error} = \frac{1}{N}\sum_{i=0}^{N}\frac{|\varepsilon_m(i) - \varepsilon_t(i)|}{\varepsilon_t(i)} \times 100\% \tag{28}$$

where i denotes different volume fractions in steps of 10% for 11 sets of data, and ε_m and ε_t are the measured and theoretical dielectric constants, respectively. The relative error at the three TPs is calculated, and the relative error of the real part is 4.32, 5.11, and 6.30%. The imaginary part is 17.80, 11.60, and 16.26%. Results indicate that the sensor can detect the real part of permittivity more precisely.

3.4 Summary

In this section, a multipole transmission-mode sensor is proposed based on a broadband fan-shaped stub and spoof-LSP resonator. The mathematical equation deduced from the higher-order coupling-oscillators model in Sect. 2 can describe the transmission spectrum of this microwave structure with three TPs. An equivalent circuit model is also proposed to qualitatively describe the mechanism. The sensor is used to detect the complex permittivity of ethanol–water and methanol–water solutions at different TPs to verify the accurate and stable sensing capability.

4 Cascaded Dual-Band Transmission-Mode Permittivity Sensor

In this section, a dual-band transmission-mode sensor is further designed based on the cascaded stub-SRR structures, an extended design based on Sect. 2. Each stage works independently. The sensor is used to measure MUTs with small complex permittivity and the position of MUTs. Experimental results show that the sensitivity of the sensor achieves 7.16%.

4.1 Structure Design

Figure 33a shows the detailed structure of the cascaded structure. The detailed geometric parameters are given in Table 5. The first stage stub-SRR on the left works at the lower frequency band of around 1.1 GHz, and the second stage stub-SRR on the right side works at the higher frequency band of around 1.8 GHz. Figure 33b shows the simulated transmission spectrum, where the TPs of two transmission modes are marked. The Q factors of the two transmission modes are 18.02 and 21.04, respectively.

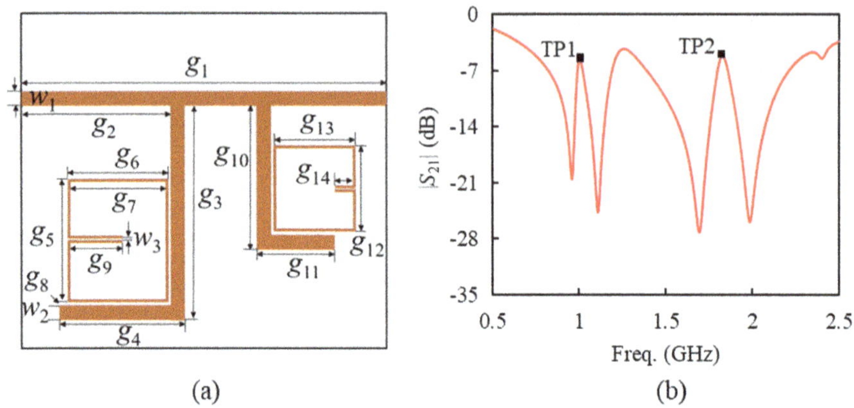

Fig. 33 **a** Cascaded dual-band transmission-mode structure. The entire structure is etched on a FR-4 dielectric substrate with permittivity of 4.5, thickness of 0.8 mm, and loss tangent of 0.025. **b** S_{21} curve of the cascade structure

Table 5 The specific parameter values of the cascade sensor

Parameters	w_1	w_2	w_3	g_1	g_2	g_3
Values	1.5 mm	1.5 mm	0.2 mm	45 mm	21.1 mm	23 mm
Parameters	g_4	g_5	g_6	g_7	g_8	g_9
Values	17.4 mm	15 mm	15 mm	14.4 mm	0.2 mm	8.7 mm
Parameters	g_{10}	g_{11}	g_{12}	g_{13}	g_{14}	
Values	15 mm	8.8 mm	10 mm	10 mm	2.7 mm	

4.2 Equivalent Circuit

As discussed in Sect. 2, a series RLC resonant circuit can describe the stub, and the SRR can be regarded as an RLC loop circuit. The cascaded equivalent circuit of the whole structure is shown in Fig. 34. We further use the curve fitting method to extract the equivalent circuit parameters. The resonant frequency and Q factor of each stub and SRR are obtained in CST simulations, which can reveal the relationship between the L, C, and R parameters. We further use the optimization tool to match the circuit and CST simulations. The final fitted curves are compared in Fig. 35, and the fitted parameter values are given in Table 6.

4.3 Experiments

The fabricated prototype and the experimental setup are shown in Fig. 36. The structure size is 4.5 cm × 5 cm × 0.8 mm. The measurement instrument is a vector network

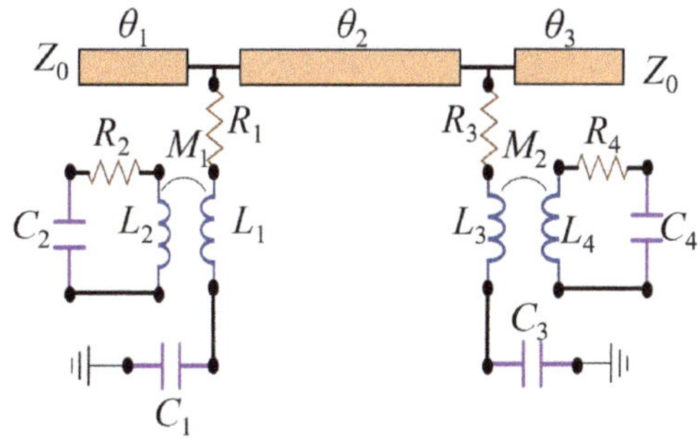

Fig. 34 The equivalent circuit of the cascade dual-band stub-SRR structure

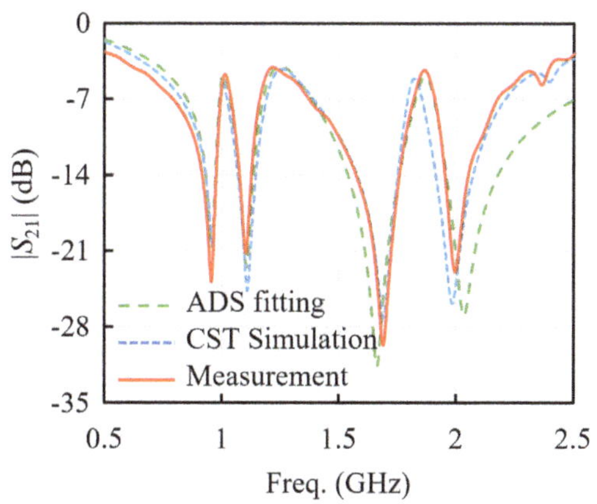

Fig. 35 The equivalent circuit model match well with simulations and measurements

Table 6 Parameters of the equivalent circuit model

Parameters	Z_0	L_1	L_2	L_3	L_4	C_1
Values	50 Ω	5.56 nH	5.94 nH	2.96 nH	2.1 nH	4.22 pF
Parameters	C_2	C_3	C_4	R_1	R_2	R_3
Values	4.17 pF	2.69 pF	3.48 pF	0.7 Ω	0.85 Ω	0.6 Ω
Parameters	R_4	M_1	M_2	θ_1	θ_2	θ_3
Values	0.55 Ω	0.8 nH	0.48 nH	47.6°	14°	36.7°

analyzer (VNA, *Agilent N5230A*). The VNA is connected to the SMA connectors at both ends of the sensor through two coaxial cables, and the number of sampling points is set to 1601 from 0.5 to 2.5 GHz.

The sensor was used to detect MUTs with small complex permittivity, where the MUTs are the 3 mm thick glass and Teflon substrates, which are directly placed on the surface of the structure. The real part of the permittivity of the Teflon medium is about 2, and that of the glass is about 4. Corresponding S_{21} curves are given in Fig. 37. The resonant frequency of TP1 decreases from 1.01 GHz to 797 MHz, and

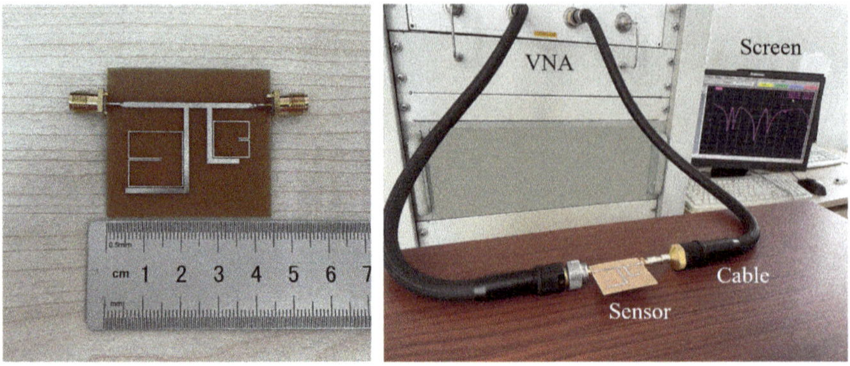

(a) (b)

Fig. 36 **a** Photograph of the fabricated prototype. **b** Experimental setup

Fig. 37 Measured S_{21} curves when detecting different MUTs

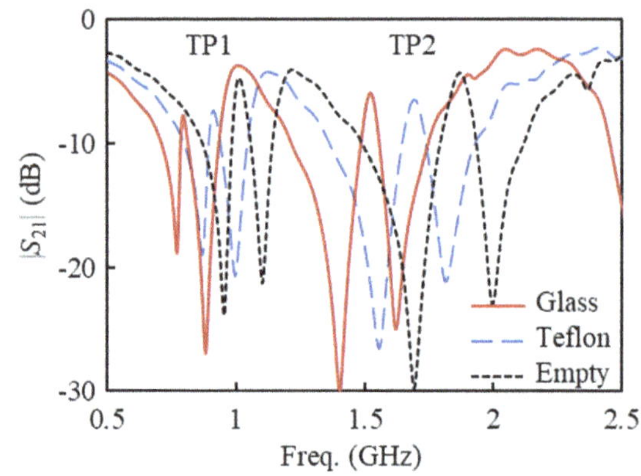

that of TP2 decreases from 1.82 to 1.52 GHz. After calculating, the sensitivity at the two frequency points is 7.16% and 6.17%, respectively.

The cascaded configuration is featured by independent detecting ability in two working bands, meaning we can test different samples on the surface of the different stub-SRR structures. To verify this idea, we covered the two stub-SRR structures separately with oil-soaked test papers, as shown in Fig. 38a, b. Figure 38c, d show the measured S_{21} curves. The two TPs have independent spectral shifts, indicating that the cascaded configuration can test the permittivity and the location of samples.

4.4 Summary

This section proposes a dual-band transmission-mode sensor based on the cascaded stub-SRR structures introduced in Sect. 2. The sensitivity of the sensor can reach

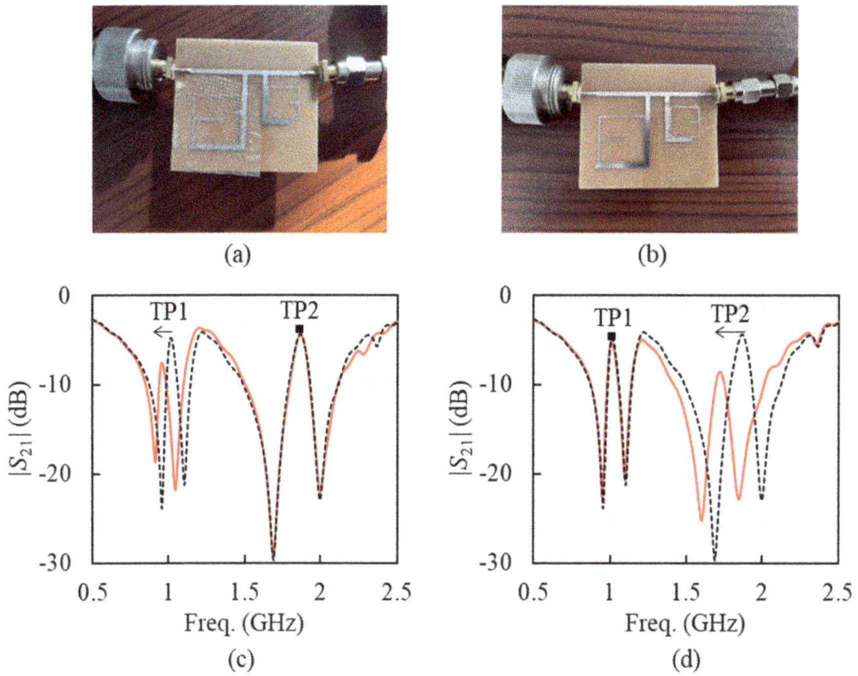

Fig. 38 Oil-soaked test papers placed **a** on the first stage and **b** on the second stage. Corresponding measured S_{21} curves of **c** the first stage sensing and **d** the second stage sensing

7.16%. The sensor can detect MUTs with different permittivity at different working bands, and it can also detect the position of MUTs. We can design more complex networks or arrays of the stub-SRR structures to realize two-dimensional surface detections.

5 Conclusions

In this chapter, we have investigated the theory and application of transmission mode sensors based on strong coupling between bright and dark mode resonators. We have used coupled-mode oscillators to derive the general mathematical model of transmission mode and its dependence on the resonant frequency and quality factor of the coupled resonators. We have designed and implemented three prototypes of transmission-mode sensors using different types of resonators. We have demonstrated the sensing performance of the sensors by measuring the complex permittivity of various liquids and solids using curve-fitting methods. The main contributions of this chapter are summarized as follows:

(1) We have revealed the universal origin of the high-Q resonant transmission modes induced by strong coupling between bright and dark mode resonators. This provides valuable insight for designing high-sensitivity and high-precision sensors.
(2) We have presented a single-band transmission-mode sensor based on a stub and an SRR as the bright and dark mode resonators. Based on a least square fitting method, the sensor can measure the permittivity of water–ethanol mixture liquids.
(3) We have developed a multipole transmission-mode sensor based on a broadband fan-shaped stub and a multipole spoof-LSP resonator. Based on a group-delay-based curve fitting method, the sensor can support multiple transmission modes over a wide frequency range and measure the permittivity of water–ethanol and water–methanol mixtures.
(4) We have designed a dual-band transmission-mode sensor based on two cascaded stub-SRR structures. The sensor can operate at two distinct frequency bands and detect the permittivity of low-permittivity solids and liquids with high sensitivity.

References

1. Chen LF, Ong CK, Neo CP, Varadan VV, Varadan VK (2004) Microwave electronics: measurement and materials characterization. Wiley, Hoboken, NJ
2. Petosa A (2007) Dielectric resonator antenna handbook. Artech House Publishers, Boston
3. Knox RM (1976) Dielectric waveguide microwave integrated circuits—an overview. IEEE Trans Microw Theory Tech 24(11):806–814. https://doi.org/10.1109/TMTT.1976.1128965
4. Ren Z, Qi S-S, Hu Z, Shen Z, Wu W (2019) Wideband water helical antenna of circular polarization. IEEE Trans Antennas Propag 67(11):6770–6777. https://doi.org/10.1109/TAP.2019.2922846
5. Jahani S, Jacob Z (2016) All-dielectric metamaterials. Nat Nanotechnol 11(1):23–36
6. Xu Z, Cui X, Shi J, Liu S, Li S, Yin X (2019) Experimental demonstration of lattice-tailored scattering features of dielectric particle arrays at microwave frequencies. Appl Phys Express 12(1):012006. https://doi.org/10.7567/1882-0786/aaf5c3
7. Muhammed Shafi KT, Jha AK, Akhtar MJ (2017) Improved planar resonant RF sensor for retrieval of permittivity and permeability of materials. IEEE Sens J 17(17):5479–5486. https://doi.org/10.1109/JSEN.2017.2724942
8. Jha AK, Tiwari NK, Akhtar MJ (2021) Accurate microwave cavity sensing technique for dielectric testing of arbitrary length samples. IEEE Trans Instrum Meas 70:1–10. https://doi.org/10.1109/TIM.2021.3073438
9. Wang Y, Xu Z, Feng Y, Fang S (2023) Coaxial resonant cavity for measuring complex permittivity of liquids. Frequenz 77(5–6):229–234. https://doi.org/10.1515/freq-2022-0115
10. Martín F, Vélez P, Muñoz-Enano J, Su L (2022) Planar microwave sensors, 1st edn. Wiley-IEEE Press
11. Velez P, Su L, Grenier K, Mata-Contreras J, Dubuc D, Martin F (2017) Microwave microfluidic sensor based on a microstrip splitter/combiner configuration and split ring resonators (SRRs) for dielectric characterization of liquids. IEEE Sens J 17(20):6589–6598. https://doi.org/10.1109/JSEN.2017.2747764

12. Baena JD et al (2005) Equivalent-circuit models for split-ring resonators and complementary split-ring resonators coupled to planar transmission lines. IEEE Trans Microw Theory Tech 53(4):1451–1461. https://doi.org/10.1109/TMTT.2005.845211
13. Xu Z, Shi J, Davis RJ, Yin X, Sievenpiper DF (2019) Rainbow trapping with long oscillation lifetimes in gradient magnetoinductive metasurfaces. Phys Rev Appl 12(2):024043. https://doi.org/10.1103/PhysRevApplied.12.024043
14. Hajiaghajani A, Afandizadeh Zargari AH, Dautta M, Jimenez A, Kurdahi F, Tseng P (2021) Textile-integrated metamaterials for near-field multibody area networks. Nat Electron 4(11), Art. no. 11. https://doi.org/10.1038/s41928-021-00663-0
15. Ebrahimi A, Withayachumnankul W, Al-Sarawi S, Abbott D (2014) High-sensitivity metamaterial-inspired sensor for microfluidic dielectric characterization. IEEE Sens J 14(5):1345–1351. https://doi.org/10.1109/JSEN.2013.2295312
16. Padilla WJ, Taylor AJ, Highstrete C, Lee M, Averitt RD (2006) Dynamical electric and magnetic metamaterial response at terahertz frequencies. Phys Rev Lett 96(10):107401. https://doi.org/10.1103/PhysRevLett.96.107401
17. Ebrahimi A, Scott J, Ghorbani K (2020) Dual-mode resonator for simultaneous permittivity and thickness measurement of dielectrics. IEEE Sens J 20(1):185–192. https://doi.org/10.1109/JSEN.2019.2941753
18. Rowe DJ, al-Malki S, Abduljabar AA, Porch A, Barrow DA, Allender CJ (2014) Improved split-ring resonator for microfluidic sensing. IEEE Trans Microwave Theory Technol 62(3):689–699. https://doi.org/10.1109/TMTT.2014.2299514
19. Xu Z, Wang Y, Liu S, Ma J, Fang S, Wu H (2023) Metamaterials with analogous electromagnetically induced transparency and related sensor designs—a review. IEEE Sens J 23(7):6378–6396. https://doi.org/10.1109/JSEN.2023.3249743
20. Alzar CLG, Martinez MAG, Nussenzveig P (2002) Classical analog of electromagnetically induced transparency. Am J Phys 70(1):5
21. Boller K, Imamolu A, Harris SE (1991) Observation of electromagnetically induced transparency. Phys Rev Lett 66(20):2593–2596. https://doi.org/10.1103/PhysRevLett.66.2593
22. Abdumalikov AA Jr, Astafiev O, Zagoskin AM, Pashkin YA, Nakamura Y, Tsai JS (2010) Electromagnetically induced transparency on a single artificial atom. Phys Rev Lett 104(19):193601. https://doi.org/10.1103/PhysRevLett.104.193601
23. Xu Z, Liu S, Li S, Yin X (2019) Analog of electromagnetically induced transparency based on magnetic plasmonic artificial molecules with symmetric and antisymmetric states. Phys Rev B 99(4):041104(R). https://doi.org/10.1103/PhysRevB.99.041104
24. Xu Z, Wang Y, Fang S (2021) Dielectric characterization of liquid mixtures using EIT-like transmission window. IEEE Sens J 21(16):17859–17867. https://doi.org/10.1109/JSEN.2021.3085954
25. Xu Z, Li S, Liu Y, Zhao H, Yin X (2018) Characteristic mode analysis of complex spoof localized surface plasmon resonators. IEEE Access 6:2871–2878. https://doi.org/10.1109/ACCESS.2017.2785809
26. Xu Z, Kong X, Chang J, Sievenpiper DF, Cui TJ (2022) Topological flat bands in self-complementary plasmonic metasurfaces. Phys Rev Lett 129(25), Art. no. 25. https://doi.org/10.1103/PhysRevLett.129.253001
27. Liu S, Xu Z, Yin X, Zhao H (2021) High-*Q*-value classical electromagnetically induced transparency based on dipoles overlapping at spoof localized surface plasmons. J Opt Soc Am B 38(4):1156–1162. https://doi.org/10.1364/JOSAB.417496
28. Xu Z, Wang Y, Chang J, Cui TJ (2022) Multiple spoof plasmonically induced transparency for sensing applications. Phys Rev Appl 18(2):024035. https://doi.org/10.1103/PhysRevApplied.18.024035
29. Bao J, Swicord ML, Davis CC (1996) Microwave dielectric characterization of binary mixtures of water, methanol, and ethanol. J Chem Phys 104(12):4441–4450. https://doi.org/10.1063/1.471197
30. Withayachumnankul W, Jaruwongrungsee K, Tuantranont A, Fumeaux C, Abbott D (2013) Metamaterial-based microfluidic sensor for dielectric characterization. Sens Actuators A 189:233–237. https://doi.org/10.1016/j.sna.2012.10.027

31. Liu S, Xu Z, Yin X, Zhao H (2020) Analog of multiple electromagnetically induced transparency using double-layered metasurfaces. Sci Rep 10(1):8469. https://doi.org/10.1038/s41598-020-65418-x
32. Huidobro PA et al (2014) Magnetic localized surface plasmons. Phys Rev X 4(2):021003. https://doi.org/10.1103/PhysRevX.4.021003
33. Garcia-Vidal FJ et al (2022) Spoof surface plasmon photonics. Rev Mod Phys 94(2), Art. no. 2. https://doi.org/10.1103/RevModPhys.94.025004

Analytical Methods to Retrieve the Permittivity by Means of Coupled Resonators

Alonso Corona-Chávez and José Luis Olvera-Cervantes

Abstract Conventional single-resonator sensors perform dielectric properties measurements by means of the well-known perturbation technique, where the resonant frequency and the quality factor are used to extract the complex permittivity values. However, coupled resonators exhibit two resonance frequencies, one corresponding to the odd mode, and the other one to the even mode. This characteristic can be used to measure dielectric properties by correlating the coupling factor of the two resonances to the permittivity. In addition, this can also be used to perform anisotropic measurements, since the geometry of the electric field lines is different for odd and even modes. Therefore, by utilizing the cavity perturbation technique for the different modes, anisotropy can be extracted. In this chapter, coupled-resonator techniques will be addressed for isotropic and anisotropic measurement.

1 Introduction

Measurements of dielectric properties can be divided in resonant and non-resonant methods [1]. Resonant methods normally offer greater accuracy, but they are limited to only a single frequency (the frequency of resonance) or to a few frequencies when harmonic frequencies of the fundamental resonance are used.

Conventionally, these measurements have been performed using waveguide cavities, where the sample under test (SUT) is placed inside the cavity. From the resonant frequency and quality factor variations, the complex permittivity $\tilde{\varepsilon}$ can be determined using the cavity perturbation technique (CPT) [2]. In addition, to measure anisotropy, the sample can be rotated inside the cavity to measure the interaction of the electric field E in different SUT directions [3]. Other methods involve the use of more than

A. Corona-Chávez (✉) · J. L. Olvera-Cervantes
Coordinación de Electrónica, Instituto Nacional de Astrofísica, Óptica y Electrónica, 72000 Puebla, México
e-mail: alonsocorona@inaoep.mx

J. L. Olvera-Cervantes
e-mail: jolvera@inaoep.mx

F. Martín and E. Bronchalo (eds.), *Coupled Structures for Microwave Sensing*, Lecture Notes in Electrical Engineering 1150,
https://doi.org/10.1007/978-3-031-53861-2_7

one resonator. For example, in [4] a two-resonator method is presented, where two resonant cavities operated in the TE_{011} and TM_{010} modes are used to measure the longitudinal and transversal dielectric constant and dielectric loss of the SUTs.

As opposed to conventional single-resonator sensors, with one fundamental resonant frequency, a pair of coupled resonators present two resonant frequencies that appear due to the even and odd mode components, also known as magnetic and electrical frequencies (f_m and f_e), respectively. In this configuration (Fig. 1a), for the odd mode (Fig. 1b), an electric wall is formed at the line of symmetry *T-T´*, which derives in strong E-fields oriented in the x direction. This E field is highly sensitive to the x-direction permittivity values. However, for the even mode, a virtual magnetic wall is formed at the symmetry line, so there is a weak E field in the x direction. Therefore, the even mode frequency is less sensitive to changes in the dielectric characteristics in x direction and is manly sensitive to changes in the z direction. This principle of operation has been applied to microwave sensing. For example, in [5] the x and z components of the dielectric constant of the substrate are measured using a pair of coupled resonators, where the SUT dielectric constant is measured in x and z directions to extract anisotropy.

On the other hand, isotropic measurements have been performed in [6], where a SUT is placed on top of a microstrip coupled-resonators sensor, and the coupling coefficient is correlated to the SUT permittivity (Fig. 1c).

2 Permittivity Tensor and Anisotropy

The complex permittivity of any linear material is generally defined as a tensor that describes the relation between the electric displacement (D) and the electric field (E) as expressed by (1):

$$\vec{D} = \tilde{\varepsilon}\vec{E} \tag{1}$$

When the permittivity of a medium is dependent on the direction of the electric field, it is said to be anisotropic, and the following tensor is used (2)

$$\tilde{\varepsilon} = \begin{bmatrix} \widetilde{\varepsilon_{xx}} & \widetilde{\varepsilon_{xy}} & \widetilde{\varepsilon_{xz}} \\ \widetilde{\varepsilon_{yx}} & \widetilde{\varepsilon_{yy}} & \widetilde{\varepsilon_{yz}} \\ \widetilde{\varepsilon_{zx}} & \widetilde{\varepsilon_{zy}} & \widetilde{\varepsilon_{zz}} \end{bmatrix} \tag{2}$$

If the coordinate axes are properly chosen, the permittivity tensor $\tilde{\varepsilon}$ adopts a diagonal form, being the non-zero components $\tilde{\varepsilon}_{xx}$, $\tilde{\varepsilon}_{yy}$ and $\tilde{\varepsilon}_{zz}$ the eigenvalues of $\tilde{\varepsilon}$. The material is said to have uniaxial anisotropy when two of these three components are equal, as shown in (3). Figure 2 shows a planar material that exhibits uniaxial anisotropy where the parallel component is $\varepsilon_{par} = \varepsilon_{xx} = \varepsilon_{yy}$ and the perpendicular component is $\varepsilon_{prp} = \varepsilon_{zz}$.

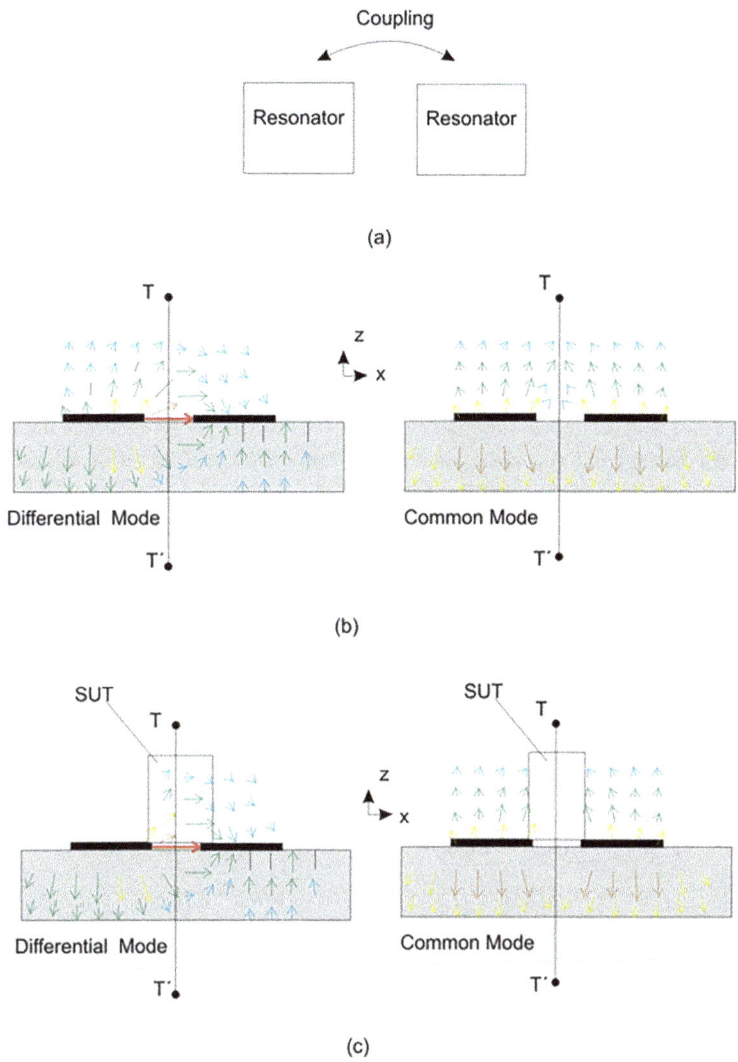

Fig. 1 **a** Coupled resonators. **b** E-Field distribution in odd and even modes for the coupled resonators without sample. **c** E-Field distribution for the coupled resonators with sample

$$\tilde{\varepsilon} = \begin{bmatrix} \varepsilon_{par} & 0 & 0 \\ 0 & \varepsilon_{par} & 0 \\ 0 & 0 & \varepsilon_{prp} \end{bmatrix} \tag{3}$$

where

$$\varepsilon_{par,prp} = \varepsilon'_{par,prp} - j\varepsilon''_{par,prp} = \varepsilon'_{par,prp}\left(1 - jtan\delta_{par,prp}\right) \tag{4}$$

Fig. 2 Dielectric anisotropy

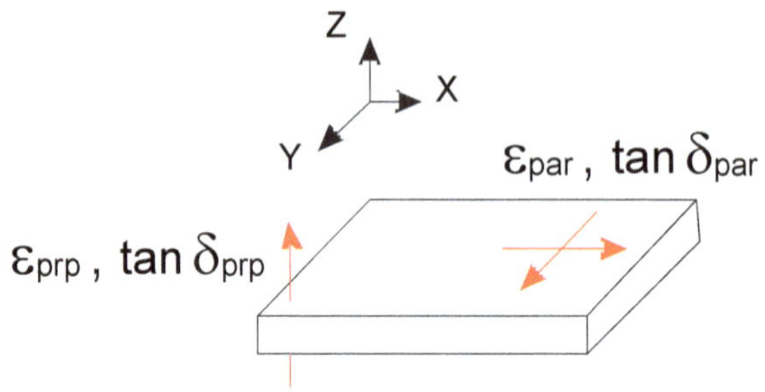

and where the loss tangent is given by $tan\delta = \varepsilon''/\varepsilon'$.

The real part of the complex permittivity normalized to ε_0 is known as the dielectric constant and it is related to the material ability to store energy, whereas the relative imaginary part of the complex permittivity known as loss factor is related to the capacity to absorb energy from the *E*-field.

The dielectric anisotropy can be expressed as ratios of the parallel and perpendicular components:

$$A = \varepsilon'_{par}/\varepsilon'_{prp} \tag{5a}$$

$$A_{loss} = \varepsilon''_{par}/\varepsilon''_{prp} \tag{5b}$$

Or, in normalized form:

$$\Delta A_{\varepsilon} = \frac{2\left|\varepsilon'_{par} - \varepsilon'_{prp}\right|}{\varepsilon'_{par} + \varepsilon'_{prp}} \tag{6}$$

Uniaxial anisotropy can be present in modern substrates used for printed circuit boards. In optical glasses [7], microwave ceramics [8] and liquid crystals [9], uniaxial anisotropy is caused mainly by their crystallography properties.

For most polycrystalline materials, glasses, plastics, and crystals having a cubic crystallographic structure, all diagonal elements are identical, and the complex permittivity becomes a scalar quantity [1].

For high frequency printed circuits, previous knowledge of the anisotropic characteristics of the substrates can lead to greater design accuracy as exhibited in [10]. In this work, two printed circuits are fabricated using FR4, a well-known anisotropic substrate, with height of 1.48 mm. The first circuit consists of a patch antenna at 2.4 GHz, and the second, two coupled resonators at the same frequency (Fig. 3). A full wave simulator [11] is used to model the structures under three different conditions:

(1) Isotropic substrate with $\varepsilon_r = 4.4$.
(2) Anisotropic substrate with $\varepsilon_{par} = 4.57$ and $\varepsilon_{prp} = 4.19$, according to the measurements presented in [9].

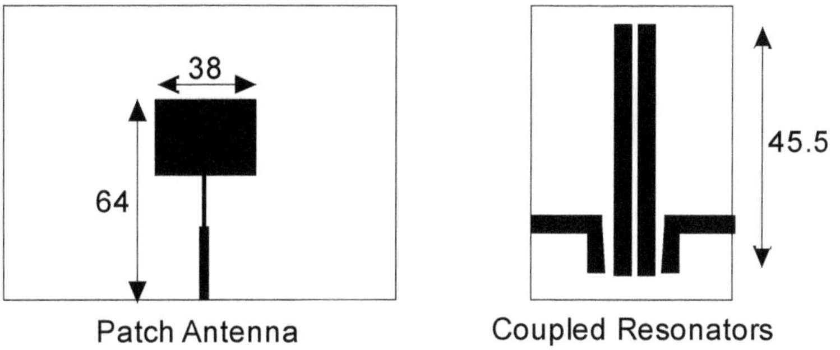

Fig. 3 Patch antenna and coupled resonators. Dimensions in mm. Adapted with permission from [10]; copyright (2020) IEEE

(3) Anisotropic substrate with $\varepsilon_{par} = 4.538$ and $\varepsilon_{prp} = 4.159$, according to [5].

The results are summarized in Table 1. Under anisotropic conditions, modeling accuracy increases. The following measure can be used to quantify the accuracy of the modeled printed circuits:

$$Error(\%) = 100\frac{f_{sim} - f_{exp}}{f_{exp}}$$

where f_{sim} and f_{exp} are, respectively, the resonant frequencies of the simulated model and measured experimentally. For the antenna, the error in resonant frequency (f_0) is 1.5% for the isotropic substrate, and, with anisotropic conditions, it decreases to 0.53%. In the same manner, for the odd (f_e) and even (f_m) resonance frequencies of the coupled resonators (see Sect. 3 below), the error in f_e decreases from 0.5% with

Table 1 Comparison of experimental results and simulations using isotropic and anisotropic assumptions [9]

Condition		Antenna		Coupled resonators			
		f_0	Error (%)	f_m	Error f_m (%)	f_e	Error f_e (%)
Anisotropic	$\varepsilon_{rpar} = 4.57$ $\varepsilon_{rprp} = 4.19$	2.468	0.53	1.753	0.17	1.938	0.2
Anisotropic	$\varepsilon_{rpar} = 4.538$ $\varepsilon_{rprp} = 4.159$	2.478	0.93	1.757	0.06	1.94	0.1
Isotropic	$\varepsilon_r = 4.4$	2.418	1.5	1.733	1.31	1.932	0.5
Measurement	–	2.455	0	1.756	0	1.942	0

f_0: antenna center frequency; f_e: coupled resonators even-mode resonant frequency; f_o: coupled resonators odd-mode resonant frequency

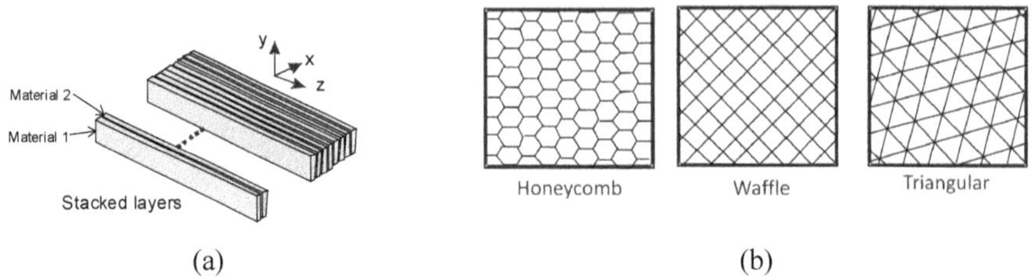

Fig. 4 Artificial materials: **a** stack of layered materials. Standard PLA patterns of 3D printed substrates

isotropic substrate to 0.1% under anisotropic conditions, whereas for f_m it decreases from 1.3% (isotropic) to 0.06% (anisotropic).

In [7], a dielectric resonator rod based on sapphire is proposed where the anisotropy of sapphire is reported to be $\varepsilon'_{par} = 9.389$ and $\varepsilon'_{prp} = 11.478$ at 9.7 GHz. For rutile at 5 GHz, $\varepsilon'_{par} = 165$ and $\varepsilon'_{prp} = 85$ [12]. In [13], several liquid crystal samples are reported, where a large anisotropy is shown for GT3-24,002 ($\varepsilon'_{par} = 3.3$ and $\varepsilon'_{prp} = 2.5$ at 8 GHz). For K15 (5CB) it is $\varepsilon'_{par} = 2.72$ and $\varepsilon'_{prp} = 2.9$ at 5 GHz.

In fiber-based substrates, anisotropy is mainly due to their inhomogeneous structure formed by reinforcing fibers, and irregular filling resins, where air balloons can be formed. Generally, the nonwoven substrates are less anisotropic than woven substrates [14]. For example, FR4 has been reported with $\varepsilon'_{par} = 4.57$ and $\varepsilon_{prp} = 4.19$, Rogers 4350B with $\varepsilon'_{par} = 3.88$ and $\varepsilon'_{prp} = 3.48$, and Diclad 880 with $\varepsilon'_{par} = 2.33$ and $\varepsilon'_{prp} = 2.17$ at 2.4 GHz [15].

Another area where anisotropy has been widely studied is "artificial materials" such as engineered composites, and 3D printed structures. In [16] a stack of layered alumina–barium titanate ($Al_2O_3/BaTiO_3$) pellets are fabricated, giving $\varepsilon'_{par} = 17.2$ and $\varepsilon'_{prp} = 45.8$ at 7.5 GHz (Fig. 4). In addition, 3D printed substrates have been analyzed in [17], where printed PLA substrates are shown with anisotropies varying depending on the printing filling (Fig. 4).

On the other hand, the anisotropic dielectric properties are required to be accurately known for heating, cooking or drying effects [18, 19]. For instance, it has been reported that anisotropic composition in rubber wood is affected by moisture content and structural direction of the fibers [20]. Table 2 shows a summary of the results.

3 Theory of Coupled Resonators

Figure 5a shows a circuit schematic of two identical lossless coupled resonators, where L and C are the resonators capacitance and inductance. C_m is the mutual coupling capacitance [21]. By analyzing the resonator's loop, the following equations can be derived.

Table 2 Different materials and their reported anisotropic dielectric properties

Material	ε'_{par}	ε'_{prp}	$tan\delta_{par}$	$tan\delta_{prp}$	*Freq.* (GHz)	Refs.
Sapphire	9.389 10.002	11.478 11.553	–	–	10	Kobayashi and Senju [12]
Rutile (25C)	85	165	10^{-4}	$8 \cdot 10^{-5}$	4.6–5.6	Tobar et al. [13]
K15(5CB)[a]	2.9	2.72	0.03	0.03	1–10	Yaghmaee et al. [14]
BL037[a]	2.61	2.35	0.06	0.06	1–10	Yaghmaee et al. [14]
BL006[a]	3.11–3.12	2.62–2.69	0.035−0.036	0.015–0.007	5–9	Yaghmaee et al. [14]
E7[a]	3.17	2.72	0.02	0.12	5–6	Yaghmaee et al. [14]
GT3-23,001[a]	2.28−2.3	2.56−2.5	0.0038−0.004	0.0143	13–19	Yaghmaee et al. [14]
GT3-24,002[a]	3.3	2.5	0.0032	0.0123	6–10	Yaghmaee et al. [14]
Diclad 880	2.32	2.15	0.0016	$9 \cdot 10^{-4}$	12	Morales-Lovera et al. [16]
RO 4350	3.88	3.48	–	–	2.4	Morales-Lovera et al. [16]
PLA	2.79	2.73	–	–	2.4	Morales-Lovera et al. [16]
$Al_2O_3/BaTiO_3$[b]	17.2	45.8	$6 \cdot 10^{-4}$	$1 \cdot 10^{-3}$	7	Mumcu et al. [17]
3D printed PLA:						
Honeycomb 50% infill	2.6	2.48			2.4	Morales-Lovera et al. [18]
Honeycomb 30% infill	2.47	2.32			2.4	Morales-Lovera et al. [18]
Waffle 50% infill	2.54	2.43			2.4	Morales-Lovera et al. [18]
Waffle 30% infill	2.4	2.26			2.4	Morales-Lovera et al. [18]
Triangle 50% infill	2.62	2.53			2.4	Morales-Lovera et al. [18]
Triangle 30% infill	2.5	2.35			2.4	Morales-Lovera et al. [18]
Dry wood	**Radial direction**	**Longitudinal direction**	ε''_{par}	ε''_{prp}		
	2.1	2.4	0.09	0.11	2	Kabir et al. [21]

(continued)

Table 2 (continued)

Material	ε'_{par}	ε'_{prp}	$tan\delta_{par}$	$tan\delta_{prp}$	*Freq.* (GHz)	Refs.
	2	2.2	0.16	0.2	10	Kabir et al. [21]
	1.8	2.1	0.13	015	17	Kabir et al. [21]

[a] *Liquid crystal;* [b] *Artificial Material*

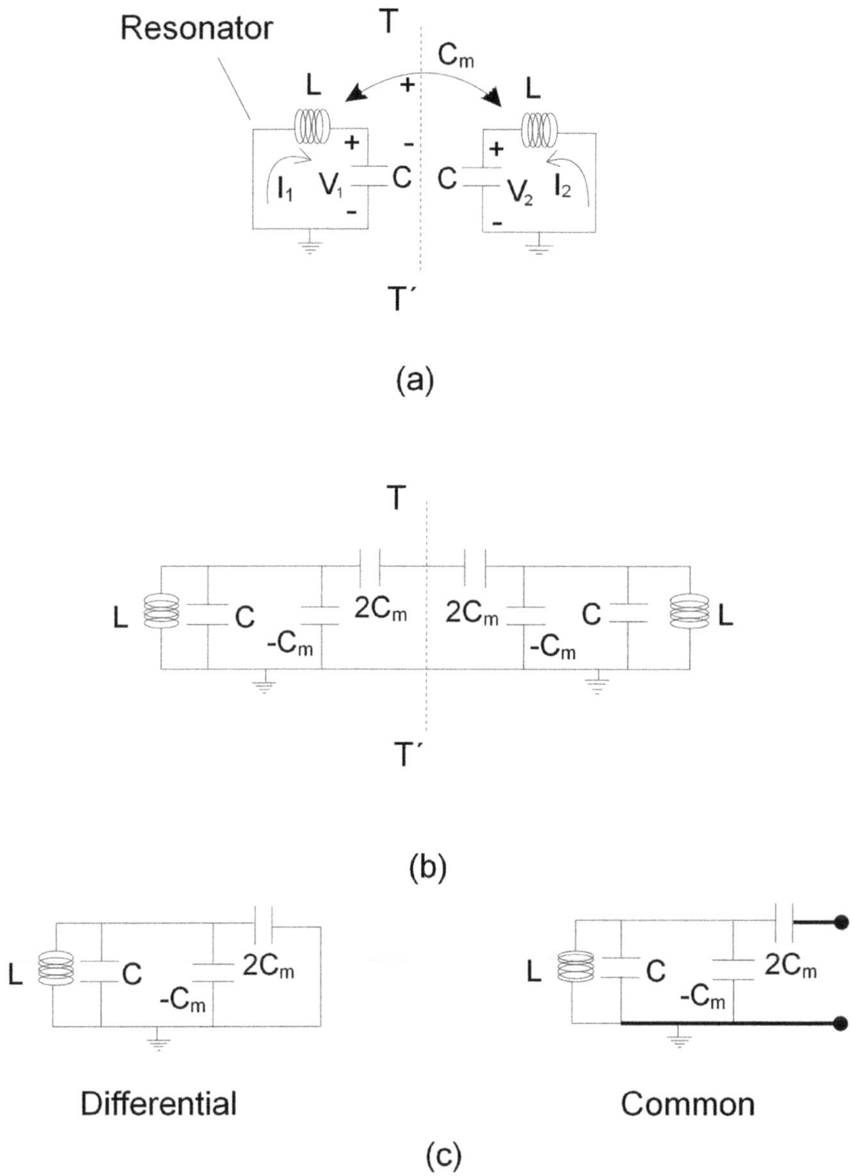

Fig. 5 **a** Coupled resonators circuit; **b** Equivalent circuit; **c** Odd and even modes

$$I_1 = j\omega_0 C V_1 - j\omega_0 C_m V_2 \tag{7}$$

$$I_2 = j\omega_0 C V_2 - j\omega_0 C_m V_1 \tag{8}$$

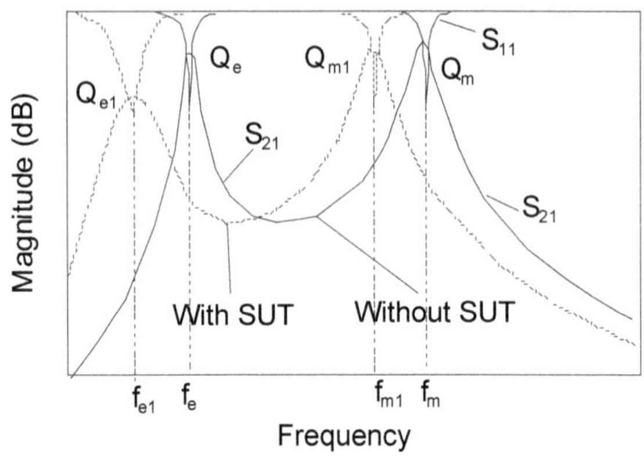

Fig. 6 Frequency response of coupled resonators showing electric and magnetic resonances

where $\omega_0 = 1/\sqrt{LC}$ is the resonance angular frequency of the uncoupled resonators.

Moreover, an equivalent circuit can be expanded as shown in Fig. 5b. From here, we can apply the odd and even mode analysis. As discussed previously, for the odd mode, an electric wall (short circuit) exists at the symmetry plane *T-T´*, whereas for the even mode, a magnetic wall (open circuit) exists at the symmetry plane, as shown in Fig. 5c. For the odd mode, the resulting resonant frequency is

$$f_e = \frac{1}{2\pi\sqrt{L(C + C_m)}} \tag{9}$$

and for the even mode

$$f_m = \frac{1}{2\pi\sqrt{L(C - C_m)}} \tag{10}$$

From these equations, the coupling coefficient k becomes:

$$k = \frac{f_m^2 - f_e^2}{f_m^2 + f_e^2} = \frac{C_m}{C} \tag{11}$$

A typical coupled-resonators response in shown in Fig. 6, where the magnitude of the scattering parameters, S_{11} and S_{21} are plotted versus frequency.

4 Isotropic Dielectric Constant Measurements with Coupled Resonators

In [5], two coupled hairpin resonators in microstrip technology are used for permittivity sensing. The structure is shown in Fig. 7, where the sample is placed on top of the resonators so that the equivalent circuit can be modified to take into account

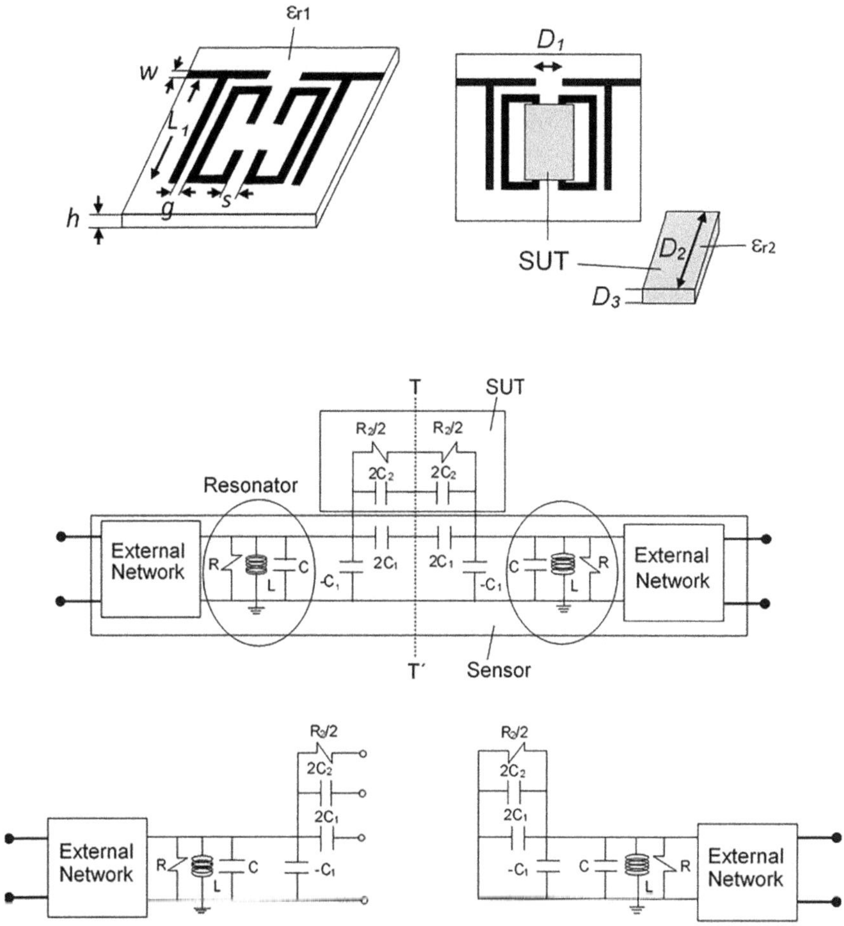

Fig. 7 Equivalent circuit of coupled resonators. Dimensions (mm): $s = 0.2$, $h = 0.787$, $w = 2.39$, $L_1 = L_2 = 21.2$, $g = 0.4$, $L = 45.44$. Substrate properties: $\varepsilon_r = 2.2$ and $\tan\,\delta = 0.0009$. Adapted with permission from [6]; copyright (2020) IEEE

the mutual capacitance associated to the sample (C_2) and the mutual capacitance associated to the substrate (C_1), such that:

$$C_m = C_1 + C_2 \tag{12}$$

In order to measure the real part of the permittivity, a lossless network is assumed, therefore, $R = R_2 = \infty$.

When a sample is placed on top of the substrate, the capacitances C_1 and C_2 can be expressed as [22]:

$$C_i = \frac{\varepsilon_0 \varepsilon'_{ri}}{\pi} ln\left[coth\left(\frac{\pi s}{4h}\right)\right] + 0.65 C_f \left(\frac{0.02h\sqrt{\varepsilon'_{ri}}}{s} - \frac{1}{\varepsilon'^2_{ri}} + 1\right) \tag{13}$$

where the subindex i can be 1 or 2 and the fringing field capacitance C_f is

$$C_f = \frac{\sqrt{\varepsilon'_{r1eff}}}{2cZ_0} - \frac{w\varepsilon_0\varepsilon'_{r1}}{2h} \tag{14}$$

where ε_{r1eff} is the effective dielectric constant of the substrate, c the speed of light, ε_{r1} the substrate dielectric constant, Z_0 the microstrip line characteristic impedance, h the substrate height and w the separation between the coupled lines (Fig. 7).

By taking Eq. (12), we can approximate C_m to:

$$C_m \approx \frac{0.013C_f\sqrt{\varepsilon'_{r1}} + 0.013C_f\sqrt{\varepsilon'_{r2}}}{s/h} \tag{15}$$

By replacing (15) in (11), the following quadratic equation is found in terms of the coupling coefficient k

$$\varepsilon'_{r2} = \alpha k^2 + \beta k + \gamma \tag{16}$$

where

$$\gamma \approx \varepsilon'_{r1} \tag{17}$$

Therefore, in order to obtain the α, β and γ constants, it is necessary to perform a calibration with three known-samples and solve the following equation system

$$\varepsilon'_{r2a} = \alpha k_a^2 + \beta k_a + \gamma \tag{18}$$

$$\varepsilon'_{r2b} = \alpha k_b^2 + \beta k_b + \gamma \tag{19}$$

$$\varepsilon'_{r2c} = \alpha k_c^2 + \beta k_c + \gamma \tag{20}$$

Herrera et. al. [6] calculate α, β and γ by using air ($\varepsilon_{ra} = 1$), FR4 ($\varepsilon_{rb} = 4.4$) and Rogers 3010 ($\varepsilon_{rc} = 11.2$) as known materials. The samples are placed on the sensor. The respective measured coupling coefficients are $k_{a1} = 0.1343$, $k_{b2} = 0.198$, and $k_{c3} = 0.2709$. By solving (18), (19), (20) we obtain α = 290.1, β = -42.37, and γ = 1.514, giving the following expression

$$\varepsilon'_{r2} = 290.1k_2^2 - 42.4k_2 + 1.5 \tag{21}$$

Table 3 Measured permittivity values reported in [6] using coupled hairpin resonators compared with the manufacturer's values

Material	Reported by manufacturer	Measured
Air[a]	1	1
Diclad880	2.2	2.217
Rogers 5870	2.33	2.311
Rogers 4003C	3.38	3.395
FR4[a]	4.4	4.4
TMM10i	9.8	9.573
Rogers 6010	10.7	10.55
Rogers 3010[a]	11.2	11.2

[a] Materials used for calibration

The dielectric constants of several samples have been measured by this method (Diclad 880, Rogers 5870, Rogers 4003C, TMM10i and Rogers 6010), showing good agreement between the manufacturers reported values and the measured results [6], as shown in Table 3.

5 Isotropic Dielectric Loss Measurements with Coupled Resonators

By considering the circuit of Fig. 7, it can be seen that R_2 contributes only to the odd mode, not to the even mode [23]. This is exemplified in Fig. 6, where Q_e and Q_m are the odd and even quality factors without SUT respectively, and Q_{e1}, Q_{m1} are the quality factors with sample. As it can be seen, only the odd quality factor is affected by the addition of the sample, whereas the even mode quality factor remains practically unaltered. For the odd mode, the unloaded quality factor can be calculated as shown in Eq. (22) (see Fig. 8)

$$\frac{1}{Q_{Te}} = \frac{1}{Q_{EXTe}} + \frac{1}{Q_{0e}} + \frac{2}{Q_2} \tag{22}$$

where Q_{Te} is the total quality factor in the odd mode, Q_{EXTe} is the external quality factor, Q_{0e} is the unloaded quality factor, and Q_2 is the SUT quality factor. These can also be calculated as:

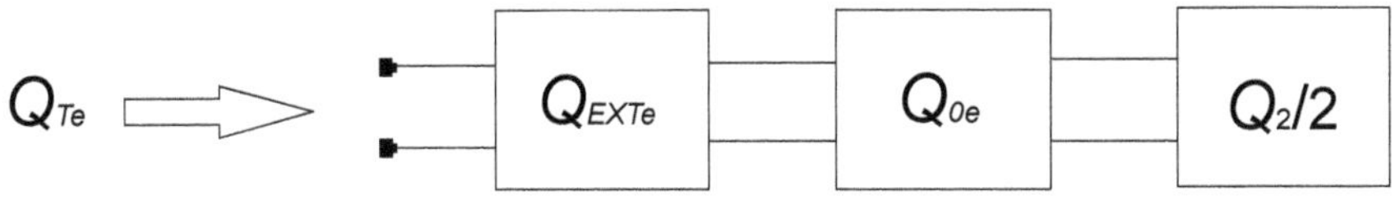

Fig. 8 Total Q_{Te} value in the odd case and its components: Q_{0e} unloaded, Q_{EXTe} external and Q_2 of the sample

$$\frac{1}{Q_{EXTe}} = \frac{\omega_e L}{R_{EXT}} \tag{23}$$

$$\frac{1}{Q_{0e}} = \frac{\omega_e L}{R} \tag{24}$$

where ω_e is given by

$$\omega_e = \frac{1}{\sqrt{L(C + C_m + 2C_2)}} \tag{25}$$

By taking into consideration that *tanδ* is the inverse of the quality factor, we can express (22) in terms of two unknown constants ρ_1 and ρ_2.

$$\frac{1}{Q_2} = tan\delta_2 = \rho_1\left(\frac{1}{Q_{Te}}\right) - \rho_2 \tag{26}$$

Therefore, in order to measure the *tanδ* losses, it is necessary to realize a calibration with two known samples and solve (26) to obtain ρ_1 and ρ_2. In [23], Rogers 4003 (RO4003, $f_{e3} = 2.143$ GHz) and polyvinyl chloride (PVC, $f_{e2} = 2.236$ GHz) are taken for calibration. Figure 9 shows the insertion losses of the sensor loaded with different SUTs. The resonant frequency of the sensor without SUT is $f_e = 2.333$ GHz. The reported losses for these samples are *tan* $\delta = 0.0027$ for RO4003, and *tan* $\delta = 0.01$ for PVC. Solving (26) for the two known samples yields 5.218 for ρ_1 and 53.352 $\times\ 10^{-3}$ for α_2, obtaining the following equation

$$tan\delta_2 = 5.218\left(\frac{1}{Q_{Te}}\right) - 53.35 \cdot 10^{-3} \tag{27}$$

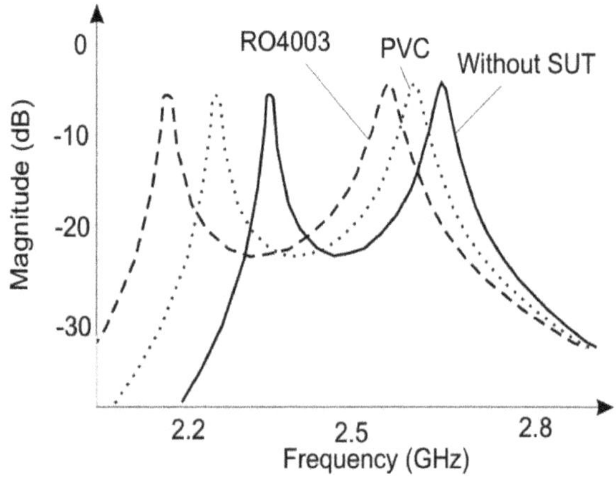

Fig. 9 Measured insertion losses for different SUTs. Adapted with permission from [23]; copyright (2020) Taylor & Francis

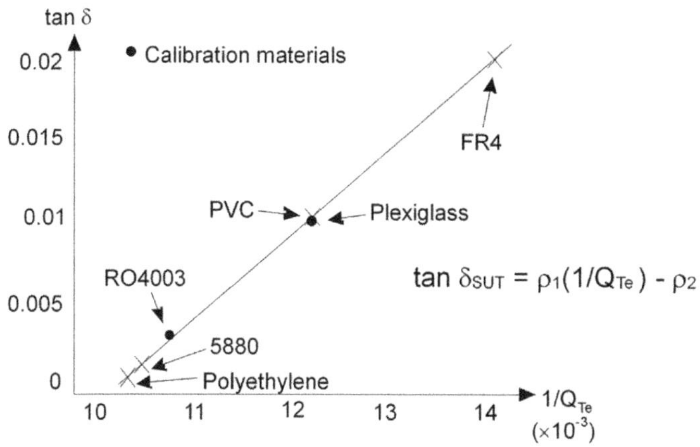

Fig. 10 Measured results of several unknown samples. Adapted with permission from [23]; copyright (2020) Taylor & Francis

In addition, four more samples (Polyethylene, Rogers 5880, Plexiglas, and FR4) were measured, giving the following Q values for the odd mode: 97.252, 96.135, 82.163 and 71.095, respectively. By solving (27), the loss tangents are 0.000303 for Polyethylene, 0.00092 for Rogers 5880, 0.0101 for Plexiglas, and 0.02 for FR4. These results are plotted in Fig. 10, where a linear trend is clearly observed between $1/Q_{Te}$ and $tan\delta$.

6 Uniaxial Anisotropy Measurements with Coupled Resonators

6.1 Method for Measuring the Dielectric Constant

In [5, 10, 17] and [24], coupled resonators are presented together with Space-Mapping techniques to measure dielectric properties. In these works, a series of simulations are utilized combined with experimental results for extracting anisotropic values of the permittivity.

As shown in Fig. 1, when performing anisotropy measurements in x–z plane directions, the x component of the E field is mainly affected by the odd mode and the z component by the even mode. The first step is to consider a proportionality constant A between the resonant frequency and the permittivity value (28).

$$A \cdot \varepsilon = f^{-2} \tag{28}$$

Since the electric field has x and z components, we can expand this equation as

$$A_{par_m}\varepsilon_{par} + A_{prp_m}\varepsilon_{prp} = f_m^{-2} \tag{29}$$

$$A_{par_e}\varepsilon_{par} + A_{prp_e}\varepsilon_{prp} = f_e^{-2} \tag{30}$$

And in matrix form we have

$$\begin{bmatrix} \varepsilon_{par} \\ \varepsilon_{prp} \end{bmatrix} = \begin{bmatrix} A_{par_m} & A_{prp_m} \\ A_{par_e} & A_{prp_e} \end{bmatrix}^{-1} \begin{bmatrix} f_m^{-2} \\ f_e^{-2} \end{bmatrix} \tag{31}$$

where it is clear that the even and odd mode frequencies depend on the parallel and perpendicular components of the permittivity. In addition, the proportionality constant is also composed of parallel and perpendicular components in the odd and even modes. Therefore, in order to calculate these constants, it would be necessary to measure two known samples to solve this two-equation system.

By applying a Space-Mapping technique, Rautio et al. [5] use a series of simulations to perform the calculations. The first step is to measure the unknown sample experimentally with a given sensor. From here, two resonant frequencies $f_{m\,exp}$ and $f_{e\,exp}$ are obtained. The second step is to simulate the sensor using two samples with different permittivities (ε_a, ε_b), obtaining f_{ma}, f_{ea}, f_{mb} and f_{eb}. From here, A_{ij1} can be calculated by solving Eq. (32):

$$\begin{bmatrix} A_{par_m} & A_{prp_e} \\ A_{par_e} & A_{prp_m} \end{bmatrix}^{-1} = \begin{bmatrix} \varepsilon_{par_a} & \varepsilon_{par_b} \\ \varepsilon_{prp_a} & \varepsilon_{prp_b} \end{bmatrix}^{-1} \begin{bmatrix} f_{ma}^{-2} & f_{mb}^{-2} \\ f_{ea}^{-2} & f_{eb}^{-2} \end{bmatrix}^{-1} \tag{32}$$

Then $A_{\text{ij}1}$, f_{mexp} and f_{eexp} are used in Eq. (31) to extract ε_{par1} and ε_{prp1}.

Subsequently, a simulation is carried out with $\varepsilon_{\text{par}1}$ and $\varepsilon_{\text{prp}1}$ to obtain f_{msim1} and f_{esim1}. A comparison between f_{mexp} and f_{eexp}, and between f_{msim1} and f_{esim1}, is performed. If the resonant frequencies do not fall within the specified tolerance error, then another A_{ij2} is extracted using $f_{m\,sim1}$ and $f_{e\,sim1}$, ε_{par1} and ε_{prp1}. This procedure is performed until the required error tolerance is achieved, as shown in Fig. 11. It is important to note that the simulations must be performed with great accuracy, and experimental errors should be minimized.

This technique has been applied in stripline (Fig. 12a) and shielded microstrip (Fig. 12b) sensors [5, 24], where the coupled resonators are printed on a substrate of unknown anisotropy value. Therefore, the sensor substrate becomes the SUT. However, in [10] the sensor is printed on a known substrate, and the SUT is placed over the sensor, with the advantage that the same sensor can be used for multiple SUTs measurements (Fig. 12c). In this latter case, air gaps between the SUT and the sensor mainly due to surface roughness, and errors in sample placement are of great concern. For multiband measurements, several harmonics can be measured, as shown in [5] and [24].

By following the above-mentioned method, in [10], a microstrip sensor with a fundamental frequency of 2.4 GHz is designed. Figure 13 shows the microstrip sensor layout using RT/Duroid 5880 substrate of Rogers Corporation, with ε_r of

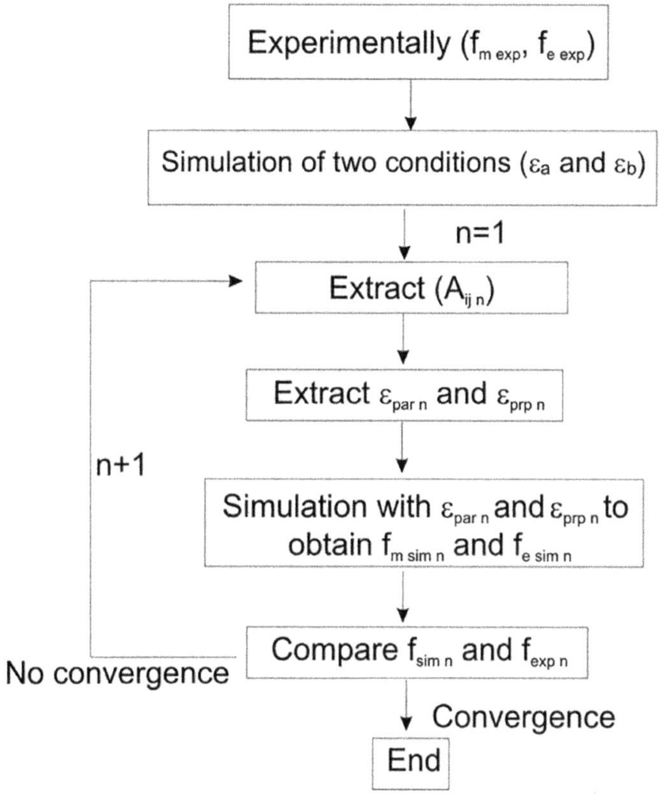

Fig. 11 Methodology for measuring the dielectric anisotropy

2.2, loss tangent of 0.0009, copper cladding thickness t of 18 μm, and a dielectric thickness h of 0.79 mm. The structure dimensions are shown in Fig. 13.

The first step is to fabricate and measure the sensor with different SUTs: FR4, Rogers 4350B, Arlon Diclad 880 and PTFE, all with the following dimensions: $L_s = 2.3$ mm, $W_s = 10$ mm, and with thickness h varying, depending on the sample (Table 4). Subsequently, the values of A_{ij} are extracted from two simulations with:

(1) $\varepsilon_{par} = \varepsilon_{prp} = \varepsilon_{iso} = 4.4$

and

(2) $\varepsilon_{par} > \varepsilon_{iso}$ and $\varepsilon_{prp} < \varepsilon_{iso}$ ($\varepsilon_{par} = 4.5$; $\varepsilon_{prp} = 4.3$)

In the simulation, a 2.5 um air gap between SUT and sensor was considered to take into account surface roughness. Then, the recursive method of Fig. 11 is applied until convergence is reached. The final results are shown in Table 4.

As it can be seen, PTFE is confirmed to be practically isotropic with a normalized anisotropy (6) of $\Delta\varepsilon = 0.39\%$, whereas FR4 has a $\Delta\varepsilon$ of 8.67%, Rogers-4350B has $\Delta\varepsilon = 10.86\%$ and Diclad-880 has $\Delta\varepsilon = 7.11\%$.

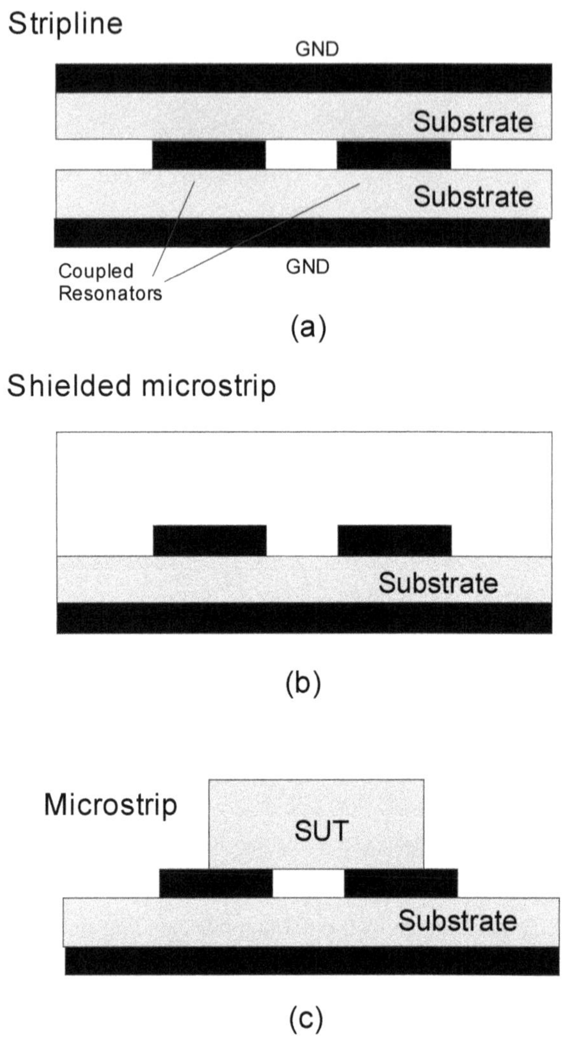

Fig. 12 Three methods for measuring anisotropy: stripline with unknown substrate anisotropy; shielded microstrip with unknown substrate anisotropy and microstrip on known substrate and unknown SUT

6.2 Method for Measuring Dielectric Constant and Dielectric Loss

The previous method is able to retrieve the real part of the permittivity. However, for calculating the real and imaginary parts, the cavity perturbation technique (CPT) may be employed.

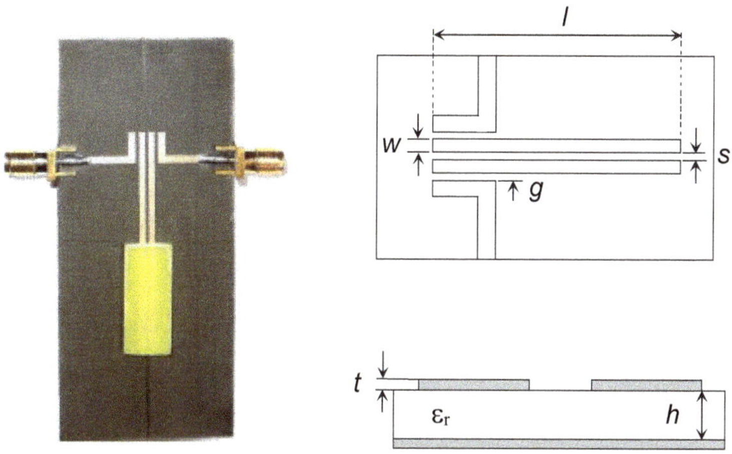

Fig. 13 Coupled resonators for anisotropic measurements. Dimensions: $w = 1.2$ mm, $s = 0.7$ mm, $l = 45.63$ mm, $g = 0.8$ mm, $h = 0.79$ mm and $t = 18$ μm. Adapted with permission from [10]; copyright (2020) IEEE

Table 4 Anisotropic measurements of different dielectrics [10]

Material	h (mm)	ε_{par}	ε_{prp}	$\Delta\varepsilon$ (%)
FR4	1.48	4.57	4.19	8.67
Rogers-4350B	0.76	3.88	3.48	10.86
Diclad-880	3.18	2.33	2.17	7.11
PTFE	1.38	2.05	2.04	0.39

The resonator of Fig. 13 can be modified to accommodate a sample substrate inside the slot as shown in Fig. 14. With this configuration, and using the odd and even mode analyses, it is seen that, for the odd mode, the x component of the E field is directed transversely to the SUT; however, for the even mode, the E field is directed mainly in the z direction. Therefore, it is possible to obtain ε_{prp} in the odd mode, and ε_{par} in the even mode (Fig. 14).

By using the CPT, the real and imaginary parts of the permittivity can be measured. The CPT equations are as follows:

$$\varepsilon_r' = 1 + A\left[\frac{V_c(f_0 - f_s)}{V_s \cdot f_s}\right] \quad (33)$$

$$\varepsilon_r'' = B\frac{V_c}{V_s}\left[\frac{1}{Q_s} - \frac{1}{Q_0}\right] \quad (34)$$

where $\varepsilon_r{}'$ and $\varepsilon_r{}'$ are the real and imaginary parts of the permittivity, A and B are calibration constants, f_0 and f_s are the resonant frequencies before and after the

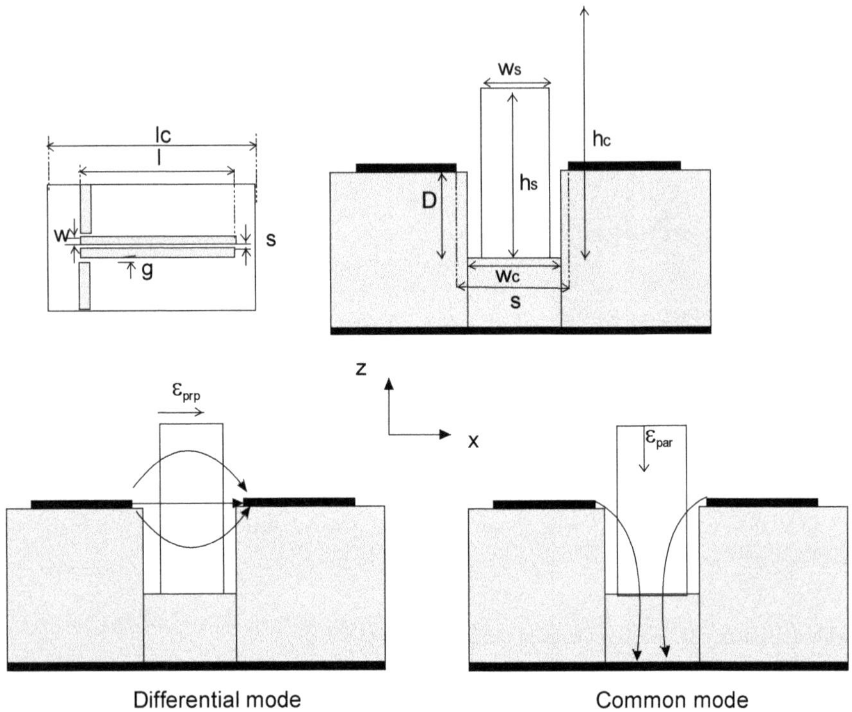

Fig. 14 Coupled resonators with middle slot to measure real and imaginary parts of the permittivity using cavity perturbation technique [15]. Dimensions (mm): $w = 2.5$, $g = 1$, $l = 45$, $s = 1.2$, $w_c = 1.07$, $D = 0.62$, $l_c = 65$ and $h_c = 3.12$. Adapted from [15] under a creative commons attribution 4.0 international license

perturbation, V_s and V_c are the volumes of the sample and cavity, respectively, and Q_0 and Q_s are the quality factors before and after the perturbation [25]. The parameters A and B are extracted experimentally by measuring known samples of similar size and shape of the SUT. In addition, greater accuracy is obtained when the permittivities of the known samples are close to the unknown SUT permittivity [26].

For determining the uniaxial anisotropy, the even-mode S parameters are used to obtain ε_{par} and the odd-mode S parameters are used for ε_{prp}.

Table 5 Materials used for calibration as shown in [15]

Material	*ws* (mm)	*hs* (mm)	$f_{2\mathrm{m}}$ (GHz)	$Q_{2\mathrm{m}}$	$\varepsilon_{\mathrm{par}}$	$\tan\delta_{\mathrm{par}}$	$f_{2\mathrm{e}}$ (GHz)	$Q_{2\mathrm{e}}$	$\varepsilon_{\mathrm{prp}}$	$\tan\delta_{\mathrm{prp}}$
PTFE	0.96	1.54	2.3518	60.84	2.051	0.0012	2.5045	103.58	2.051	0.0012
RT5870	0.97	1.51	2.3462	61.36	2.35	0.0016	2.5049	102.43	2.5	0.0023
RO4003	1.03	0.79	60.145	2.4951	3.37	0.0029	2.4951	98.29	3.67	0.0037

In [15], a sensor was fabricated on a 3D printed substrate, using PLA with a 50% honeycomb pattern with $\varepsilon'_r = 2.47$, and two substrates ($h = 0.127$ mm, $\varepsilon'_r = 2.2$ and $tan\delta = 0.0009$) attached to the top and bottom of the PLA substrate. The volume of the cavity (Fig. 14) is $V_c = w_c \times l_c \times h$. h_c is the effective cavity height, and it is defined as the maximum height for sensor sensitivity (approximately $5D$), which would mean that at any height greater than h_c, the coupling E-field is negligible. By simulations with HFSS, h_c was established as 3.12 mm. l_c can be chosen to be the length l plus about 5 times the substrate thickness per side. The sample volume is $V_s = w_s \times l_s \times h_s$, where $w_s \le w_c$, and $l_s \le l_c$, h_s should be smaller than $3D$. All dimensions are shown in Fig. 14.

The sensor was measured with a VNA (Keysight FieldFox N9918A), giving the following resonant frequencies and quality factors: for the even mode, $f_{1m} =$ 2.3543 GHz and $Q_{1m} = 63.372$, and for the odd mode, $f_{1e} = 2.52425$ GHz and Q_{1e} = 105.432. It is important to note that, given that the sensor is weakly coupled (S_{21} at resonance < -25 dB), the measured quality factor is equivalent to the unloaded quality factor.

Morales et al. [15] have used PTFE, RT5870 and RO4003 as known materials to determine the proportionality constants A and B. All measurements were repeated 12 times and the results averaged. By using Eqs. (33) and (34), the resulting values are shown in Table 5.

After calculating the proportionality constants, the sensor was used to characterize 3 anisotropic samples (PLA, Diclad 880 and RO4350B). Table 6 shows the SUT characteristics and measured parameters. As it can be seen, both real and imaginary parts present anisotropy. The maximum anisotropy for the real part is observed for RO4350 $\Delta\varepsilon = 11.47\%$, whereas PLA presents only $\Delta\varepsilon = 1.47\%$. For the loss tangent, anisotropy can be as large as $\Delta A_{tan\delta} = 131\%$, and the lowest value was $\Delta A_{tan\delta} =$ 50%, for PLA.

Table 6 Different permittivity values as reported in [15]

Material	ws (mm)	hs (mm)	f_{2m} (GHz)	Q_{2m}	ε_{par}	$tan\delta_{par}$	f_{2e} (GHz)	Q_{2e}	ε_{prp}	$tan\delta_{prp}$	ΔA_{ε} (%)	ΔA_{tand} (%)
PLA	0.95	1.91	2.3457	60.98	2.7	0.0012	2.5	101.73	2.74	0.002	1.47	50
Diclad 880	0.92	1.51	2.3486	61.44	2.13	0.0011	2.5037	102.62	2.28	0.0022	6.8	66.7
RO4350B	0.96	0.75	2.3459	60.212	3.55	0.003	2.496	99.51	3.73	0.0034	11.47	131

References

1. Krupka J (2006) Frequency domain complex permittivity measurements at microwave frequencies. Meas Sci Technol 17(6).
2. Lobato-Morales H, Corona-Chavez A, Murthy DV, Olvera-Cervantes JL (2010) Complex permittivity measurements using cavity perturbation technique with substrate integrated waveguide cavities. Rev Sci Instr 81(6):064704
3. Chen L, Ong CK, Tan BT (1999) Cavity perturbation technique for the measurement of permittivity tensor of uniaxially anisotropic dielectrics. IEEE Trans Instrum Measur 48(6):1023–1030
4. Dankov PI (2006) Two-resonator method for measurement of dielectric anisotropy in multilayer samples. IEEE Trans Microw Theory Techn 54(4):1534–1544
5. Rautio JC, Arvas S (2009) Measurement of planar substrate uniaxial anisotropy. IEEE Trans Microw Theory Techn 57(10):2456–2463
6. Herrera-Sepulved LV, Olvera-Cervantes JL, Corona-Chavez A, Kaur T (2019) Sensor and methodology for determining dielectric constant using electrically coupled resonators. IEEE Microw Wireless Compon Lett 29(9):626–628
7. Kobayashi Y, Senju T (1993) Resonant modes in shielded uniaxial-anisotropic dielectric rod resonators. IEEE Trans Microw Theory Techn 41(12):2198–2205
8. Fritsch U, Wolff I (1992) Characterization of anisotropic substrate materials for microwave applications. In: 1992 IEEE MTT-S microwave symposium digest. IEEE, pp 1131–1134
9. Mueller S, Koeberle M, Goelden F, Penirschke A, Gaebler A, Lapanik A, Haase W, Jakoby R (2008) W-band characterization of anisotropic liquid crystals at room temperature. In: 2008 38th European microwave conference. IEEE, pp 119–122
10. Morales-Lovera HN, Olvera-Cervantes JL, Corona-Chavez A, Kataria TK (2020) Dielectric anisotropy sensor using coupled resonators. IEEE Trans Microw Theory Techn 68(4):1610–1616
11. HFSS v.12 Ansoft.
12. Tobar ME, Krupka J, Ivanov EN, Woode RA (1998) Anisotropic complex permittivity measurements of mono-crystalline rutile between 10 and 300 K. J Appl Phys 83(3):1604–1609
13. Yaghmaee P, Karabey OH, Bates B, Fumeaux C, Jakoby R (2013) Electrically tuned microwave devices using liquid crystal technology. Int J Antenn Propag. 824214
14. Dankov IP (2020) Dielectric anisotropy of modern microwave substrates. In: Microwave and millimeter wave technologies from photonic bandgap devices to antenna and applications. IntechOpen
15. Morales-Lovera HN, Olvera-Cervantes JL, Pérez-Ramos AE, Corona-Chavez A, Saavedra CE (2022) Microstrip sensor and methodology for the determination of complex anisotropic permittivity using perturbation techniques. Sci Rep 12(1):2205
16. Mumcu G, Sertel K, Volakis JL (2008) A measurement process to characterize natural and engineered low-loss uniaxial dielectric materials at microwave frequencies. IEEE Trans Microw Theory Techn 56(1):217–23
17. Morales-Lovera HN, Olvera-Cervantes JL, Corona-Chavez A, Kataria TK (2021) Dielectric characterization of anisotropic 3D-printed biodegradable substrates based on polylactic acid [application notes]. IEEE Microwave Mag. 22(9):18–20
18. Bodakian B, Hart FX (1994) The dielectric properties of meat. IEEE Trans Dielectr Electr Insul 1(2):181–187
19. Sun E, Datta A, Lobo S (1995) Composition-based prediction of dielectric properties of foods. J Microwave Power Electromagn. Energy 30(4):205–212
20. Kabir MF, Khalid KB, Daud WM, Aziz SH (1997) Dielectric properties of rubber wood at microwave frequencies measured with an open-ended coaxial line. Wood Fiber Sci 319–324
21. Hong JS, Lancaster MJ (2004) Microstrip filters for RF/microwave applications. John Wiley & Sons
22. Garg R, Bahl IJ (1979) Characteristics of coupled microstrip lines. IEEE Trans Microw Theory Techn 27(7):700–705

23. Herrera-Sepúlveda LV, Olvera-Cervantes JL, Corona-Chavez A, Kaur T (2020) New methodology to determine the loss tangent of dielectric planar samples by using electrically coupled resonators. J Electromagn Waves Appl 34(18):2410–2418
24. Rautio J, Brian J, El Sabbagh M, Rautio JC (2011) Detailed error analysis and automation of the RA Resonator technique for measurement of uniaxial anisotropic permittivity. In: 78th ARFTG microwave measurement conference. IEEE, pp 1–5
25. Subramanian V, Sobhanadri J (1994) New approach of measuring the Q factor of a microwave cavity using the cavity perturbation technique. Rev Sci Instr 65(2):453–455
26. Kaur T, Olvera-Cervantes JL, Rojas-Laguna R, Sosa-Morales ME, Corona-Chavez A (2023) Dielectric properties of Sphenarium purpurascens at 2.4 GHz. J Microwave Power E E 15(1):1

Planar Microwave Sensors Based on Coupled Ring Resonators and Applications

Mohammad H. Zarifi, Omid Niksan, Ali Maleki Gargari, and Dima Kilani

Abstract The potential of planar microwave resonators for sensing applications has been signified, when a single resonant structure is exposed to dielectric properties-variant environments. However, enhancing sensing parameters such as sensitivity, selectivity and dynamic range remains a continuing challenge. To address these challenges, coupled microwave resonant structures have been introduced in various microwave sensing applications. It is demonstrated that the quality factor of a coupled resonator system can be improved, and the dynamic range of sensing can be extended before the sensitivity saturation range is reached. In this chapter, the electromagnetic coupling between planar split ring resonators will be discussed. Their performances will be presented in simulation, mathematical analysis and experimental forms, and their outstanding performance in the introduced unique applications will be studied in wired and wireless readout methods. Moreover, it will be shown that not only the electromagnetic coupling of split ring resonators in wired systems enhances the dynamic range of the sensors, but also, in wireless readout systems, it increases the communication distance between an array of coupled passive split ring resonators and an interrogator antenna.

1 Introduction

Coupled microwave resonators offer a promising solution to the limitations of single microwave resonators within the framework of sensing and detection. While single resonators have demonstrated potential, they often suffer from selectivity issues, low resolution, and a limited dynamic range [1]. However, by employing a configuration of multiple coupled resonators, these limitations can be effectively addressed. In a coupled arrangement of resonators, each element can be designed to detect certain target substances, or an environmental parameters (e.g., displacement, temperature),

M. H. Zarifi (✉) · O. Niksan · A. M. Gargari · D. Kilani
Faculty of Applied Sciences, School of Engineering, University of British Columbia, Kelowna, BC V1V 1V7, Canada
e-mail: Mohammad.zarifi@ubc.ca

F. Martín and E. Bronchalo (eds.), *Coupled Structures for Microwave Sensing*, Lecture Notes in Electrical Engineering 1150,
https://doi.org/10.1007/978-3-031-53861-2_8

thereby improving the selectivity of sensing. The interplay between the coupled resonators can amplify their field interactions with test materials, leading to sensitivity enhanced responses. Moreover, the coupled configuration of resonators allows for an extension in the range of measured magnitude, such as dielectric constant, effectively enhancing the dynamic range of detection for the microwave sensor. Coupled resonators enable unique directions in the advancing field of microwave sensing, with applications in biomedical, industrial, safety, and security sectors [2].

The present chapter begins with a classification of coupling phenomena in microwave split ring resonators (SRRs). The majority of this chapter discusses how coupling between SRRs can enhance the dynamic range, selectivity, and detection range in microwave sensors. Additionally, the chapter will illustrate how utilizing coupled SRR configurations can further enable the detection of physical quantities, such as UV radiation, or mass of material samples.

2 Classification of Electromagnetic Coupling in Planar Microwave Resonators

Coupling phenomena in microwave resonators is often referred to as electromagnetic coupling, which involves the transfer of energy (and information) between resonators through their electromagnetic fields. The coupling between resonators can be addressed through both electrical and magnetic interactions, depending on the dominant field involved and the relative strength. The resonators can exhibit increased sensitivity (which is amplified at the resonant frequency of the system) to the external stimuli, i.e., the sensing target. Therefore, the coupling mechanisms between resonators can be carefully designed and optimized to achieve the desired interaction depending on the requirements of a specific sensing application.

The concept of coupled resonators has been previously investigated by Hong et al. for cross-coupled planar microwave filters [3]. Hong's research focused on open loop coupled resonators, commonly known as SRRs, where a comprehensive theoretical framework based on filter theory was provided for understanding these structures. Based on Hong's classification, coupled resonators can be categorized into three primary types: electric coupling, magnetic coupling, and mixed (electromagnetic) coupling. The illustration of these configurations is depicted in Fig. 1. These coupled structures are formed by arranging a pair of folded half-wavelength open-loop resonators in different orientations while maintaining a separation distance denoted as S. The couplings observed in these structures are primarily proximity coupling which occurs through fringe fields. The nature and strength of the couplings are determined by the characteristics of these fringing fields. At fundamental resonance mode, each SRR exhibits maximum electric field intensity on the split section (open gap), while the maximum magnetic field intensity is observed on the opposite thru section. Due to the exponential decay of the fringe field beyond this region, the

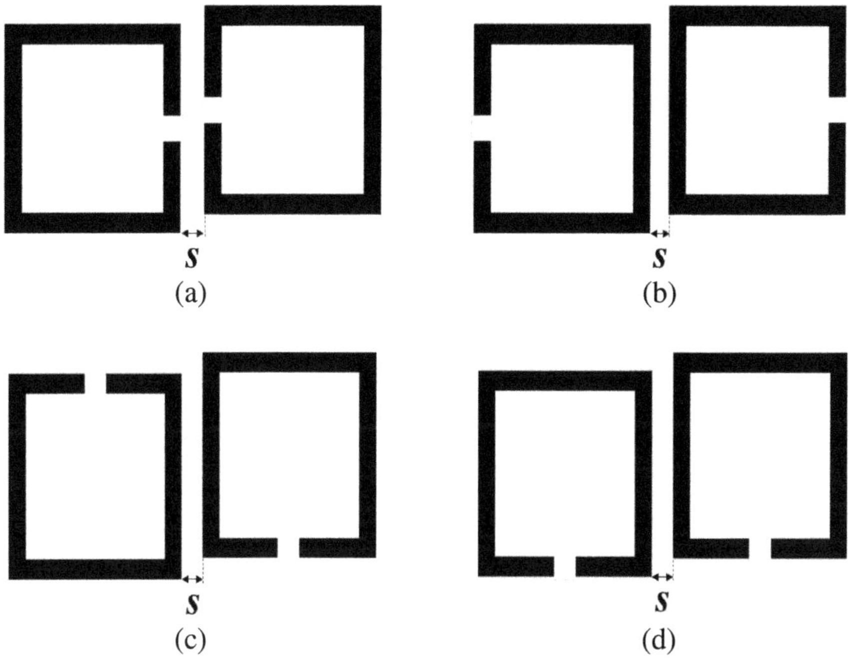

Fig. 1 Typical structures of planar coupled resonators with **a** electric coupling **b** magnetic coupling, and **c–d** mixed coupling. Reprinted with permission from [3]; copyright (1996) (IEEE)

electric fringe field is more evident near the side with the highest electric field distribution, whereas the magnetic fringe field is stronger near the side with the maximum magnetic field distribution. Therefore, when the open sides of two coupled resonators are positioned in proximity, as depicted in Fig. 1a, electric coupling is prevailing. Alternatively, if the sides with the highest magnetic field of the coupled resonators are closely positioned, as shown in Fig. 1b, magnetic coupling is achieved. In the coupling configurations depicted in Fig. 1c and d, the electric and magnetic fringe fields at the coupled sides may exhibit comparable distributions, resulting in both electric and magnetic couplings. This combined coupling is referred to as "electro-magnetic" or "mixed" coupling.

An equivalent lumped-element circuit model for electrically and magnetically coupling of the resonators with equal dimensions (i.e., synchronously tuned coupled resonators) is given in Figs. 2a and b. The value of L and C are the self-inductance and self-capacitance of single resonator, so that $(LC)^{-1/2}$ equals the angular resonant frequency of uncoupled resonators. C_m and L_m represent the mutual capacitance and inductance indicating the coupling factor in the electrically- and magnetically-coupled modes. A coupling coefficient can be defined as a ratio of the coupled electric (or magnetic) energy to the stored energy of uncoupled single resonator, and simplified as follows:

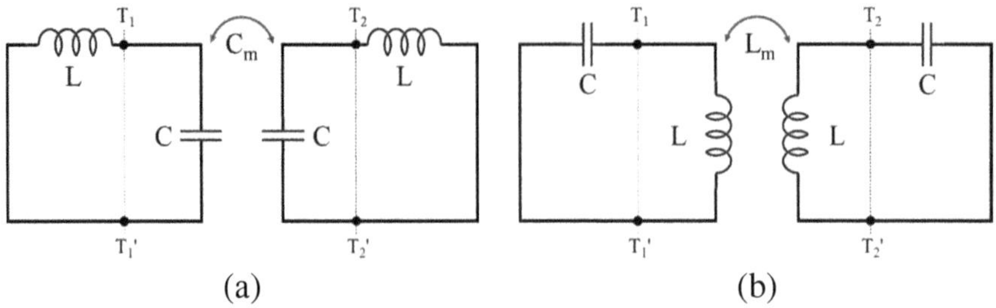

Fig. 2 Equivalent circuit models of coupled resonators with **a** electric and **b** magnetic coupling. Reprinted with permission from [3]; copyright (1996) (IEEE)

$$K_E = \frac{C_m}{C} \tag{1}$$

$$K_M = \frac{L_m}{L} \tag{2}$$

Among various design parameters of coupled structures, spacing (S) between coupled resonators plays a crucial role in determining the coupling behavior of the system. Reducing S between coupled resonators not only leads to an increase in the mutual inductance or capacitance but also results in stronger coupling coefficients. This enhanced coupling can be understood by examining the eigenvalues of the eigenequation associated with the coupled circuit models. It has been shown in [3] that the natural resonance mode of an individual resonator was split into two distinct modes when coupling occurs. The coupled structures can be analyzed by deriving the two eigenvalues, which correspond to the two natural resonant frequencies of the system. For electric coupling, these eigenvalues can be obtained using (1) and (2). On the other hand, for magnetic coupling, (3) and (4) provide the derived resonant frequencies. An alternative approach to extract the resonant frequencies is by analyzing the even and odd symmetry modes of the structure. By applying electric and magnetic wall conditions to the center of the equivalent circuit model, parallel to the T_1-T_1' and T_2-T_2' planes, the resonant frequencies can be determined. It is worth noting that reducing the spacing S increases spectral separation of the low and high resonant frequencies. This effect is significant in the design and optimization of coupled resonator structures, allowing for precise control over frequency responses and tailored resonant characteristics.

$$f_{\text{E, low}} = \frac{1}{2\pi}\frac{1}{\sqrt{L(C+C_m)}} \tag{3}$$

$$f_{\text{E, high}} = \frac{1}{2\pi}\frac{1}{\sqrt{L(C-C_m)}} \tag{4}$$

$$f_{\text{M,low}} = \frac{1}{2\pi}\frac{1}{\sqrt{(L+L_m)C}} \tag{5}$$

$$f_{\text{M, high}} = \frac{1}{2\pi} \frac{1}{\sqrt{(L - L_m)C}} \tag{6}$$

3 Coupled Split Ring Resonators with Enhanced Sensing Performance

In a work conducted by Zarifi et al. [4], the concept of coupled SRRs was investigated for sensing applications, and the performance was thoroughly investigated for both magnetic (i.e., inductive) and electric (i.e., capacitive) couplings. The sensing platform utilized in the study consisted of two planar SRRs that were either capacitively or inductively coupled, as depicted in Fig. 3. To evaluate and compare the sensitivity and dynamic range of the coupled resonator configurations, the differentiation of various complex permittivity values was used as figure of merit. The investigation by Zarifi et al. demonstrated that the capacitive coupled structure exhibited a significant advantage over the inductive coupled form in terms of sensing non-magnetic, low-loss, dielectric materials. The sensitivity of a microwave sensor can be defined as the ability of detecting changes in the physical magnitude of interest (e.g., permittivity, conductivity). In the context of permittivity sensing, the capacitive coupled structure exhibited higher sensitivity, implying that it could accurately detect and quantify variations in the permittivity of the target material. Dynamic range, on the other hand, refers to the range of input signals or physical quantities that a sensor can effectively measure. A larger dynamic range indicates that the sensor can accommodate a broader range of input variations without losing accuracy or reaching saturation. The capacitive coupled structure outperformed the inductive coupled one by offering a wider dynamic range, making it more suitable for applications that involve a broad range of complex permittivity values.

In Fig. 3a, each resonant ring is shown magnetically coupled to the probing transmission lines (TLs), while electric coupling is established between the two adjacent rings. For the inductively coupled structure (Fig. 3b), the two rings are magnetically coupled to each other while a capacitive coupling is established between the resonators and the TLs. To study the frequency splitting phenomena and the impact of the coupling gap (denoted as g_r in here) on the separation between the two resonant frequencies of the coupled system, Fig.4 shows the results of a set of full wave simulations in Ansys HFSS environment. The transmission coefficients of the sensor ($|S_{21}|$) were plotted in Fig. 4a and b when g_r varies from 0.3 to 0.9 mm with steps of 0.2 mm. In both coupling cases, increasing the coupling gap reduces the coupling strength, and consequently, the separation between the resonant frequencies of the coupled system. The equivalent circuit models of the proposed sensing structure are shown in Fig. 5a and b for electrical and magnetic coupling forms, respectively. These models are consistent with the model described in the previous section and

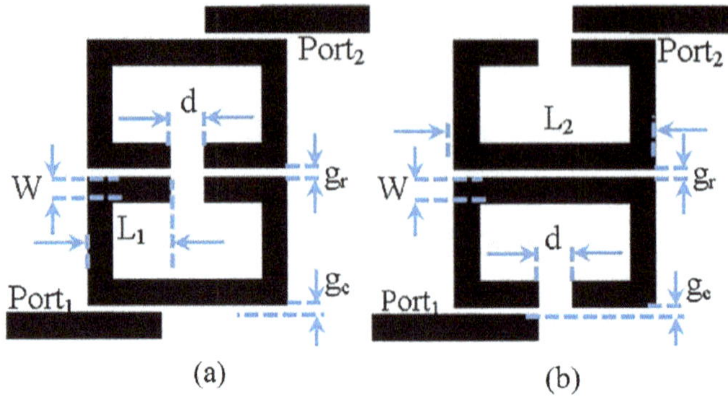

Fig. 3 Planar coupled resonator sensor structure: **a** capacitive coupling and **b** inductive coupling with the following dimensions (mm): $W = 2.4$, $g_r = 0.7$, $g_c = 0.6$, $d = 3$, $L_1 = 5.4$, $L_2 = 14$. Reprinted with permission from [4]; copyright (2016) (American institute of physics)

highlight that the separation between low and high resonant frequencies are reduced by increasing the coupling factors of k_E or k_M.

In planar microwave resonator sensors, the characteristics of the sensor are influenced by changes in the permittivity of the surrounding environment, which in turn can affect the electric field. The electric field distribution is governed by the electrical properties of the substrate and the materials above the resonator's surface, referred to as the material under test (MUT). To visualize this, Fig. 6 presents the simulated distribution of the electric field within a 2-mm distance above the sensor's surface. Notably, the regions experiencing capacitive coupling exhibit stronger electric fields, indicating their heightened sensitivity, and being referred to as "hot-spots".

To assess the effectiveness of the proposed sensor structures, a simulation was conducted by placing the MUT on the sensitive hot spots of both resonators and

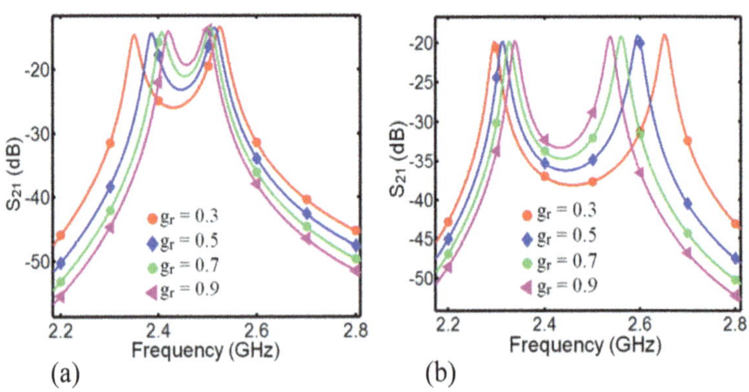

Fig. 4 Effect of varying coupling gap for **a** capacitive and **b** inductive coupled resonator with the corresponding equivalent circuit models of **a** capacitive and **b** inductive coupled structures. Reprinted with permission from [4]; copyright (2016) (American institute of physics)

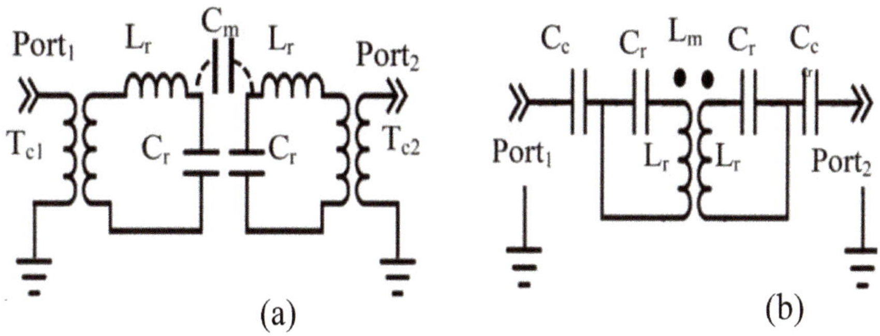

Fig. 5 The corresponding equivalent circuit models of **a** capacitive and **b** inductive coupled structures. Reprinted with permission from [4]; copyright (2016) (American institute of physics)

Fig. 6 HFSS simulations, probing the electric field intensities 2 mm above the surface of the capacitive and inductive coupled resonators. Capacitive coupled areas demonstrate higher field intensity than other regions. Reprinted with permission from [4]; copyright (2016) (American institute of physics)

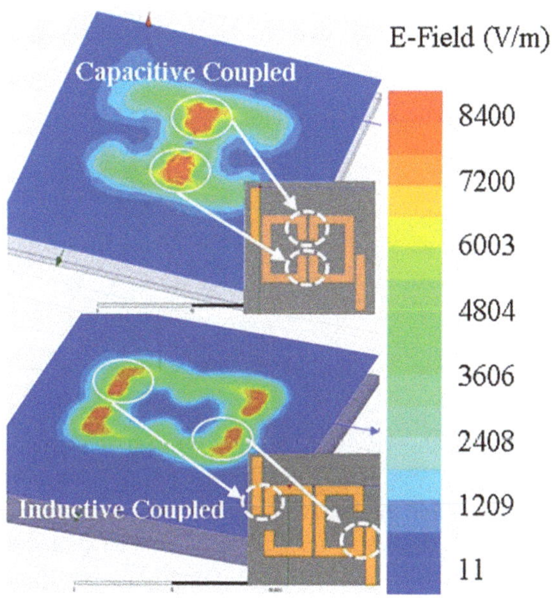

monitoring the changes in the $|S_{21}|$ response of the sensor for various permittivity values of the MUT. Figure 7a and b provide an illustration of the simulated structure with the MUT positioned precisely in these critical areas. Figure 7c clearly shows that in the capacitive coupled structure, the higher resonant frequency (f_H) of the coupled configuration suffers less variation than the lower resonant frequency (f_L). Conversely, in the inductive coupled resonator, the higher resonant frequency experience larger variations than the lower resonant frequency, leading to a distorted frequency response, as shown in Fig. 7d. As these two resonant frequencies show larger variations with the MUT permittivity. It is essential to study their behavior more in detail.

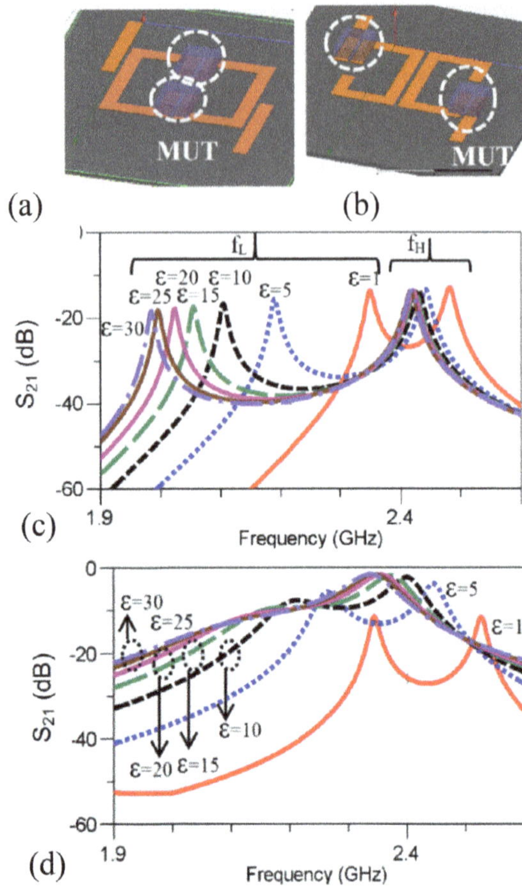

Fig. 7 HFSS simulation of capacitive and inductive coupled resonators for MUTs of varying permittivity: **a** capacitively coupled; **b** inductively coupled; **c** $|S_{21}|$ for capacitively coupled structure; and **d** $|S_{21}|$ for inductively coupled structure. Reprinted with permission from [4]; copyright (2016) (American institute of physics)

To examine the behavior of the lower resonant frequency in capacitive coupling (f_{LCC}) and the higher resonant frequency in inductive coupling (f_{HIC}), as shown in Fig. 8, a sensitivity study was carried out. The simulation results revealed that f_{L} in capacitive coupled resonator structures was more sensitive to changes in permittivity compared to f_{H} in inductive coupled structures. In addition, the quality factor (loaded) of the resonance (*i.e.*, the ratio of resonant frequency to the 3 dB bandwidth) demonstrated a 20% increase in the capacitive coupled resonator, while the corresponding value in the inductive coupled resonator experienced a substantial decline (see Fig. 8a and b). The observed changes in the quality factor of simulated models were attributed to the permittivity variations (real part), with the loss factor set to have a constant value. To analyze the impact of loss-factor variation at a constant permittivity of 2, the response of the resonators was studied and depicted in Fig. 8c and d. Increasing the loss-factor from 0 to 0.2 resulted in a 0.5% variation in the resonant frequency for both capacitive and inductive coupled resonators. Nonetheless, it was observed that the capacitive coupled resonator experienced a substantial

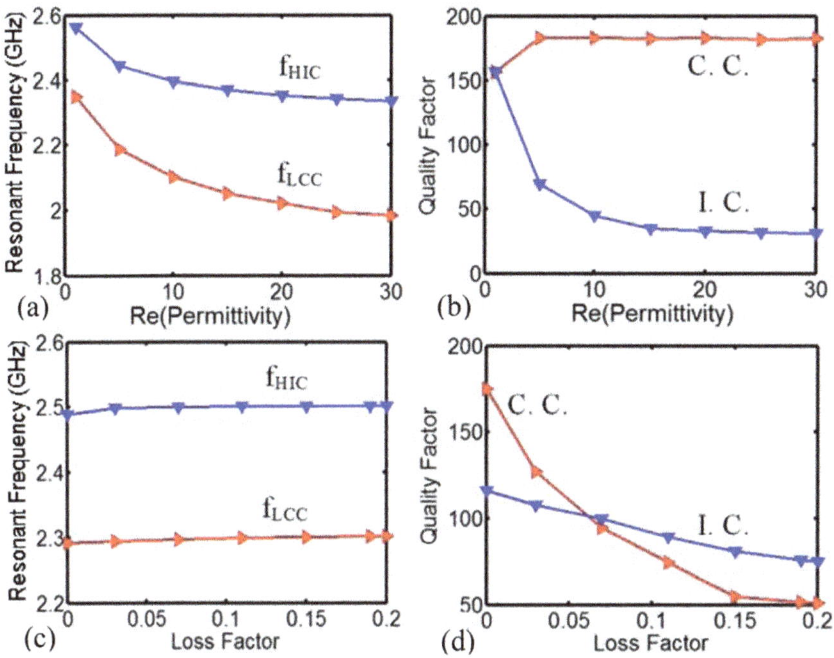

Fig. 8 Comparison of simulated capacitive (C.C.) and inductive coupled resonators (I.C.). **a** Resonant frequency versus variant permittivity at zero loss factor; **b** quality factor for the resonant frequencies versus variant permittivity at zero loss-factor; **c** resonant frequency versus variant loss-factor at permittivity of 2; and **d** quality factor versus variant loss-factor at permittivity of 2. Reprinted with permission from [4]; copyright (2016) (American institute of physics)

decrease in quality factor at its corresponding resonant frequency (f_L), in comparison to the inductive coupled resonator at its corresponding resonant frequency (f_H) (Fig. 8d). The findings of this work indicated that the capacitively coupled resonator demonstrated a sensitivity to permittivity variations that was nearly double that of the inductive coupled resonator within the same range. Additionally, the capacitively coupled resonator maintained a high and a relatively constant quality factor.

The fabricated resonators on a 0.79-mm thick 5880 Rt/Druid substrate are shown in Fig. 8. To investigate their sensing performance, two sets of experiments were conducted by introducing different MUTs on the hotspots. Two standard MUT samples of 5880 Rt/Druid substrate with equal cross-sectional area of 7 × 7 mm and different heights of 0.79 mm and 1.57 mm were prepared and placed on the hotspots. The $|S_{21}|$ response of the sensors with and without the MUTs, measured from 2.2 to 2.5 GHz, are plotted in Fig. 9. The bare results (i.e., sensor without MUT) demonstrated low and high resonant frequencies of 2.29 and 2.44 GHz for the capacitive coupled resonator, and 2.28 and 2.46 GHz for the inductive coupled resonator. When a 0.79 mm-thick MUT was introduced, the measured results showed a 0.64% variation in the high resonant frequency (f_H) of the inductive mode and a

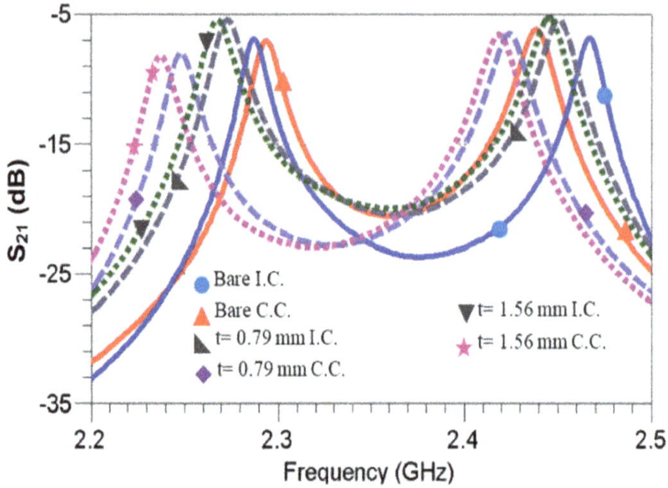

Fig. 9 Measured $|S_{21}|$ parameters of the implemented resonators for bare and different MUTs with permittivity of 2.2 and dimensions of $7 \times 7 \times 0.79$ mm^3 and $7 \times 7 \times 1.56$ mm^3. Reprinted with permission from [4]; copyright (2016) (American institute of physics)

1.96% variation in the low resonant frequency (f_L) of the capacitive mode. Furthermore, increasing the thickness of the MUT to 1.76 mm led to increased variations in both the inductive and capacitive coupling modes, with values of 0.87% and 2.48%, respectively. Comparing the changes of the resonant frequency for the inductively and capacitively coupled resonators, the variations of the capacitively coupled resonator were 200% higher than for the inductively coupled counterpart.

To demonstrate the enhanced dynamic range of capacitively coupled structures, an additional set of experiments was conducted using deionized water (DI) and high concentrations of sodium chloride (NaCl) in DI water (saline solution). The results, depicted in Fig. 8, revealed that when 20 μL of DI water was added, the inductive coupled resonator lost its second resonant frequency due to a significant decrease in quality factor. In contrast, the capacitive coupled resonator maintained its lower resonant frequency and exhibited distinguishable variations among different NaCl concentrations in DI water. The measurements confirmed that increasing the NaCl concentration elevated the conductivity of the solution, degrading the quality factor and increasing the low resonant frequency (f_L) (see inset of Fig. 11c). These findings emphasized the potential of the capacitively coupled structures for sensing applications (Fig. 10).

As stated in the previous section of this chapter, the coupling phenomena occurring between two (or more) resonators can be leveraged to adjust or enhance the dynamic range of deduction in microwave resonator sensors. The dynamic range, for instance, can be associated with the range of complex permittivity for materials under test. Various materials of different complex permittivity values can be differentiated

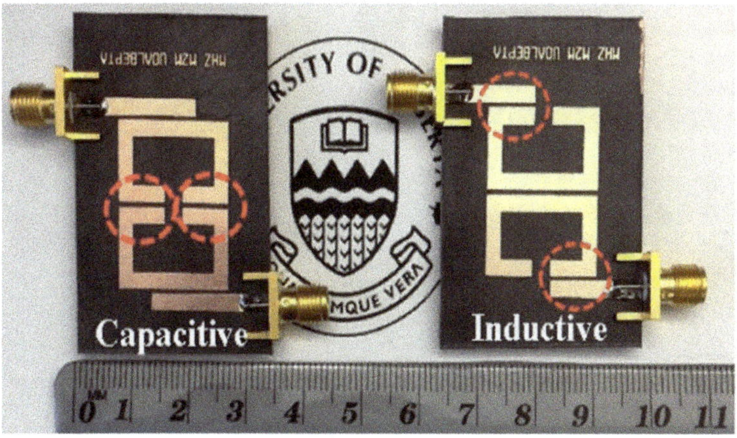

Fig. 10 Implemented capacitive and inductive coupled resonators to investigate and compare dielectric sensing performance. Reprinted with permission from [4]; copyright (2016) (American Institute of physics)

by tailoring the coupling mechanism between resonators, mainly by changing the geometrical disposition of the resonators.

One of the prominent real-world demands for detecting extreme variations in the complex permittivity is ice formation from sea sprays in cold conditions. Salty water can demonstrate significant changes in the complex permittivity (especially the loss tangent/conductivity of the water) compared to purified water (equivalent to reduced concentrations of ions). The difference in the complex permittivity of liquid water phase can also occur in the icing phase, as salt inclusion as part a solid ice body can change the effective permittivity and loss factor of the medium. Such variations in the complex permittivity can go beyond the dynamic range of detection in microwave resonator sensors, as extremely conductive materials or high permittivity values can degrade the resonant quality (peak sharpness of $|S_{21}|$ dB) and signal amplitudes in microwave sensor readouts. Thereby, the coupling mechanism between two adjacent resonators can be used to adjust dynamic range for such extreme variations of the complex permittivity of the sample.

In an approach for adjusting and enhancing the dynamic range of microwave sensors for detecting salt-water (brine) ice detection, Luckasavitch et al. [5] proposed the dual configuration of the SRRs to utilize magnetic couplings and electric couplings for differentiating saltwater, pure water, saltwater ice, and pure water ice. The phenomenon of saltwater icing is mainly observed in industrial marine environments, and the mitigation strategies for marine ice accretion is gaining momentum due to potential devastating effects on maritime vehicles, and the disruption on marine establishments. As a side note, it should be mentioned that visual techniques for identification of icing are frequently effective, yet the demand for a non-optical ice detection technique persists, with the goal of minimizing/eliminating human involvement in accurate identification of icing [5].

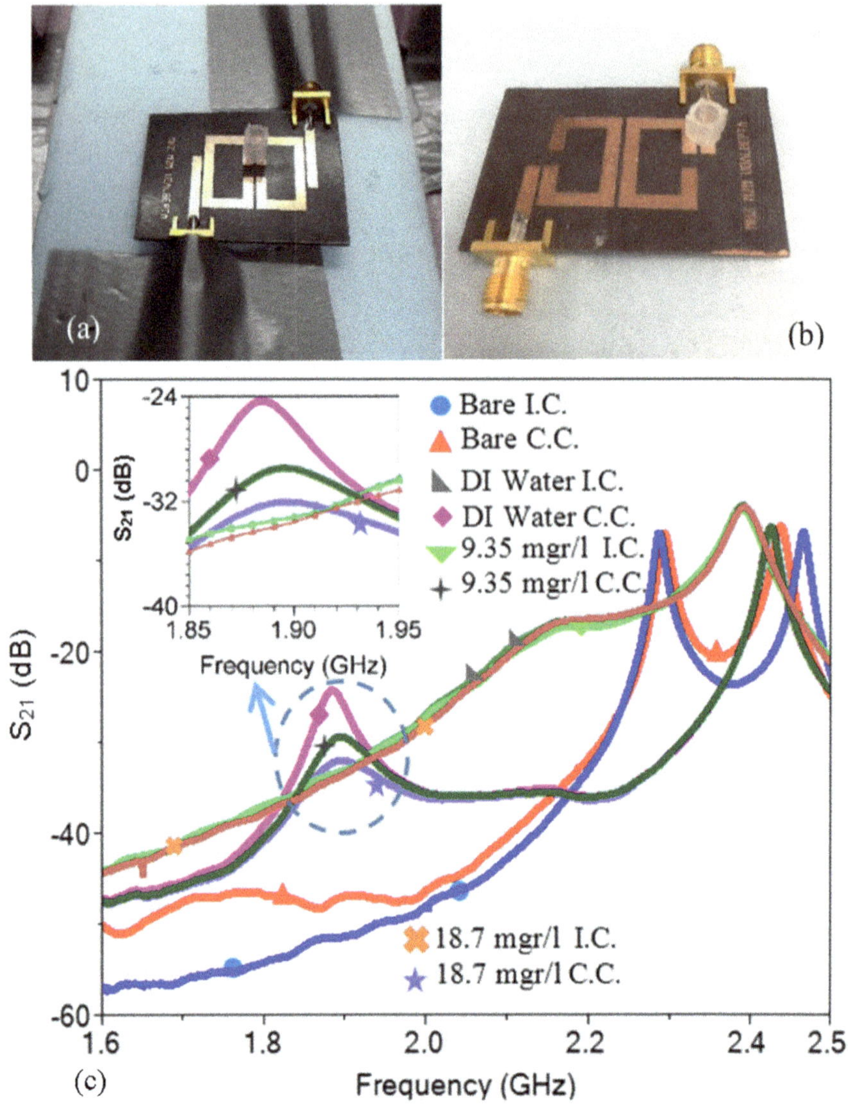

Fig. 11 **a** Capacitive coupled ring resonator with liquid container. **b** Inductive coupled ring resonators with liquid container. **c** Comparison of the measured S_{21} parameter for the inductive coupled (I.C.) and capacitive coupled (C.C.) resonator sensors for DI-water and different concentrations of NaCl in DI-water. Reprinted with permission from [4]; copyright (2016) (American Institute of physics)

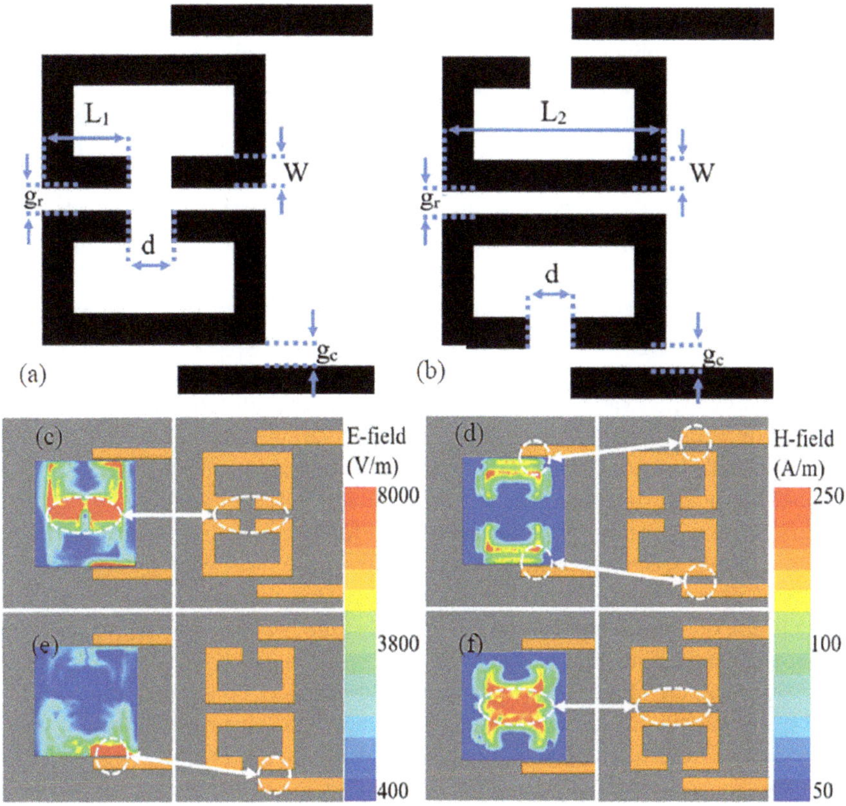

Fig. 12 Design topology and HFSS simulations for E and H field distribution over the coupled arrangements of the resonators. Electrically coupled **a** and magnetically coupled **b** configurations. **c** Electric field 0.2 mm above the electrically coupled resonator surface, **d** magnetic field 0.2 mm above the electrically coupled resonator surface, **e** electric field 0.2 mm above the magnetically coupled resonator surface, and **f** magnetic field 0.2 mm above the magnetically coupled resonator. Reprinted with permission from [5]; copyright (2022) (Elsevier)

The proposed microwave sensor by Luckasavitch et al. was based on coupled SRRs in two different configurations with predominant electrical coupling, and predominant magnetic coupling, respectively. The designed and fabricated resonator pair with electrical coupling demonstrated two resonant peaks at frequencies of 4.726 and 4.942 GHz. The resonator pair which exhibited primarily magnetic coupling showed two resonant peaks at 4.647 GHz and at 4.960 GHz. The calculated quality factors for the electrically coupled resonators were 260 ± 2 and 175 ± 2, for the lower and higher frequencies, respectively. For the magnetically coupled resonators, the corresponding quality factors were reported to be 195 ± 2 and 260 ± 2, respectively.

Figure 12 shows the planar topology of the electrically and magnetically coupled resonator sensors, along with the simulated E and H field distributions on a surface slightly (=0.2 mm) above the plane of the resonators. As shown in Fig. 11a the electrically coupled resonator configuration was comprised of a pair of SRR, oriented such that the ring splits were in proximity of each other (splits facing inwards). On the other hand, as shown by Fig. 12b the design of magnetically coupled resonators involved positioning the thru section of the rings in proximity to each other (splits facing outwards). Both the electrically coupled resonators (ECR) and magnetically coupled resonators (MCR) were excited using open-ended microstrip transmission line (TL), and a secondary TL was utilized as a probe for obtaining the scattering parameter S_{21} of the resonator configuration. Details and the design parameters of the resonator configuration can be found in [5].

Figure 12c-f further demonstrates the simulated E and H-field distribution over the ECR and MCR configuration. As shown in Fig. 12c, the simulated E-field distribution over ECR configuration showed a maximum field intensity in the region surrounding the two splits. This was an indication that for material differentiation with pure permittivity variations (real part), the ECR configuration can provide a high spatial resolution and sensitivity, when the sample was placed on the split region. Noting the simulated H-field distribution over the thru section (MCR configuration) in Fig. 12d, it is indicative that the marked area can be used for detecting disturbances in the H-field due to magnetic (material) perturbation or samples with varying conductivity values. These results suggest that the coupling of the two resonators allows the design and tailoring of sensitive spots for characterizing both the real and imaginary part of permittivity in the microwave regime.

To showcase the effectiveness of engineering resonator couplings for modifying sensitive hot spots, Luckasavitch et al. performed repetitive experimental investigation and validation based on various samples of water, salt water, ice, and salt-water ice. The ice formation was achieved by utilizing a Peltier cooling stage where different concentrations of saltwater were cooled down until freezing on the electrically coupled resonators (ECR) and magnetically-coupled resonators (MCR) configurations. Results of Fig. 13a and b show S_{21} spectra of the ECR and MCR configurations after the placement of material samples containing reverse osmosis (RO) water, ice, 5% and 10% saltwater, and 5% and 10% saltwater ice mixtures. As indicated by Fig. 13a, the S_{21} magnitude for the ECR configuration with samples of 5% and 10% saltwater concentrations was under −35 dB threshold, which can be considered as a minimum detectable signal level in practical applications. The ECR, therefore, was unable to differentiate between two concentrations of salt water and salt/ice mixtures. Not only the ECR configuration was demonstrated ineffective for differentiating salt/water concentrations, but also the resonant quality degraded due to the presence of filtered RO water. In contrast, looking at the $|S_{21}|$ results of Fig. 13b, the MCR sensing configuration could differentiate all the samples containing RO water, ice, 5% and 10% saltwater, and 5% and 10% saltwater ice mixtures. Comparing the $|S_{21}|$ results for pure ice and salt-ice mixtures in Fig. 13b, it is evident that the increased conductivity of the materials samples (due to the inclusion of salt in ice) results in higher electromagnetic power dissipation, consequently reducing the maximum magnitude of the $|S_{21}|$ spectra from ~−10 dB for pure ice to ~−20 dB for 5% saltwater ice. The time dynamics of resonant frequency upon freezing and thawing sequences were also assessed for samples with pure water, 5% saltwater and 10% salt water (see

Fig. 13c and d). It was observed that an increase in material conductivity (due to salt inclusion) led to a decreasing trend for the resonant frequency. The findings of Luckasavitch et al. concluded that by engineering the coupling mechanism in microwave SRRs, the dynamic range for differentiating materials with extreme difference in complex permittivity values (e.g., saltwater ice) can be enhanced. Based on the complex permittivity range for samples of interest, MCR, ECR or a combination of both, couplings can be formed to enable the microwave sensing capability.

Following the investigations of coupled SRRs for enhancing the selectivity of sensing, an alternative approach would be to have a resonator with a tunable resonant frequency. By reconfiguring the resonant frequency over a wide range, the frequency-dependence of complex permittivity can be exploited to differentiate materials that demonstrate similar permittivity values. In this regard, a modified SRR was integrated with a microfluidic channel filled with liquid metal to develop a frequency-tunable resonator structure [6, 7]. This novel design, depicted in Fig. 14 and introduced by Rafi et al. can expand the capabilities of resonator-based sensing systems by enabling a dynamic control over the resonant frequency through the manipulation of the liquid metal within a microfluidic channel. The design methodology tackles

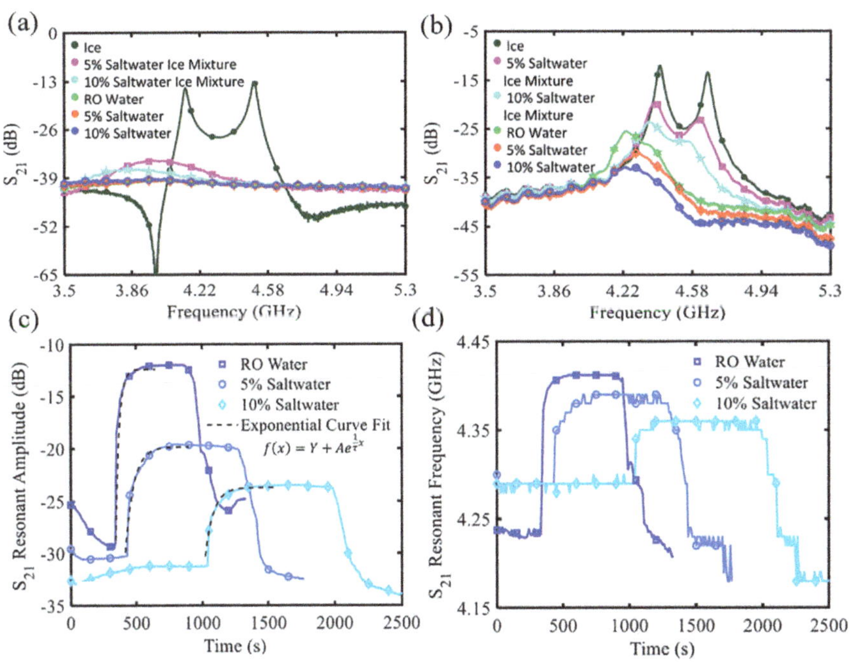

Fig. 13 S_{21} response of the sensor for loadings with RO water, 5% and 10% saltwater and saltwater ice mixture for **a** electrically and **b** magnetically coupled resonator configurations. Time dynamics of the monitored **c** resonant amplitude and **d** resonant frequency in the MCR configuration for RO (pure) water, 5% saltwater, and 10% saltwater samples. Reprinted with permission from [5]; copyright (2022) (Elsevier)

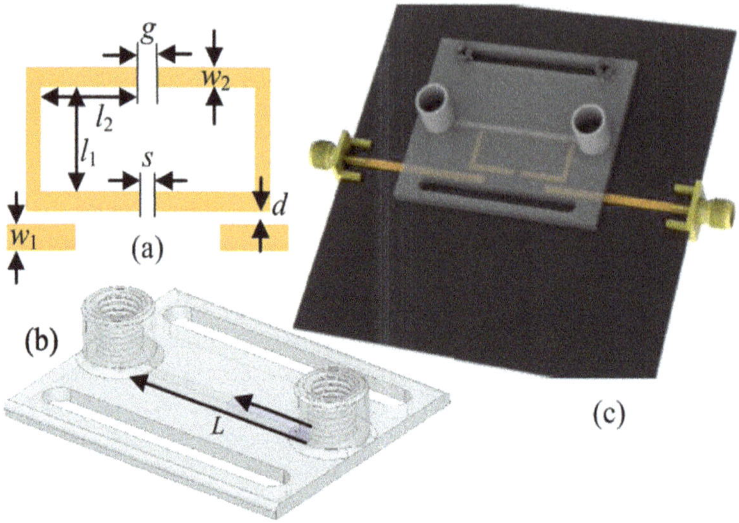

Fig. 14 **a** Highly coupled ring resonator structure ($l_1 = 8.0$ mm, $l_2 = 9.75$ mm, $g = 2.0$ mm, $s = 1.5$ mm, $w_1 = 2.0$ mm, $w_2 = 1.5$ mm, $d = 1.0$ mm). **b** 3-D printed microfluidic channel ($L = 3$ cm, channel height and width are 1.5 and 2 mm, respectively). Implementation as a tunable resonator. Reprinted with permission from [6]; copyright copyright (2020) (IEEE)

a key limitation of traditional microstrip resonators, which lack tunability once they are fabricated. The proposed structure comprises a double-gap SRR, as illustrated in Fig. 14a, accompanied by a 3-D printed microfluidic channel, depicted in Fig. 14b. To achieve tunability, the microfluidic channel was positioned above the resonator and subsequently filled with liquid metal, as shown in Fig. 14c. Having double-gap structures resulted in two highly coupled resonances in the transmission response ($|S_{21}|$) of the structure. Introducing liquid metal alters the electrical parameters of the double gap ring structure, thereby enabling the tuning of the resonant frequencies. The dimensions of the microchannel, including its width and height, play a role in influencing the resonant frequency and quality factor to some extent. These details, along with comprehensive insights into the proposed structure, are provided in [6].

The proposed structures were excited using an edge-coupling technique with a discontinuous microstrip transmission line, where a gap was intentionally introduced to break the connection between one end and the other end. This discontinuity led to the observation of resonant peaks in the $|S_{21}|$ response. Simulation results using Ansys HFSS demonstrated the successful operation of the highly coupled resonant structures, exhibiting two resonant peaks in the measured $|S_{21}|$ parameters as shown in Fig. 15. The unloaded resonator had center frequencies at 3.52 and 3.57 GHz with a resonant amplitude of −24.06 and −23.00 dB, respectively. The changes in the resonant frequencies were shown to be influenced by the length of the liquid metal in the microfluidic channel. To explain the tuning operation, a lumped element equivalent circuit model using ADS is shown in Fig. 16, where the lumped elements represented

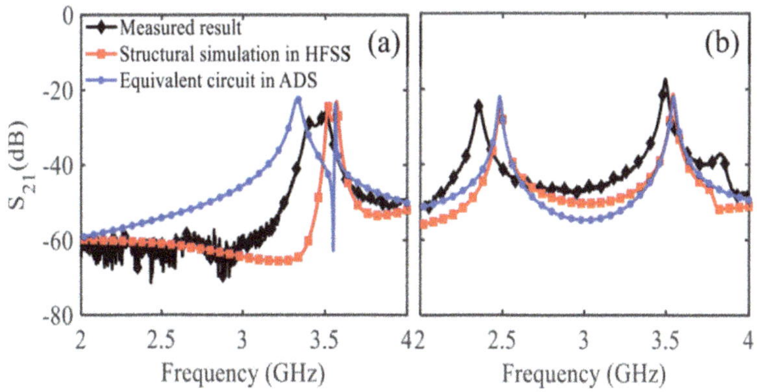

Fig. 15 Comparison between the simulated (HFSS and ADS) and measured S_{21} frequency spectra for the resonator with the microfluid channel in a **a** empty state and **b** filled state. Reprinted with permission from [6]; copyright (2020) (IEEE)

the effects of liquid metal volume on electromagnetic coupling and capacitive loading between the coupled resonators. The ADS simulation results were also plotted in Fig. 15a for the empty state and b for the filled stated (liquid metal) (Fig. 17).

The effectiveness of the proposed design by Rafi et al. was validated through experimental results. The resonator and probing TLs were fabricated using Rogers RT/duroid 5880 high-frequency substrate, while a polyester material was 3-D printed to create the microchannel. Galinstan, a liquid metal alloy, was injected into the channel to achieve tunability. Measurements of $|S_{21}|$ were conducted using a vector network analyzer (VNA), capturing the resonant frequency shifts for different volumes of liquid metal. The experimental results demonstrated that increasing the length of the liquid metal led to a downward shift in the resonant frequency.

The developed tunable resonator structure exhibited an impressive tuning range of 1.04 GHz (Fig. 7), allowing the resonant frequency to be adjusted between 2.48 GHz and 3.52 GHz. The average sensitivity of the resonator, in terms of the liquid metal length, was found to be 375 MHz/cm. Notably, the experimental results showcased significant resonant frequency variation with minimal impact on the resonant amplitude, affirming the effectiveness of the proposed design. This structure holds promise for different applications, including microwave frequency spectroscopy, microwave filters, and complex permittivity sensing.

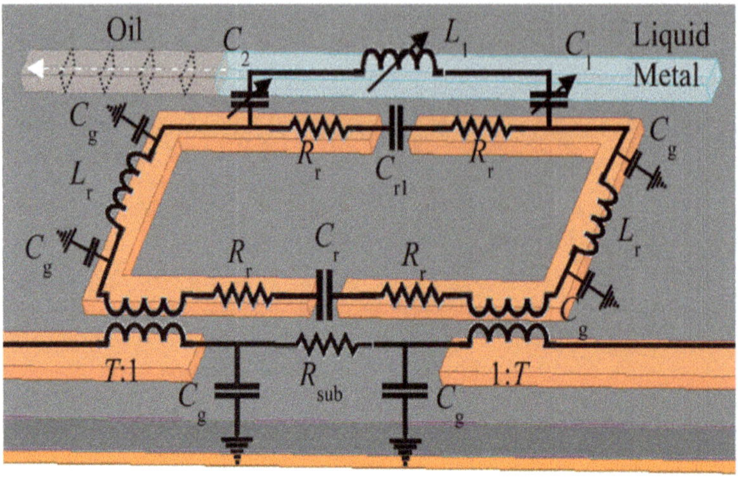

Fig. 16 Lumped element equivalent circuit model overlaying the substrate, resonator, and liquid metal ($R_r = 5\Omega$, $L_r = 31.485$ nH, $C_r = 0.0357$ pF, $C_{r1} = 0.031$ pF, T = 7.66, $C_1 = 5 - C_2$ pF, C_2 = 0.28–2.90 pF, $L_1 = 9.60$–30.21 nH, $C_g = 0.02$ pF, $R_{sub} = 54.3$ kΩ). Reprinted with permission from [6]; copyright (2020) (IEEE)

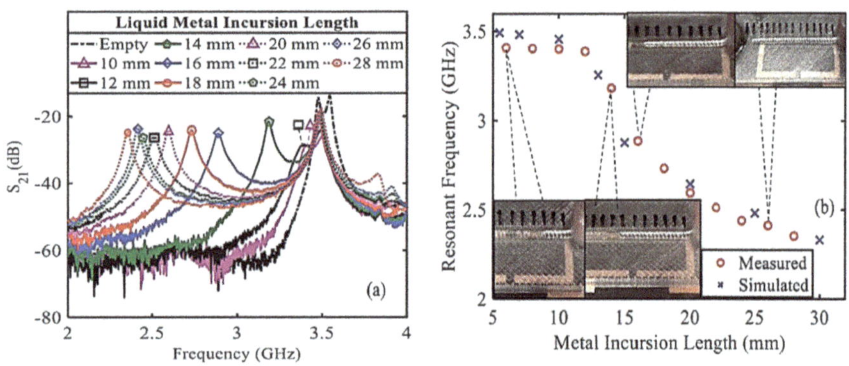

Fig. 17 **a** Resonant S_{21} spectra recorded by the VNA as the volume of metal inside the channel is varied and **b** resonant frequencies recorded for different metal lengths with selected images included. Reprinted with permission from [6]; copyright (2020) (IEEE)

4 Coupled Split Ring Resonators with Enhanced Monitoring Range in Wireless Implementations

The coupling mechanism of microwave resonators can be utilized to increase the range (distance) of excitation and probing in antenna-based (wireless) readouts of resonator sensors. Unlike the previous implementations where microstrip SRRs were probed and excited using microstrip TLs in their vicinity, a wireless, antenna-based approach is also a practical way of excitation and probing resonators. To achieve antenna-based excitation of SRRs, it is necessary to increase the effective surface area (i.e., most often coming with the increase of the physical area) that interacts with incident electromagnetic waves. This increase in surface area is crucial for ensuring the scattered signal power from the resonators is detectable at the input of the receiver antenna. In order to tackle this challenge, resonators were arranged in a two-dimensional configuration to effectively enhance the surface area of interaction with the incident electromagnetic waves.

The proof of concept for SRR coupling in antenna-based excitation was introduced by Wiltshire et al. [8], where four SRR were arranged in a plane, such that individual resonances of each SRR were affected by the nearby SRRs. This work was subsequent to the industrial demand for microwave sensors encompassing wireless readouts, miniaturized footprints, and non-contact assessment functionalities, in addition to having durable sensor characteristics capable of enduring potentially harsh environmental conditions [8]. Not to mention, the implementation of coupled resonators excited by a readout antenna can be also considered as a chipless RFID sensing system if the resonating elements are not integrated with microchips nor lumped elements.

The resonator topology with the antenna-based readout which was demonstrated in [8], consisted of a coupled configuration of SRRs on a single substrate with no ground plane backing it, a microfluidic channel on top of the resonator elements (for bringing liquid samples), and finally a readout (RX/TX) antenna. Subsets of Fig. 18a and b show the array of (four) resonators that were oriented in two different directions, where an incident E-field with a steady direction was incident on the resonator elements. The simulations of Fig. 18a and b show that the resonator elements have a specific polarizability, directly associated with the polarization of the incident waves. For the electric field bridging the splits of the resonators, the array of four resonators generates a resonant profile highly like the dominant mode resonance of a single SRR. Yet, observing the E-field distribution and intensities in Fig. 18b, the dominant mode resonance is not visible from the excited resonator elements. This observation is in accordance with the experimental investigations in [9] on the polarizability of SRRs.

Upon confirming the polarizability of the four-element coupled resonator configuration with an incident electromagnetic wave, the experimental investigation (setup shown in Fig. 19) was performed to assess the functionality of the configuration for liquid sample differentiation. As shown in Fig. 19a the coupled resonator elements were excited by a standard gain horn antenna kept a 24 cm distance from the plane of the resonators. A fluidic channel with width of 1.5 mm, length of 30 mm, and height of 500 μm was 3D printed and mounted on the resonator configuration, and a substrate to channel fixture was implemented to prevent mechanical drifts during the experiments. Four different liquid samples (with measured dielectric properties given in Table 1), water, isopropranol, methanol and ethanol, were used as test samples to evaluate the sensing performance of the wirelessly excited coupled resonator configuration. The liquid samples were injected into the channel using a syringe pump

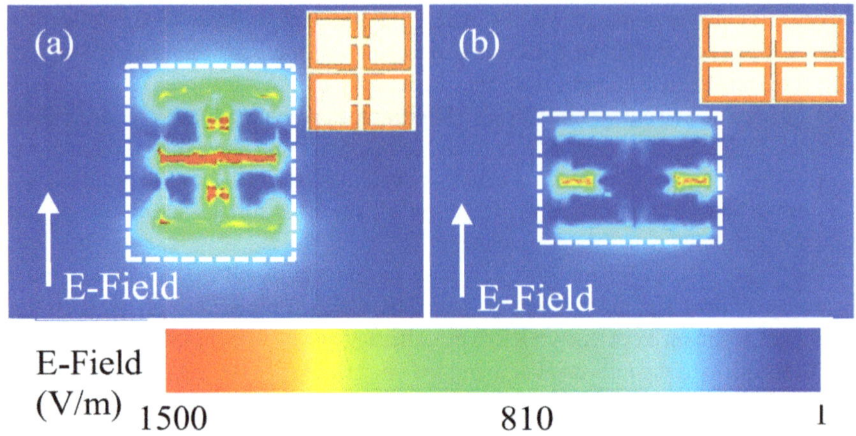

Fig. 18 Design topology and HFSS simulations for E-field polarizability of the coupled arrangements of four SRRs. **a** Excitation with electric field along the split of each SRR. **b** Excitation with electric field perpendicular to the split of each SRR. Reprinted with permission from [8]; copyright (2020) (IEEE)

Table 1 Dielectric properties of sample materials (liquids) measured at 2 GHz and 21°C (room conditions)

Sample	Water	Isopropranol	Methanol	Ethanol
ε_r	80	4	22	7
tanδ	0.1	1	0.3	0.7

and the real time data of the resonant characteristics were monitored using a VNA connected to the TX/RX antenna.

Figure 19b illustrates the reflection coefficient at the input of the antenna, following the injection of various liquids into the vicinity of coupled resonator configuration. For these set of experiments, the resonator elements were placed inside a plastic box to demonstrate the capability of wireless readout in case of having non-metallic obstacles. The S_{11} results of the antenna indicated that loading the resonator configuration with liquids of different permittivity values affected the resonant characteristics and therefore, the scattering profile of the 4-element resonator configuration. Such changes in the scattering of the resonator elements were monitorable through the alterations of antenna's reflection coefficient (Fig. 19b), where the interaction of illuminating waves with a material-sensitive resonator configuration resulted a unique spectrum for each liquid sample. To characterize these variations, the frequency of the S_{11} notch (in the monitored spectra) was monitored and plotted versus the permittivity of each liquid sample (see Fig. 19). Similarly, the same analysis was applied to the amplitude of the notch point in the monitored S_{11} spectra, and the results are presented in Fig. 19d. These experiments were conducted for two

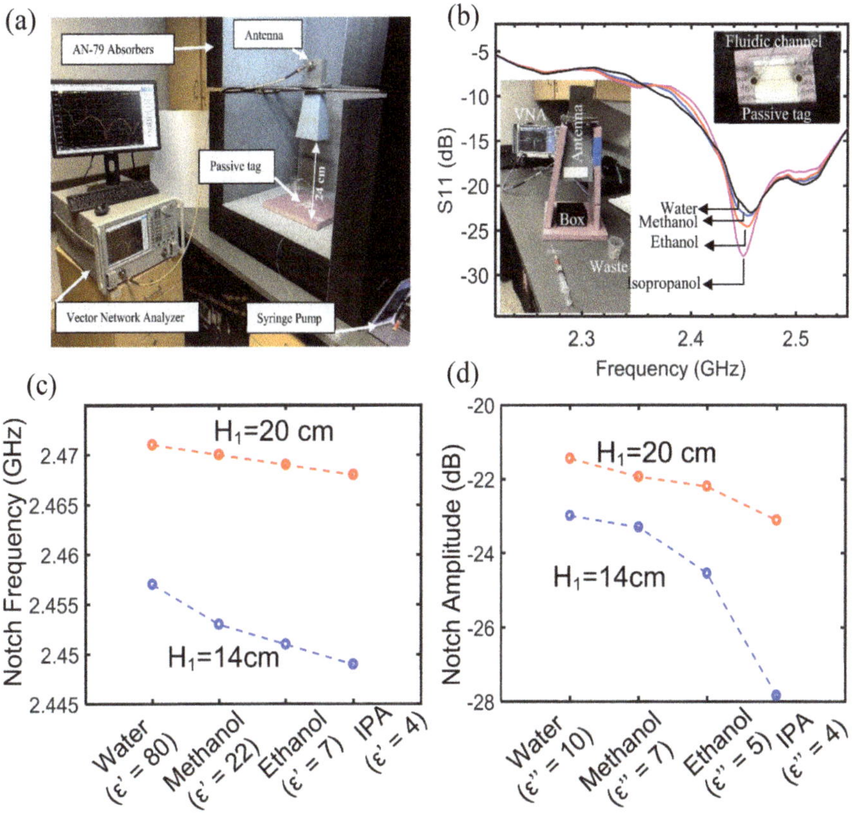

Fig. 19 **a** Experimental setup included a VNA, an interrogator RX/TX antenna, material under test, and the sensor tag in the line of sight of the antenna. **b** S_{11} responses for various sample liquids on the sensor, where there was an obstacle in the line of sight of the antenna. **c** Frequency of the S_{11} notch for different sample liquids at antenna to tag distances of 14 cm and 20 cm. **d** Amplitude of the S_{11} notch for different liquid samples at two antennas to tag distances of 14 cm and 20 cm. Reprinted permission with from [8]; copyright (2020) (IEEE)

antenna-to-resonator distances, 14 and 20 cm. The findings depicted in Fig. 19 effectively showcased the benefits of utilizing coupled microwave resonators to increase the range of readout in antenna-based readout method for microwave resonator sensing applications, such as the classification of liquids.

The concept of coupled microwave resonators can be visualized in an array of inter-coupled microwave resonators forming a frequency selective surface. In such two-dimensional topology, the resonator cell is coupled to adjacent cells, and the behavior of the entire array can be modeled as an infinitely periodic structure. The premise of frequency selective surfaces with intercoupled resonator cells holds immense potential for microwave sensing applications, where the increased resonant surface area can aid in surface interrogation and detection of materials of interest, particularly over a larger surface area in comparison to a single SRR. A notable application related to the wireless detection of materials on interactive microwave surfaces can be the monitoring of ice accretion in the wind turbine and aerospace industries. The formation of ice on aircraft wings and turbine blades is a significant issue in these industries. Even a small accumulation of ice, approximately 0.8 mm, on aircraft wings can lead to a substantial decrease in lift, up to 25% [10]. Similarly, the accumulation of ice on turbine blades can result in a loss of power up to 19 MWh over a period of 29 months [11].

To accomplish the aforementioned premise of wirelessly detecting ice accretion on distant surfaces, Niksan et al. [12] proposed a microwave system involving a frequency-selective array of intercoupled microwave resonators, wireless interrogation, and AI-assisted signal processing. Figure 20 illustrates an abstract of this microwave system, where the array of intercoupled resonator cells can be mounted on rotary blades or aircraft wings, and the reflection coefficient of the surface can be

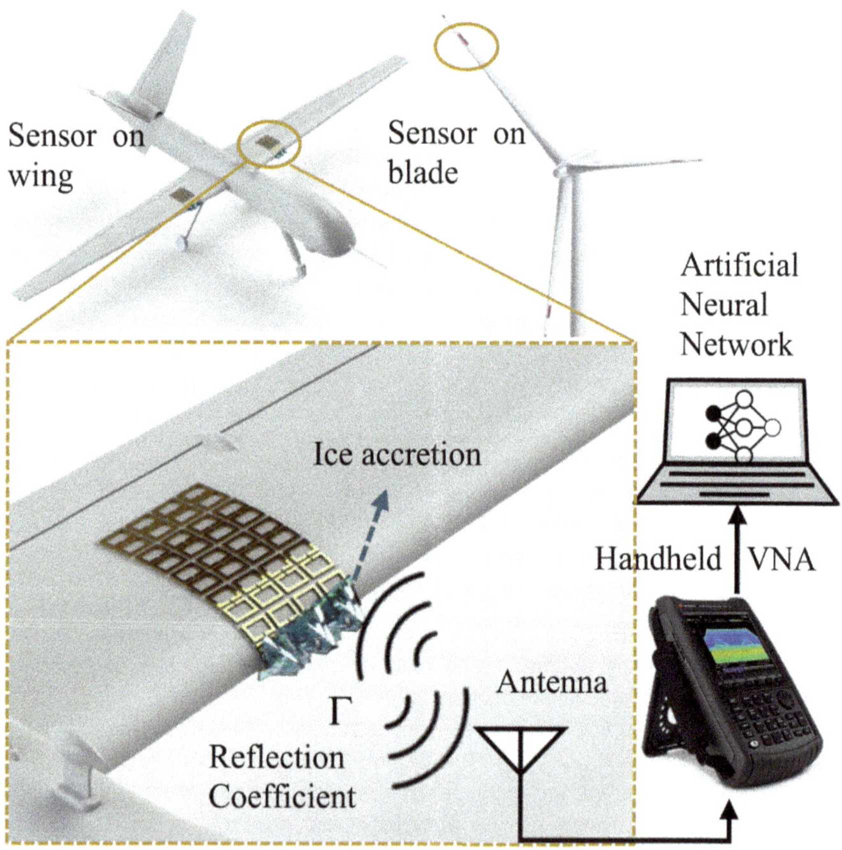

Fig. 20 An abstract illustration of the ice detection system, comprising the frequency selective surface (consisted of intercoupled resonator cells), reader antenna, and ANN-based post-processing unit. Reprinted with permission from [12]; copyright (2023) (IEEE)

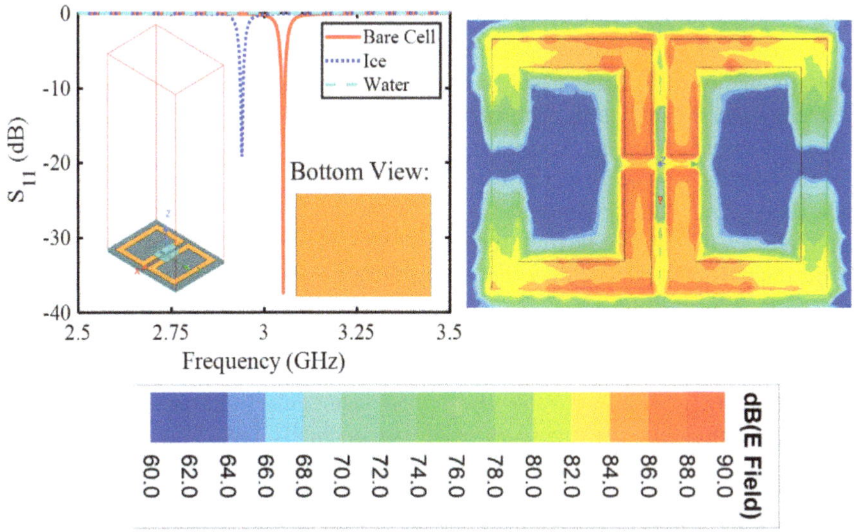

Fig. 21 Simulated model of the resonant cell (top and bottom views), S_{11} (dB) results for different materials on the main gap, and E-field intensities over each resonant unit cell. Reprinted with permission from [12]; copyright (2023) (IEEE)

monitored using a VNA, with the measurement results being post-processed using an Artificial Neural Network.

The material sensing mechanism for the designed array of resonators was based on the changes in reflection coefficient from its surface. The resonant array surface shown in Fig. 20 was designed for impedance matching with the air (~377 Ω) at the resonant frequency of each intercoupled resonator/unit cell. Loading the surface of the array with arbitrary (non-magnetic, dielectric) materials can affect the resonant characteristics of the array, translated in its reflection coefficient (i.e., air-to-array impedance matching). This correlation can be described by (7):

$$\Gamma = \frac{Z_{in} - 377}{Z_{in} + 377} \tag{7}$$

where Z_{in} denotes the input impedance of the array at a particular frequency. Subset of Fig. 21 shows a unit cell of the designed resonant cell, where the cell was simulated assuming an infinitely periodic array of intercoupled (adjacent) resonators. Each resonator cell and the entire array was backed by a ground plane, which acted as a shielding layer for maintaining the resonant characteristics intact from the effect of background installation platform in a real-world implementation. Moreover, each resonator cell itself consisted of two coupled SRRs. This topology, compared to a single SRR unit cell, was chosen to increase the effective permittivity-sensitive area for each resonant cell that interacted with the material loading its surface. As seen by the S_{11} (dB) results in Fig. 21, a bare resonator cell consists of two resonators, operated at 3.05 GHz, with ~−40 dB depth of notch as the indicator of an excellent air-to-array impedance matching. Loading a layer of material with a permittivity of ~3.15 (mimicking ice) onto the split section of the resonator affects both the resonant frequency and the reflection coefficient (i.e., impedance matching). As an extreme case for material perturbation on the resonator cell, loading water (with a permittivity of ~80) onto the split section completely degrades the impedance matching, with no indication of resonant behavior in the monitored spectra and the detectable signal amplitude levels. Moreover, from the plot of the simulated E-field distribution over the resonant cell, it was evident that the split section, along with the side arms of each SRR, show the highest E-field intensities (up to 30 dB higher than in the central part of the SRR, which confirms the sensitivity of these regions to non-magnetic dielectric material loadings and perturbations.

The sensitivity of the fabricated array of resonator cells to ice and water droplets formed on different coupling spots was experimentally investigated in the work proposed by Niksan et al. [12]. Figure 22a shows a snapshot of frozen water droplets on different coupling spots of the fabricated array, where the spots, based on the location, were denoted as M, Main, and X gaps. This investigation was performed to utilize the coupling mechanism of SRRs for determining the regions of ice formation. The reflection coefficient of the reader antenna interacting with the resonant surface is shown in Fig. 22b. The results indicated the resonance of the array at 3.05 GHz, and further drop casting water on the main and X gap disrupted the resonant response. The reflection coefficient of the array with water on the Main and X gaps was highly similar to the case of bare antenna response (Fig. 22). Yet, when water droplets were placed on the M gap, the array still resonated in the dominant mode. This observation follows the findings of Luckasavitch et al., where the gap with predominant magnetic coupling is more effective in the detection of loss factor and conductance variations of the samples [5]. For ice droplets on the three different regions, the main gap demonstrated the highest change in resonant frequency, indicating its relatively higher spatial sensitivity to permittivity perturbation and material loading.

To assess the real-time response of the shown microwave system to ice formation on the surface, the array of resonators was cooled down, and the resonant frequency of the surface was monitored. Figure 23 shows the time dynamics of the resonant frequency over a span of cooling, freezing, and thawing for the placed water droplets on the main, X, and M, gaps, respectively. The results of Fig. 23a, b, and c show the changes in the resonant frequency in a time window of 2000s, and the right Y-axis plots the monitored temperature of the cooling stage. Upon reaching sub-zero temperatures after ~800 s, formation of ice recovers a resonant notch at ~2.9 GHz. Figure 23a and b demonstrate similar changes in the resonant frequency after ice formation. Yet, as seen in Fig. 23c, when ice forms on the M-gap of the resonators, a minute 9 MHz change in the resonant frequency was seen. This again confirms the hypothesis of higher spatial sensitivity to permittivity variations in the regions where higher E-field intensities are distributed. Accordingly, the main and X gaps show 150 MHz and 120 MHz changes in the resonant frequencies, respectively.

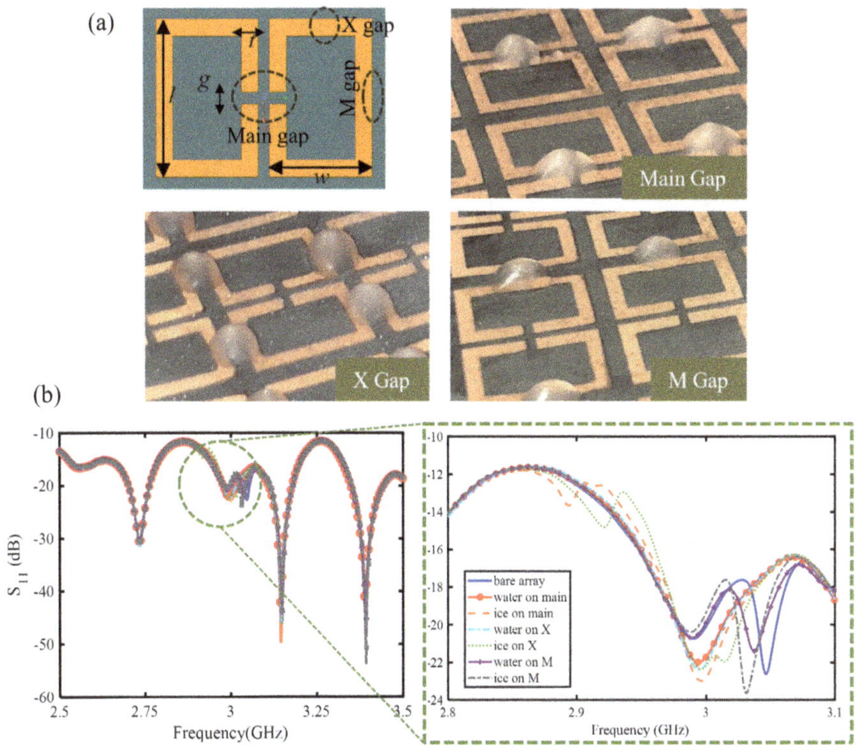

Fig. 22 **a** Designation of three different coupling gaps over each resonant unit cell and the placements of water/ice droplets on the fabricated array. **b** S_{11} (dB) results of the reader antenna for ice and water placed over different spots. Reprinted with permission from [12]; copyright (2023) (IEEE)

The findings of Niksan et al. successfully showcased a frequency-selective array of intercoupled resonators for the detection of ice and water on the surface, and a wireless microwave sensing system, in which ice can be detected on passive, battery-free, and chip less resonant surfaces, and where, the detection accuracy can be significantly ensured using ANNs.

5 Coupled Split Ring Resonators for Microwave Sensing of Different Physical Magnitudes

It was shown that the coupling between SRRs enhances the dynamic range of detection for the sensors. Surpassing this advantage, the coupling phenomena can also be useful for extracting additional information about the materials under test, or measuring physical quantities, such as mass, or light intensity. In a work conducted

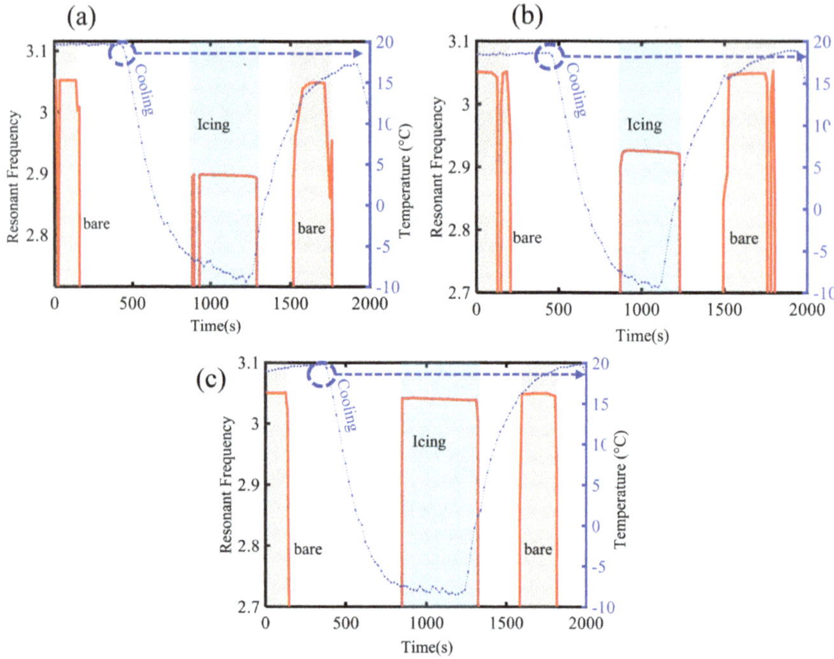

Fig. 23 Time dynamics of the monitored resonant frequency over the sequence of cooling-freezing–thawing for **a** ice/water droplets on the main gap, **b** ice/water droplets on the X gap, and **c** ice/water droplets on the M gap (magnetic coupling gap). Reprinted with permission from [12]; copyright (2023) (IEEE)

by Zarifi et al. [13], a coupled configuration of sensor resonator and tag resonator was proposed to integrate the effects of the mass and the dielectric properties of the materials in the sensing process, consequently improving the sensitivity and selectivity of the ring resonator for precise detection.

Figure 24 depicts the sensor structure consisting of a reader resonator, a sensing resonator (tag), and a cantilever to hold the tag. The cantilever structure exhibited bending toward the main resonator, when loaded by a mass or a force. The deformation of the cantilever depended on the applied vertical force on its top surface. As is well known from Elasticity Theory, the displacement of a point at the free edge of the cantilever is described by (8).

$$\delta_{max} = \frac{Pl^3}{3EI} \tag{8}$$

with P being a point force at a distance l from the anchor point, E the young modulus of the cantilever, and I the moment of inertia of the cantilever section.

The distance between the tag and the reader changed due to the deflection of the cantilever, and affects the S_{21} parameter of the two coupled resonators at a frequency range of 2.4 to 3.2 GHz. Simulations shown in Fig. 25 validate the sensor performance and the impact of the mechanical deflection on the S_{21} response. A sample material with relative permittivity $\varepsilon_r = 6$ was placed on the gap of the sensing tag that was carried by the cantilever. Since the electric field intensity across the gap of the sensor tag is higher, this region is sensitive to changes in permittivity. The deflection of the cantilever was applied by adjusting the angle θ between zero and four degrees, changing the distance between the sensing tag and reader resonator, as shown in Fig. 25a. Figure 25b depicts the simulation results of the sensor's structure with two resonant peaks at low and high resonant frequencies of f_L and f_H, respectively. The coupling between the two resonators cause further splitting of the resonant frequencies, which increases the sensor response of $\Delta f_{sense} = f_H - f_L$. f_H changes only when θ changes, while it remains constant when the permittivity of the test sample changes. On the other hand, f_L varies when both θ and permittivity changes. As θ, which corresponds to the effect of mass on the cantilever beam, increases, both frequency separation of f_L and f_H increased. It is evident that a significant variation in the resonant frequency shift occurs at $\theta = 4°$ compared to the resonant frequency at $\theta = 0°$.

The sensor was fabricated using two microwave substrates, as shown in Fig. 26a. The reader resonator was implemented on RT 5880 with a dielectric layer thickness of 0.79, $\varepsilon_r = 2.2 \pm 0.2$, and loss tangent of 0.0009. The sensing resonator was fabricated on a flexible substrate from Rogers with $\varepsilon_r = 3$, a loss factor of 0.003, and a thickness of 400 μm. The flexible cantilever length and width were 39 mm and 24 mm, respectively. The center of the sensing resonator was 20 mm away from the non-clamped edge of the cantilever beam. As test sample, a stack of 4 layers of 5×5 mm^2 Rogers 5880 copper etched substrate with a thickness of 1.56 mm and weight of 0.022 g, was placed on the cantilever. Figure 27 depicts the sensor response under two different test conditions, including the bent and flat of the cantilever. The sensor with a bending cantilever had more sensitivity than the same sensor with a non-bending structure. As shown in Fig. 27, the sensor's sensitivity with a bent cantilever was improved by 20% compared to the flat case. This improvement was achieved due to the addition of test sample's mass, which exerts an effect independent from the dielectric properties.

Coupled resonators have the potential to facilitate the detection of physical quantities, including electromagnetic disturbances beyond the microwave regime, such as light or UV radiation. To accomplish this, an interface material transducing the effect of light intensity to complex permittivity variations is required. Yet as mentioned previously in this chapter, the placement of lossy/conductive materials on the split region of the SRRs can degrade the resonant quality, disabling the sensing performance of the device. This issue can be tackled by introducing another tag SRR coupled to a primary SRR so that the variations of conductivity (due to the external stimuli) can be monitored from the coupling parameters of the two resonators. Moreover, this approach can allow for the integration of an active feedback circuitry that amplifies the effect of mentioned conductivity variations, leading to an enhanced resolution of sensing.

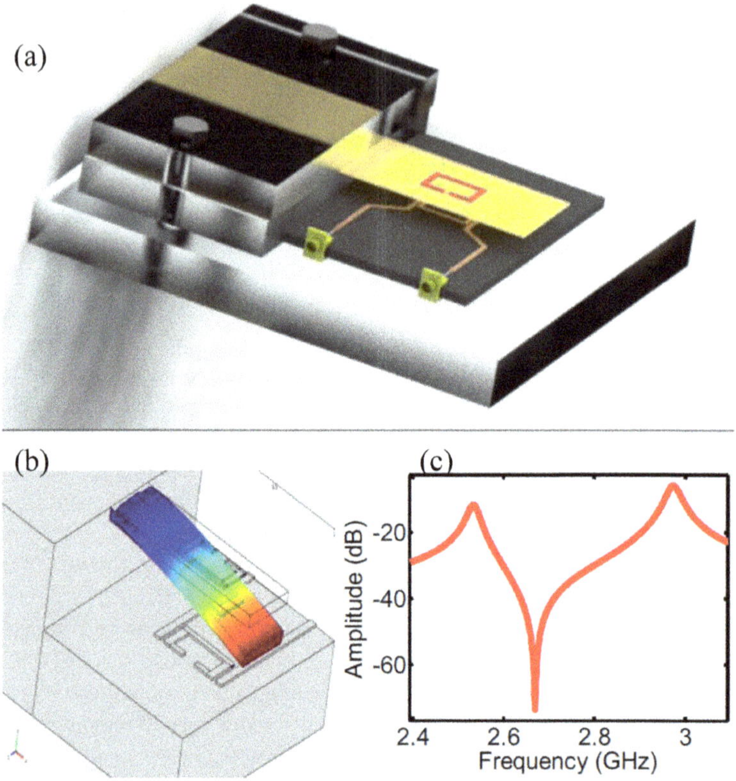

Fig. 24 Sensor structure with a fixed reader resonator and a sensing resonator on a cantilever structure. Reprinted with permission from [13]; copyright (2018) (IEEE)

The work introduced by Jain et al. [14] implemented a coupled sensing structure to detect ultra-violet (UV) radiation using a wireless passive microwave resonator integrated with TiO_2 nanotubes (TNT) with high sensitivity. In this approach, a flexible passive SRR tag was integrated with a TNT membrane for the detection and measurement of UV radiation. Figure 28 shows the UV radiation light detection system consisting of a wireless passive microwave resonator integrated with TNT and coupled to an SRR reader. The main sensing element of the device was a passive SRR tag combined with TNT. The TNT was placed on the passive tag to facilitate its interaction with the electric field, which was positioned 3 cm above the active feedback assisted SRR reader. The strong coupling between the TNT-integrated passive tag and the SRR reader allowed the SRR reader to interpret changes in the dielectric properties of the TNT. These changes were caused by the generation of charges through UV absorption by TNTs and were detected by the SRR reader as variations in both the resonant frequency and amplitude. The system demonstrated capability to detect UV radiation with an improved sensitivity promoted its suitability for various industrial and biomedical applications.

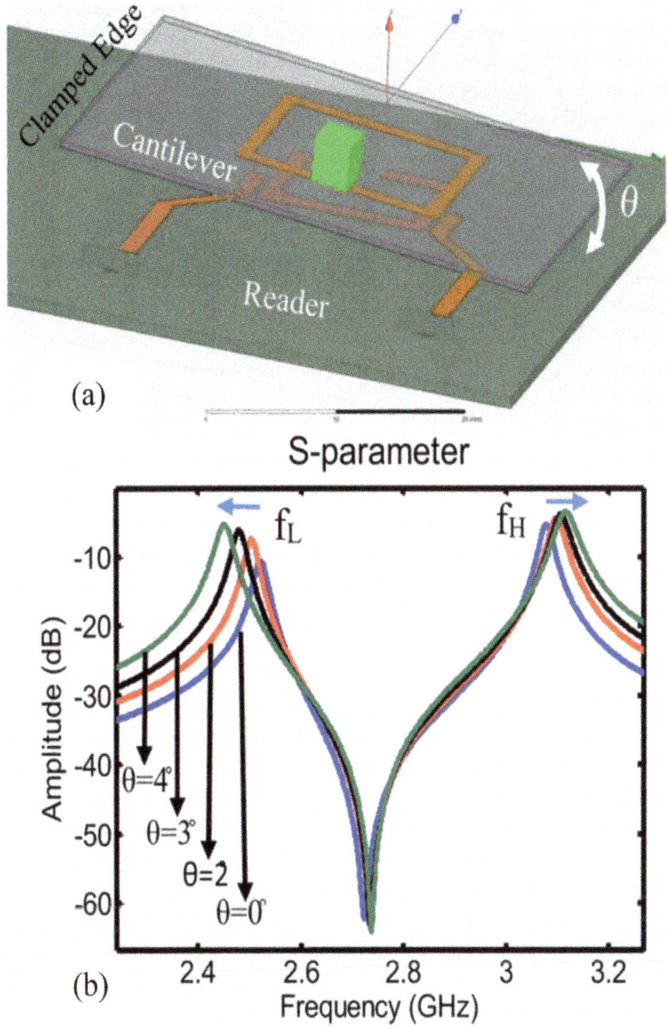

Fig. 25 Implemented sensor structure in HFSS with a test sample ($\varepsilon_r = 6$) and moving angle θ. **b** S_{21} simulation profile for four different angles with a constant test sample. Reprinted with permission from [13]; copyright (2018) (IEEE)

The lumped element model of the sensor was implemented in ADS simulation software, as shown in Fig. 29. The sensor operated over a frequency band of 2.5 GHz to 3.2 GHz. By incorporating an active feedback loop consisting of a phase shifter and low noise variable gain amplifier, the SRR reader compensated for the losses in the coupled sensor, effectively increasing the quality factor of the resonance and enhancing the resolution of detecting minute perturbations to the sensing system. The active configuration was implemented by generating a positive feedback loop

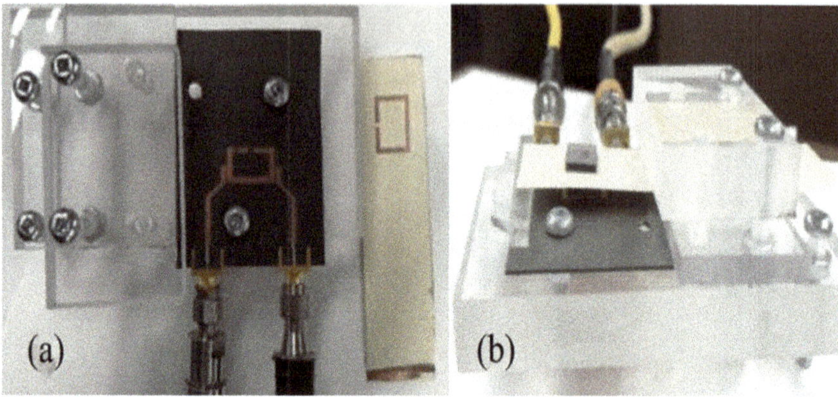

Fig. 26 **a** Implemented sensor, reader, and sensing tags. **b** Cantilever-based coupled microwave resonator sensor with standard samples. Reprinted with permission from [13]; copyright (2018) (IEEE)

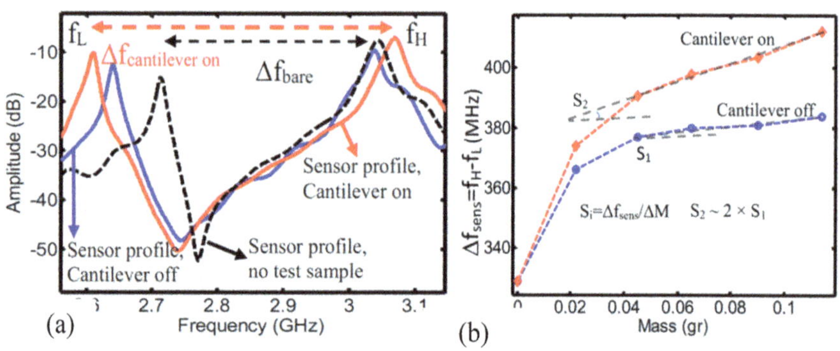

Fig. 27 **a** Resonant profiles for bare sensor (sensor with no test-sample), sensor with no bending cantilever (cantilever off), and sensor with bending cantilever (cantilever on). **b** Sensor sensitivity for the mass on cantilever for bending and fixed cantilevers. Reprinted with permission from [13]; copyright (2018) (IEEE)

based on a constructive phase matching at the input/output couplings of the reader resonator.

The SRR reader was fabricated on a Rogers 5880 substrate with a thickness of 0.79 mm, a permittivity of 2.20, and a loss tangent of 0.0009. The resonator tag was fabricated on a flexible substrate, with a thickness of 0.4 mm, a permittivity of 2.9, and a loss tangent of 0.0025. Figure 30a and 30b show the fabricated sensors. To create self-organized TNT membranes, Ti foils were subjected to anodization using the recipe of [15]. The TNT membranes were annealed in the air in a tube oven at 400 °C for 1 h to achieve the desired anatase structure.

Figure 31 shows the simulated and measured transmission coefficient response and the E-field plots obtained by simulation with HFSS. As depicted in Fig. 31a, the

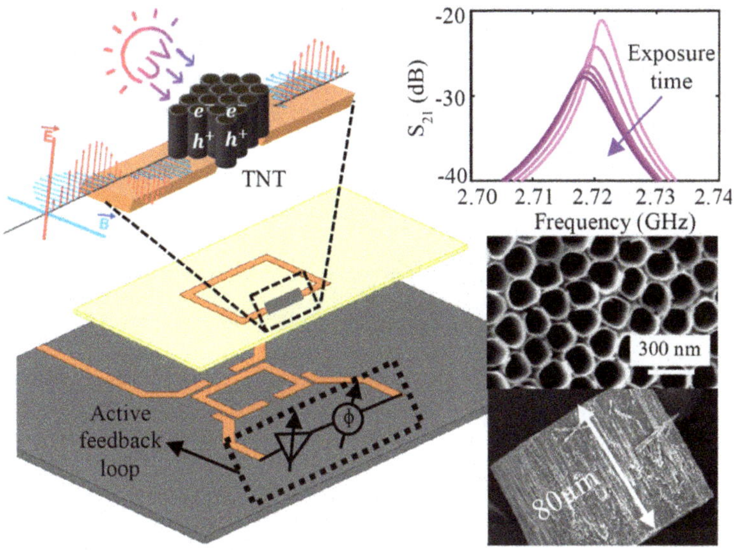

Fig. 28 High-resolution ultra-violet radiation detection schematic using a TNT-integrated wireless passive microwave resonator. Reprinted with permission from [14]; copyright (2022) (IEEE)

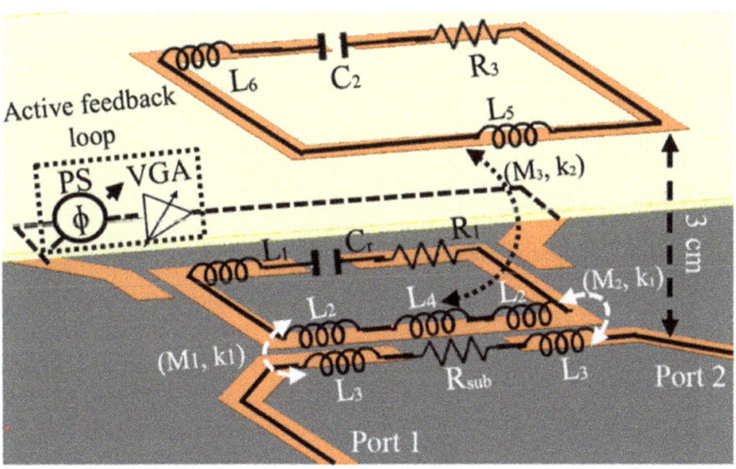

Fig. 29 Lumped circuit mode of the sensor. PS is the phase shifter and VGA is a variable gain amplifier. Reprinted with permission from [14]; copyright (2022) (IEEE)

resonator quality factor was significantly improved from 87 to 700 due to the loss compensation (active feedback) circuitry incorporated into the system. Furthermore, the field simulation shown in Fig. 31b demonstrates that the coupling between the reader and the tag in the active configuration was substantial compared with the

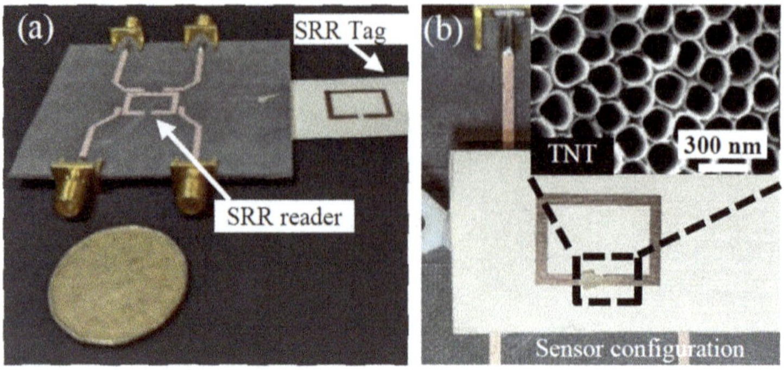

Fig. 30 **a** Fabricated SRR reader and passive tag. **b** Sensor configuration with 80 μm TNT (SEM image in the inset) placed on the tag. Reprinted with permission from [14]; copyright (2022) (IEEE)

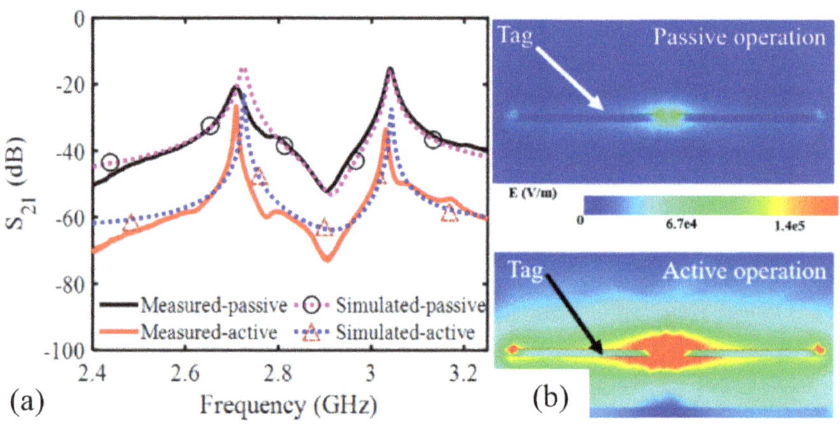

Fig. 31 **a** S_{21} simulation and measurement profile. **b** E-field intensity mapping for passive and active operation. Reprinted with permission from [14]; copyright (2022) (IEEE)

passive, improving the signal-to-noise ratio and enabling efficient detection even with small variations. The coupled sensor configuration was exposed to UV LED with light intensities of 18 $\mu W/cm^2$ and 38 $\mu W/cm^2$ for 10 min while the S_{21} was measured, as shown in Fig. 2. The S_{21} response of Fig. 32a changed due to the variation of the TNT dielectric properties under the illumination, with the active mode demonstrating higher amplitude variation compared to the passive mode. In addition, since the Q-factor and loss compensation were significantly improved in the active coupled sensing structure, the achieved sensitivity was five times higher than for the passive mode, as depicted in Fig. 32b.

The advantage of coupled resonator structures has been also explored by Abbasi et al. [16] for monitoring volatile organic compounds (VOC). The sensor structure utilized two passive SRRs electro-magnetically coupled to each other. The sensing

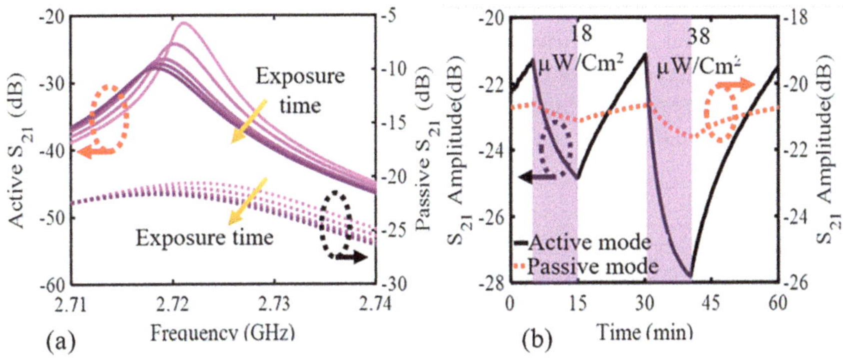

Fig. 32 **a** Measured active (solid lines) and passive (dotted lines) transmission coefficient (S_{21}) profile during UV illumination. **b** Time-resolved S_{21} amplitude (dB) for active and passive operation under UV (purple) illumination. Reprinted with permission from [14]; copyright (2022) (IEEE)

(tag) resonator was implemented on a flexible RF substrate, the wireless coupling between the reader and the tag allowed for both contactless operation and high sensitivity. The microwave sensor operated at a frequency of 4 GHz, and the distance between the reader and the tag could be extended up to 1.5 cm. The sensitivity was enhanced by placing the VOC polymeric adsorbent beads inside a cylindrical quartz reactor, where the tag was capable of monitoring gas adsorption. The sensor successfully detected various concentrations of VOCs, including methyl ethyl ketone (MEK) and cyclohexane, within the range of 250 to 1000 parts per million (ppm), displaying distinct responses for each compound. The sensor demonstrated a sensitivity of 40 kHz/ppm for MEK and 2 kHz/ppm for cyclohexane.

It has also been demonstrated in [17], that the coupling mechanism between two resonators can be engineered for enhancing the sensitivity of microwave resonator sensors, upon their implementation as a reflection-mode sensor (rather than a transmission mode configuration). In their work, Casacuberta et al. proposed a reflection-mode resonator sensor, in which phase variations of the reflection coefficient were correlated with the dielectric constant of materials under test (MUT)s, establishing the primary sensitivity mechanism. Based on the conducted sensitivity analysis, adjusting the coupling coefficients between two resonators affected the sensitivity of phase to the permittivity of the tested samples. Based on the method proposed in [17], the sensitivity of the sensor was (comparatively) increased by ~3 times, compared to a resonator with stronger, unadjusted coupling. The engineering of coupling phenomena in this work, consequently lead to enhanced figure of merit (Max sensitivity/sensing area) for the resonator sensor. The advantages of coupled, symmetrically positioned resonators can also be utilized for detecting material samples with spatial non-uniformities, where the appearance of two transmission notches rather than one can be used as an indicator for material defects [18]. Ebrahimi et al. demonstrated another instance of tuning coupling phenomena for detecting spatial defects in material samples by imposing an electric wall on the coupled region of two ring resonators for enabling the "balanced" and "unbalanced" sensing principle based on the symmetric and asymmetric loadings of the resonator, respectively [19]. Operating at 2.18 GHz, the sensor exhibited an (exclusively defined) average sensitivity of 88 MHz, where the occurrence of two transmission notches acted as an indicator for detecting spatial defects in the tested material sample.

References

1. Ebrahimi A, Withayachumnankul W, Al-Sarawi SF, Abbott D (2014) Metamaterial-inspired rotation sensor with wide dynamic range. IEEE Sens J 14(8):2609–2614
2. Ebrahimi A, Withayachumnankul W, Al-Sarawi S, Abbott D (2014) High-sensitivity metamaterial-inspired sensor for microfluidic dielectric characterization. IEEE Sens J 14(5):1345–1351
3. Hong J-S, Lancaster MJ (1996) Couplings of microstrip square open-loop resonators for cross-coupled planar microwave filters. IEEE Trans Microw Theory Tech 44(11):2099–2109
4. Zarifi MH, Daneshmand M (2016) Wide dynamic range microwave planar coupled ring resonator for sensing applications. Appl Phys Lett 108(23):232906
5. Luckasavitch K, Kozak R, Golovin K, Zarifi MH (2022) Magnetically coupled planar microwave resonators for real-time saltwater ice detection. Sens Actuators A Phys 333
6. Rafi MA, Wiltshire BD, Zarifi MH (2020) Wideband tunable modified split ring resonator structure usingliquid metal and 3-D printing. IEEE Microw Wirel Components Lett 30(5):469–472
7. Wiltshire BD, Rafi MA, Zarifi MH (2021) Microwave resonator array with liquid metal selection for narrow band material sensing. Sci Rep 11(1):8598
8. Wiltshire BD, Zarifi T, Zarifi MH (2020) Passive split ring resonator tag configuration for RFID-based wireless permittivity sensing. IEEE Sens J 20(4):1904–1911
9. Katsarakis N, Koschny T, Kafesaki M, Economou EN, Soukoulis CM (2004) Electric coupling to the magnetic resonance of split ring resonators. Appl Phys Lett 84(15):2943–2945
10. Rasmussen R, Cole J, Moore RK, Kuperman M (2000) Common snowfall conditions associated with aircraft takeoff accidents. J Aircr 37(1):110–116
11. Pinar Pérez JM, García Márquez FP, Ruiz Hernández D (2016) Economic viability analysis for icing blades detection in wind turbines. J Clean Prod C(135):1150–1160
12. Niksan O, Colegrave K, Zarifi MH (2023) Battery-free, artificial neural network-assisted microwave resonator array for ice detection. IEEE Trans Microw Theory Tech 71(2):698–709
13. Zarifi MH (2018) Sensitivity and selectivity enhancement in coupling ring resonator sensors using splitting resonant frequencies. IEEE MTT-S Int Microw Symp Dig 2018-June:36–39
14. Jain MC, Alijani M, Wiltshire BD, MacAk JM, Zarifi MH (2022) High resolution ultra-violet radiation detection using TNT-integrated wireless passive microwave resonator. IEEE MTT-S Int Microw Symp Dig 2022-June:133–136
15. Alijani M, Sopha H, Ng S, Macak JM (2021) High aspect ratio TiO_2 nanotube layers obtained in a very short anodization time. Electrochim Acta 376:138080
16. Abbasi Z, Shariaty P, Nosrati M, Hashisho Z, Daneshmand M (2019) Dual-band microwave circuits for selective binary gas sensing system. IEEE Trans Microw Theory Tech 67(10):4206–4219
17. Casacuberta P, Vélez P, Munoz-Enano J, Su L, Gil M, Martin F (2022) Reflective-mode phase-variation permittivity sensors based on coupled resonators. Proc IEEE Sensors 2022-Octob:1–4
18. Naqui J, Su L, Mata J, Martín F (2015) Analysis of transmission lines loaded with pairs of coupled resonant elements and application to sensors. J Magn Magn Mater 383:144–151
19. Ebrahimi A, Beziuk G, Scott J, Ghorbani K (2020) Microwave differential frequency splitting sensor using magnetic-LC resonators. Sens (Switz) 20(4)

Solute Concentration Sensing in Aqueous Solutions with Coupled Microstrip Resonators

Benjamin Potelon, Enrique Bronchalo, Carlos G. Juan, Cédric Quendo, and Alexis Chevalier

Abstract This chapter presents the application of planar microwave sensors based on coupled resonators for the detection of solute concentration in aqueous solutions. First, the dielectric and conductive properties of these solutions, based on broadband experimental measurements, are presented. Then, different aspects of the sensor design, as well as a sensor circuit model, are exposed. For this model, the theoretical response and the influence of the mutual and self-capacitances of the resonators on the sensitivity are studied. The last section presents two practical examples of sensors applied, respectively, to binary (water-glucose) and ternary (water-sucrose-sodium chloride) solutions.

Sensing and monitoring of physico-chemical properties of aqueous solutions is of high interest for a variety of fields, including chemistry, environmental monitoring, medical or food and drinks industries. Amongst the numerous available technologies, microwave sensing appears as interesting thanks to its non-invasiveness and its ability to be continuously operated. The principle of microwave sensing is based on the detection of changes in the electromagnetic properties of the materials, which are related to the physico-chemical evolution of the samples (viscosity, temperature, concentrations...) [1]. The use of coupled-resonator sensor topologies leads to great

B. Potelon (✉)
IMT Atlantique, Lab-STICC, UMR CNRS 6285, 29238 Brest, France
e-mail: benjamin.potelon@imt-atlantique.fr

B. Potelon · C. Quendo · A. Chevalier
Univ Brest, CNRS, Lab-STICC UMR 6285, 29238 Brest, France

E. Bronchalo
Institute for Engineering Research, Miguel Hernández University of Elche, 03202 Elche, Spain

C. G. Juan
Neuroengineering Biomedical Research Group, Institute of Bioengineering, Miguel Hernández University of Elche, 03202 Elche, Spain

Electronic Design and Signal Processing Techniques Research Group, Department of Electronics, Computer Technology and Projects, Technical University of Cartagena, 30202 Cartagena, Spain

F. Martín and E. Bronchalo (eds.), *Coupled Structures for Microwave Sensing*, Lecture Notes in Electrical Engineering 1150,
https://doi.org/10.1007/978-3-031-53861-2_9

sensitivities of the measured electrical response of the sensors with respect to changes in the electromagnetic properties of the samples [2, 3].

This chapter deals with the design of microwave sensors based on coupled-resonator topologies applied to aqueous solutions. In order to achieve the design and optimization of microwave sensors, it is paramount to identify the underlying physic behind electromagnetic properties of materials. Consequently, the first part of this chapter is devoted to the presentation of the dielectric and conductive properties of materials, and in particular those of aqueous solutions in the DC to microwave range, including their definition, measurement and interpretation. The second part will focus on the exploitation of these electromagnetic properties in the design of microwave sensors and, in particular, in the design of coupled-resonator sensors. Finally, some examples featuring advanced monitoring of solute concentrations in aqueous solutions are detailed and commented.

1 Electromagnetic Properties of Sugars and Salts in Aqueous Solutions at Microwave Frequencies

1.1 Definition of Magnetic, Dielectric and Conductive Properties

The magnetic properties of materials refer to their response to an applied magnetic field. The magnetic behavior of a material is determined by its atomic structure and the interactions between neighboring atoms. The magnetic behavior of a material is expressed through its magnetic permeability μ, measured in Henry per meter (H/m), defined as the ability to conduct the magnetic flux, relative to the vacuum. The relative magnetic permeability of a material is $\mu_r = \mu/\mu_0$, where $\mu_0 = 4\pi \times 10^{-7}$ H/m is the magnetic permeability of vacuum. Most aqueous solutions are diamagnetic, i.e., they have a relative magnetic permeability μ_r very close to 1. In diamagnetic materials, all electrons are paired and their magnetic moments cancel out.

The dielectric properties are expressed through the dielectric permittivity, ε, which represents the material response when an electric field is applied. It is defined as the ratio of the electric flux density to the electric field strength in the material. Dielectric permittivity is expressed in Farads per meter (F/m). The relative dielectric permittivity is $\varepsilon_r = \varepsilon'/\varepsilon_0$, where $\varepsilon_0 = 8.854 \times 10^{-12}$ F/m is the dielectric permittivity of vacuum.

For time-varying electric excitations, the dielectric permittivity of materials is usually expressed as a complex number, $\varepsilon = \varepsilon' - j\varepsilon''$ [4]. The real part of the dielectric permittivity, ε', is a measure of the polarization of the material in response to the applied electric field. It also represents the ability of a material to store electric energy. The imaginary part of the dielectric permittivity, ε'', is a measure of the energy dissipation in the material when an electromagnetic field is applied. Both parameters are related through the loss tangent, defined, in the absence of conductivity, as $\tan\delta = \varepsilon''/\varepsilon'$. Materials with high loss tangents, called lossy materials, show a high dissipation rate of electromagnetic energy. Aqueous solutions generally present variations of their relative dielectric permittivity when the solute concentrations change, and thus complex permittivity is one of the key-features used to track concentrations within solutions [5–7].

Finally, the electric conductivity of a material is a measure of its ability to conduct electric current. The conductivity, denoted by the symbol σ, is defined as the ratio of the current density (J) in the material to the applied electric field strength (E), and it is consequently measured in siemens per meter (S/m). Conductivity is related to the density of free charges within the materials and their mobility. Regarding aqueous solutions, although deionized water has a very low conductivity because of its very low ionization (H_3O^+ and OH^- ions), some solutes can introduce charges, and consequently increase the conductivity [8].

The properties μ, ε and σ, together, allow for the full description of the electromagnetic behavior of a material. It is important to note that, for the sake of simplicity, materials are sometimes considered to have ideal properties, which does not necessarily represent the complexity of the reality. The following assumptions are usually adopted:

- *Linearity*: materials are considered as linear, which means that their electromagnetic properties do not depend on the magnitude of the applied fields.
- *Homogeneity*: materials are considered as homogenous, which means that their properties are the same everywhere inside the material.
- *Isotropy*: materials are considered as isotropic, which means their electromagnetic properties do not depend on the direction of the applied field.

Each of the hereinbefore properties of materials are defined with respect to the applied fields, that can be static or dynamic. In this latter category, the harmonic time-dependence is of particular interest. In general, electromagnetic properties of materials are dispersive, i.e., frequency-dependent, and thus the three above parameters should rather be defined as functions of the frequency f of the harmonic applied field. Consequently, any material, and especially complex ones like solutions or composite materials, is described by $\mu_r(f)$, $\varepsilon_r(f)$ and $\sigma(f)$.

1.2 Characterization and Interpretation of Electromagnetic Properties of Materials

In order to design microwave sensors to monitor or sense the properties of aqueous solutions, a previous knowledge of the electromagnetic signatures of each tracked parameter is necessary. The characterization of materials consists in extracting their permeability, permittivity and conductivity over a wide frequency band. As the aqueous solutions considered in this chapter are diamagnetic ($\mu = \mu_0$ to a very good approximation), their characterization is achieved through exploitation of permittivity and conductivity changes.

The dielectric properties of aqueous solutions may be measured using a variety of techniques such as parallel plate capacitor, resonant cavity, or coaxial lines [9], to cite a few. These techniques enable the extraction of the real and imaginary parts of the permittivity at various frequencies. The conductivity can be measured through different techniques or instruments, such as conductometer [10], but it can also be seen on the imaginary part of the permittivity extracted by means of dielectric characterization techniques [4]. Conductive effects can be considered as losses from the dielectric point of view, and thus they can be related to the imaginary part of the relative dielectric permittivity. This relation can be made explicit through the Ampère-Maxwell law:

$$\nabla \times \vec{H} = \vec{J} + \frac{\partial \vec{D}}{\partial t} \tag{1}$$

For linear materials, and assuming harmonic time dependence, (1) can be expressed as:

$$\nabla \times \vec{H} = \left(\sigma + j\omega(\varepsilon' - j\varepsilon'')\right)\vec{E} = \left(\sigma + \omega\varepsilon'' + j\omega\varepsilon'\right)\vec{E} \tag{2}$$

indicating the somehow equivalent effect of σ and $\omega\varepsilon''$ (ω is the angular frequency). In the rest of this section, dielectric and conductive properties of some aqueous solutions in the DC to microwave ranges are presented, analyzed and discussed.

1.2.1 Dipolar Effects

Dipolar effects occur in materials composed of polar molecules, and can have a significant impact on the permittivity. In a polar molecule, the centers of positive and negative charges are separated, thus creating a dipole moment. When an electric field is applied to the material, the dipoles tend to align with the field, resulting in an overall polarization of the material, and contributing to its permittivity. In aqueous solutions, the polarization from DC to microwave frequencies originates from rotation of the dipole moment of water molecule, and the large value of ε_r' (close to 80 at low frequency) comes not only from the interaction of each molecule with the applied field, but also from cooperative motions due to intermolecular hydrogen bonds [11–13].

Dipolar effects are usually frequency-dispersive. As the frequency increases, the applied field inversion time decreases, impacting the ability of the material to be polarized. In fluids (liquids or gases), the intermolecular forces lead to micro-mechanical resistance. This resistance to flow, also known as viscosity, reduces molecules rotational mobility. At molecular level, it prevents the instantaneous alignment of the polar molecule with the biasing electric field. The associated delay is called relaxation time, and is consequently closely dependent on the viscosity of the fluid [14]. As a result, high viscosities lead to high relaxation times [15].

The relaxation time, intrinsic property of the material related to its viscosity [16], thus impacts the frequency-dependence of the real part of the permittivity, ε' [17]. Accordingly, the real part of the permittivity of a polar liquid decreases with increasing frequency, since the time available for molecular polarization to fully occur or to be totally released decreases. It is important to note that any parameter impacting viscosity, such as temperature or pressure [18], will also impact the relaxation time, and thus the dielectric behavior of the material.

Regarding the imaginary part of the permittivity, ε'', it represents the energy dissipation within the material when an electric field is applied. In polar materials, it is closely related to the frictional effects appearing as the applied electrical field induces the alignment of the polar molecules [19]. Regarding its frequency dependence, ε'' presents a maximum when the exciting field period is close to the relaxation time, because of the high level of mechanical frictions implying high dielectric losses [20].

Real and imaginary parts of the dipolar permittivity are related by the Kramers–Kronig relations [21]. These relations establish a fundamental relationship between the dispersions of the real and imaginary parts. Knowing one part over the whole frequency spectrum makes it possible to calculate the other part:

$$\varepsilon'(\omega) = 1 + \frac{2}{\pi}\int_0^\infty \frac{\nu\varepsilon''(\nu) - \omega\varepsilon''(\omega)}{\nu^2 - \omega^2} d\nu \tag{3a}$$

$$\varepsilon''(\omega) = \frac{-2\omega}{\pi}\int_0^\infty \frac{\varepsilon'(\nu) - \varepsilon'(\omega)}{\nu^2 - \omega^2} d\nu \tag{3b}$$

Hence, the typical behavior of the permittivity of polar materials is a decrease of the real part and a peak in the imaginary part. Debye proposed a model for this behavior [22], which can be expressed as:

$$\varepsilon_r = \varepsilon_\infty + \frac{\varepsilon_{stat} - \varepsilon_\infty}{1 + j\omega\tau} \tag{4}$$

where ε_{stat} and ε_∞ are the limits of the relative dielectric permittivity at low and high frequencies, respectively, and τ is the relaxation time. A semi-empirical evolution of the Debye is the Cole–Cole model [23]:

$$\varepsilon_r = \varepsilon_\infty + \frac{\varepsilon_{stat} - \varepsilon_\infty}{1 + (j\omega\tau)^{1-\alpha}} \tag{5}$$

where α is a dimensionless exponent parameter taking into account the dispersion in the relaxation time.

Water is a polar molecule and, as such, its permittivity function has the typical shape for dipolar materials shown in Fig. 1. At microwave frequencies, it can be

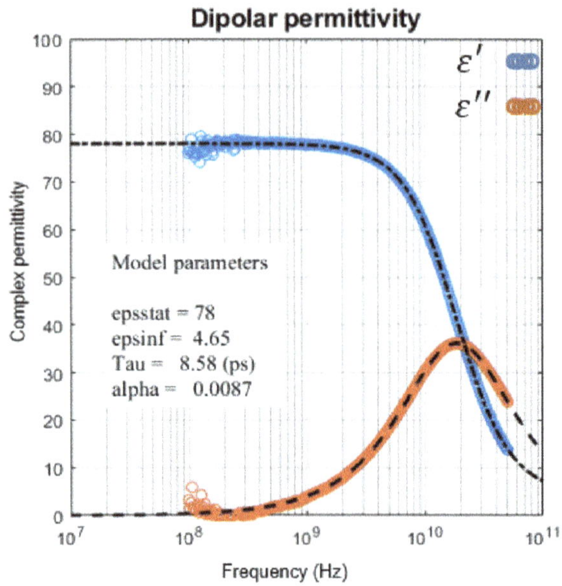

Fig. 1 Complex permittivity of an aqueous solution: measured (o) and theoretical (–) values. Determination of Cole–Cole parameters using fitting of the dipolar contribution (logarithmic scale for *x*-axis). Reprinted from [46] under a Creative Commons Attribution 4.0 International License

modelled with the Debye equation or Cole–Cole model. It has been widely investigated and many parametrizations of this model can be found in the literature [24–26].

1.2.2 Electrical Conductivity

Electrical conductivity in fluids is mainly due to ionic charges. These free charges move by the effect of the applied electric field, thus realizing the conductive effect. The static ionic conductivity σ_s of a solute can be described by the Kohlrausch law [27]:

$$\sigma_s = \Lambda \times C \tag{6}$$

where Λ is the molar electrical conductivity and C is the concentration of ions.

Electrical conductivity is frequency-dispersive, as it depends not only on the number of ions per unit volume but also upon their drift velocity [11]. The drift velocity of an ion varies with the electric field intensity, with the mass of the ion, and with other factors such as micro-mechanical resistances occurring within the liquid (viscosity). As for the case of the dipolar effect, viscosity is thus impacting electrical conductivity. At high frequencies, the conductivity decreases because the ions are not able to change their movement as fast as the field inversion would require. Hence, under high-frequency fields, the movement of ions along a particular direction deceases with respect to static or low-frequency conditions, resulting in a conductivity continuously decreasing with frequency.

As explained previously, conductivity can be related to the permittivity (a poor conductor could be seen as a lossy dielectric). Therefore, the permittivity model proposed by Cole–Cole for a polar material can be modified to integrate the conductivity contribution, if any ionic compounds are present in the solution. It becomes:

$$\varepsilon = \varepsilon_\infty + \frac{\varepsilon_{stat} - \varepsilon_\infty}{1 + (j\omega\tau)^{1-\alpha}} + \frac{\sigma_s}{j\omega\varepsilon_0} \tag{7}$$

with the same parameters as in (5), and where σ_s is the static ionic conductivity. The ionic conductivity term is added to the imaginary part of the permittivity and increases the loss tangent, its effect being decreasingly important for increasing frequencies.

Again, any environmental condition impacting the viscosity (pressure, temperature…) would impact the electrical conductivity and should be taken into account during the characterization process.

1.2.3 Examples

In this section, some examples of dielectric characterizations are proposed. They were performed with the broadband coaxial probe technique [28, 29] applied onto several solutions, exhibiting either dipolar effects exclusively or aggregate dipolar and conductive effects. The section will be divided into subsections depending on the kind of solutions under analysis.

i. Binary solutions

   a. Deionized water and glucose

   In the frequency range of the dielectric relaxation of water, centered at around 18 GHz at room temperature, and ranging roughly from 0.5 to 100 GHz, the chief effect of glucose and fructose (both molecules have the same chemical composition, although different chemical structure) in aqueous solution is seen as an increase of the dielectric relaxation time [30, 31]. This effect is associated to an increment of the viscosity of the solution linked to the molecular interactions between glucose and water. Apart from their influence over water molecules, the effect of glucose molecules as participants in the relaxation process is comparatively weak and hardly detectable. Another relevant effect is the increase in the parameter α of the Cole–Cole relaxation function, related to the width of the relaxation peak [15, 31, 32]. As the glucose concentration c_g increases, the dielectric spectrum of the solution shifts from a Debye function ($\alpha = 0$) for pure water to a Cole–Cole function, with α values around 0.1 for c_g around 1 mol/l.

   While high-concentration glucose solutions have importance for the food and beverage industry [33, 34], low-concentration solutions are especially relevant in the biomedical field [35]. The typical c_g values in human blood oscillates around 100 mg/dl, ranging from 50 to 400 mg/dl for diabetic people [36]. Glucose concentrations in interstitial fluid and in blood are highly correlated, and their values are approximately equal except for a time delay of 10–20 min in interstitial fluid with respect to blood [37–39]. Owing to the practical inconveniences of experimenting with blood, water–glucose solutions have been widely used as a first-step medium to test microwave glucose sensors. For this reason, the effects of glucose at low concentrations in the dielectric spectrum of water have been studied by several authors [40–44]. The influence of glucose at low c_g values on the dielectric spectrum has been expressed by parameterizing the Debye parameters ε_{stat}, ε_∞ and τ as second-order polynomials of c_g:

$$\varepsilon_\infty = A_\infty + B_\infty c_g + C_\infty c_g^2 \tag{8a}$$

$$\varepsilon_{stat} = A_{stat} + B_{stat} c_g + C_{stat} c_g^2 \tag{8b}$$

$$\tau(ps) = A_\tau + B_\tau c_g + C_\tau c_g^2 \tag{8c}$$

For high glucose concentrations, the validity of this parameterization is limited, as the dielectric spectrum is described by a Cole–Cole function rather than by a Debye function. The coefficients of this parameterization, following [43, 44], are given in Table 1 for c_g expressed in mg/dl. The parameterization (8a)–(8c) allows estimating the variation $\Delta\varepsilon_r'/\Delta c_g$ at a given frequency by using the Debye model. The small values of the linear and quadratic coefficients (B's and C's) in Table 1 indicate the technical difficulty of measuring physiological glucose levels with enough accuracy by using sensing principles based on dielectric permittivity.

b. Deionized water and sucrose
Sucrose, widely known as table sugar, is a disaccharide formed by the association of the monosaccharides glucose and fructose, whose molecular formula is $C_{12}H_{22}O_{11}$ [45]. Although sucrose is a polar molecule, it is not possible to identify its own dipolar signature in aqueous solution due to its great molecular size relatively to that of water. However, sucrose has a great impact on water dielectric properties [46]. The imaginary permittivity of the solution exhibits

Table 1 Coefficients of the parameterization (8a)–(8c) for c_g in mg/dl and τ in ps

Refs.	A_∞	$10^3 \times B_\infty$	$10^8 \times C_\infty$	A_{stat}	$10^4 \times B_{stat}$	$10^9 \times C_{stat}$	A_τ	$10^4 \times B_\tau$	$10^9 \times C_\tau$
[43]	8.722	2.148	−8.214	81.015	−2.793	2.318	8.776	5.150	−8.370
[44]	5.076	1.395	−6.28	77.692	−2.741	−4.1	7.851	3.737	−10.7

a single, wide peak at high frequencies corresponding to the water relaxation process, as shown in Fig. 2. As expected, no conductivity effects can be seen, as sucrose does not introduce ions in the solution.

Effects can be seen both on the imaginary and real parts of the permittivity. The shift towards low frequencies of the maximum of the imaginary part shows the effect of viscosity brought by sucrose. Moreover, the maximum values of both real and imaginary parts of the permittivity significantly decrease as sucrose concentration increases. Table 2 summarizes the values of Cole–Cole model parameters extracted from measurements at each concentration of sucrose.

The static permittivity ε_{stat} experiences a significant decrease with increasing sucrose concentration. This is due to a lower concentration of water molecules as sucrose concentration increases. Additionally, the large size of the sucrose molecule results in over 37 water molecules being included in its first hydration shell [47], leading to an abundance of non-rotational bonds of water molecule that further contribute to the decrease in static permittivity [15]. The measured

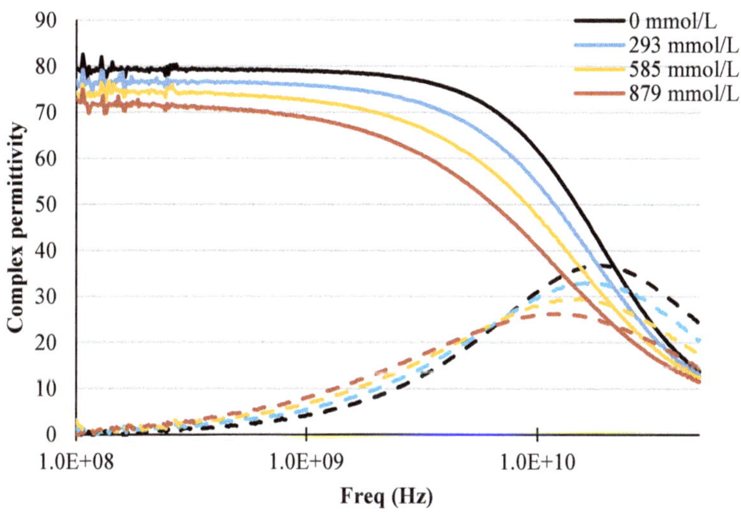

Fig. 2 Real part (solid lines) and imaginary part (dashed lines) of the complex permittivity of aqueous solutions at various sucrose concentrations as a function of frequency (logarithmic scale). Reprinted from [46] under a Creative Commons Attribution 4.0 International License

Table 2 Extracted Cole–Cole model parameters for water–sucrose solutions

$\frac{C_{C_{12}H_{22}O_{11}}}{(\text{mmol/L})}$	ε_∞	ε_{stat}	τ (ps)	α	σ (mS/cm)
0	4.60	79.20	8.71	0.011	0
293	2.14	76.90	9.55	0.075	0
585	1.43	74.80	1.12	0.130	0
879	1.55	72.30	1.36	0.180	0

values of static permittivity are consistent with those reported in the literature [48].

Sucrose concentration has a significant impact on the relaxation time τ due to its strong influence on viscosity [49]. As sucrose concentration increases, the relaxation time increases significantly. The static conductivity is zero because sucrose does not add any ionic contribution to aqueous solutions. The parameter α experiences a substantial increase with sucrose concentration, indicating a widening of the imaginary permittivity spectra associated to the increase in viscosity.

c. Deionized water and sodium chloride

Because of its polar character, and in addition to the specific features of this property upon the permittivity, water is an excellent solvent. It means that it is able to separate anions and cations and to have specific separate interactions with these ions. When sodium chloride (NaCl) is diluted in water, it dissociates into anions and cations and generates ionic conductivity. As a result, water–sodium chloride solutions are interesting as they exhibit a conductive effect brought by the ions along with the dipolar effect brought by water. It has been shown that varying the concentration of NaCl in an aqueous solution leads to the variation of the conductivity (the higher the concentration, the higher the number of ionic charges, the higher the conductivity) [46]. This effect is clearly visible at the low frequencies in Fig. 3. Table 3 summarizes the values of Cole–Cole model parameters extracted from measurements at each concentration of NaCl.

The imaginary part of the permittivity for non-zero NaCl concentrations exhibits two regions of losses. At low frequencies (up to approximately 1 GHz), there is an inverse proportionality between the imaginary part and the frequency,

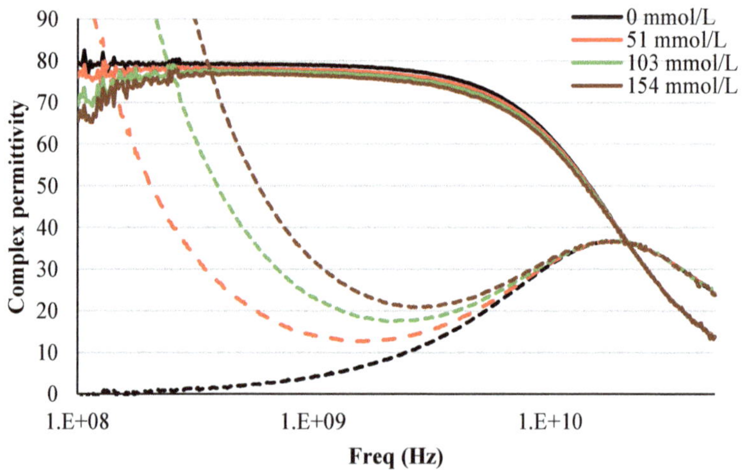

Fig. 3 Real part (solid lines) and imaginary part (dashed lines) of the complex permittivity of aqueous solutions at various sodium chloride concentrations as a function of frequency (logarithmic scale). Reprinted from [46] under a Creative Commons Attribution 4.0 International License

Table 3 Extracted Cole–Cole model parameters for water–sodium chloride solutions at 25 °C

C_{NaCl} (mmol/L)	ε_∞	ε_{stat}	τ (ps)	α	σ (mS/cm)
0	4.60	79.20	8.71	0.011	0
51	4.00	78.20	8.52	0.016	5.76
103	3.20	77.50	8.41	0.024	10.92
154	2.85	76.90	8.36	0.030	15.99

and a proportionality to the sodium chloride concentration, which corresponds to ionic conductive losses (see Eq. (6)). At high frequencies, the losses are due to dipolar relaxation of water molecules, with the maximum imaginary part occurring at around 18 GHz, regardless of the NaCl concentration.

The real part of the permittivity shows a slight decrease as the sodium chloride concentration increases. However, for high concentrations, measurement uncertainties arise at low frequency due to electrode polarization effects [50]. Hence, the first frequency points of the measured permittivity were ignored for curve fitting at high concentrations. Close scrutiny of Fig. 3 reveals that sodium chloride clearly modifies the ionic behavior of the permittivity, as expected, but hardly affects the dipolar behavior. The static permittivity, ε_{stat}, decreases slowly with increasing NaCl concentration, which is due to the lower water concentration and ion–solvent interaction leading to non-rotational bonding of water molecules near ions.

The relaxation time, τ, slightly decreases as NaCl concentration increases, which has been observed for other ionic solutes [51]. This effect could be related to the decrease of the molecular order of the dipolar network of water induced by the introduction of ions [52]. However, viscosity increases very little with NaCl concentration [53]. This leads to a slightly decreasing mean relaxation time for sodium chloride concentrations below 1 mol/L [8].

The ionic conductivity, σ_s, considerably increases with increasing NaCl concentration. Since the viscosity of NaCl solutions shows very little change, the conductivity behavior is mainly driven by the number of ions added to the DI water. The parameter α slightly increases as NaCl concentration increases. The presence of solutes in the solution leads to broadening of the relaxation losses, which is well-described by dispersion in relaxation time. The parameter ε_∞ decreases slightly as the concentration increases, but since it acts as a mathematical optimization parameter restricted to the studied frequency range, no physical phenomenon can be clearly associated with its variation. In conclusion, NaCl diluted in water mainly provides a conductive character due to dissolved ions and does not affect the dipolar character brought by water molecules.

d. Deionized water and sodium lactate

Sodium lactate ($NaC_3H_5O_3$) is the sodium salt of L(+)-lactic acid. It is a polar molecule, like glucose, sucrose or water, but it ionizes when dissolved in water. Thus, the solution of sodium lactate in water exhibits simultaneously a dipolar behavior and a conductive effect. The evolution of the complex permittivity of water–sodium lactate solutions according to the sodium lactate concentration is shown in Fig. 4, whilst the extracted Cole–Cole model parameters are gathered in Table 4.

The characterization of dielectric properties indicates that as the concentration of sodium lactate increases, there is a reduction in relaxation time and an increase in the α parameter. This can be attributed to the viscosity changes caused by the sodium lactate, which affects the dipolar relaxation of the solution. In addition, the static conductivity increases rapidly with increasing sodium lactate concentration up to 2.848 mol/L, and then decreases. This non-monotonic effect is due to two opposing factors: the concentration of free ionic charges and their ability to move. At low concentrations, the number of free charges increases with increasing sodium lactate concentration, leading to an increase in conductivity. However, at higher concentrations, the increase in viscosity restricts the

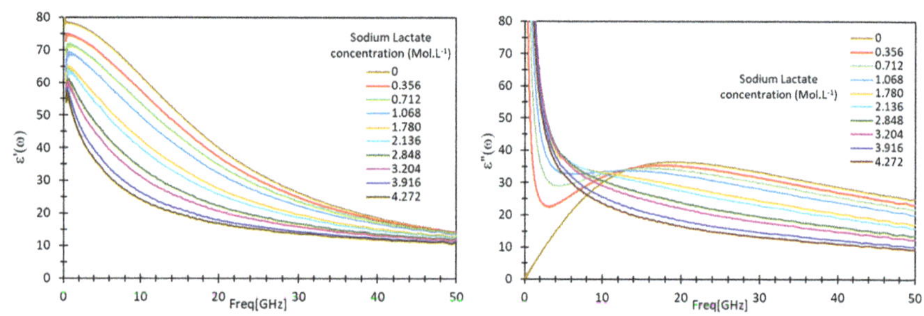

Fig. 4 Real part (left) and imaginary part (right) of the complex permittivity of aqueous solutions at various sodium lactate concentrations as a function of frequency (logarithmic scale). Reprinted from [97] under a Creative Commons Attribution 4.0 International License

Table 4 Extracted Cole–Cole model parameters for aqueous solutions of sodium lactate

Sodium lactate concentration (mol/L)	ε_∞	ε_{stat}	τ (ps)	α	σ (mS/cm)
0 (DI water)	4.60	79.2	8.71	0.011	0.0
0.356	4.47	74.8	8.48	0.027	19.0
0.712	4.41	72.0	8.77	0.046	32.1
1.068	3.39	69.5	9.08	0.079	42.1
1.780	2.58	66.0	10.4	0.14	53.6
2.136	2.33	65.3	11.3	0.17	54.8
2.848	2.85	63.5	14.6	0.23	55.7
3.204	3.38	63.2	17.1	0.25	54.1
3.916	2.89	62.9	23.8	0.32	48.5
4.272	3.51	62.0	28.9	0.33	44.9

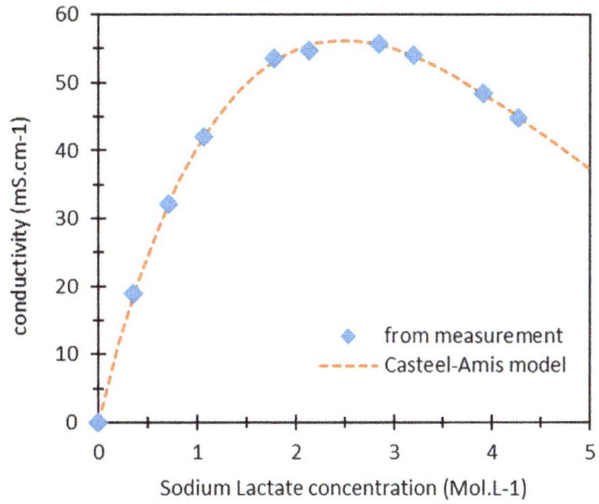

Fig. 5 Static conductivity extracted from permittivity measurement of sodium lactate/water solutions at 25 °C. Reprinted from [97] under a Creative Commons Attribution 4.0 International License

mobility of the free charges, resulting in a decrease in conductivity as depicted in the Casteel-Amis model [54] of Fig. 5.

ii. Ternary solutions: deionized water, sodium chloride and sucrose
In ternary solutions made from sodium chloride, sucrose and water, several physical phenomena occur and interactions between them appear. Especially, the viscosity brought by sucrose significantly affects the ionic conductivity brought by NaCl. Permittivity measurements by means of coaxial probe were made in [46] with several sets of water–NaCl–$C_{12}H_{22}O_{11}$ solutions at room temperature (23 °C) and with a broadband approach. The resulting complex permittivities are plotted in Fig. 6, where each curve is labelled with a letter associated with the sodium chloride concentration followed by a number associated with the sucrose concentration. The Cole–Cole model parameters resulting from curve fitting, as well as the equivalence between the curves labels and the sodium chloride and sucrose concentrations, can be seen in Table 5.

It can be seen that the static permittivity (ε_{stat}) decreases as the concentration of NaCl or sucrose increases, as it happens with binary mixtures. This effect is due, first, to the lower water concentration when other solutes are present, and second, to the solute–solvent interaction which leads to non-rotational bonding of water molecule in the vicinity of solute. The relaxation time (τ) of ternary mixtures is also greatly influenced by the sucrose concentration, which has a strong impact on viscosity. As sucrose concentration increases, the relaxation time significantly increases, whereas the influence of NaCl is relatively minor.

Viscosity effects due to sucrose are particularly noticeable on ionic conductivity (σ_s), which decreases as sucrose concentration increases. For example, at a fixed sodium chloride concentration of 154 mmol/L, σ_s decreases from 15.99 to 8.12 mS/cm as sucrose concentration increases from 0 to 877 mmol/L, respectively. The conductivity of ternary solutions is mainly related to the amount of solvated ions, with a higher concentration of sodium chloride leading to greater

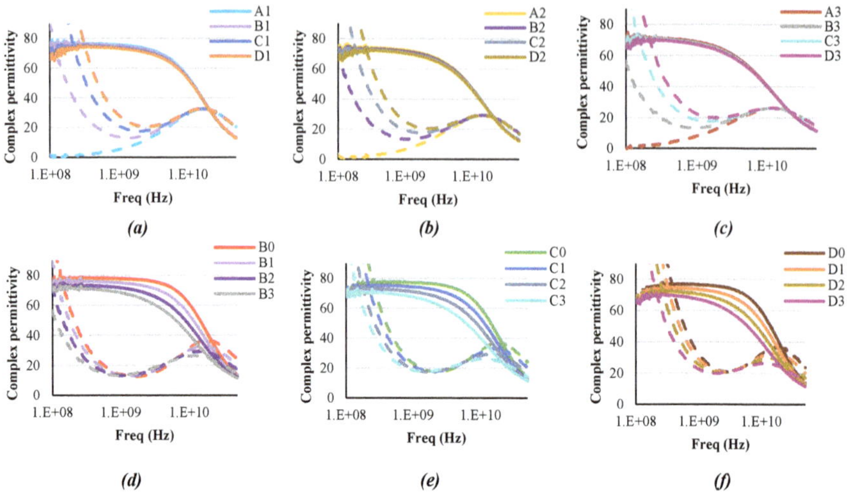

Fig. 6 Real part (solid lines) and imaginary part (dashed lines) of the complex permittivity of samples for different sucrose concentrations (**a**–**c**) and for different sodium chloride concentrations (**d**–**f**) as a function of frequency (logarithmic scale). Reprinted from [46] under a Creative Commons Attribution 4.0 International License

Table 5 Extracted Cole–Cole parameters for aqueous solutions of sucrose and sodium chloride

Solution samples	C_{NaCl} (mmol/L)	$C_{C_{12}H_{22}O_{11}}$ (mmol/L)	ε_∞	ε_{stat}	τ (ps)	α	σ_s (mS/cm)
A0	0	0	4.60	79.2	8.71	0.011	0
B0	51	0	4.00	78.2	8.52	0.016	5.76
C0	103	0	3.20	77.5	8.41	0.024	10.92
D0	154	0	2.85	76.9	8.36	0.030	15.99
A1	0	293	2.14	76.9	9.55	0.075	0
B1	52	294	1.88	76.3	9.56	0.084	4.77
C1	104	293	1.30	75.4	9.45	0.085	9.11
D1	154	292	0.70	74.8	9.15	0.096	13.22
A2	0	585	1.43	74.8	11.2	0.130	0
B2	51	585	1.09	73.9	11.0	0.140	3.80
C2	103	586	0.38	73.4	10.8	0.150	7.26
D2	154	585	0.10	73.0	10.8	0.150	10.53
A3	0	879	1.55	72.3	13.6	0.180	0
B3	51	877	0.98	72.0	13.5	0.190	2.92
C3	106	877	0.30	71.5	13.2	0.200	5.82
D3	154	877	0.10	71.1	13.0	0.200	8.12

conductivity. However, viscosity reduces the mobility of ions, and thus, the higher the sucrose concentration, the lower the conductivity.

Similarly to binary mixtures, the parameter α in ternary mixtures also increases with sucrose concentration. Additionally, α slightly increases with sodium chloride concentration. Therefore, the wider the loss curves, the higher the solute concentration in ternary mixtures. The parameter ε_∞ decreases with an increase in the concentration of sucrose. Furthermore, ε_∞ slightly decreases with an increase in sodium chloride concentration. One should note that ε_∞ is the parameter used to fine tune the model in order to make it fit the measured characteristics. For this reason, it may sometimes reach values lower than one, which is neither physically significant nor possible.

Conclusions

This section has shown how dissolved species could impact the dielectric permittivity and the electrical conductivity of aqueous solutions. Relying on these phenomena, it is now possible to monitor the concentration of these dissolved species by designing sensors sensitive to these changes. To obtain optimal sensitivities, two ways should be simultaneously driven: on the one hand, it is overarching to identify which feature of the electromagnetic properties will be sensed (frequency, real or imaginary part of the permittivity, conductivity), and on the other hand, but taking into account the first point, sensor should be optimally designed and its sensing indicators should be adequately chosen. The next sections are devoted to address these points.

2 Design of Sensors Based on Electromagnetic Features of Liquids Devoted to the Monitoring of Aqueous Solutions Containing Sugars and Salts

Now that underlying physical phenomena impacting electromagnetic properties associated to salts and sugars diluted into water have been presented, there is a real interest to build sensors exploiting these features to track evolutions of their concentrations. The aim is then to optimize the sensor design in order to maximize interactions between the device and the sample to be monitored. This ability is referred to as sensitivity, which can be defined as the ratio between the variation of the response of the sensor over the variation of the parameter at the origin of the modification. As an example, for a sensor with voltage as output, its sensitivity to the parameter p is expressed as:

$$S(p) = \frac{\Delta V}{\Delta p} \tag{9}$$

where the voltage variation ΔV is the consequence of the variation Δp in the input parameter. Sensitivity has a dimension that depends on the input and output parameter dimensions. The ideal sensor has a high sensitivity, which is constant over the whole measuring range. This latter feature is called linearity and is defined as the variation of the sensitivity along the sensing range.

Multiple choices have to be properly made to obtain a highly sensitive device, and these choices can be categorized as related to the environment where the system is going to operate, to the sensor driving electronics and the associated data processing, to the material or sample itself, and finally to the sensor design and its implementation. Of course, many choices are simultaneously related to several of these points and lead to trade-offs between aspects that are sometimes antagonist.

2.1 Environmental Constraints

Sensors are usually purposed to be used in various environments and it is of paramount importance to take them into account for the design as they can have impact on the material, on the device itself, or on the interactions between them. Here are a few examples of each of these aspects.

1. Environment impact on material properties

Regarding the environmental parameters that can impact the material electromagnetic behavior, every element that influences viscosity of the liquid would also modify its electromagnetic properties. This is notably the case, amongst others, for pressure or temperature. In particular, this latter one has a strong impact upon the dielectric characteristics of the material, as it is reflected through the expression of the relaxation time τ, which was theoretically expressed by Debye as [22]:

$$\tau = \frac{4\pi \eta a^3}{kT} \tag{10}$$

η being the viscosity, T the temperature, a the molecular radius and k the Boltzmann constant. This is also directly related to changes in the viscosity of dielectric liquids with temperature, as evidenced by Arrhenius [55]:

$$\eta = A \times exp\left(\frac{E_a}{kT}\right) \tag{11}$$

being A a frequency-related factor and E_a the activation energy. Temperature is thus an important parameter impacting the dielectric behavior of aqueous solutions and subsequently the ability of the sensors to extract concentrations of the solutions under study.

2. Impact of the surrounding environment upon the sensor

Sensors are devices optimized so that they are sensitive to the tracked parameter, and ideally only to that parameter. Other parameters affecting its response could therefore be considered as spurious, and for this reason, cross-sensitivities are usually investigated to identify potential interferences [56]. Ideally, every parameter, compound, or change in the environment should be tested to identify the specific impact it can have on the sensor sensitivity.

The ability of a sensor to react and be sensitive to one parameter solely is defined as selectivity. It is an essential feature of sensors and can be extrapolated from the cross-sensitivity of a sensor. Cross-sensitivity of a sensor is defined as the impact a parameter b has on the sensor sensitivity with regard to the parameter a. It describes how the sensor response is subject to possible interferences from other parameters. Because they are not highly selective (contrarily to electro-chemical sensors for instance), microwave-based sensors are particularly liable to be affected by other compounds, that can hinder their sensitivity to the tracked parameter [57]. In this case, the cross-sensitivity is a relevant parameter to identify the impact of environment over the sensor sensitivity.

Furthermore, operating environment of microwave sensors is, in most cases, neither protected from nor protecting other microwave sources or detectors, being all the devices thereby prone to interferences. Consequently, frequencies and powers should be adequately chosen so that they respect international regulations on electromagnetic spectrum to avoid any possible interferences of the sensor upon any other microwave-based device. Reversely, using some frequencies out of the free ISM (Industrial, Scientific and Medical) frequency bands exposes the system to the reception of unwanted signals that can blur or even deteriorate the system if the detection floor is particularly low, which is generally required to achieve high sensitivities.

3. Environmental aspects impacting sensor/sample interactions

Finally, an essential aspect to obtain high sensitivities for microwave sensors relies upon the optimum interactions between the electromagnetic waves and the media or samples to be monitored. One should note that the ability to operate on a remote basis is a key feature of electromagnetic sensors (often a differential one) and as such it appears as a considerable asset for non-invasive sensing. Notwithstanding that, the environment between the sensor and the material under test may have a considerable impact over the sensitivity.

In particular, to prevent the devices from any sensitivity hindering due to the environment, all the materials that are located in between the sensor and the sample should be as electromagnetically-transparent as possible, which signifies that their losses should be low, and their thicknesses thin. Depending on the sensing configuration (in vivo, ex vivo, in vitro…) the surrounding environment can be either chosen or mastered by the sensor designer (sample holder, container, tubes…) or undergone when the sensor is targeted to operate remotely. In this latter case, and aiming at non-invasive measurement of biomedical variables, biological tissues, skin or other materials can affect interactions between the samples to be monitored and the electromagnetic waves emitted by the sensor. For this reason, many works have been carried out to characterize electromagnetic properties of biological materials or tissues [58, 59].

2.2 System and Architecture Considerations in the Design of Microwave-Based Sensors

A microwave sensor is usually composed of several parts: the transducer itself, which directly interacts with the material under test, and the system that feeds and analyzes the signals to/from the transducer. This latter part, although it is not directly interacting with the samples is crucial to the monitoring process. Indeed, the RF system architecture should ensure the signal generation and the transducer electrical response analysis.

Regarding signal synthesis, the challenge is to be able to generate the waves at the required frequencies, with a sufficient power level. Signals are achieved thanks to oscillators or synthesizers, which are able to generate rather narrowband signals. Although solutions to obtain wideband spectrums exist, they require the use of numerous devices such as mixers and filters, which makes them difficult to implement at low costs. On the other hand, generating single frequency signals is simpler [60] but it implies constraints on the sensing scenario. Finally, intermediate solutions allowing to trade-off between complexity of the signal generation and sensing possibilities rely on the generation and use of moderately-wide band or multiple narrow-band signals [61]. As for any other applications (telecommunications, defense…) signal generation for sensing purposes is quantified thanks to features regarding frequency and spectral purity (phase noise, frequency accuracy…) and amplitude (dynamic range and amplitude accuracy).

The generated signal is feeding the transducer, which modifies the signal, further analyzed by the receiver. Regarding signal detection and quantification, the same difficulties arise in relation to bandwidth, and special attention should also be paid to sensitivity, linearity, dynamic range and noise figure. Several architecture options exist to implement signal generation and detection—direct conversion, homodyne, heterodyne, superheterodyne—each one addressing differently the receiver paradigms. However, it appears that, from the system point of view, narrowband systems, at rather low-to-moderate frequencies (up to a few GHz), with relatively high-power levels, are easier to implement, which usually means lower costs for their design and manufacturing.

Additionally, several choices are possible regarding the electrical parameter to be sensed. Between the input and output signals, changes induced by the transducer could be included in resonance frequency modifications [62], phase shifts [63], magnitude level changes [64], among others [65, 66]. Although there are existing possibilities to converse from one to another, magnitude levels are probably simpler to detect, as long as their powers are sufficiently high.

Finally, the transducer, which is the sensor part interacting with the sample, can provide either a reflection-based response (S_{11} parameter only, one port, e.g. [67, 68]) or a transmitted and reflected response (S_{21} and S_{11}, two ports, e.g. [69, 70]). Obviously, the difference is significant in terms of collected data but also regarding complexity of the analysis, which consequently affects not only the system architecture but also the measurement accuracy and simplicity (simple or more complex calibration procedure).

Sensor design, and particularly the transducer, is the main topic addressed in this chapter, however, because of its location, between the samples to be monitored and the RF system that generates and analyzes the signal, it has to cope with paradigms inherited from the system together with those coming from the material described in the first part of this chapter. Consequently, the optimal sensor design results in trade-offs between all these constraints. Next sections are dedicated to detailing some considerations regarding sensor design optimization with respect to the impact of the material properties.

2.3 Optimization of the Sensor Regarding Material Properties

In the first part of this chapter, impact of some kind of solutes on electromagnetic properties of water has been presented. These electromagnetic properties were extracted thanks to wideband characterizations made with network analyzer (VNA) and commercial probes, performed in dedicated laboratories. However, for practical applications such as industrial process monitoring, biomedical, or food and drug industry for example, deployment of turnkey, low-cost solutions is vital. As detailed hereinbefore, this implies discarding wideband solutions and developing narrow-band, or possibly multiband sensors. Consequently, and bearing in mind the frequency-dependent evolution of the electromagnetic properties of water, in order to achieve sensors with high sensitivity, it is crucial to choose the best sensing frequencies.

Sensing frequencies choice obviously depends on the solutes to be tracked and their electromagnetic signatures, but from the general properties presented, it appears that ionic solutes (salts, for instance) concentrations can be sensed through conductivity modifications at rather low frequency. On the other hand, non-ionic compounds that have effects over dipolar relaxations (notably through the viscosity changes) may be sensed at various frequencies. Amongst them, several antagonist effects lead to requirement for trade-off arbitration.

Indeed, as the sensing principle is based on dielectric properties changes, a high concentration of the electric field into the samples is sought [71]. This is optimally obtained if the real part of the permittivity is rather high. From this point of view, the lower the frequency the better, apart from undesirable effects that could occur at very low frequencies, near DC (such as electrode polarization, for example). However, if it is necessary to choose frequencies with a high value of the real part of the permittivity to obtain a high density of electric field within the sample, it is also important to choose a frequency where the relative variations of the permittivity induced by concentration changes are maximal in order to obtain high sensitivities. It appears that the highest relative variations of the dielectric permittivity usually appear around the relaxation phenomena.

The permittivity is made of a real and an imaginary part, and these two parts are somehow related (Kramers–Kronig relations (3a), (3b) and through underlying physical phenomenon related to viscosity). Subsequently, variations of dielectric permittivity imply variations on both real and imaginary parts.

Changes in the real part of the dielectric permittivity modify field distributions. They can also change propagating or resonant modes, which mean phase, magnitude and/or resonance frequency variations of the signal, depending on the microwave devices. All these effects have to be taken into account for the design and optimization of the sensor in order to magnify electromagnetic wave–sample interactions, and thus reach the highest sensitivity for the sensor.

2.4 Design of the Sensor

Regarding the design and implementation of the sensor, there are plenty of possibilities, spanning from very simple to more complex configurations. In the following lines we will review the main aspects that should be considered.

1. Technological implementation

A large variety of implementation techniques are available for microwave components, from the electrically-performant waveguides to the highly integrable planar configurations. Each solution exhibits assets and drawbacks, which means that each application particularities would lead to a specific choice. For instance, in vivo measurements usually require high integration, which means compact structures and thus generally discards waveguide implementations, at the cost of lower electrical performances. On the other hand, waveguide solutions raise interest as far as high performances are sought and when no constraints on size or weight arise.

Another important feature to consider is the way the monitored liquids are brought in the neighborhood of the RF device. In situ monitoring means that the environment around the solutions is not controlled and the technological configuration of the sensor should be adapted (e.g. [72–74]). If the liquids are extracted from their environment, they can be flowing through bespoke tubes or fluidic channels (e.g. [75–81]), thus allowing for continuous monitoring, or sampled and punctually placed within containers (e.g. [82–85]). Obviously, designers have to cope with these aspects when making their technological choices.

Additionally, aqueous solutions monitoring usually takes place in an environment with strong requirements regarding the integrity of the samples. For instance, for food and drink industry, materials in contact with the samples should be certified so as not to contaminate them. Similarly, biocompatibility is needed for biomedical applications. Nowadays, a wide range of materials and processes are available for the fabrication of microwave devices. For instance, dielectric resins processed thanks to biocompatible 3-D printing appear as potentially interesting for the realization of waveguide-based solution sensing. Regarding planar or multilayer configurations, non-organic solutions exist, such as those based on ceramic substrates, including low-temperature cofired ceramics (LTCC) and high-temperature cofired ceramics (HTCC) [86].

2. Sensing parameter

Taking a brief look at the literature of microwave resonant sensors, it appears that several sensing parameters can be considered. Resonance frequencies, quality factors, insertion/return loss or phase are amongst the most often used. However, criteria for selecting them are not evident, and a consensus on which one (or ones) to use for each case does not seem to appear. In a recent publication [65], authors have proposed to use a common indicator named Relative Sensitivity (*RS*) to evaluate and compare performances of various sensors dedicated to the monitoring of a given compound. Indeed, choice of the sensing parameter is closely related to the sensor design and reversely, optimization of the sensor topology depends on the chosen indicator. In this sense, choice of the sensing parameter should be done jointly with the optimization of the sensor design.

3. Topological implementation

Many sensors dedicated to monitoring solute concentration in aqueous solutions have been developed and are presented in the scientific literature. As explained in the previous sections, for simplicity of the signal generation and analysis, most of them are intended to work on relatively narrow frequency bands. Furthermore, to reach high sensitivities and good precisions and resolutions while remaining narrowband, resonant sensors topologies are usually preferred. The principle is to use a microwave device at frequencies around its resonance frequency, and to benefit from the intrinsic sensitivity of resonances to the electromagnetic properties of the immediate vicinity in order to achieve sensors [87]. Amongst the paradigms to be addressed in the design of a resonant sensor, the main ones deal with optimizing location of the sample with regard to the field distribution on the one hand, and resonator feeding scheme on the other hand.

Regarding the choice of the most appropriate location for the sample, it is related to the field distribution, which is usually quite straightforward to identify for simple resonant structures. However, complexity relies in the modifications of the field distribution around resonance frequency related to the introduction of the sample and its effects upon field distribution, which can be considerable if the dielectric permittivity contrast between the sensor, its environment (air, substrate, tube/container layers, biological tissue layers…) and the liquid to be sensed is high enough, but this effect could be somehow masked by the environment if a proper configuration is not considered [71, 88].

With regard to the couplings, input/output (I/O) coupling levels should be high enough to provide enough power to the device (and particularly if the sample is lossy around the resonance frequency) but also low enough to take full benefit of the unloaded quality factor (Q_u) of the resonator (as it will be shown in the next section), which allows for an easier detection on most of the sensing parameters (this is particularly true for resonance frequency-, phase- and Q-factor-based approaches). Hence, I/O coupling design usually needs to trade off powering capabilities (and thus sensing signal readability) against detection capabilities [84].

Recently, the possible topologies for planar sensors have been painstakingly studied [89], and a novel one based on coupled resonators has started to raise interest. This strategy enables to address the same issues as those of resonant sensors (relative simplicity of signal generation/analysis related to limited bandwidth), but it seems to exhibit higher sensitivities [3, 65, 90, 91]. The next section proposes a comprehensive approach together with the modelling of these topologies applied to the monitoring of liquids. For compacity, simplicity, and cost sake, planar technologies are retained although most of the content presented hereafter would remain valid for waveguide implementations.

3 Coupled-Resonator Sensors

In this section, we present a theoretical model of a sensor made up of two mutually coupled resonators. The coupling is considered to be essentially electric, since the magnitudes to be sensed are related to the permittivity of a sample. Although the model is based on lumped elements, and it therefore lacks of complexity or sophistication, it can reproduce rather realistically the behavior of a two-port coupled-resonator sensor.

3.1 Circuit Model of Two Identical Resonators Electrically Coupled

Each resonator is considered as a lumped-element parallel resonator, which is a valid assumption for half-wave (open-loop, for instance) or quarter-wave (short-ended line) conventional resonators. Electric coupling is chosen to concentrate the electric interaction with the sample. The I/O coupling networks can be modeled by admittance inverters [92]. Unless the response bandwidth is too wide (strong inter-resonator coupling), the inverter parameter J can be taken as constant (frequency-independent). The full lumped-element model is depicted in Fig. 7.

In the model in Fig. 7, Z_0 is the input/output ports impedance and L_r is the inductance of the resonators. The sample has an impact on the resonators capacitance (C_r)

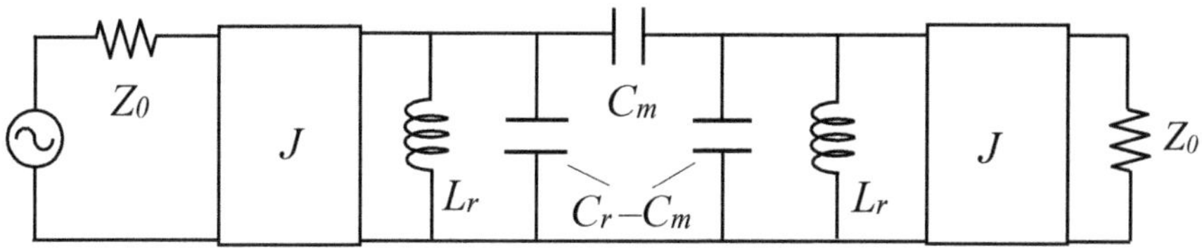

Fig. 7 Lumped-element electric model of a two-port coupled-resonator sensor

and the coupling or mutual capacitance (C_m). Both capacitances can have a contribution which is independent of the sample (C_{r0} and C_{m0}) along with a contribution linked to the sample (C_{rs} and C_{ms}). In the case of aqueous solutions, the sample is highly lossy (loss tangent ranging from 0.2 to 0.4 in the frequency region of 4–8 GHz [26]), so that the model capacitances can be expressed as:

$$C_r = C_{r0} + C_{rs}(1 - j \times tg\delta) \tag{12a}$$

$$C_m = C_{m0} + C_{ms}(1 - j \times tg\delta) \tag{12b}$$

The model could incorporate losses not related to the sample, but, for highly lossy samples like water, other contributions to the model losses can be neglected in a first approximation. The resonance frequency of the uncoupled resonators is:

$$\omega_0 = \frac{1}{\sqrt{L_r C_r}} \tag{13}$$

The network in Fig. 7 may be considered as a cascade of different two-port sub-networks: inverter–resonator–coupling network–resonator–inverter. Therefore, its response can be calculated by using transmission (*ABCD*) matrixes. The different matrixes are the following ones:

(a) Inverters:

$$T_J = j\begin{pmatrix} 0 & 1/J \\ J & 0 \end{pmatrix} \tag{14}$$

(b) Parallel resonators:

$$T_R = \begin{pmatrix} 1 & 0 \\ Y_r & 1 \end{pmatrix} \tag{15}$$

where:

$$Y_r = j\left(\omega(C_r - C_m) - \frac{1}{\omega L_r}\right) \tag{16}$$

(c) Capacitive coupling:

$$T_C = \begin{pmatrix} 1 & 1/Y_c \\ 0 & 1 \end{pmatrix} \tag{17}$$

where:

$$Y_c = j\omega C_m \tag{18}$$

The transmission matrix of the global network is:

$$T = T_J T_R T_C T_R T_J = -\begin{pmatrix} 1 + Y_r/Y_c & \frac{2Y_r + Y_r^2/Y_c}{J^2} \\ J^2/Y_c & 1 + Y_r/Y_c \end{pmatrix} \tag{19}$$

The transmission parameter is then calculated as:

$$S_{21} = \frac{2}{A + \frac{B}{Z_0} + CZ_0 + D} = \frac{-Y_c}{Y_c + Y_r + \frac{Y_r Y_c + Y_r^2/2}{G_{ex}} + \frac{G_{ex}}{2}} \tag{20}$$

where:

$$G_{ex} \equiv J^2 Z_0 \tag{21}$$

is the equivalent conductance of the ports, as seen by the resonators.

The conductance associated to the sample losses are:

$$G_r = \omega C_{rs} tg\delta \tag{22a}$$

$$G_m = \omega C_{ms} tg\delta \tag{22b}$$

If the resonance is not very wide, these conductance can be approximated by their values at ω_0. Within this approximation, also Y_c can be taken as constant, and then the denominator of the transmission parameter S_{21} around ω_0 yields a second order polynomial in $\Omega = \omega/\omega_0 - \omega_0/\omega$. The poles of S_{21} can be obtained maximizing $|S_{21}(\Omega)|^2$ with respect to Ω, leading to an equation of degree 3 in Ω. If the inter-resonator coupling is high enough, this third-degree equation leads to two maximums (poles) and a minimum of $|S_{21}(\Omega)|^2$, with the minimum between the poles. For resonators coupled by a lossless coupling mechanism, the condition for the existence of a minimum is:

$$k \geq 1/Q_L \tag{23}$$

where k is the coupling factor, that for capacitive coupling is simply the ratio of mutual to self capacitance:

$$k = \frac{C_{m0} + C_{ms}}{C_{r0} + C_{rs}} \tag{24}$$

and where Q_L is the loaded quality factor of each single resonator:

$$\frac{1}{Q_L} = \frac{1}{Q_u} + \frac{1}{Q_{ex}} \tag{25}$$

being Q_{ex} the external quality factor, i.e. the contribution to the measured (loaded) quality factor related to elements not strictly belonging to the resonator, and Q_u the unloaded quality factor, i.e. the quality factor without considering any possible effects of the external loads from the measurement equipment. In the approximate range $1 < k \times Q_L < 5$, the frequencies of the two poles in the S_{21} response vary with Q_L. For $k \times Q_L > 5$, these frequencies remain essentially unchanged, and the coupling factor can be estimated by the expression (see Chap. 2) [93]:

$$k \cong \frac{f_2^2 - f_1^2}{f_2^2 + f_1^2} \tag{26}$$

where f_1 and f_2 are the resonance frequencies (poles) of the coupled system.

The above considerations indicate that, for coupled resonators with high losses, the coupling intensity needs to be moderate or even strong to assure the presence of two separate peaks in the response. As an example, the graph of Fig. 8 shows the transmission response of two identical, lossless, coupled resonators with resonance frequency $f_0 = 5$ GHz for $k = 0.025$ and three $k \times Q_L$ values.

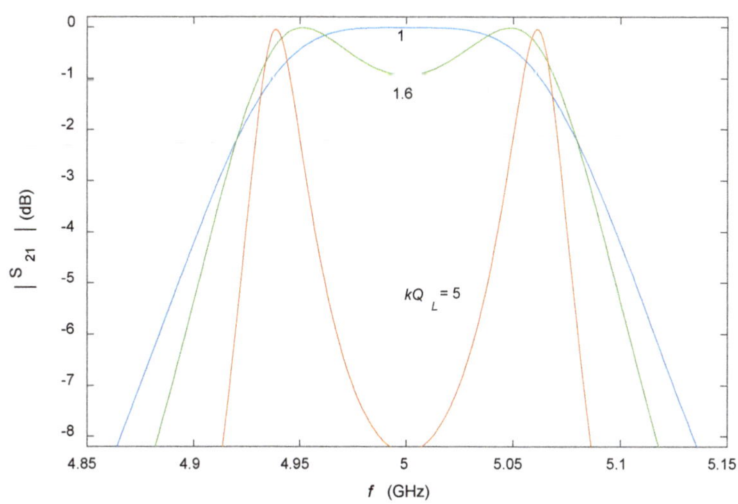

Fig. 8 Transmission response of two identical, lossless, coupled resonators depending on the $k \times Q_L$ value

When considering losses in the resonators (but not in the coupling mechanism), the maximum S_{21} levels at the peaks are:

$$|S_{21}(poles)| = \frac{Q_u}{Q_u + Q_{ex}} \tag{27}$$

This expression shows that very high external quality factors (i.e. very weak I/O couplings) lead to very low S_{21} levels, which could be difficult to detect by the instrumentation equipment, and thus they should to be avoided.

3.2 Design Considerations for Aqueous Solutions

Considering more specific applications, for solutes like glucose, sucrose and other sugars, the more relevant effect in the dielectric relaxation of water, as stated in Sect. 1, is a shift of the whole spectrum (real and imaginary permittivity functions) towards lower frequencies. In the 1–10 GHz region, this effect generates a decrease in the real permittivity and an increase in the imaginary permittivity, and therefore in the loss tangent as well. This intrinsic sensitivity of the solution dielectric properties to solute concentration is higher in the 4–10 GHz range, approximately. In this range, the upper frequencies (8–10 GHz) show higher losses: the loss tangent of water increases from ~0.2 at 4 GHz to ~0.4 at 8 GHz, at room temperature [94]. Therefore, the range between 4 and 8 GHz seems to raise as a reasonable trade-off between sensitivity of the dielectric properties of the aqueous solution to glucose concentration and not excessively high losses.

For sensors based on the inter-resonator coupling, the resonance frequencies of the coupled system (poles) are the measurable magnitudes providing the coupling factor. In order to have acceptable peaks in the transmission response, the condition $k \times Q_L > 1$ must be met. Both the coupling factor and the loaded quality factor are influenced by the sample, and the strength of this influence is determined by the strength of the resonator–sample interaction. The relative magnitude of this interaction can be roughly estimated through the relative weights of the resonator and coupling capacitances dependent on the sample (C_{rs} and C_{ms}) to the independent ones (C_{r0} and C_{m0}). For highly lossy samples, such as aqueous solutions, a strong resonator–sample interaction implies high insertion losses and S_{21} peak widening. On the other hand, a weak interaction implies a weak dependence of the response on the sample properties, i.e. a low sensitivity for the sensor. Therefore, a trade-off between both effects needs to be sought in the sensor design.

For diluted aqueous solutions, the capacitances linked to the sample may be approximated as linear functions of the concentration. Then, the coupling factor can be expressed as

$$k = \frac{C_{m0} + C_{ms}(1 + ac_s)}{C_{r0} + C_{rs}(1 + bc_s)} \tag{28}$$

where c_s is the solute concentration in the sample and the parameters a and b are the relative sensitivities of the capacitances influenced by the sample to c_s:

$$a = \left.\frac{1}{C_{ms}} \frac{dC_{ms}}{dc_s}\right|_{c_s=0} \tag{29a}$$

$$b = \left.\frac{1}{C_{rs}} \frac{dC_{rs}}{dc_s}\right|_{c_s=0} \tag{29b}$$

The parameters a and b depend on the sample geometry, specifically on the way that the sample influences the inter- and intra-resonator electric fields. For coupled-resonator sensors tracking the changes in the solute concentration by means of the variations in k, the sensitivity of the device to the solute concentration is defined as:

$$S_k \equiv \frac{dk}{dc_s} \tag{30}$$

For low concentrations, from (30), S_k can be easily shown to be:

$$S_k \cong \frac{1}{C_{r0} + C_{rs}} \left(\frac{dC_{ms}}{dc_s} - k \frac{dC_{rs}}{dc_s} \right) \tag{31}$$

Therefore, the two capacitances influenced by the sample (C_{ms} and C_{rs}) exert opposite effects on the sensitivity. In order to obtain a high sensitivity in magnitude, there are two options: a mutual capacitance highly influenced by the sample whilst having a resonator capacitance weakly influenced by the sample, or vice versa. The signs of the sensitivities for both options are opposite.

For water–glucose solutions in the 1–10 GHz frequency range, an increment of the concentration leads to an increment in the loss tangent and a decrement in the real permittivity. Therefore, the two derivatives in the right side of (29a), (29b) are negative. If the sample affects more the mutual capacitance than the resonator capacitance, so that the first term dominates, the resulting sensitivity of k will be negative, i.e. the coupling factor decreases for increasing concentration.

As a matter of fact, as mentioned above, the resonance frequencies of the coupled system are influenced by the loaded quality factor, Q_L. If, in a first approximation, the losses associated to the coupling are disregarded, the resonance frequencies f_1 and $f_2 > f_1$ of the coupled system fulfil (see Chap. 2):

$$\frac{f_2 - f_1}{f_0} = \sqrt{k^2 - \left(\frac{1}{Q_{ex}} + \frac{1}{Q_u} \right)^2} \tag{32}$$

When the glucose concentration increases, the loss tangent of the sample increases, and therefore Q_u decreases. If the sensitivity of k is negative (first term dominant in the right-side of (31)), both effects combine jointly to reduce the bandwidth $f_2 - f_1$. In that case, the parameter $\Delta f = f_2 - f_1$ is more sensitive than k to the glucose concentration, and it seems therefore a good candidate as sensing parameter. As previously noted, the influence of Q_L (and thus Q_u) in the positions of the peaks is small for $k \times Q_L > 5$, so that Q_L fulfilling $1 < k \times Q_L < 5$ is compatible with: (i) having two separate peaks in the response; and (ii) adding the influences of the sample in k and in Q_u to obtain a high sensitivity of Δf. If the sensitivity of k is positive (second term dominant in the right-side of (31)), the product $k \times Q_L$ must be kept greater than 5 to avoid the influence of the losses in Δf.

3.3 Design Example

Let us consider, as a typical example of planar microwave resonator, a microstrip open-loop resonator as those shown in Fig. 9. These are indeed the type of resonators used in the practical realizations presented in the last sections of this chapter. Around its fundamental resonance, an open-loop resonator can be described by an equivalent parallel lumped-element resonator. The lumped-model equivalent parameters L_r and C_r can be easily found by imposing that both resonators (lumped and distributed) have the same resonance frequency and susceptance slope parameter. Supposing that the resonator is not loaded with a sample, these parameters are:

$$L_r = \frac{2L}{\pi^2} l \tag{33}$$

$$C_r = \frac{C}{2} l \tag{34}$$

where L and C are the per-unit-length inductance and capacitance of the microstrip lines, and l is their length. The inter-resonator coupling factor, supposing that the coupling is not as high as to alter significantly the voltage and current distributions of the uncoupled resonators, is also easy to evaluate. For the electric-coupling configuration of Fig. 9 (see Chap. 2), it yields:

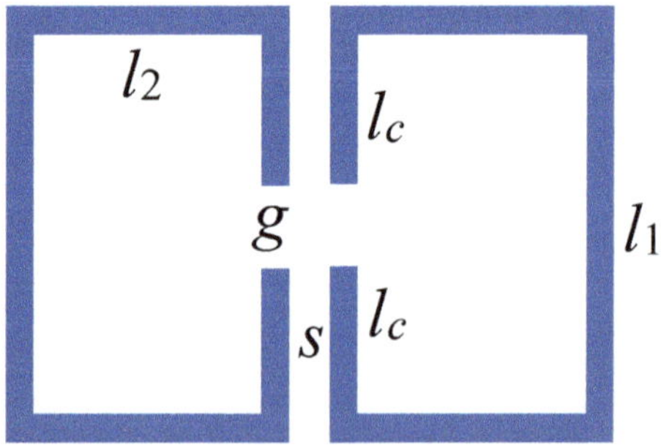

Fig. 9 Outline of two coupled open-loop microstrip resonators electrically coupled

$$k \cong \frac{|k_L(2\pi l_c/l - \sin(2\pi l_c/l)) + k_C(2\pi l_c/l + \sin(2\pi l_c/l))|}{\pi} \tag{35}$$

where l_c is the coupling length in each extreme of the resonator, and $l = \lambda_r/2$ is the length of the resonating lines, being λ_r the guided wavelength at resonance.

When loading the sensor with a sample, especially if the sample is water or an aqueous solution (high permittivity, high losses), several features of the sensor response change with respect to the unloaded sensor:

- The voltage distribution changes, and the region of dominant electric energy shifts toward the sample, due to its high permittivity. The equivalent capacitance of the resonators increases.
- Given that voltage and current distributions in the resonators are linked, the current distribution changes as well, and then the equivalent inductance of the resonators changes.
- The resonator losses increase significantly due to the high loss tangent of the sample. The unloaded quality factor of the resonators decreases.
- If the sample is placed in the coupling region, the inter-resonator coupling increases due to the increase in the mutual capacitance, and losses associated to the coupling appear.

It is difficult to establish a theoretical model that includes the effects linked to the sample, as the geometry in the region loaded with the sample is highly inhomogeneous. On the other hand, if from the outset the design is based on the system without sample, the changes in the response after placing the sample will be, in general, important, unless the interaction is very weak.

Assuming that the effects described in the preceding paragraphs are not too strong for a moderate resonator–sample interaction, we can start a design in the 5 GHz region, which combines a reasonable intrinsic sensitivity of the sample to glucose variations with not too high dielectric loss tangent. Consider a low-loss substrate TLX-8 ($\varepsilon_r = 2.55$, tan $\delta = 0.0017$) with thickness $h = 825$ μm. For a line width of w = 1 mm (not an excessively wide line to keep a proper width/length ratio), the line impedance is $Z_c \cong 80\ \Omega$ and the effective relative permittivity is $\varepsilon_{re} \cong 2.014$. The resulting line length for resonance (without sample) at 5 GHz is $l = 21.14$ mm, and the equivalent parameters of the resonator are:

$$L_r = 1.35\, nH$$
$$C_r = 0.52\, pF$$

Considering the design in Fig. 9, the total length l can be distributed, for instance, as $g = 1.6$ mm, $l_1 = 6.6$ mm, $l_2 = 4.77$ mm with a gap wide enough to neglect its capacitance, and a length $l_c = 2.5$ mm for each coupling region. For a separation between the coupled lines $s = 0.8$ mm the coupling factor gives $k = 0.0612$, a moderate value compatible with the condition $k \times Q_L > 1$, unless the sample–resonator interaction is very strong.

Let us suppose that the sample is placed in the coupling region, and its geometry is designed to affect more the mutual capacitance than the resonators capacitances. For instance, let us suppose that:

$$\frac{C_{rs}}{C_r} = 0.035$$
$$\frac{C_{ms}}{C_m} = 0.35$$

indicating that the sample affects strongly the coupling and weakly the resonators capacitances. The effect of the glucose concentration on the dielectric properties of water can be estimated from published experimental data at 5 GHz and 25 °C as [41, 42, 94]:

$$\varepsilon_r'(c_g) \cong 73.6 - 0.6c_s(\%) \tag{36}$$

$$\varepsilon_r''(c_g) \cong 17.7 + 0.3c_s(\%) \tag{37}$$

Using these expressions, the capacitances in (12a, b) and the conductance in (22a, b) can be estimated as:

$$\left.\frac{1}{C_{ms}}\frac{dC_{ms}}{dc_s}\right|_{c_s=0} = \left.\frac{1}{C_{rs}}\frac{dC_{rs}}{dc_s}\right|_{c_s=0} \cong \left.\frac{1}{\varepsilon_r'}\frac{d\varepsilon_r'}{dc_s}\right|_{c_s=0} \cong -8.15 \times 10^{-3}/\%$$
$$\left.\frac{1}{G_{ms}}\frac{dG_{ms}}{dc_s}\right|_{c_s=0} = \left.\frac{1}{G_{rs}}\frac{dG_{rs}}{dc_s}\right|_{c_s=0} = \left.\frac{1}{\varepsilon_r''}\frac{d\varepsilon_r''}{dc_s}\right|_{c_s=0} \cong +16.9 \times 10^{-3}/\%$$

It must be emphasized that these are quite rough estimations, since the interaction of a liquid sample with the sensor is usually influenced by the container walls, that have not been considered in this approximate analysis. Moreover, when loading the sensor with an aqueous sample, the whole response shifts towards low frequencies, depending on the strength of the sensor–sample interaction, and the numbers in (36) and (37) may change.

Once the elements of the circuit in Fig. 9 have been calculated, the transmission response of the coupled system can be calculated with (20). Figure 10 shows the S_{21} magnitude for the cases $Q_{ex} = 78$ and $Q_{ex} = 130$. The Q_u without sample has been estimated to be roughly 280 by evaluating the dielectric and ohmic losses of a half-wave resonator with the dimensions given above. In any case, the influence of the initial Q_u (without sample) is weak, since the Q_u of the loaded sensor is chiefly determined by the sample losses.

The results of the model have been evaluated for three values of Q_{ex} (78, 130 and 194) and two glucose concentrations, 0% and 10%. The following absolute and relative sensitivities have been calculated from the results:

Absolute sensitivity in the bandwidth:

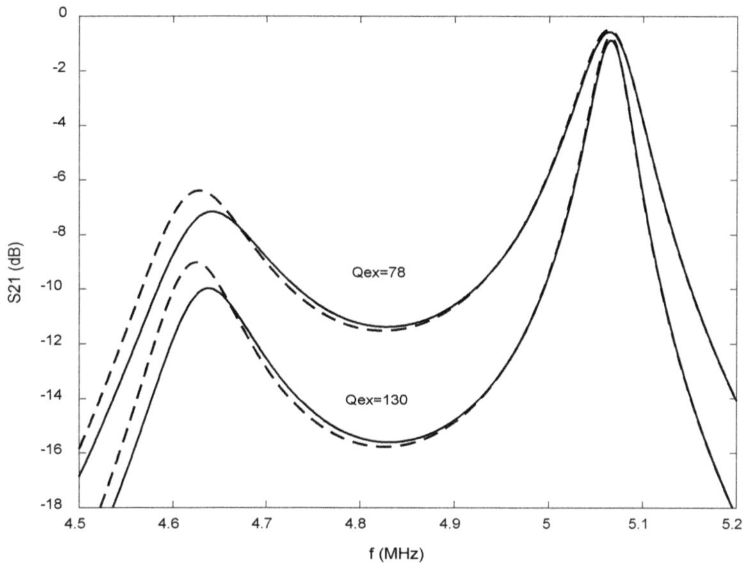

Fig. 10 Transmission response of the sensor modeled in this section for water–glucose solutions at 0% (dashed lines) and 10% (solid lines). Upper curves: $Q_{ex} = 78$. Lower curves: $Q_{ex} = 130$

$$S_{BW} = \frac{\Delta(f_2 - f_1)}{\Delta c_g} \frac{\text{MHz}}{\%} \tag{38}$$

Relative sensitivity in the bandwidth:

$$RS_{BW} = 100 \frac{\Delta(f_2 - f_1)}{\Delta f_0 \times \Delta c_g} \frac{\%}{\%} \tag{39}$$

Absolute sensitivity in the coupling factor:

$$S_k = \frac{\Delta k}{\Delta c_g} \frac{1}{\%} \tag{40}$$

Relative sensitivity in the coupling factor:

$$RS_k = 100 \frac{\Delta k}{k \times \Delta c_g} \frac{\%}{\%} \tag{41}$$

The obtained sensitivities are collected in Table 6.

The coupling factors of Table 6 have been calculated from the peak frequencies by using (26). The sensitivity of k calculated that way, as mentioned in previous section, is slightly increased by the decrease of Q_u (see (32)). In fact, the sensitivity of k estimated from the model by (31), and not affected directly by the quality factor,

Table 6 Sensitivities to the glucose concentration obtained with the sensor model of Fig. 9 for the design example above

Q_{ex}	S_{BW} (MHz/%)	RS_{BW} (%/%)	S_k (1/%)	RS_k (%/%)
78	−1.24	−0.32	-2.6×10^{-4}	−0.31
130	−1.20	−0.30	-2.5×10^{-4}	−0.30
194	−1.14	−0.29	-2.3×10^{-4}	−0.28

is slightly lower (-2.4×10^{-4}/%). In the following section, two practical examples of experimental works involving this kind of sensors will be reviewed.

4 Examples of Realization of Coupled-Resonator Sensors

The purpose of this section is to demonstrate the particular features of sensors based on coupled open-loop resonators. In the following subsections, two state-of-the-art experimental approaches will be reviewed. Both of these works rely upon the coupled-resonator sensing strategy to face solute concentration measurements in aqueous solutions. Specifically, a glucose concentration sensor will be shown and discussed, as well as a sensor devoted to extract simultaneously the concentrations of sodium chloride and sucrose from ternary solutions.

4.1 Glucose Coupled-Resonator-Based Sensor

Aiming at glucose concentration sensing in aqueous solutions, a highly-sensitive sensor exploiting the coupled-resonator paradigm was introduced in [90]. The sensor was designed for glucose concentration sensing of small volume, low-concentration water–glucose solutions. It was conceived as an early version of future sensors intended for biomedical applications, like monitoring of glucose in blood or interstitial fluid. In this preliminary step, water–glucose solutions were chosen as a simplified medium to test the sensor features.

The working frequency region was selected to be in the range 4.0–4.5 GHz (once the sample holder and the samples were already placed), combining good intrinsic sensitivity to glucose (see Sect. 3.2) and moderate sensor size with reasonable penetration depth in tissues, having in mind future evolutions of the sensor. The sensor concept evolves from prior designs based on a single microstrip resonator, adding a second resonator in a coupled, two-port network. In previous works [84], the unloaded quality factor (Q_u) and the maximum transmission level (S_{21max}) were explored as sensing magnitudes. In this design, the coupling factor (k) is added to these parameters as a new sensing magnitude.

The design of the sensor is shown in Fig. 11, and a picture of the fabricated device, including an ad hoc sample holder, can be seen in Fig. 12. Open-loop topology was chosen for the resonators to benefit from the strong electric fields in the open-end region of the resonators. As it can be seen in Fig. 11, the microstrip lines are narrower in the central part than in the open ends, so that each resonator is actually a stepped-impedance resonator (SIR). The line width was smaller in the central part to reinforce its magnetic character, thus reinforcing as well the electric character of the extremes. The resonators are coupled in the open-end region, so that the coupling is essentially electric.

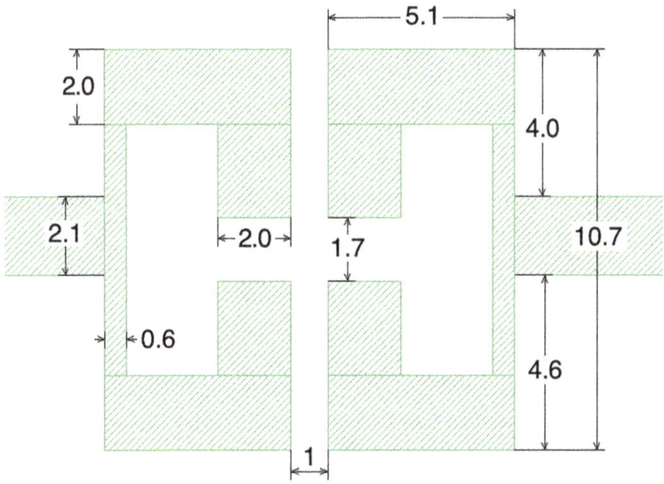

Fig. 11 Design of the glucose concentration coupled-resonator-based sensor (all dimensions are in mm). Reprinted with permission from [88]; copyright 2019 EuMA

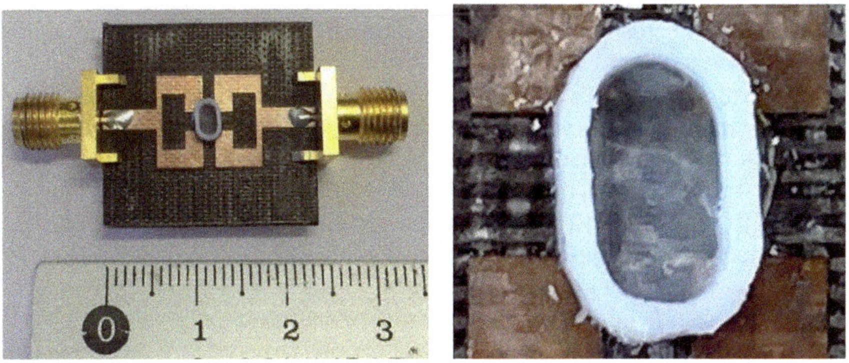

Fig. 12 Fabricated sensor (left, scale in cm) and zoom-up of the sample holder (right). Reprinted with permission from [88]; copyright 2019 EuMA

The external coupling was implemented by means of direct (tapped) connection of the I/O lines. The external quality factor is controlled by the position of the tapped line with respect to the symmetry line across the coupled resonators. The connection is slightly separated from this symmetry line, giving a moderate external quality factor, leading to $S_{21\max} \cong -10.8$ dB. A low external quality factor was avoided to prevent too wide S_{21} peaks that would lead to some degree of merging. This would result in poor resolution in the peak positions (resonance frequencies) and in the unloaded quality factors. It is worthy to note that the unloaded quality factors contain relevant information about the dielectric losses of the Material Under Test (MUT), closely related to the glucose concentration [95].

Regarding fabrication, the sensor was manufactured on the TLX-8 substrate (from Taconic), which combines low loss tangent (0.0017) with low relative permittivity (2.55). The low permittivity value helps to reduce the influence of the substrate in the sensing area. Having the MUT a relative permittivity much higher than the substrate permittivity (around 74 at the working frequencies of the sensor), the electric energy in the sensing region is concentrated in the MUT. The substrate thickness is 825 μm.

The sample holder was fabricated by micromachining into a plate with the form of an oval. This shape is intended to avoid corners, where air bubbles tend to form when filling the holder with the solutions. The material used for the sample holder was polytetrafluoroethylene (PTFE), which has low permittivity and low loss tangent at microwave frequencies. In the fabrication process, the thickness of the bottom wall was made as thin as mechanically possible looking to enhance the sensor–MUT interaction. To prevent the risk of cracks, and to avoid excessive inhomogeneity—the bottom was taken in the simulations as a flat homogeneous layer—, a moderately small thickness of ~60 μm was obtained. A close look of the sample holder already integrated in the sensor is shown in Fig. 12. A very thin layer of adhesive was used to glue the sample holder onto the substrate.

Concerning the experimental assessment, eight samples of water–glucose solutions at equally spaced concentrations between 1.25 and 10% in mass were prepared. A volume of 20 mL of each solution was prepared, which is a much greater volume than necessary, in order to reduce the concentration error. A precision micropipette was used to take sample volumes of 5 μL for each measurement. The scattering parameters (S-parameters) were recorded with a duly calibrated VNA for the different samples. During the measurements, the room temperature remained stable at roughly 22 °C. After each measurement, the sample holder was emptied and cleaned with ethanol and the S-parameters of the sensor in air-loaded state (i.e. without sample), frozen in the screen from the outset of the experiment, were reached again. The obtained transmission responses for the different concentrations are plotted in Fig. 13. The glucose concentrations are given in the box as glucose percentage in mass.

For each response, corresponding to a particular concentration, several parameters were extracted: the resonance frequencies of the two peaks, $S_{21\max}$ values for each peak, and the peak widths at a 2 dB fall of the S_{21} level from its the maximum. The unusual choice of peak width at 2 dB fall is due to the partial merging between both peaks, which makes the usual 3 dB width influenced by the other peak. These widths are related to the Q_u of each peak through the following expression (see Chap. 2 and [96]):

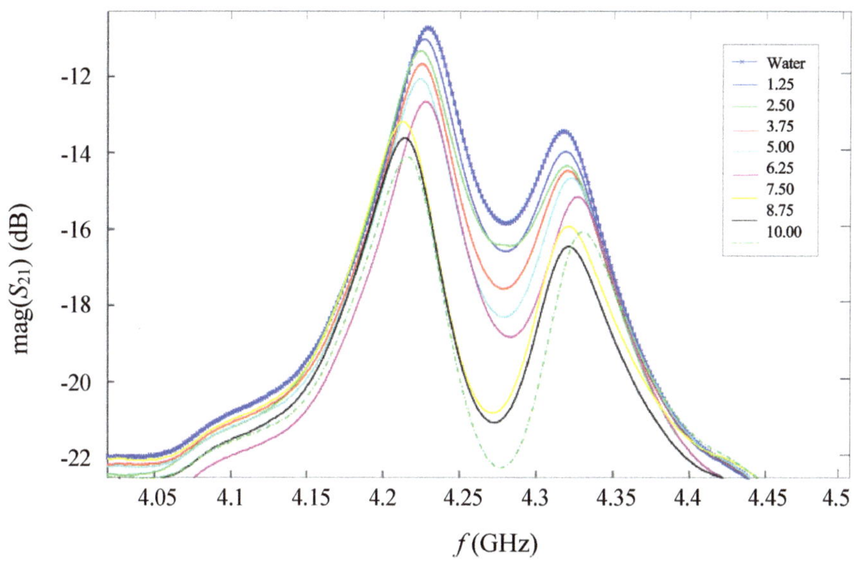

Fig. 13 Transmission responses obtained for the different measurements. Reprinted with permission from [88]; copyright 2019 EuMA

$$Q_L = \frac{f_{peak}}{\Delta f_x}\sqrt{\frac{1-x}{x}} \tag{42}$$

where x is the power fraction of the maximum power (for 2 dB fall, $x = 0.631$) and Δf_x is the peak bandwidth at this power fraction of the maximum. The unloaded quality factor of the peaks is then given by (see Chap. 2):

$$Q_u = \frac{Q_L}{1 - S_{21\,\max}} \tag{43}$$

where $S_{21\max}$ must be in linear scale. As a reference, the Q_{u} values of the two peaks of water (0% glucose) are 126.2 and 104.9 for the lower and higher resonance, respectively.

Three sensing magnitudes were evaluated for each curve: the $S_{21\max}$ of the lower-frequency peak, which showed a more regular evolution with the glucose concentration than the upper-frequency peak, the unloaded quality factor of the lower-frequency peak according to (42) and (43), and the overall coupling factor according to (26). The results of these three sensing magnitudes are shown in Figs. 14, 15 and 16. In all cases, the difference respect to the reference curve (water) is shown as a function of the glucose concentration. The results of electromagnetic simulations are also included in these plots (red points). These simulations were performed with *Ansys HFSS* full-wave simulation software. A good agreement between measurements and simulations can be seen in these graphs.

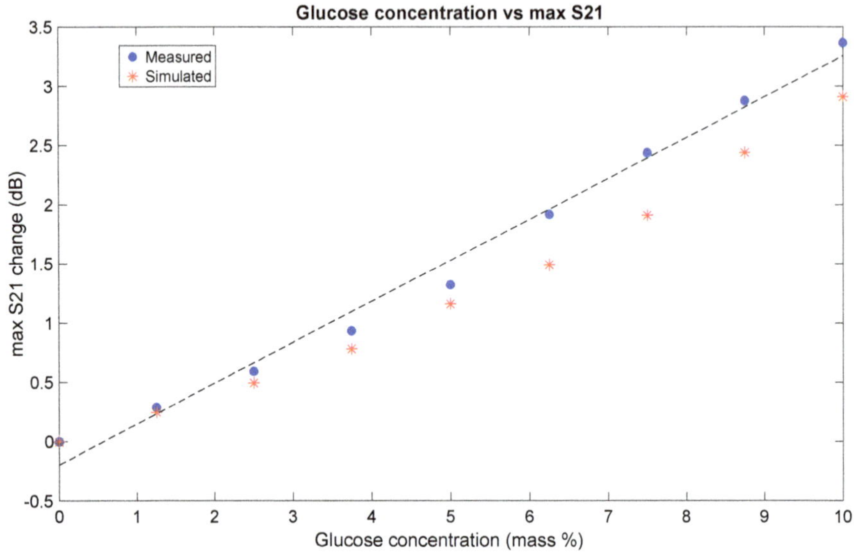

Fig. 14 Evolution of measured and simulated S_{21max} depending on the glucose concentration. Reprinted with permission from [88]; copyright 2019 EuMA

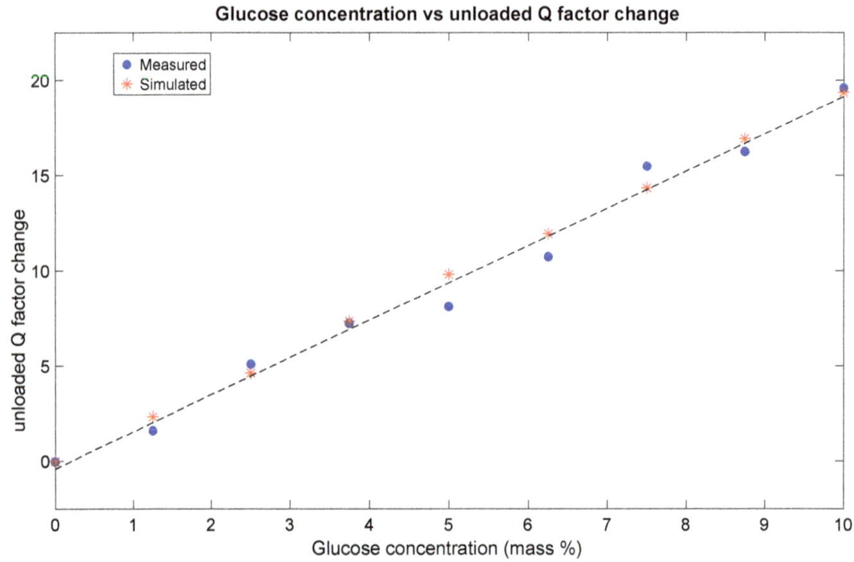

Fig. 15 Evolution of measured and simulated Q_u depending on the glucose concentration. Reprinted with permission from [88]; copyright 2019 EuMA

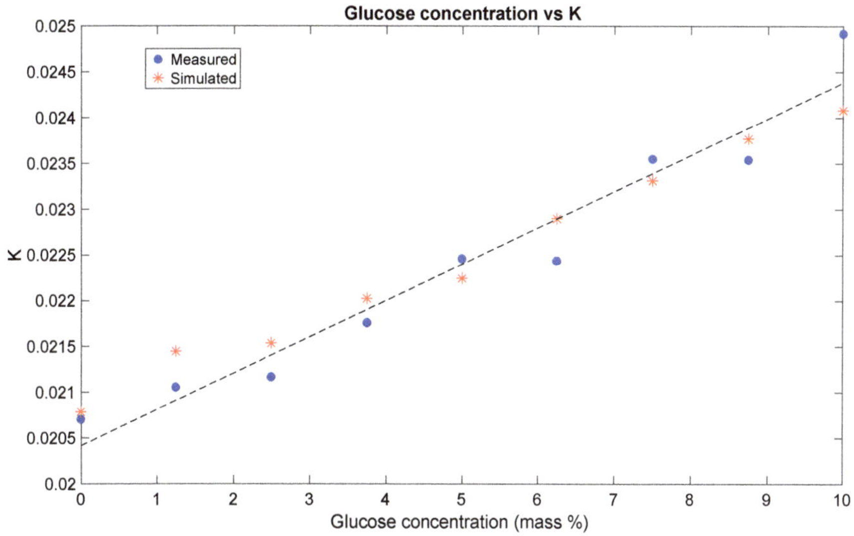

Fig. 16 Evolution of measured and simulated k depending on the glucose concentration. Reprinted with permission from [88]; copyright 2019 EuMA

The three graphs show a positive correlation between the sensing magnitude ($S_{21\max}$, Q_u and k) and the glucose concentration. The respective absolute sensitivities to glucose are:

$$S(G)_{S_{21\max}} = -0.34 \text{ dB}/\%$$
$$S(G)_{Q_u} = -1.96/\%$$
$$S(G)_k = 4.0 \times 10^{-4}/\%$$

The minus signs simply indicate that both Q_u and $S_{21\max}$ decrease for increasing glucose concentration. The relative sensitivities to glucose (i.e. the percentage change of the sensing magnitude relative to the percentage change of the glucose concentration, as described in [65]), taking as reference for the sensing magnitude the measured value for water, are:

$$RS(G)_{S_{21\max}} = -3.24\%/\%$$
$$RS(G)_{Q_u} = -4.12\%/\%$$
$$RS(G)_k = 1.95\%/\%$$

The sensitivities obtained with Q_u and $S_{21\max}$ can be comped directly with the same parameters for the one-resonator sensor of [84]. In that case, the best results from the three single-resonator sensors were absolute sensitivities of −0.658/% for Q_u and −0.084 dB/% for $S_{21\max}$, as well as relative sensitivities of 0.978%/% for Q_u and 0.329%/% for $S_{21\max}$. It should be noted that the experiments in [84] were performed with samples of 25 μL, a fivefold volume. Therefore, the proposed coupled-resonator topology outperforms the single-resonator one with resonators of the same type.

With respect to the coupling factor as sensing magnitude, it should be pointed out that the sensitivities $S(G)_k$ and $RS(G)_k$ are positive, i.e. the coupling factor increases when the glucose concentration increases. Having in mind the discussion in Sect. 3 of this chapter, this positive sign means that the self-capacitances of the coupled resonators of the sensor are more strongly influenced by the properties of the MUT than the mutual capacitance. As the permittivity of the solution decreases with glucose concentration, when the glucose concentration increases the self-capacitances decrease, and therefore the coupling factor increases. This can be understood by considering the geometry in the MUT region (Fig. 12, right). The fringing fields of the resonators open-ends entail an important capacitive effect, amplified by the high permittivity of the MUT. The effect in the small region in between the two resonant microstrip lines seems to be comparatively lower.

From this point, there are two possible lines to follow for further increasing the sensitivity. On the one hand, it could be achieved by optimizing the geometry of the MUT, something that has to be done with the help of an electromagnetic simulation software rather than analytically. On the other hand, another possibility is to reduce the thickness of the container bottom wall to get the MUT closer to the fringing fields of the microstrip lines. The research on these and other possible sensitivity improvement techniques raises as an interesting line for the future years.

4.2 Sodium Chloride and Sucrose Coupled-Resonator-Based Sensor

In addition to the ability to extract glucose concentrations in aqueous solutions with remarkable sensitivities, this sensor topology, because of its intrinsic properties, has certain features that could be potentially exploited for measurement purposes. For example, it also allows for the simultaneous extraction of both an ionic solute (such as sodium chloride) concentration together with a dipolar molecule (such as sucrose) concentration exhibiting a great impact over water dipolar relaxation. The effective simultaneous extraction of the concentration of both solutes in ternary aqueous solutions by means of a two-port coupled-resonator sensor was demonstrated in [46]. The goal of this example is thus to demonstrate that such a topology is interesting not only regarding sensitivity, as shown in the previous subsection, but also regarding sensor selectivity.

As demonstrated previously, in Sect. 1.2.3.ii.a, concentration of solutes in ternary mixtures can be determined by analyzing the dielectric properties of the solutions at different microwave frequencies. However, to minimize sample handling and reduce the number of components, it is recommended to use a single sensor and a single sample for both RF and low-frequency measurements. The objective in [46] was to develop a unique sensor capable of extracting both low-frequency conductivity and complementary RF-range information to simultaneously determine the concentrations of sodium chloride and sucrose in ternary mixtures with high sensitivity. This was rendered possible with the coupled-resonator topology involving open-loop resonators thanks to the location of the sample, which was placed between the input and the output of the device, being the resonators not connected to each other in any other way.

Conductivity measurements at low frequencies were thus achieved by incorporating a serial capacitance between the two ports, with the Material Under Test (MUT) placed on it in contact with the electrodes. The presence and movement of ions in the conductive MUT affect the serial capacitance, resulting in short-circuit behavior. Measuring the transmitted or reflected parameters at low frequencies allows for the extraction of conductivity. At low frequencies, it can be modelled as shown in Fig. 17.

At higher frequencies, as explained hereinbefore, the coupled-resonator device can be modeled as shown in Fig. 18. This model includes lumped elements G_1, C_1, L_1, G_2, C_2, and L_2, representing the first and second resonators, as well as the coupling (C) between them. The dielectric contributions of the sample (MUT) to each element of the sensor model are denoted as $G_{1Sample}$, $C_{1Sample}$, $G_{2Sample}$, $C_{2Sample}$, R_{Sample}, and C_{Sample}. The conductive effect of the sample is modeled by R_σ (as previously shown in Fig. 17).

The sensor consists of two resonators coupled to each other, fed by I/O lines directly connected to the resonators. The sample is placed in the coupling area, where the electric field is maximized, leading to significant effects on the inter-resonator coupling (R_{Sample} and C_{Sample} contributions). Additionally, the sample's proximity to the open-end tips of the resonators influences the inner capacitance and resistance of each resonator, which are represented by $G_{1Sample}$, $C_{1Sample}$, $G_{2Sample}$, and $C_{2Sample}$.

If the physical implementation is symmetrical, resulting in identical resonators, the values with indices 1 and 2 are equal ($G_1 = G_2$, $C_1 = C_2$, $L_1 = L_2$, $G_{1Sample} = G_{2Sample}$, and $C_{1Sample} = C_{2Sample}$). For very low frequencies, the resonators do not resonate, allowing us to consider a simplified model similar to the one depicted in Fig. 17. Conversely, the model at high frequencies (Fig. 18) is relevant for a microwave sensor, assuming that R_σ is sufficiently high (indicating open-circuit behavior) to ensure good sensitivity with respect to R_{Sample} and C_{Sample}. As it can be seen, variations in the dielectric properties of the samples (ε' and ε'') related to

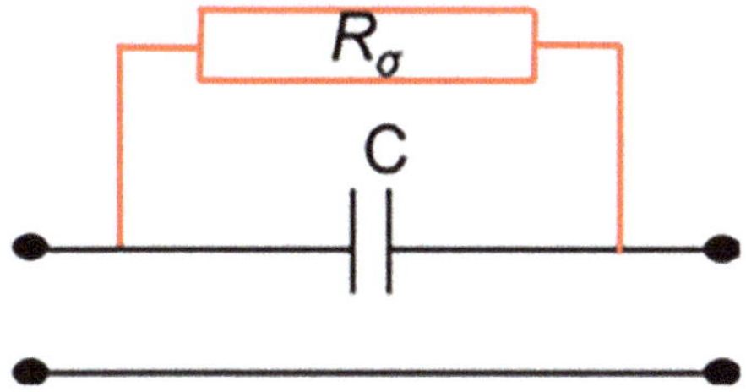

Fig. 17 Lumped-element model of a conductivity sensor (low frequencies). Reprinted from [46] under a Creative Commons Attribution 4.0 International License

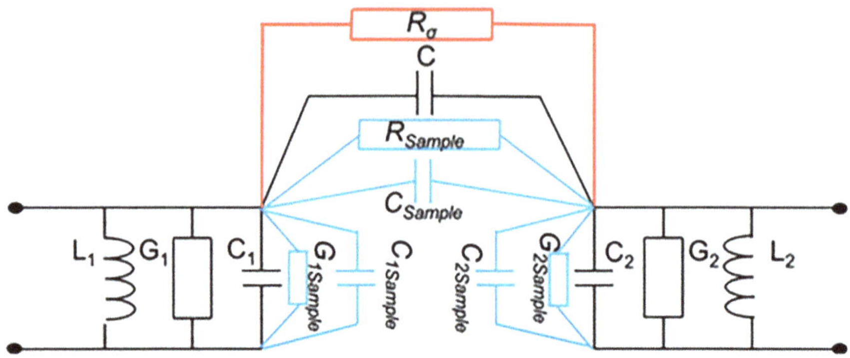

Fig. 18 Lumped-element model of the sensor. Reprinted from [46] under a Creative Commons Attribution 4.0 International License

sodium chloride and sucrose concentrations will alter $G_{1Sample}$, $C_{1Sample}$, $G_{2Sample}$, $C_{2Sample}$, R_{Sample}, and C_{Sample}, consequently affecting the electrical response of the sensor, observed through its scattering parameters. However, the optimal frequency range should be identified to account for all the aspects described beforehand.

After having selected the sensing topology, the next step in sensor design is to determine the optimal frequency range for the microwave portion of the operation. Several factors need to be considered in this process. On the one hand, for the dual detection, the microwave operating frequency should be sufficiently high to mitigate the conductive effects of sodium chloride, as described in Sect. 1.2.3.ii. Frequencies below 5 GHz should be excluded to ensure accurate sensing. On the other hand, very high frequencies (>10 GHz) are not ideal due to the behavior of the real part of permittivity for water-based solutions. At high frequencies, the real part of the permittivity decreases, as shown in Sect. 1.2.3.ii. This leads to a reduced concentration of the electric field within the sample, which is unfavorable for sensing purposes. Moreover, the imaginary part of the dielectric permittivity reaches a peak around 12 GHz, resulting in increased losses at these frequencies.

Considering the proposed model, the sensitivity of the sensor is primarily associated with the variations in $C_{1Sample}$, $C_{2Sample}$, and C_{Sample}, which are chiefly influenced by changes in the real part of the permittivity of the solution. To identify the most suitable frequency range for sensing, the maximum variations of the real part of the dielectric permittivity were computed using the wideband electromagnetic characterizations presented in Sect. 1.2.3.ii. Specifically, to determine the optimal operating frequency, the difference between the real part of the permittivity for various solutions and that of deionized water (DIW) was measured at each frequency for all the solutions. The resulting difference ($\Delta\varepsilon'(f) = \varepsilon_{DIW}(f) - \varepsilon_{(Samples)}(f)$) is plotted against frequency for three solutions in Fig. 19.

This analysis helps to identify the frequency range where the greatest differences in permittivity occur for the same variations of concentrations in the samples, indicating higher sensitivity for the sensor. By selecting the frequency range with significant

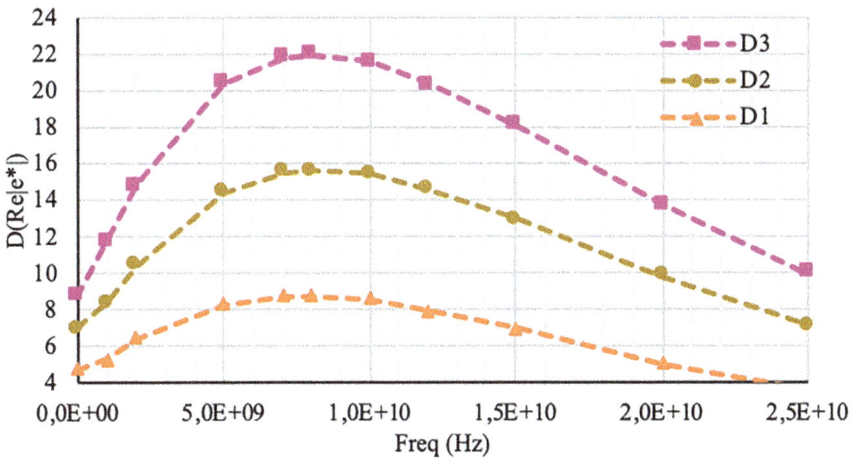

Fig. 19 Maximum difference in the real part of the permittivity as a function of frequency. Reprinted from [46] under a Creative Commons Attribution 4.0 International License

variations in permittivity—around 7–8 GHz—, the sensor can effectively detect and differentiate the concentrations of the targeted solutes in the ternary mixtures.

Further to the definition of the optimal operating frequency range, the next step is to proceed with the physical implementation. Open-loop resonator topology was chosen for convenience and ease of design, similar to the one in [90]. To optimize field distribution and sensitivity of the sensor, fundamental mode and first order harmonic field distributions were investigated, and particularly in the odd and even modes, which stand for the uncoupled and coupled cases, respectively. They are depicted in Fig. 20, where plus and minus signs represent the maximum and minimum of E-field, respectively.

By considering these configurations, and based on the theoretical aspects detailed in the previous sections, the 1st harmonic field distribution enables a better separation between effects of the sample upon the sensor mutual and coupled capacitances. Indeed, in the even mode configuration, for the fundamental mode (Fig. 20 top right), the two tips of a single resonator are at opposite voltages, inducing a strong electrical field between them, and this E-field strongly flows through the sample because of the value of the real part of the permittivity with respect to that of the substrate. The resonator capacitances are thus strongly impacted by the sample properties in the fundamental case whereas in the 1st harmonic configuration, for the even mode configuration (Fig. 20 top left), the electrical field is mainly located between the microstrip and the ground plane, which means that the resonator capacitance is weakly affected by the sample properties.

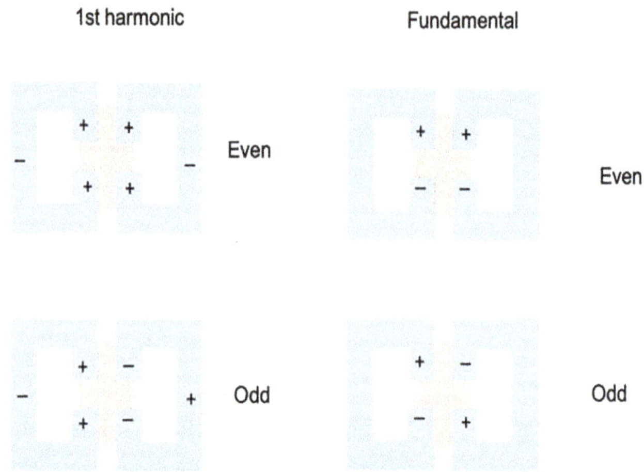

Fig. 20 Representation of voltage distribution along resonators for fundamental and 1st harmonic, even and odd modes

On the other hand, the odd modes exhibit different electrical field distributions between the fundamental (Fig. 20 bottom right) and the 1st harmonic (Fig. 20 bottom left) modes but they are both exhibiting strong densities in the sample, which means high dependence on the sample properties. To put it in a nutshell, 1st harmonic mode field distribution shows lower dependence of the resonator capacitance with respect to the sample properties compared to the fundamental mode. Given that mutual capacitances have the same order of magnitude in both configurations, the 1st harmonic mode appears as more suitable. The sensor was thus developed so as to work around 7–8 GHz, on the 1st harmonic, when loaded with the sample. Bearing all these aspects in mind, the layout of the proposed sensor design is shown in Fig. 21.

The sensor was implemented on a 0.8 mm-thick RT Duroid 5880 substrate with a relative dielectric permittivity of 2.2 and a loss tangent of 0.0009. Following the conclusions in Sect. 4.1, and with a conductivity measurement aim, a bottomless sample holder was made thanks to a 3-D printer with RS-F2-GPWH-04 resin from FormLabs®. Its expected relative dielectric permittivity is 3.55 and loss tangent is 0.01. This piece was stuck to the device by mean of an epoxy adhesive. The

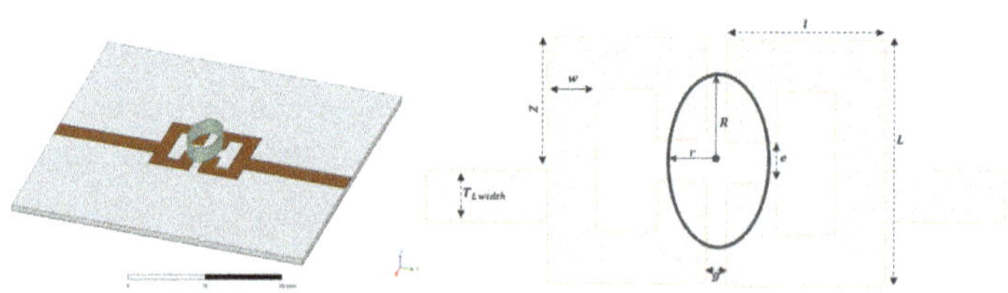

Fig. 21 D model (left) and layout (right) of the designed sensor. Reprinted from [46] under a Creative Commons Attribution 4.0 International License

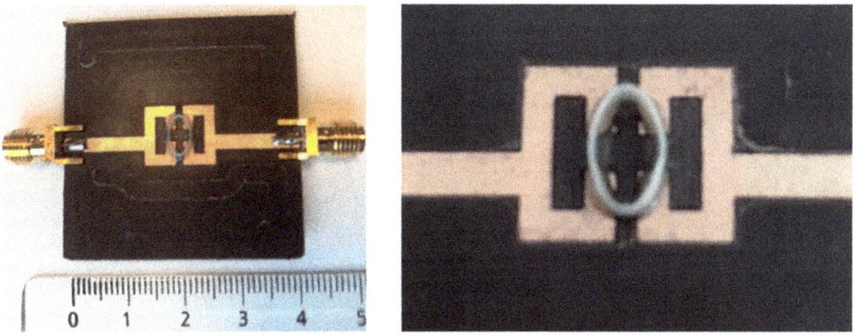

Fig. 22 Fabricated sensor (with a centimetric scale) and zoomed-in view of the sensing area. Reprinted from [46] under a Creative Commons Attribution 4.0 International License

dimensions of the sensor, according to Fig. 21, are $L = 10$ mm, $l = 5.4$ mm, $w = 1.8$ mm, $g = 1.2$ mm, $T_{Lwidth} = 2.474$ mm, $Z = 5$ mm and $e = 2$ mm. Bottomless sample holder dimensions are $R = 3.4$ mm and $r = 2.04$ mm, which correspond to an estimated sample volume of 43.56 μL. The device was fabricated, and a picture of the manufactured sensor is presented in Fig. 22.

Once the sensor was implemented, measurements with aqueous sucrose–sodium chloride solutions were made. The recorded responses of the sensor are plotted in Fig. 23. The labels associated to each curve are referring to the concentrations detailed in Table 5 (Sect. 1.2.3.ii). As explained in the previous parts, several parameters could be chosen to extract relevant information from the sensor electrical response. It is worthwhile at this point to mention that although electrical responses are shown over a wide frequency band, the sensor was designed to allow for the extraction of the sought concentrations thanks to narrow bands analysis only.

Given that this sensor is purposed to extract concentrations of two compounds, one impacting conductivity and the other one related to dipolar relaxation, two narrow bands were identified, low frequencies on one hand and around resonances of the coupled resonators on the other hand. One of the chosen indicators was the magnitude of the reflected *S*-parameter (S_{11}) (in natural value) at the frequency of 100 MHz: $|S_{11}(100 \text{ MHz})|$. Around the resonances, because the two peaks are not clearly identifiable, it was not possible to extract the coupling coefficient, consequently, Sensing Ratio (*SR*) computed as follows was chosen:

$$SR = \frac{\left(\frac{f_c}{BW}\right)}{(1 - |S_{21}(f_c)|)} \tag{44}$$

where BW is the bandwidth computed from 1 dB fall from S_{21} maximum (instead of the commonly-used 3 dB fall, which cannot be computed for high sucrose concentrations, due to losses), f_c is the central resonance frequency and $|S_{21}(f_c)|$ is the magnitude (in linear scale) of the transmission *S*-parameter at the central frequency.

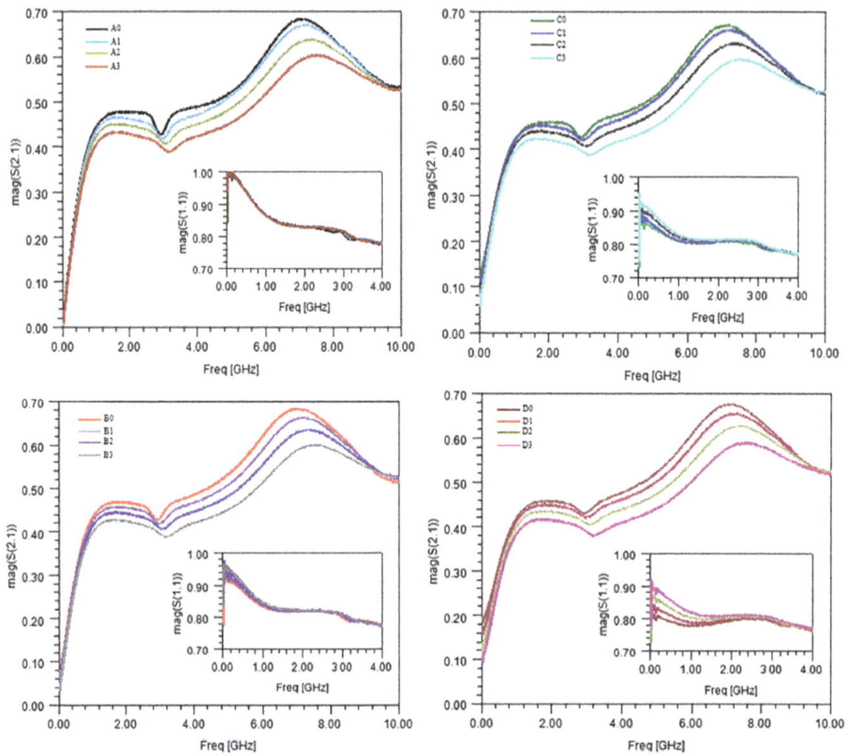

Fig. 23 Measured transmitted (S_{21}) and reflected (S_{11}) parameters of the sensor with various samples. Reprinted from [46] under a Creative Commons Attribution 4.0 International License

It is then possible to retrieve concentrations by solving the inverse problem from measured values of these parameters and the sensitivity law. There are several ways to obtain the equations relating sensing parameters together with the measured features ($|S_{11}(100\,\text{MHz})|$ and *SR*). Indeed, these equations can be extracted from full-wave simulations integrating the above presented Cole–Cole model for the samples or they can be directly extracted from measurements if it is possible to "calibrate" the sensor electrical response with enough reference solutions. Choice of the way to obtain the sensitivity law constituted by these equations relating concentrations with the sensing parameters mainly depends on the possibility to calibrate the sensor electrical response.

In the present case, the extraction was made thanks to equations derived from measurements. These equations are given hereafter:

$$|S_{11}(100\ \text{MHz})| = |S_{11}(100\ \text{MHz})|_{\text{H}_2\text{O}} - \left[a_i C_{\text{NaCl}}^2 + b_i C_{\text{NaCl}}\right] \tag{45a}$$

with:

$$a_i = 2.29 \times C_{\text{C}_{12}\text{H}_{22}\text{O}_{11}} - 2.94 \tag{45b}$$

and:

$$b_i = -0.93 \times C_{\text{C}_{12}\text{H}_{22}\text{O}_{11}} + 1.65 \tag{45c}$$

and:

$$SR_{Samples} = SR_{DIW} - (-3.71 \times C_{\text{NaCl}} \times C_{\text{C}_{12}\text{H}_{22}\text{O}_{11}} + 3.65 \times C_{\text{C}_{12}\text{H}_{22}\text{O}_{11}} + 5.08 \times C_{\text{NaCl}}) \tag{46}$$

with sodium chloride concentration (C_{NaCl}) and sucrose concentration $\left(C_{\text{C}_{12}\text{H}_{22}\text{O}_{11}}\right)$ expressed in mol/L. A summary of the extracted values for all the measurements is presented in Table 7. Relative differences for each compound concentration together with the quadratic error were also computed and reported in the same table. The quadratic error is computed thanks to the formula:

$$\sigma_{quadratic} = \sqrt{(\Delta C_{\text{NaCl}})^2 + \left(\Delta C_{\text{C}_{12}\text{H}_{22}\text{O}_{11}}\right)^2} \tag{47}$$

From Table 7, it can be seen that relative differences on concentration are reaching 5.1% at maximum when they are considered individually and quadratic error, representing the error over the two sought parameters, is lower than 5.5%.

This second application example demonstrates that it is possible to effectively monitor several compounds simultaneously with a single coupled-resonator sensor. In addition to the great sensitivities reached with this topology, additional extractions are thus possible thanks to the great flexibility offered by this kind of sensors. The exploitation of all the benefits brought by these sensors calls for optimism when trying to push the limits of sensing science in the near future.

Table 7 Summary of extracted concentrations using equations derived from measured results ((45) and (46))

Sample	Sodium chloride			Sucrose			Quadratic error (%)
	Real concentration (mmol/L)	Extracted concentration (mmol/L)	Relative difference (%)	Real concentration (mmol/L)	Extracted concentration (mmol/L)	Relative difference (%)	
A0	0	0	0	0	0	0	0
B0	51	53	+3.9	0	0	0	3.9
C0	103	103	0	0	11	N/A	0
D0	154	156	+1.3	0	6	N/A	1.3
A1	0	0.0	0	293	280	−4.4	4.4
B1	51	52	+2.0	294	279	−5.1	5.4
C1	104	100	−3.8	293	284	−3.1	4.9
D1	154	154	0	292	287	−1.7	1.7
A2	0	0.0	0	585	588	+0.5	0.5
B2	51	53	+3.9	585	591	+1.0	4.0
C2	103	101	−1.9	586	585	−2.0	2.8
D2	154	153	−0.6	585	580	−0.9	1.1
A3	0	0.0	0	879	874	−0.6	0.6
B3	51	52	+1.9	877	878	+0.1	1.9
C3	106	107	+0.9	877	883	+0.7	1.1
D3	154	156	+1.3	877	882	+0.6	1.4

References

1. Raicu V, Feldman Y (2015) Dielectric relaxation in biological systems: physical principles, methods and applications. Oxford Univ. Press, Oxford, U.K.
2. Zarifi MH, Daneshmand M (2016) Wide dynamic range planar coupled ring resonator for sensing applications. Appl Phys Lett 108, Art no 232906
3. Herrera Sepulveda LV, Olvera Cervantes JL, Saavedra CE (2021) Multifrequency coupled-resonator sensor for dielectric characterization of liquids. IEEE Trans Instrum Meas 70, Art no 8005507
4. Böttcher CJF, Bordewijk P (1992) Dielectrics in time-dependent fields. Elsevier, Amsterdam, The Netherlands
5. Tura A, Sbrignadello S, Barison S, Conti C, Pacini G (2007) Impedance spectroscopy of solutions at physiological glucose concentrations. Biophys Chem 129(2–3):235–241
6. Olarte O, Barbé K, Van Moer W, Van Ingelgem Y, Hubin A (2014) Measurement and characterization of glucose in NaCl aqueous solutions by electrochemical impedance spectroscopy. Biomed Signal Process Control 14:9–18
7. Yoon G (2011) Dielectric properties of glucose in bulk aqueous solutions: influence of electrode polarization and modeling. Biosens Bioelectron 26(5):2347–2353
8. Peyman A, Gabriel C, Grant EH (2007) Complex permittivity of sodium chloride solutions at microwave frequencies. Bioelectromagnetics 28(4):264–274
9. Kaatze U (2010) Techniques for measuring the microwave dielectric proper- ties of materials. Metrologia 47(2):S91–S113
10. Mohammadi T, Kaviani A (2003) Water shortage and seawater desalination by electrodialysis. Desalination 158:267–270
11. Coelho R (1979) Physics of dielectrics. Elsevier, Amsterdam, The Netherlands
12. Sharma M, Resta R, Car R (2007) Dipolar correlations and the dielectric permittivity of water. Phys Rev Lett 98, Art no 247401
13. Kirkwood JG (1939) The dielectric polarization of polar liquids. J Chem Phys 7(10):911–919
14. Grant EH (1957) Relationship between relaxation time and viscosity for water. J Chem Phys 26(6):1575–1577
15. Levy E, Puzenko A, Kaatze U, Ben Ishai P, Feldman Y (2012) Dielectric spectra broadening as the signature of dipole-matrix interaction. I. Water in nonionic solutions. J Chem Phys 136(11), Art no 114502
16. Rötger H (1974) Relaxation times and viscosity. J Non-Cryst Solids 14(1):201–217
17. Barón M, Buep AH (1997) Empirical relation between permittivity (ε) and viscosity (η) of polar and non-polar liquids. Speculations Sci Technol 20:45–50
18. Swindells JF, Snyder CF, Hardy RC, Golden PE (1958) Viscosities of sucrose solutions at various temperatures: tables of recalculated values. United States Department of Commerce, National Bureau of Standards, Supplement to National Bureau of Standards Circular 440, Washington D. C., U. S. A.
19. Fröhlich H (1958) Theory of dielectrics: dielectric constant and dielectric loss. Clarendon Press, Oxford, U. K.
20. Hill NE, Yaughan WE, Price A, Davis M (1969) Dielectric properties and molecular behaviour. Van Nostrand Reinhold, London, U. K.
21. Landau LD, Lifshitz EM, Pitaevskii LP (1984) Electrodynamics of continuous media, 2nd edn. Pergamon Press, London, U. K.
22. Debye P (1929) Polar molecules. The Chemical Catalog Company, Inc., New York, NY, U. S. A.
23. Cole KS, Cole RH (1941) Dispersion and absorption in dielectrics I. Alternating current characteristics. J Chem Phys 9(2):341–451
24. Wyman J (1930) Measurements of the dielectric constants of conducting media. Phys Rev 35:623–634
25. Kaatze U (2007) Reference liquids for the calibration of dielectric sensors and measurement instruments. Meas Sci Technol 18(4):967–976

26. Ellison WJ (2007) Permittivity of pure water, at standard atmospheric pressure, over the frequency range 0–25 THz and temperature range 0–100 °C. J Phys Chem Ref Data 36(1):1–18
27. Picálek J, Kolafa J (2007) Molecular dynamics study of conductivity of ionic liquids: The Kohlrausch law. J Mol Liq 134:29–33
28. Marsland TP, Evans S (1987) Dielectric measurements with an open-ended coaxial probe. Proc IEEE 134:341–349
29. Blackham DV, Pollard RD (1997) An improved technique for permittivity measurements using a coaxial probe. IEEE Trans Instrum Meas 46(5):1093–1099
30. Saito A, Miyawaki O, Nakamura K (1997) Dielectric relaxation of aqueous solution with low-molecular-weight nonelectrolytes and its relationship with solution structure. Biosci Biotechnol Biochem 61(11):1831–1835
31. Fuchs K, Kaatze U (2001) Molecular dynamics of carbohydrate aqueous solutions. Dielectric relaxation as a function of glucose and fructose concentration. J Phys Chem B 105(10):2036–2042
32. Shiraga K, Suzuki T, Kondo N, Tajima T, Nakamura M, Togo H, Hirata A, Ajito K, Ogawa Y (2015) Broadband dielectric spectroscopy of glucose aqueous solution: analysis of the hydration state and the hydrogen bond network. J Chem Phys 142(23):234504
33. Cao R, Komura F, Nonaka A, Kato T, Fukumashi J, Matsui T (2014) Quantitative analysis of D-(+)-glucose in fruit juices using diffusion ordered-1H nuclear magnetic resonance spectroscopy. Anal Sci 30:383–388
34. Galant AL, Kaufman RC, Wilson JD (2015) Glucose: detection and analysis. Food Chem 188:149–160
35. Shokrekhodaei M, Quinones S (2020) Review of non-invasive glucose sensing techniques: optical, electrical and breath acetone. Sensors 20(5), Art no 1251
36. Carr JH (2021) Clinical hematology atlas, 6th edn. Elsevier
37. Thennadil SN, Rennert JL, Wenzel BJ, Hazen KH, Ruchti TL, Block MB (2001) Comparison of glucose concentration in interstitial fluid, and capillary and venous blood during rapid changes in blood glucose levels. Diabetes Technol Ther 3(3):357–365
38. Harman-Boehm I, Gal A, Raykhman AM, Zahn JD, Naidis E, Mayzel Y (2009) Noninvasive glucose monitoring: a novel approach. J Diabetes Sci Tech 3(2):253–260
39. Kottmann J, Rey JM, Luginbühl J, Reichmann E, Sigrist MW (2012) Glucose sensing in human epidermis using mid-infrared photoacoustic detection. Biomed Opt Expr 3(4):667–680
40. Potelon B, Quendo C, Carré J-L, Chevalier A, Person C, Queffelec P (2014) Electromagnetic signature of glucose in aqueous solutions and human blood. In: Proceedings MEMSWAVE conference, La Rochelle, France, pp 4–7
41. Choi H, Luzio S, Porch A (2016) Dielectric properties of aqueous glucose solutions using microwave cavity and coaxial probe. In: Proceedings of the IEEE 2nd Australian microwave symposium (AMS), Adelaide, SA, Australia, pp 7–8
42. Lin T, Gu S, Lasri T (2017) Highly sensitive characterization of glucose aqueous solution with low concentration: application to broadband dielectric spectroscopy. Sens Actuator A Phys 267:318–326
43. Turgul V, Kale I (2018) Permittivity extraction of glucose solutions through artificial neural networks and non-invasive microwave glucose sensing. Sens Actuator A Phys 277:65–72
44. Zhang YJ, Yong JL, Tian Y, He Y (2023) High sensitivity detection method for liquid concentrations based on coupled microwave resonators. IEEE Trans Instrum Meas 72, Art no 8003011
45. Queneau Y, Jarosz S, Lewandowski B, Fitremann J (2007) Sucrose chemistry and applications of sucrochemicals. In: Horton D (ed) Advances in carbohydrate chemistry and biochemistry, vol 61. Elsevier, pp 217–292
46. Bakam Nguenouho OS, Chevalier A, Potelon B, Benedicto J, Quendo C (2022) Dielectric characterization and modelling of aqueous solutions involving sodium chloride and sucrose and application to the design of a bi-parameter RF-sensor. Sci Rep 12, Art no 7209
47. Engelsen SB, Pérez S (1996) The hydration of sucrose. Carbohydr Res 292:21–38

48. Malmberg CG, Maryott AA (1956) Dielectric constant of water from 0 to 100 C. J Res Natl Bur Stand 56(1):1–8
49. Chenlo F, Moreira R, Pereira G, Ampudia A (2002) Viscosities of aqueous solutions of sucrose and sodium chloride of interest in osmotic dehydration processes. J Food Eng 54:347–352
50. Bobowski JS, Johnson T (2012) Permittivity measurements of biological samples by an open-ended coaxial line. Prog Electromagn Res B 40:159–183
51. Chen T, Hefter G, Buchner R (2003) Dielectric spectroscopy of aqueous solutions of KCl and CsCl. J Phys Chem A 107(20):4025–4031
52. Levy E, Puzenko A, Kaatze U, Ben Ishai P, Feldman Y (2012) Dielectric spectra broadening as the signature of dipole-matrix interaction. II. Water in ionic solutions. J Chem Phys 136(11), Art no 114503
53. Herbert AW, Jackson CP, Lever DA (1988) Coupled groundwater flow and solute transport with fluid density strongly dependent upon concentration. Water Resour Res 24(10):1781–1795
54. Casteel JF, Amis ES (1972) Specific conductance of concentrated solutions of magnesium salts in water-ethanol system. J Chem Eng Data 17(1):55–59
55. Arrhenius S (1889) Über die Reaktionsgeschwindigkeit bei der Inversion von Rohrzucker durch Säuren. Z Physik Chem 4:226–248
56. Juan CG, Bronchalo E, Potelon B, Quendo C, Sabater-Navarro JM (2019) Glucose concentration measurement in human blood plasma solutions with microwave sensors. Sensors 19(17), Art no 3779
57. Juan CG, Bronchalo E, Potelon B, Quendo C, Muñoz VF, Ferrández-Vicente JM, Sabater-Navarro JM (2023) On the selectivity of planar microwave glucose sensors with multicomponent solutions. Electronics 12(1), Art no 191
58. Gabriel S, Lau RW, Gabriel C (1996) The dielectric properties of biological tissues: II. Measurements in the frequency range 10 Hz to 20 GHz. Phys Med Biol 41(11):2251–2269
59. Gabriel S, Lau RW, Gabriel C (1996) The dielectric properties of biological tissues: III. Parametric models for the dielectric spectrum of tissues. Phys Med Biol 41(11):2271–2293
60. Vélez P, Paredes F, Casacuberta P, Elgeziry M, Su L, Muñoz-Enano J, Costa F, Genovesi S, Martín F (2023) Portable reflective-mode phase-variation microwave sensor based on a rat-race coupler pair and gain/phase detector for dielectric characterization. IEEE Sens J 23(6):5745–5756
61. García H, Juan CG, Ávila-Navarro E, Bronchalo E, Sabater-Navarro JM (2019) Portable device based on microwave resonator for noninvasive blood glucose monitoring. In: Proceedings of the 41st annual international conference of the IEEE engineering in medicine and biology society (EMBC), Berlin, Germany, pp 1115–1118
62. Govind G, Akhtar J (2020) Design of an ELC resonator-based reusable RF microfluidic sensor for blood glucose estimation. Sci Rep 10, Art no 18842
63. Casacuberta P, Vélez P, Muñoz-Enano J, Su L, Martín F (2023) Highly sensitive reflective-mode phase-variation permittivity sensors using coupled line sections. IEEE Trans Microw Theory Technol, Early Access 1–15
64. Zidane MA, Rouane A, Hamouda C, Amar H (2021) Hyper-sensitive microwave sensor based on split ring resonator (SRR) for glucose measurement in water. Sens Actuator A Phys 321, Art no 112601
65. Juan CG, Potelon B, Quendo C, Bronchalo E (2021) Microwave planar resonant solutions for glucose concentration sensing: a systematic review. Appl Sci 11(15), Art no 7018
66. Mayani MG, Herraiz-Martínez FJ, Matanza Domingo J, Giannetti R (2021) Resonator-based microwave metamaterial sensors for instrumentation: Survey, classification, and performance comparison. IEEE Trans Instrum Meas 70, Art no 9503414
67. Casacuberta P, Vélez P, Muñoz-Enano J, Su L, Gil M, Ebrahimi A, Martín F (2022) Circuit analysis of a coplanar waveguide (CPW) terminated with a step-impedance resonator (SIR) for highly sensitive one-port permittivity sensing. IEEE Access 10:62597–62612
68. Juan CG, Bronchalo E, Potelon B, Álvarez-Pastor J, Sabater-Navarro JM (2020) Use of coplanar quarter-wave resonators for glucose sensing in aqueous solutions. In: Proceedings of the international microwave biomedical conference (IMBioC), Toulouse, France

69. Kiani S, Rezaei P, Fakhr M (2021) Dual-frequency microwave resonant sensor to detect noninvasive glucose-level changes through the fingertip. IEEE Trans Instrum Meas 70, Art no 6004608
70. A. Aquino, C. G. Juan, B. Potelon, and C. Quendo, “Dielectric permittivity sensor based on planar open-loop resonator,” *IEEE Sens. Lett.*, vol. 5, no. 3, Art. no. 3500204, 2021.
71. Turgul V, Kale I (2018) Sensitivity of non-invasive RF/microwave glucose sensors and fundamental factors and challenges affecting measurement accuracy. In: Proceedings of the IEEE international instrumentation and measurement technology conference (I2MTC), Houston, TX, U. S. A.
72. Choi H, Luzio S, Beutler J, Porch A (2017) Microwave noninvasive blood glucose monitoring sensor: human clinical trial results. In: Proceedings of the IEEE MTT-S international microwave symposium digest, (IMS), Honololu, HI, U. S. A., pp 876–879
73. Juan CG, García H, Ávila-Navarro E, Bronchalo E, Galiano V, Moreno O, Orozco D, Sabater-Navarro JM (2019) Feasibility study of portable microwave microstrip open-loop resonator for noninvasive blood glucose level sensing: proof of concept. Med Biol Eng Comput 57(11):2389–2405
74. Srour M, Potelon B, Quendo C, Person C, Carré J-L (2020) Biosensor based on a resonant technique for aqueous glucose monitoring using standardized medical test tubes. In: Proceedings of the international microwave biomedical conference (IMBioC), Toulouse, France
75. Pinon S, Diedhiou DL, Boukabache A, Conedera V, Bourrier D, Gue A-M, Prigent G, Rius E, Quendo C, Potelon B, Favennec J-F (2012) Fabrication and characterization of a fully integrated biosensor associating microfluidic device and RF circuit. In: Proceedings of the IEEE MTT-S international microwave symposium digest, (IMS), Montréal, QC, Canada
76. Abduljabar AA, Rowe DJ, Porch A, Barrow DA (2014) Novel microwave microfluidic sensor using a microstrip split-ring resonator. IEEE Trans Microw Theory Technol 62(3):679–688
77. Rowe DJ, al-Malki S, Abduljabar AA, Porch A, Barrow DA, Allender CJ (2014) Improved split-ring resonator for microfluidic sensing. IEEE Trans Microw Theory Technol 62(3):689–699
78. Chretiennot T, Dubuc D, Grenier K (2016) Microwave-based microfluidic sensor for non-destructive and quantitative glucose monitoring in aqueous solution. Sensors 16, Art no 1733
79. Masoumabad MA, Delmonte N, Silvestri L, Perregrini L, Bozzi M (2023) A simple technique for liquid–liquid percentage determination using single-frequency amplitude measurements. IEEE Microw Wirel Technol Lett 33(7):1086–1089. https://doi.org/10.1109/LMWT.2023.3261984
80. Vélez P, Canalias X, Muñoz-Enano J, Casacuberta P, Su L, Martín F (2023) Effects of losses on the sensitivity of reflective-mode phase-variation liquid sensors. IEEE Trans Microw Theory Technol. https://doi.org/10.1109/TMTT.2023.3308586
81. Muñoz-Enano J, Vélez P, Gil M, Martín F (2020) Microfluidic reflective-mode differential sensor based on Open Split Ring Resonators (OSRRs). Int J Microw Wirel Technol 12:588–597
82. Shaji M, Akhtar MJ (2013) Microwave coplanar sensor system for detecting contamination in food products. In: Proceedings of the IEEE MTT-S international microwave and RF conference, New Delhi, India
83. Raj A, Jha AK, Ansari MAH, Akhtar MJ, Panda S (2016) Metamaterial-inspired microwave sensor for measurement of complex permittivity of materials. Microw Opt Technol Lett 58(11):2577–2581
84. Juan CG, Bronchalo E, Potelon B, Quendo C, Ávila-Navarro E, Sabater-Navarro JM (2019) Concentration measurement of microliter-volume water–glucose solutions using Q factor of microwave sensors. IEEE Trans Instrum Meas 68(7):2621–2634
85. Islam MR, Islam MT, Salaheldeen M, Bais B, Almalki SHA, Alsaif H, Islam MS (2022) Metamaterial sensor based on rectangular enclosed adjacent triple circle split ring resonator with good quality factor for microwave sensing application. Sci Rep 12, Art. no. 6792
86. Chen H, Guo L, Zhu W, Li C (2022) Recent advances in multi-material 3D printing of functional ceramic devices. Polymers 14(21), Art. no. 4635

87. Juan CG (2021) Designing microwave sensors for glucose concentration detection in aqueous and biological solutions: towards non-invasive glucose sensing. Springer, Cham, Switzerland
88. Turgul V, Kale I (2017) Simulating the effects of skin thickness and fingerprints to highlight problems with non-invasive RF blood glucose sensing from fingertips. IEEE Sens J 17(22):7553–7560
89. Martín F, Vélez P, Muñoz-Enano J, Su L (2023) Planar microwave sensors. Wiley-IEEE Press, Hoboken, NJ, U.S.A.
90. Juan CG, Potelon B, Quendo C, Bronchalo E, Sabater-Navarro JM (2019) Highly-sensitive glucose concentration sensor exploiting inter-resonators couplings. In: Proceedings of the 49th European microwave conference (EuMC), Paris, France, pp 662–665
91. Herrera-Sepulveda L-V, Olvera-Cervantes J-L, Corona-Chavez A, Kaur T (2020) New methodology to determine the loss tangent of dielectric planar samples by using electrically coupled resonators. J Electromagn Waves Appl 34(18):2410–2418
92. Pozar DM (2021) Microwave engineering, 4th edn. Wiley
93. Hong J-S (2011) Coupled resonator circuits. In: Hong J-S (ed) Microstrip filters for RF/microwave applications, 2nd edn. Wiley, pp 235–272
94. Kaatze U (1989) Complex permittivity of water as a function of frequency and temperature. J Chem Eng Data 34(4):371–374
95. Juan CG, Potelon B, Quendo C, García-Martínez H, Ávila-Navarro E, Bronchalo E, Sabater-Navarro JM (2021) Study of Q_u-based resonant microwave sensors and design of 3-D-printed devices dedicated to glucose monitoring. IEEE Trans Instrum Meas 70, Art no 8005716
96. Bray JR, Roy L (2004) Measuring the unloaded, loaded, and external quality factors of one- and two-port resonators using scattering-parameter magnitudes at fractional power levels. IEE Proc Microw Antennas Propag 151(4):345–350
97. Nguenouho OSB, Chevalier A, Potelon B, Quendo C, Benedicto J, Simon B (2023) Electromagnetic characterization of aqueous sodium lactate solutions and its application for enhanced concentration monitoring. IEEE Trans Instrum Meas 72:1–8. https://doi.org/10.1109/TIM.2023.3312490

Sensing Using Magnetoinductive Waves

Daffodil Dhayaa, Callum Long, Anna Radkovskaya, Georgiana Dima, Jiaruo Yan, Eleanor O'Hara, Laszlo Solymar, and Ekaterina Shamonina

Abstract The propagation of slow magnetoinductive (MI) waves by virtue of magnetic coupling between resonant elements (called meta-atoms in the terminology of MI waves) enables applications ranging from improved medical imaging to wireless power transfer. A recent development is the use of MI waves in contactless sensing of conductive environments, both in the frequency and in the time domain. The underlying physics in both methods is that a perturbation due to the presence of a conductive object in the vicinity of a meta-atom changes resonant characteristics of the meta-atom and causes reflections of MI waves within the array. In this chapter we focus on time-domain reflectometry, demonstrating accurate defect localisation and characterisation. We discuss potential applications for real-time contactless monitoring of inhomogeneous conductive environments ranging from in-situ 3D printing quality monitoring to medical imaging.

1 Introduction

What are Metamaterials? Two possible definitions [1] would give a good idea.

1. Metamaterials are engineered composites that exhibit superior properties not found in nature and not observed in the constituent materials.
2. A Metamaterial is an artificial material in which the electromagnetic properties as represented by the permittivity and permeability can be controlled. The means of achieving these aims is by subwavelength structuring.

Neither of these definitions is general enough. There are many other definitions in common use but the two above will be sufficient for our purpose. In the past two decades, a vast variety of electromagnetic and acoustic metamaterials has been proposed [1–7]. Subwavelength structuring, i.e. assembling a metamaterial using

D. Dhayaa · C. Long · A. Radkovskaya · G. Dima · J. Yan · E. O'Hara · L. Solymar · E. Shamonina (✉)
Department of Engineering Science, University of Oxford, Parks Road, Oxford OX1 3PJ, UK
e-mail: ekaterina.shamonina@eng.ox.ac.uk

F. Martın and E. Bronchalo (eds.), *Coupled Structures for Microwave Sensing*, Lecture Notes in Electrical Engineering 1150,
https://doi.org/10.1007/978-3-031-53861-2_10

unit cells, called 'meta-atoms' smaller than the wavelength of the waves allowed for such unusual physical phenomena as negative refraction [8–10], inverse Doppler shift [11], inverse Goos-Hanchen effect [12–14], or growing evanescent waves [15, 16]. Unusual physics led to novel applications, such as subwavelength imaging, cloaking [17–19], wireless power transfer and superdirective antennas to mention a few.

A particular area of metamaterials research is on tightly packed metamaterials, with individual meta-atoms coupled magnetically. It was shown [20, 21] that a new type of wave can propagate on such structures, opening up a possibility to guide electromagnetic signals on the subwavelength scale. The mechanism of these waves can be understood as follows. An alternating current excited in a single meta-atom via an ac voltage creates an ac magnetic field, which threads a neighbouring meta-atom, exciting an ac current in it via the mutual inductance M. This, in turn, will excite its neighbour and so on. It may be seen that this is a kind of chain reaction that embraces element after element having all the properties of a wave which we called [20] a magnetoinductive (MI) wave and that term has since been generally accepted. A magnetoinductive waveguide (MIW) can then be defined as a one-dimensional structure/array upon which MI waves can propagate. Propagation in two and three dimensions has also been discussed [21]. All these waves have been described theoretically and observed experimentally in metamaterial structures operating from low MHz frequencies to the IR and visible frequency range [1].

The passband, i.e. the frequency spectrum in which they can propagate, is determined by the dispersion of the MI waves. MI waves belong to the family of slow waves meaning that they are slow relative to the velocity of light and their wavelength is smaller than the wavelength of an electromagnetic wave of the same frequency. In this sense they are analogous to surface plasmons and acoustic waves. Note that there is another kind of analogy with acoustic waves as well: both waves manifest themselves only in discrete locations [22, 23].

Like other types of eigenmodes of a medium (e. g. plasmons, phonons etc.) MI waves can couple to electromagnetic waves forming magnetoinductive polaritons [24, 25] and should generally be taken into account when calculating effective medium parameters e. g. permeability [24–27].

A unique set of properties of MI waves open possibilities for potential applications [1, 28, 29]. Let us enumerate these briefly.

(i) MI waveguides with well-defined and controllable properties are easy to fabricate. One can use printed circuit board (PCB) technology [30], use printing on a flexible substrate [31, 32] or 3D printing [33].

(ii) The coupling strength between individual meta-atoms and hence the dispersion properties of MI waves are easily controlled via suitable choice of the geometry of individual resonators and their mutual position [23, 34] enabling design of MIWs with desired characteristics. The dispersion in diatomic metamaterials (i.e. with two meta-atoms per unit cell) feature dispersion characteristics with two branches [22, 23] which enables such novel phenomena as unidirectional waveguiding due to near-field interference [35], parametric amplification [36] and deep subwavelength imaging [37–39].

(iii) Individual meta-atoms are small relative to the wavelength of the electromagnetic wave and there is no continuous metallic path connecting the meta-atoms, which makes MIWs suitable for diagnostic applications using Magnetic Resonance Imaging (MRI) as they are MRI-safe [31, 32]. Another important direction is wireless data [40, 41] and power transfer [40, 42, 43].
(iv) The slow-wave nature of MI waves [21] enabled development of novel delay lines [44], phase shifters [45] and superdirective antennas [46–48].
(v) MI waves can propagate in a lossy environment enabling applications in underground and underwater communications as well as sensing in in-vivo medical diagnostics [31, 32, 49, 50].

A precursor for the recent upsurge of interest in employing MI waves for sensing in conductive environments is the sensitivity of a single ring probe to its environment due to eddy currents excited by the probe [51–54]. A variety of non-destructive testing (NDT) applications have been proposed using a single ring resonator probe, for conductivity and thickness measurements [55–59] and for detection of cracks, defects and foreign objects [60–70].

Wave reflectometry is a well established method used with other types of waves, from electromagnetic waves in transmission line testing and ultrasound waves in medical imaging to seismic waves in geological studies. When a wave is incident upon a discontinuity, the wave will be reflected; by detecting the reflected wave, one can retrieve information on the location and on the nature of the discontinuity. In the context of MI waves, the idea of contactless sensing the presence of a metallic object using an array of magnetically coupled resonators is based on the following simple idea: if a metallic object is placed next to the meta-atom, the resonant characteristics of the meta-atom will change due to interaction with the metallic object. Hence the presence of a conductive object would create a 'defect' in an otherwise uniform metamaterial array. MI waves incident upon the defect will experience reflections opening up a possibility to locate and to characterise the metallic object in the vicinity of the array. Contactless distributed sensors based on MIWs were proposed for structural health and sensing corrosion [71–73], contactless resonant imaging of conductive environments [30, 74–78] and RFID tag location [79].

While reflection properties of MI waves by a defect is the common key to sensing using MI waveguides, there are two different approaches, frequency-domain reflectometry (FDR) [30] and time-domain reflectometry (TDR) [73–78].

In FDR monochromatic MI waves are injected and the resulting standing wave patterns caused by reflection are analysed to locate the defect [30] whereas in TDR a MI pulse comprising many frequency components is injected and the arrival time and the magnitude of the reflected pulse are used for defect localisation [73–76, 78] and characterisation [76–78].

Trivedi and Chen et al. [73–75] were first to develop the methodology for MI TDR using an open-circuit resonator (essentially mimicking an 'ideal' defect with an infinite impedance) to create a mismatch impedance boundary of which the MI wave would reflect. Capturing these reflections and combining this information with group velocity analysis allowed them to accurately estimate the location of defects.

More recently we were able to advance the accuracy of single-defect localisation, allowing weaker defects to be located as well as characterised [76–78]. The aim of this chapter is to provide a detailed description of our method, present the results obtained, discuss the accuracy of the method and point out potential applications to enable real-time contactless monitoring of inhomogeneous conductive environments.

This chapter is organised as follows. Section 2 provides pre-requisites needed for this work. It follows a classical approach with a brief introduction of resonant properties of meta-atoms and coupling mechanisms, followed by a standard treatment of MI waves in the frequency domain. Defects caused by a metallic object in the proximity of a meta-atom are described and a frequency-domain reflection coefficient is derived. In Sect. 3 the TDR of MI waves is developed as a means for localisation of metallic objects by analysing the time of flight of the echoes caused by the defects. In Sect. 4 the TDR of MI waves is expanded to incorporate also characterisation of the metallic object by quantifying the time-domain reflection coefficient of the echo signals. Section 5 rounds up this chapter with conclusions and provides a brief discussion for future research stemming from this work.

2 Fundamentals

In this section, we provide a brief overview of the key ingredients of a circuit model for metamaterials capable of propagating magnetoinductive waves, which we will later employ to enable localisation (Sect. 3) and characterisation (Sect. 4) of conductive objects. We start with properties of individual meta-atoms comprising a metamaterial and discuss briefly how inter-element coupling enables propagation of magnetoinductive waves and how a conductive object causes reflections of the waves.

2.1 Meta-Atoms

To understand the fundamental physics behind the propagation of MI waves, we must first appreciate the behaviour of individual meta-atoms. A capacitively-loaded split-ring resonator (SRR) on a PCB substrate shown schematically in Fig. 1a is a common example of a meta-atom. A SRR has a strong resonant response when exposed to ac magnetic fields—resonating at a frequency f_0. Such meta-atoms are the building blocks of metamaterials and can be arranged into 1D arrays, 2D surfaces, or 3D structures. The overall electromagnetic response of the metamaterial arises from the properties of the individual meta-atoms and their interactions [1].

A capacitively loaded SRR can be modelled as an LCR circuit, as shown in Fig. 1b, where V represents an ac voltage used to excite the meta-atom. We note that the excitation can be either direct, e. g. from a voltage source, or inductive, e. g. from

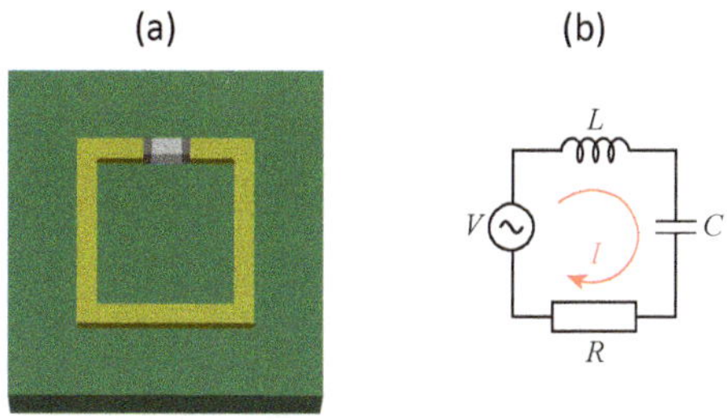

Fig. 1 A schematic of a capacitively loaded SRR meta-atom on a PCB substrate (**a**) and its equivalent LCR circuit (**b**)

a transmitter coil placed nearby. Using Ohm's law, we can compute the current in the coil, I:

$$I = \frac{V}{Z_0}, \tag{1}$$

where Z_0 denotes the self-impedance of the LCR circuit,

$$Z_0 = j\omega L + \frac{1}{j\omega C} + R = j\omega L \left(1 - \frac{\omega_0^2}{\omega^2} - \frac{j}{Q}\frac{\omega_0}{\omega}\right). \tag{2}$$

Here L is the self-inductance, C the total capacitance (the SRR self-capacitance plus the capacitance of the tuning capacitor), R the resistance, ω $(= 2\pi f)$ the angular frequency, f the frequency. ω_0 $(= 2\pi f_0)$ is the resonant frequency and Q is the quality factor defined as

$$\omega_0 = \frac{1}{\sqrt{LC}}, \quad Q = \frac{\omega_0 L}{R}. \tag{3}$$

2.2 *Coupling of Meta-Atoms*

When meta-atoms are assembled into a structure they can interact via the mutual inductance. Figure 2 provides the circuit diagram for two identical coils as in Fig. 1, carrying currents I_1 and I_2, with the first coil being driven by an external voltage V and with both coils being coupled via the mutual inductance M. Assuming the harmonic excitation of frequency ω, Kirchhoff's equations for each coil can be written in the following form:

$$\begin{aligned} V &= Z_0 I_1 + Z_M I_2 \\ 0 &= Z_0 I_2 + Z_M I_1, \end{aligned} \tag{4}$$

with

$$Z_M = j\omega M \tag{5}$$

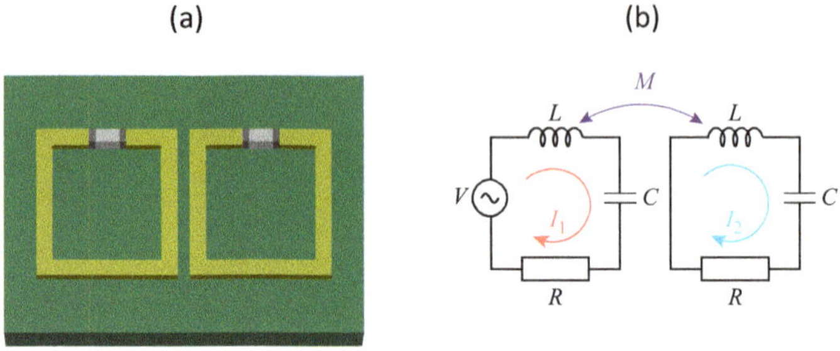

Fig. 2 A schematic (**a**) and an equivalent circuit (**b**) for two coupled meta-atoms, represented as identical LCR circuits

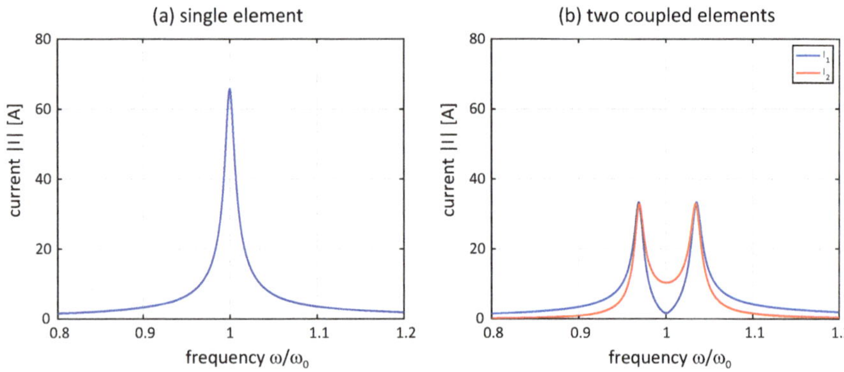

Fig. 3 Resonance curves for a single coil (**a**) and for two coupled coils (**b**). Parameters used: $f_0 = 114.76\,\text{MHz}$, quality factor $Q = 95$, coupling coefficient $\kappa = -0.13$, $L = 20\,\text{nH}$, and excitation voltage of amplitude $V = 10\,\text{V}$ applied to the first element

being the mutual impedance.

Solving Eqs. (4) we obtain explicit expressions for both currents:

$$I_1 = V \frac{Z_0}{Z_0^2 - Z_M^2}, \qquad I_2 = -V \frac{Z_M}{Z_0^2 - Z_M^2}. \tag{6}$$

As may be expected, single and coupled coils have different resonant properties. Figure 3a shows the resonance curve for a single coil with a clear single peak at the resonant frequency ω_0. Figure 3b illustrates that for two coupled coils we observe two peaks, one below ω_0 and one above ω_0. One of the modes is symmetric (with currents oscillating in phase) and the other one is antisymmetric (with currents oscillating in anti-phase) [1]. This behaviour is analogous to coupled pendulums where we would find two eigenmodes of motion, a symmetric mode and an antisymmetric mode, and two corresponding resonant frequencies.

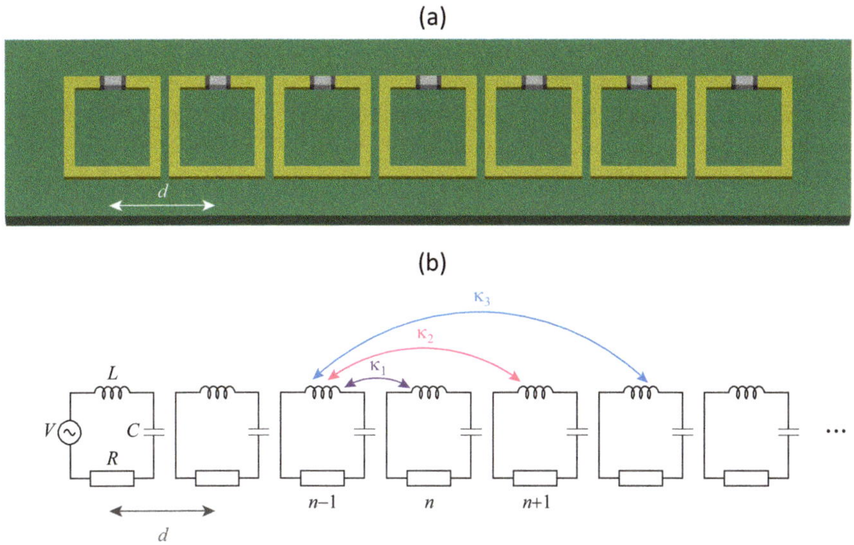

Fig. 4 A schematic (**a**) and a circuit diagram (**b**) of a MI waveguide section, represented as an array of coupled LCR circuits showing nearest-neighbour coupling κ_1 and non-nearest couplings κ_2 and κ_3

2.3 *Dispersion of MI Waves*

Let us now consider an array of identical elements spaced a distance d apart with the first element being driven by an ac voltage source V as shown in Fig. 4a. An alternating current excited in element 1 will generate a magnetic ac field that will thread element 2 and induce thereby a current in it, which, in turn, will produce a magnetic ac field that will induce a current in element 3, and so on. This coil-by-coil excitation propagates along the structure as a wave, a magnetoinductive wave. Figure 4b provides a circuit diagram of a section of the MI waveguide with non-nearest coupling taken into account. A useful parameter for our circuit model is a dimensionless coupling coefficient

$$\kappa_m = 2\frac{M_m}{L}, \tag{7}$$

where index m refers to the coupling between two elements of the array separated by $(m - 1)$ elements, so that index 1 refers to nearest-neighbour coupling. The strength of the coupling coefficients depends on the geometry of the array and the circuit parameters. In metamaterials designed for higher frequencies, the coupling coefficient is generally a complex quantity due to retardation [80–83] and can also have both a magnetic and an electric component [34, 84]. We note that in the case of low-frequency metamaterials, the subject of this chapter, the coupling coefficient is a real quantity [34, 84] and it can be positive (for axial arrays) or negative (for planar arrays, like the one shown in Fig. 4).

The next step is to consider the dispersion of MI waves. Dispersion is the relationship between wave frequency, ω, and wavenumber, k, ($k = \beta - j\alpha$, where β is the propagation coefficient and α the attenuation coefficient). Visualising the dispersion relations can be extremely insightful in understanding how an MI waveguide behaves like a bandpass filter.

Let us start with the nearest neighbour approximation (by assuming $\kappa_m = 0$ for $m > 1$ in Fig. 4) and write Kirchhoff's law for the current I_n in the nth coil as follows:

$$Z_0 I_n + Z_M (I_{n+1} + I_{n-1}) = 0\,. \tag{8}$$

Assuming a wave solution in the form

$$I_n = I_0 \mathrm{e}^{j(\omega t - kdn)}\,, \tag{9}$$

where I_0 a constant, t is the time, and d is the distance between the elements of the array, we can rewrite Eq. (8) as

$$Z_0 I_n + Z_M \left(I_n \mathrm{e}^{jkd} + I_n \mathrm{e}^{-jkd} \right) = 0 \tag{10}$$

and, rearranging further, as

$$Z_0 + 2 Z_M \cos(kd) = 0\,, \tag{11}$$

where Z_0 and Z_M have been defined in Eqs. (2) and (5), respectively. Substituting Eqs. (2) and (5) into Eq. (11) we arrive at an expression for the dispersion:

$$1 - \frac{\omega_0^2}{\omega^2} - \frac{j}{Q}\frac{\omega_0}{\omega} + \kappa \cos(kd) = 0\,, \tag{12}$$

with κ being the nearest-neighbour coupling coefficient. Extending now this analysis to non-nearest neighbour coupling the dispersion relation for an MI wave can be generalised to

$$1 - \frac{\omega_0^2}{\omega^2} - \frac{j}{Q}\frac{\omega_0}{\omega} + \sum_{m=1}^{C} \kappa_m \cos(mkd) = 0\,, \tag{13}$$

where each circuit is coupled with their nearest C neighbours.

The MI waveguide allows MI waves to propagate through the structure in a passband centred around the resonant frequency. Figure 5 provides a visualisation of the dispersion relation showing the variation for β and α with frequency ω in the nearest-neighbour case for two examples with different κ values.

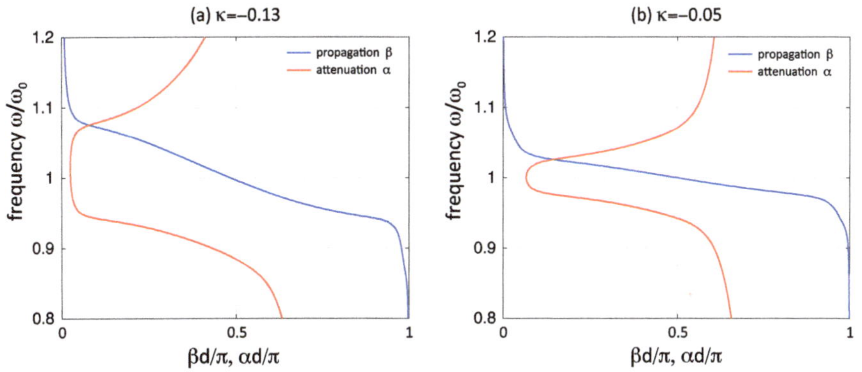

Fig. 5 Dispersion relation for nearest-neighbour coupling with $\kappa = -0.13$ (**a**) and $\kappa = -0.05$ (**b**). Parameters used: $f_0 = 114.76\,\text{MHz}$, $Q = 95$, $d = 10.5\,\text{mm}$

Neglecting losses we can derive the limits of the MI pass band, by using the values $kd = 0$ and $kd = \pi$ in Eq. (13). For negative coupling coefficients we can write:

$$\omega_{\max} = \frac{\omega_0}{\sqrt{1 + \sum_{m=1}^{C} \kappa_m}}, \qquad \omega_{\min} = \frac{\omega_0}{\sqrt{1 + \sum_{m=1}^{C} (-1)^m \kappa_m}}. \tag{14}$$

In the example of Fig. 5a with $\kappa = -0.13$ the pass band limits are $f_{\min} = 108\,\text{MHz}$ and $f_{\max} = 123\,\text{MHz}$. As the magnitude of κ decreases (i.e. as coupling gets weaker) the the passband gets narrower. In the example of Fig. 5b with $\kappa = -0.05$ the pass band limits are $f_{\min} = 112\,\text{MHz}$ and $f_{\max} = 117.7\,\text{MHz}$.

2.4 *Group Velocity*

The dispersion relation can also be used to derive the phase velocity, v_p, and the group velocity, v_g, of the MI wave. The phase velocity is the velocity at which any singular frequency component of the wave travels, and can be written as $v_p = \omega/k$. The group velocity is the velocity at which the entire wave of information travels, i.e. the overall envelope of the wave, and is defined as $v_g = d\omega/dk$. Differentiating Eq. (13) with respect to k gives the group velocity:

$$v_g = \frac{\omega^3}{2\omega_0^2} d \sum_{m=1}^{C} m\kappa_m \sin(mkd). \tag{15}$$

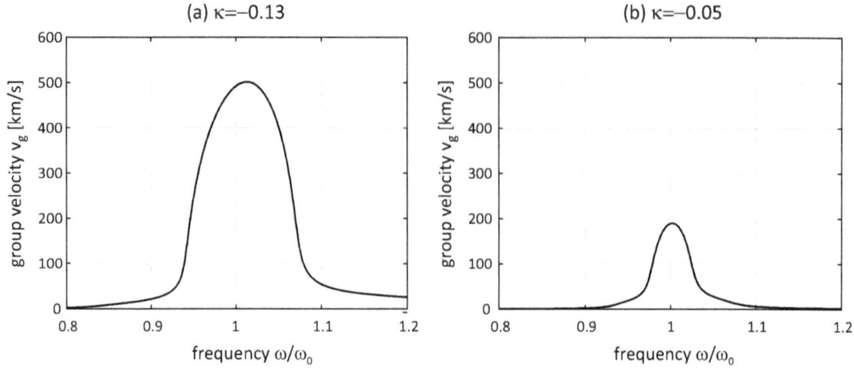

Fig. 6 MI wave group velocity for a 1D array of coils, for nearest-neighbour coupling with $\kappa = -0.13$ (**a**) and $\kappa = -0.05$ (**b**). Parameters used: $f_0 = 114.76\,\text{MHz}$, $Q = 95$, $d = 10.5\,\text{mm}$

Figure 6 illustrates a breakdown of the velocity of each frequency component using the same set of parameters with two values for the coupling constant as in Fig. 5, $\kappa = -0.13$ and $\kappa = -0.05$. We find that for $\kappa = -0.13$ the group velocity reaches 500 km/s which occurs close to the resonant frequency, whereas for frequencies outside the bandpass limits the group velocity is low—MI waves of these frequencies are essentially not propagating. For $\kappa = -0.05$ the peak group velocity is lower at 200 km/s.

The group velocity will later be useful in the time-domain analysis when we determine the distance travelled by a signal along the array by measuring the time taken between the moment the signal is injected into the array and the moment the reflection echo arrives back at the receiving probe. A conductive object placed above an element of the array causes a partial reflection of the MI wave—creating a shorter 'effective length' of the array. It may be inferred how far the MI wave had travelled and thus determine the position of the impedance discontinuity that had caused its reflection. Hence, combining the velocity and reflection arrival time one can evaluate the 'effective length' of the MI waveguide and thus locate the conductive object. Trivedi and Chen et al. [73–75] use a similar approach to locate ideal defects of infinite impedance in the time domain.

2.5 Defects

As a next step we need to quantify the resonant properties of a defect created in a MIW due to a presence of a conductor. The mechanism of the defect formation can be best understood if we consider a single meta-atom (modelled as an LCR circuit with a resonant frequency f_0 and a quality factor Q) in the vicinity of a conductor. Once the resonator is excited and an ac current flows in it, a magnetic flux proportional to the current is generated. If there is a conductive object nearby, the time-varying magnetic flux will generate eddy currents within the conductive object. This, in turn, will induce changes to the overall magnetic flux affecting the self-inductance of the meta-atom [51–54, 85]—in the context of MI waveguide we can say that the meta-atom affected by the conductive object nearby becomes a 'defect' with an impedance

$$Z_{0,\mathrm{def}} = j\omega L_{\mathrm{def}} + \frac{1}{j\omega C} + R_{\mathrm{def}} = j\omega L_{\mathrm{def}} \left(1 - \frac{\omega_{0,\mathrm{def}}^2}{\omega^2} - \frac{j}{Q_{\mathrm{def}}} \frac{\omega_{0,\mathrm{def}}}{\omega}\right), \quad (16)$$

where the resonant frequency and the quality factor of the 'defected' meta-atom are defined now as follows:

$$\omega_{0,\mathrm{def}} = \frac{1}{\sqrt{L_{\mathrm{def}} C}}, \qquad Q = \frac{\omega_{0,\mathrm{def}} L_{\mathrm{def}}}{R_{\mathrm{def}}}. \quad (17)$$

We note that within this model it is assumed that only the self-inductance and the resistance are affected by the presence of the conductive object, but not the capacitance of the meta-atom [51–54, 85].

Adopting the approach used by Yan et al. [30], we performed an experiment placing a metallic bolt above a single meta-atom and mapping how this affects $f_{0,\mathrm{def}}$ and Q_{def}. A meta-atom, a square-shaped SRR on a PCB substrate, was the same as in the array later used for TDR as will be detailed in the next section. A metal bolt was chosen for the conductive object due to its high conductivity and due to its use in the previous work on localisation using FDR [30] as well as in our work on TDR [76–78].

The measurements of the resonant frequency and the quality factor were taken using a Hewlett Packard 8753ES Vector Network Analyser (VNA). The two ports of the VNA were connected to two inductive loop probes placed either side of this single element, before measuring S_{21} as shown in Fig. 7. The resonator was placed on a balsa wood table. In order to enable accurate positioning of a conductive object a robotic 3D table was used enabling vertical movement of a custom holder arm with the tip of the bolt screwed into the holder arm. This frequency response of the S_{21} parameter was then used to calculate f_0 and Q for both an unperturbed single element and one in the presence of the metal bolt positioned at a height z above the meta-atom.

The resulting variation of the resonant frequency and of the quality factor with the bolt height is shown in Fig. 8. If the bolt is close to the meta-atom, it causes strong changes to the resonant parameters of the meta-atom, with the resonant frequency shifting upwards by up to 42% and the quality factor shifting downwards by up to 70% from their unperturbed values [85]. In the context of a MI waveguide, such a defect can be expected to cause strong reflection of the MI wave [76–78]. If the metallic object is moved further away, the defect it creates is weak (i.e. its impedance does not differ much from that of the unperturbed meta-atoms of the array) and the MI reflections it will cause can be expected to be weak and it might be difficult to detect these. We shall discuss this in more detail in the next section.

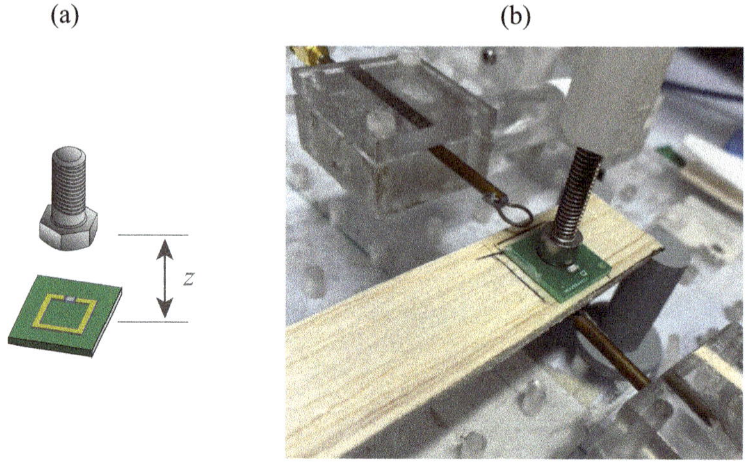

Fig. 7 A schematic of the geometry (**a**) and a photograph (**b**) of the setup used to measure $f_{0,\mathrm{def}}$ and Q_{def} of the meta-atom versus bolt height z

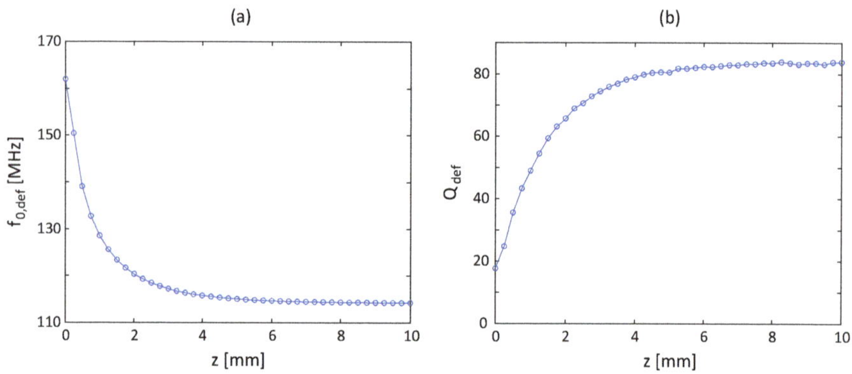

Fig. 8 Resonant frequency $f_{0,\mathrm{def}}$ (**a**) and quality factor $Q_{0,\mathrm{def}}$ (**b**) of the meta-atom for varying bolt height z

2.6 *Frequency-Domain Reflection Coefficient*

Once a MI wave encounters a defected element, it will be partly reflected and partly transmitted. Our goal is to characterise the defect, by calculating the frequency spectrum of the reflection coefficient and relate it to the resonant characteristics of the defected meta-atom. While staying in the frequency domain, this will enable the time-domain characterisation of the properties of conductive objects as shall be discussed later in Sect. 4.

The approach used is similar to that employed by Syms et al. [86] who were concerned with reflections due to perturbation to the mutual impedance, whereas we look at reflections caused by a single defect in the otherwise uniform MI waveguide, as shown schematically in Fig. 9.

We assume that the MI waveguide is infinitely long and that a MI wave of frequency ω is injected and travels from the left to the right along the waveguide. The elements are numbered such that element $n = 0$ is defected by a nearby conductive object, and all other elements are unaffected, giving for the self-impedances

$$\begin{cases} Z_{0,\mathrm{def}} & \text{for} \quad n = 0 \\ Z_0 & \text{for} \quad n \neq 0 \,, \end{cases} \tag{18}$$

where $Z_{0,\mathrm{def}}$ is defined by Eq. (16) and Z_0 is defined by Eq. (2). For simplicity we assume that there is only nearest-neighbour coupling with the mutual impedance Z_M defined in Eq. (5). We assume that to the left of the defect there are two MI waves in the line, the incident wave travelling forwards (in the direction of increasing n) and the reflected one, travelling backwards. To the right of the defect we assume that only the transmitted wave is present travelling forwards. As the waveguide is uniform (apart from the defect), all three waves (incident, reflected and transmitted) have the same wavenumber k, governed by the dispersion equation (11). The current in element n can therefore be written as:

$$I_n = \begin{cases} I_{00}(\mathrm{e}^{-jkdn} + R\mathrm{e}^{jkdn}) & \text{for} \quad n \leq 0 \\ I_{00}T\mathrm{e}^{-jkdn} & \text{for} \quad n > 0 \,, \end{cases} \tag{19}$$

where I_{00} is the amplitude of the incident wave, R is the reflection coefficient defined as the ratio of the incident and the reflected wave at element 0 and T is the transmission coefficient. To calculate the reflection coefficient R, we now write Kirchhoff's equations for element 0 (defect):

$$Z_{0,\mathrm{def}} I_0 + Z_M(I_{-1} + I_1) = 0 \tag{20}$$

and for element 1:

$$Z_0 I_1 + Z_M(I_0 + I_2) = 0 \,. \tag{21}$$

Substituting $n = -1, \ 0, \ 1, \ 2$ into Eqs. (19) we can write:

$$\begin{cases} I_{-1} & = I_{00}\left(\mathrm{e}^{jkd} + R\mathrm{e}^{-jkd}\right) \\ I_0 & = I_{00}(1 + R) \\ I_1 & = I_{00}T\mathrm{e}^{-jkd} \\ I_2 & = I_{00}T\mathrm{e}^{-2jkd} \end{cases} \,. \tag{22}$$

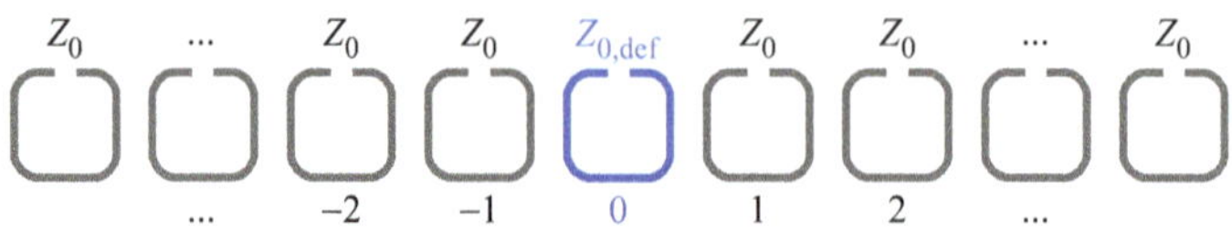

Fig. 9 Section of an infinitely long uniform waveguide with a single defect showing numbering convention used for our derivation. Element $n = 0$ is defected, with all other elements unperturbed

Using the currents from Eq. (22), Kirchhoff's equations (20) and (21) can be rewritten as

$$Z_{0,\mathrm{def}}(1+R) + Z_M\left(\mathrm{e}^{jkd} + R\mathrm{e}^{-jkd} + T\mathrm{e}^{-jkd}\right) = 0 \tag{23}$$

and

$$Z_0 T\mathrm{e}^{-jkd} + Z_M\left(1 + R + T\mathrm{e}^{-j2kd}\right) = 0\,, \tag{24}$$

respectively. Rearranging Eq. (24) we obtain:

$$T\mathrm{e}^{-jkd}\left(Z_0 + Z_M\mathrm{e}^{-jkd}\right) + Z_M(1+R) = 0\,. \tag{25}$$

This expression can be further simplified if we make use of the dispersion equation (11) written in the following form:

$$Z_0 + Z_M\mathrm{e}^{-jkd} = -Z_M\mathrm{e}^{jkd}\,. \tag{26}$$

Using (26), Eq. (25) simplifies to:

$$T = 1 + R\,, \tag{27}$$

which can be recognised as the matching boundary condition for element 0 which serves as a boundary to two sections of uniform MI waveguides. Using Eq. (27), we can find the reflection coefficient from Eq. (23) as follows:

$$R = -\frac{Z_{0,\mathrm{def}} + 2Z_M\cos(kd)}{Z_{0,\mathrm{def}} + 2Z_M\mathrm{e}^{-jkd}}\,. \tag{28}$$

We shall introduce a parameter Δ,

$$\Delta = Z_{0,\mathrm{def}} - Z_0\,, \tag{29}$$

which characterises how 'strong' the defect is, i.e. how much its impedance differs from the impedance of the unperturbed elements in the line. Then Eq. (28) can be further simplified to:

$$R = -\frac{\Delta}{\Delta - 2jZ_M\sin(kd)}\,. \tag{30}$$

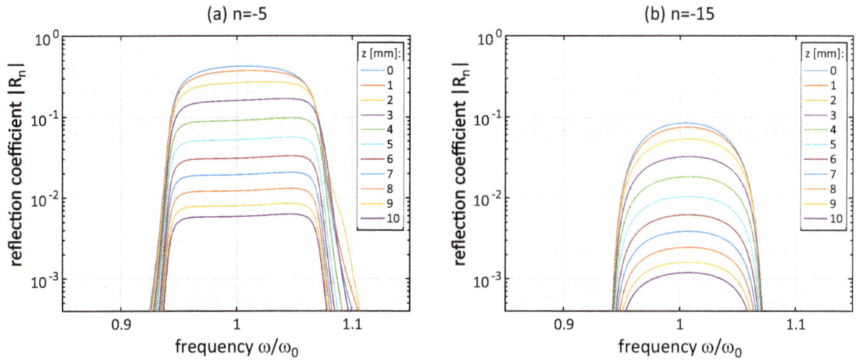

Fig. 10 Frequency response of the magnitude of the reflection coefficient R_n for $n = -5$ (**a**) and $n = -15$ (**b**), for the bolt positioned above element 0 at a height varying from 0 to 10 mm. Parameters for the uniform MI line: $f_0 = 114.76$ MHz, $Q = 95$, $\kappa = -0.13$. Parameters of the defect are as in Fig. 8

The reflection coefficient R is clearly dependent on the strength to which the element is defected and it reduces to zero when no defect is present ($\Delta = 0$) and its magnitude is equal to unity if the defect has an infinite impedance ($\Delta = \infty$).

We note that the reflection coefficient R can be treated as a local parameter defined directly at the occurrence of the reflection caused by the defected element. We can also define a general reflection coefficient R_n, as a ratio of the incident and the reflected wave at element n:

$$R_n = R\,\mathrm{e}^{-2jkd|n|} \tag{31}$$

Here we take into account that the incident wave traveled a distance nd to reach the defect and that the reflected wave traveled the same distance back. Figure 10 shows the spectra of $|R_n|$ for $n = -5$ (a) and $n = -15$ (b) for the bolt positioned above element 0 at a height varying from 0 to 10 mm. Comparing plots in (a) and (b) illustrates the role of the attenuation during the propagation of the MI wave to the defect and back—for the same vertical bolt position the reflection coefficient R_n determined for $n = -15$ is lower than for $n = 5$. It is interesting to note that, for any defect position, the reflection coefficient is maximum around the resonant frequency and remains fairly constant across the entire passband of MI waves. We shall return to the reflection coefficient in Sect. 4 when discussing characterisation of defects using time-domain reflectometry.

3 Localisation by Time-Domain Reflectometry

We are now in the position to progress onto our first main topic, contactless localisation of conductive objects. The fundamental basis of the proposed localisation model is to use time-domain reflectometry echoes of MI waves to detect the position of defects in a MIW caused by a metallic object. This involves analysing the time of flight for an MI wave, and reverse-engineering this information to determine the exact location in the array from which the MI wave was reflected. We start by setting up a circuit model, which will be later tested in experiment, and noise reduction methods will be employed demonstrating accuracy going beyond that reported previously.

3.1 Analytical Model

The setup is shown schematically in Fig. 11. The uniform MI waveguide comprises twenty identical SRR elements. The unit cell size (centre-to-centre) is $d = 10.5\,\text{mm}$. The elements of the array, except the element with a defect, are assumed to have the resonant frequency $f_0 = 114.76\,\text{MHz}$ and the quality factor $Q = 95$. Coupling coefficients up to the 3rd neighbour are taken into account, with the values $\kappa_1 = -0.13$, $\kappa_2 = -0.062$, $\kappa_3 = -0.005$ [35].

The array is directly excited by applying voltage to element 1 and a metal bolt can be placed at a height z above any of the elements creating a defect of impedance $Z_{0,\text{def}}$, see Eq. (16). To simulate the defect, we use the experimentally found values for $f_{0,\text{def}}$ and Q_{def} at a selected height shown in Fig. 8.

So far our analysis assumed monochromatic excitation of frequency ω. The process of calculating time-domain signals is outlined schematically in Fig. 12. We shall now assume a periodic sampled square-wave voltage excitation, $V(t)$, in the time domain t, as shown in Fig. 13a. Taking the Fast Fourier Transform (FFT) of the excitation signal, we obtain its spectrum, $V(\omega)$, shown in Fig. 13b. The inset confirms that the spectrum spans the passband of MI waveguide allowing signal propagation to occur. By using Kirchhoff's equations as outlined in Sect. 3.1, we are able to find current spectra, $I_n(\omega)$, for each element n of the array. Using Inverse Fast Fourier Transform (IFFT) we can then reconstruct currents $I_n(t)$ in the time domain.

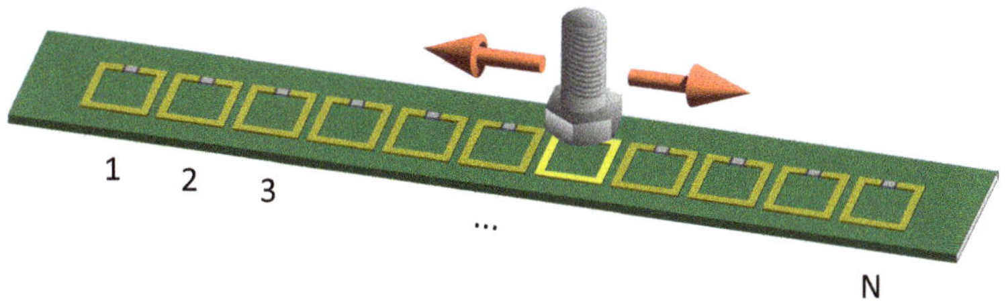

Fig. 11 A schematic of a 1D MIW with a metallic bolt creating a defect in the waveguide

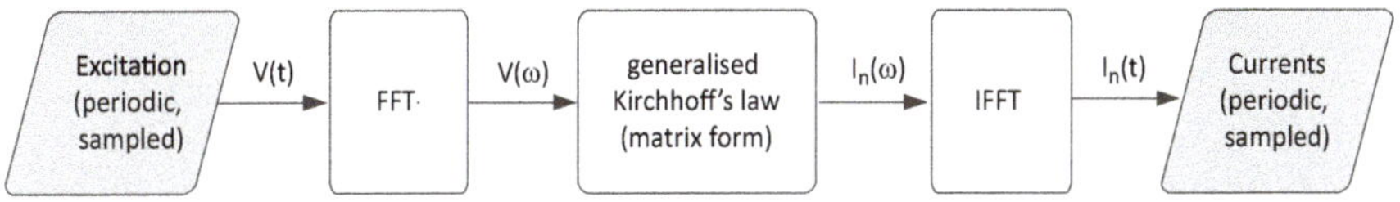

Fig. 12 A schematic for the time-domain calculations using frequency-domain Kirchhoff's equations

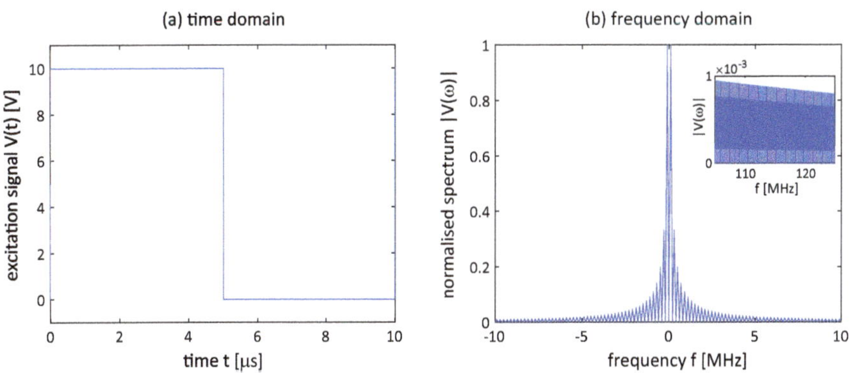

Fig. 13 Square wave pulse used to excite element 1 of the MI waveguide (**a**), alongside its frequency spectrum (**b**). Parameters of square wave are frequency $f = 100\,\text{kHz}$, amplitude $A = 10\,\text{V}_{\text{pp}}$, duty cycle $D = 0.5$, sampling frequency $f_s = 5\,\text{GSa/s}$. Inset in (**b**) shows a zoom-in onto the bandpass of the MI waveguide - these are the components of the spectrum that are able to propagate through the waveguide

3.1.1 Unperturbed Signal Propagation

To enable accurate localisation we need to first look at the behaviour of the unperturbed waveguide, without any defects present. The time-domain signal for the unperturbed waveguide ('unperturbed signal') will provide a reference point; comparing the signal in the presence of a conductive object to the unperturbed signal we will be able to isolate the changes caused by the object and use these for localisation and characterisation of the object.

In the frequency domain, assuming that element 1 is driven by a harmonic voltage, $V(\omega)\exp(j\omega t)$, Kirchhoff's equations for the currents in each element of the array of N elements can be written in a compact matrix form as

$$\mathbf{V} = Z\mathbf{I}\,, \tag{32}$$

where

$$\mathbf{V} = \begin{bmatrix} V(\omega) \\ 0 \\ 0 \\ \vdots \\ 0 \end{bmatrix} \qquad \text{and} \qquad \mathbf{I} = \begin{bmatrix} I_1 \\ I_2 \\ I_3 \\ \vdots \\ I_N \end{bmatrix} \tag{33}$$

are the vectors of length N for the voltages and the currents in all elements of the array, respectively, and

$$Z = \begin{bmatrix} Z_0 & Z_{M,1} & Z_{M,2} & Z_{M,3} & \cdots & 0 \\ Z_{M,1} & Z_0 & Z_{M,1} & Z_{M,2} & \cdots & 0 \\ Z_{M,2} & Z_{M,1} & Z_0 & Z_{M,1} & \cdots & 0 \\ Z_{M,3} & Z_{M,2} & Z_{M,1} & Z_0 & \cdots & 0 \\ \vdots & \vdots & \vdots & \vdots & \vdots & \vdots \\ 0 & 0 & Z_{M,3} & Z_{M,2} & Z_{M,1} & Z_0 \end{bmatrix} \tag{34}$$

is the $N \times N$ impedance matrix, with the main diagonal representing the self-impedances of each resonator and the off-diagonal elements representing the mutual impedances $Z_{M,m} = j\omega M_m$ between any pair of elements separated by $(m-1)$ elements. Note that, for the nearest-neighbour coupling case, the impedance matrix reduces to a tridiagonal form, with just the main diagonal, subdiagonal and superdiagonal having non-zero entries:

$$Z = \begin{bmatrix} Z_0 & Z_M & 0 & 0 & \cdots & 0 \\ Z_M & Z_0 & Z_M & 0 & \cdots & 0 \\ 0 & Z_M & Z_0 & Z_M & \cdots & 0 \\ 0 & 0 & Z_M & Z_0 & \cdots & 0 \\ \vdots & \vdots & \vdots & \vdots & \vdots & \vdots \\ 0 & 0 & 0 & 0 & Z_M & Z_0 \end{bmatrix}. \tag{35}$$

If the voltage excitation vector is known, the frequency spectra of the currents may be obtained by inverting the Z matrix,

$$\mathbf{I} = Z^{-1}\mathbf{V}\,, \tag{36}$$

before applying the IFFT to the total frequency response to determine the time-domain behaviour of the currents in the MI waveguide.

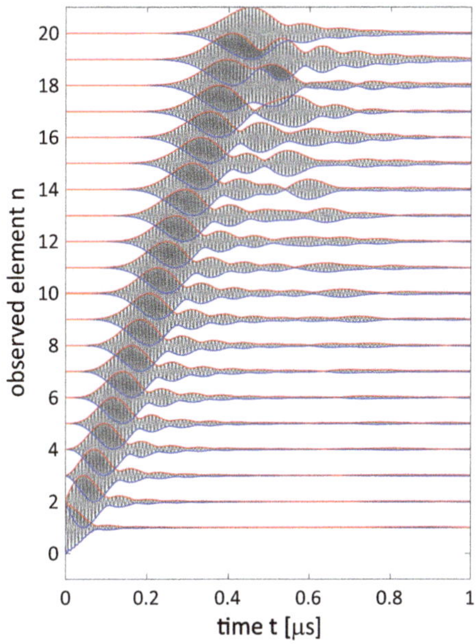

Fig. 14 Normalised currents predicted by the analytical model through elements 1 to 20 of the waveguide (grey). Also shown are the upper (red) and the lower (blue) envelopes to serve as a guide to the eye

Figure 14 shows the time evolution of the currents I_n in each element of the array calculated using the analytical model (grey) for the excitation by a periodic voltage as in Fig. 13. For a better visualisation, each current is normalised to its maximum value and shifted vertically by an integer number corresponding to the element position in the array, n, varying from 1 to 20. Also shown are the upper (red) and the lower envelopes (blue).

The shape of the initial time evolution of the currents $I_n(t)$ in each element can be seen to be fairly similar. In keeping with the slow-wave property of MI waves, there is a measurable difference between the arrival times of the signal in each element as it traverses through the waveguide moving away from the driven element 1. The time of flight of the signal between two neighbouring elements of the array can be evaluated as the difference between the peak values of the current envelopes in the elements. Analysing the data from Fig. 14 the mean time of flight is found to be approximately equal to $\tau = 22$ ns. Using the unit cell size, $d = 10.5$ mm, we find the average velocity with which the signal travels along the array to be $v = d/\tau = 480$ km/s, which is in agreement with the peak value in the group velocity spectrum $v_g(f)$ in Fig. 6.

Once the signal reaches the end of the line after approximately 0.45 μs, we can observe in Fig. 14 a reflection signal returning back down the waveguide. This reflection is caused by the impedance discontinuity created by the end of the line, in effect with a hypothetical (non-existing) element $N + 1$ acting as a perfect defect of infinite impedance. The signal makes a round trip along the waveguide and returns to element 1 appearing in the signal I_1 as a weak echo with a peak at 0.9 μs. We note that the speed of travel of the injected signal and of the reflected signal remains constant—a crucial property used for the localisation technique.

3.1.2 Reflection Due to a Conductive Object

In Sect. 2.5 we have already described the mechanism by which the proximity of a conductive object affects the self-impedance of a resonant element, modifying its resonant frequency and its quality factor, thus creating a defect in the MI waveguide. For the analytical model of the MI waveguide, we assume that only the element nearest to the conductive object (i.e. the one positioned directly beneath it) is affected by its presence. To give an example, for a defect above element 3, the impedance matrix of the MI waveguide has the form

$$Z = \begin{bmatrix} Z_0 & Z_{M,1} & Z_{M,2} & Z_{M,3} & \cdots & 0 \\ Z_{M,1} & Z_0 & Z_{M,1} & Z_{M,2} & \cdots & 0 \\ Z_{M,2} & Z_{M,1} & Z_{0,\text{def}} & Z_{M,1} & \cdots & 0 \\ Z_{M,3} & Z_{M,2} & Z_{M,1} & Z_0 & \cdots & 0 \\ \vdots & \vdots & \vdots & \vdots & \vdots & \vdots \\ 0 & 0 & Z_{M,3} & Z_{M,2} & Z_{M,1} & Z_0 \end{bmatrix}, \tag{37}$$

with the modified self-impedance $Z_{0,\text{def}}$ in the defected element given by Eq. (16) and with the values of its resonant frequency and its quality factor dependent on the vertical position of the conductive object as shown in Fig. 8.

When a signal is injected into the array, we can now expect reflections not only from the end of the array but also from the defected element. The amount of reflection caused by the defect is obviously higher when the conductive object is nearer to the element as the defect is stronger—i.e. its self-impedance differs more from the unperturbed impedance in all other elements of the array. Following the same procedure as in the unperturbed case, the currents in the defected array are found by substituting the defected impedance matrix into Eq. (36) for every frequency component of the voltage signal, and re-assembling the time-domain signals by using IFFT.

To investigate the reflection echoes from different defects in the array we shall, from now on, monitor only the current in element 2 of the MI waveguide—the same approach as later employed in the experiment. We shall also consider defects in elements 3 to 17 only, allowing no defects in the last, idle, section of the MI waveguide, to avoid harmful interference of reflections from the defects and from the end of the array.

Figure 15 shows the theoretical currents in element 2 for a defect element in positions 3–17 at a height $z = 0$. For more convenient visualisation, each signal in Fig. 15 is normalised to its maximum and vertically shifted by an integer value corresponding to the location of the defected element. We observe the reflections caused by the presence of the conductive element in the form of a secondary maximum in the

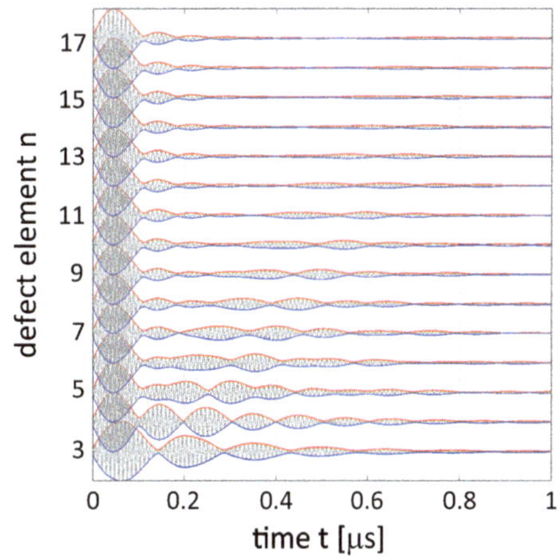

Fig. 15 Normalised theoretical current in element 2 for a single defect element located at different positions in the waveguide at a height $z = 0$. Grey: actual signal. Red: upper envelope. Blue: lower envelope

envelope of each signal. As the defect element moves further down the waveguide, this reflection takes longer to arrive back at element 2 due to the longer overall time of flight. We are now in the position to isolate this reflection signal, and use the time of flight to infer the distance along the line that the conductive object is placed, but also use the amplitude of the reflection signal to characterise the height of a conductive object above the waveguide.

3.1.3 Isolating Reflection Signals

Accurate isolation of this reflection signal is achieved by exploiting the superposition of MI waves as a combination of forward and backward travelling signals. By subtracting the unperturbed signal from the signal received in the presence of a defect, it is possible to isolate purely the reflection signal due to the defected element. Figure 16 shows the unperturbed (a) and the perturbed signal (b) for a defect in element 7. It can be clearly seen that in the time interval between 0 and 0.2 μs both signals are identical, showing the injected signal. The difference between the unperturbed and the perturbed signal becomes noticeable after 0.2 μs—this is when the reflection from element 7 starts to arrive at element 2.

Figure 17 shows that subtracting the unperturbed signal from the perturbed signal does indeed remove the injected signal while leaving intact key features of the reflected signal, namely the arrival time of the first peak and its amplitude.

A family of the isolated reflection signals in element 2 for different defected elements is depicted in Fig. 18a. It shows clearly that the reflected signal arrives later for defect elements further down the waveguide enabling localisation of the defect from the arrival time.

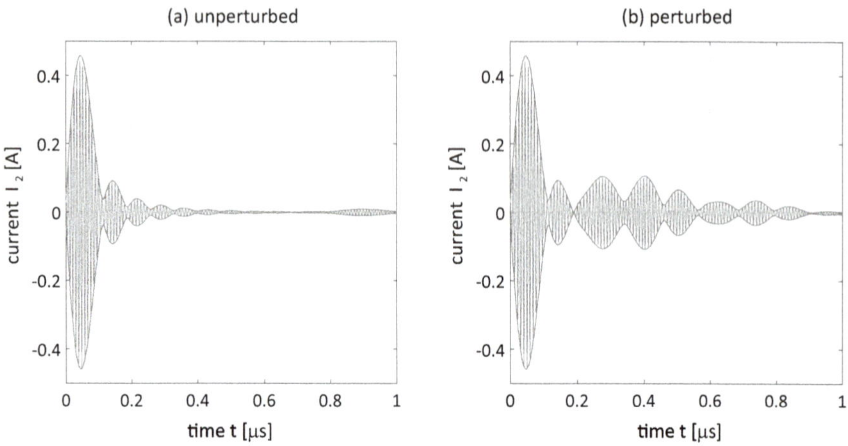

Fig. 16 Theoretical current in element 2 for an unperturbed array (**a**) and for a defect at element 7 at height $z = 0$ (**b**)

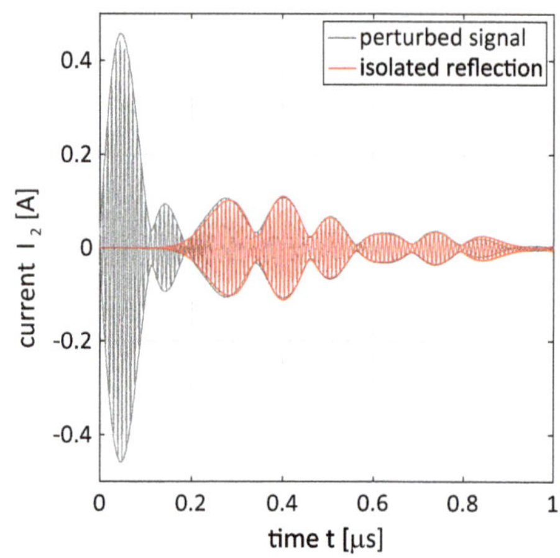

Fig. 17 Theoretical current in element 2 for a single defect at element 7 (grey), with the reflection signal isolated from it (red)

3.2 *Experimental Verification*

We have performed experiments verifying our localisation method. The remainder of this section describes the experimental setup, discusses the data post-processing, presents a detailed comparison between the theory and the experiment and evaluates the accuracy of our method.

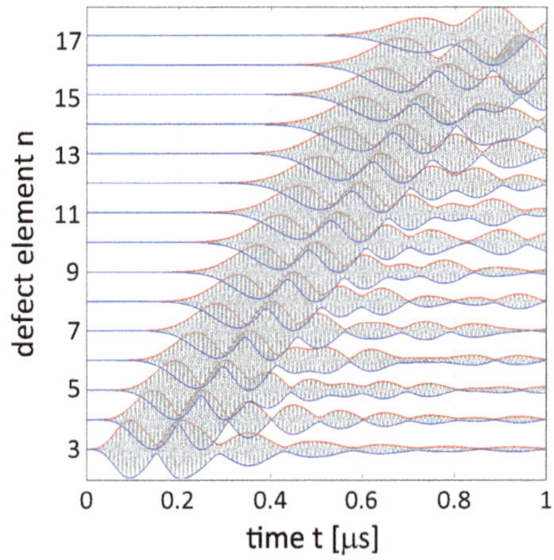

Fig. 18 Normalised reflection signals isolated from the theoretical model for different locations of the defected element with a fixed vertical bolt position $z = 0\,\text{mm}$

3.2.1 Setup

The MI waveguide used in the experiment is an improved but geometrically identical structure to that used in the research conducted by Yan et al. [30] on MI waves in the frequency domain. It comprises 20 single-turn square-shaped SRRs, measuring $d = 10.5\,\text{mm}$ apart from centre to centre, loaded with $C = 100\,\text{pF}$ capacitors, held in place by a PCB (FR4), see Fig. 19a. The resonant frequency and the quality factor of each element of the array have the values $f_0 = 114.76\,\text{MHz}$ and $Q = 95$, the same as used in the analytical modelling in the previous section.

A square wave periodic voltage of amplitude $A = 10\,\text{V}_{\text{pp}}$, frequency $f = 100\,\text{kHz}$, duty cycle $D = 0.5$, and sampling frequency $f_s = 5\,\text{GSa/s}$ (see Fig. 13) is injected from a KeySight 33600A signal generator via a direct feed soldered onto element 1, allowing excitation of the waveguide as shown in Fig. 19b. For modelling of this system, we make an adjustment to the properties of element 1: we assume it to have the resonant frequency and the quality factor that are different from the rest of the array, $f_{01} = 131.5\,\text{MHz}$ and $Q_1 = 20$, due to the internal resistance of the signal generator and due to the presence of the metallic connector which lowers the quality factor and increases the resonant frequency of element 1 [30, 85].

(a) (b)

Fig. 19 Photograph of a section of the MI waveguide (**a**) and of a close-up showing element 1 (with the connector for the direct feed) and a element 2 with a receiver probe above it (**b**)

A small receiving loop probe connected to a KeySight MSOX3034G oscilloscope is measuring the signal above element 2 as shown in Fig. 19b; it is assumed that the signal is proportional to the current in element 2, I_2, and we shall in the following refer to the signal as 'current in element 2'. To guarantee that the receiver probe is picking up the signal in element 2 only, the receiver probe is placed at such a height above element 2 that there is no coupling between the antenna and the nearest neighbouring elements (elements 1 and 3) [76, 78].

Elements 3 to 17 comprise the section of the array used for sensing of metallic objects, whereas the last three elements of the array (elements 18–20) remain an idle termination annex. As discussed in the previous section, this approach enables us to avoid reflection echoes from the end of the array to obscure reflections caused by the metallic object.

The MI waveguide is placed on a balsa wood table. In order to enable accurate positioning of a conductive object a robotic 3D table was used which is capable of lateral and vertical movement of a custom holder arm with a metallic bolt with micrometer precision. The holder arm was 3D printed using high-impact polystyrene (HIPS). A metal bolt was chosen for the conductive object due to its high conductivity and due to its use in our previous work on frequency-domain localisation [30] which we seek to compare to and improve upon. The photograph of the complete setup is shown in Fig. 20, with the tip of the conductive bolt screwed into the holder arm, and the block diagram of the setup is shown in Fig. 21.

The effect of background noise was reduced by averaging the signal over 1024 samples. Measurements of the signal above element 2, for both the unperturbed case (no bolt present) and the perturbed case (with the metal bolt held, in turn, above elements 3–17, for heights ranging from 0 to 10 mm) were taken, post-processed and compared to the results from the analytical model.

3.2.2 Post-Processing Experimental Data

Our next step is to post-process the experimental data. Generally we wish to adopt the same approach as the one used in the analytical model: to retrieve the reflection signal by subtracting the unperturbed signal (in the absence of a defect) from the perturbed signal (with the defect present).

Figure 22 shows the result of this direct procedure of extracting the reflection signal for different positions of the bolt. It is apparent from Fig. 22 that the receiving antenna picks up noise that masks the useful signal, especially in the cases of weaker reflections, e. g. from defects located further down the waveguide or when the bolt is too far up above the waveguide.

To localise the horizontal position of the reflecting object we should be able to correctly identify the corresponding peak in both the analytical model and experimental signal. Then, through comparison of the time of the arrival of these two reflected peaks, the algorithm can infer how far down the line the conductive object is located. In the analytical model identifying the right peak is straightforward due to the noise-free nature of theory producing smooth and pronounced reflection sig-

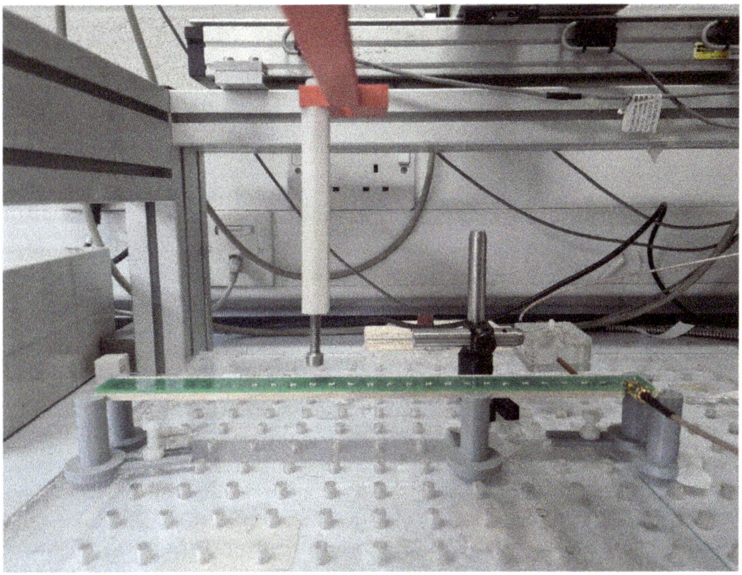

Fig. 20 Photograph of the 3D table holding the metal bolt used as a conductive object above the MI waveguide

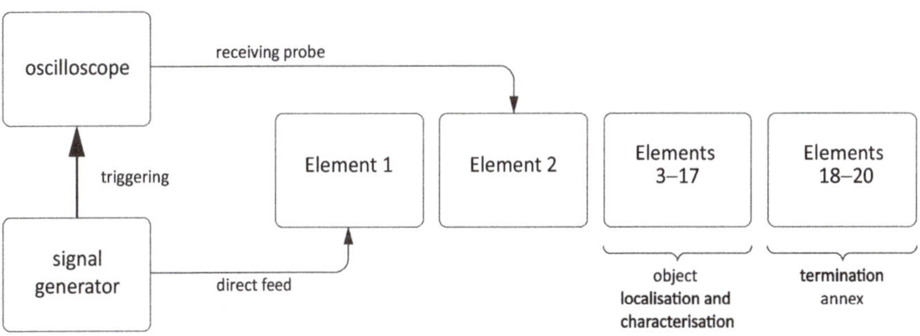

Fig. 21 Block diagram of the time-domain measurement setup

nals; however, this is not the case for experimental measurements. For bolt positions further down the line and at greater heights it becomes increasingly difficult to distinguish between the weaker reflection signal and the noise surrounding it. Whilst for $z = 0$ and $z = 4$ mm (plots (a) and (b), respectively) the peak of the reflection is clearly seen for all horizontal bolt positions, once the defect is weak or far away and the noise is of similar magnitude as the reflection signal, the accuracy of the localisation may be significantly reduced as can be seen from plot (c) for $z = 7$ mm.

To mitigate the noise issue, a zero-phase bandpass filter with the boundaries 108.2 MHz and 136.5 MHz was found to give a satisfactory result. The chosen bandpass filter covers both the passband of the MI waveguide, 108.2 MHz $\leq f \leq$

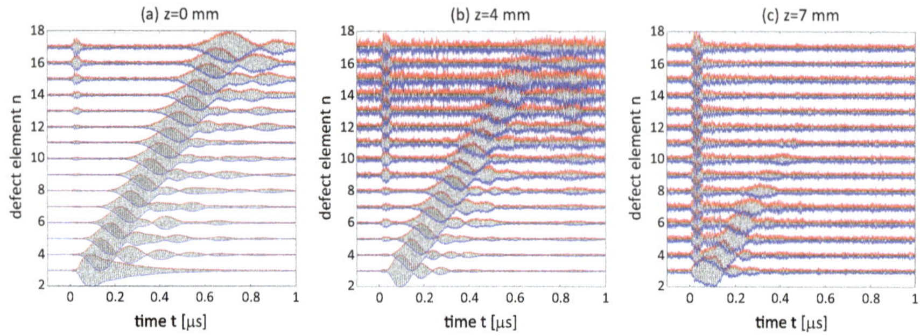

Fig. 22 Normalised experimental reflection signal (with unperturbed signal subtracted) in element 2 for varying defect position: **a** $z = 0$, **b** $z = 4$ mm and **c** $z = 7$ mm. Grey: actual signal. Red: upper envelope. Blue: lower envelope. As in Figs. 15 and 16, to aid visualisation, each curve is normalised to its maximum and shifted vertically by an integer number corresponding to the horizontal defect location

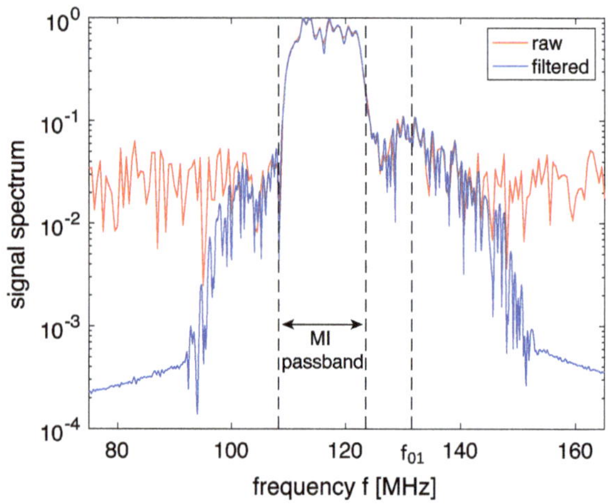

Fig. 23 Experimental spectrum of the unperturbed signal before (red) and after the filtering (blue). Vertical dashed lines show the boundaries of the MIW passband as well as the resonant frequency of element 1

123.5 MHz centred at $f_0 = 114.76$ MHz and the spectral peak at 131.5 MHz associated with the resonance of element 1. As an illustration, Fig. 23 shows the spectrum of a measured reflection signal for the bolt at a height of $z = 5$ mm above element 10, before and after filtering.

The impact of the filtering is illustrated in Fig. 24, in which both unperturbed and perturbed signals were filtered prior to the subtraction. A direct comparison of Figs. 22 and 24 shows that filtering has reduced noise while keeping main features of the reflection signals still intact.

In Fig. 24 another experimental obstacle becomes apparent—moving cables introduce some minor variation to the overall time travel of the signals through the cables.

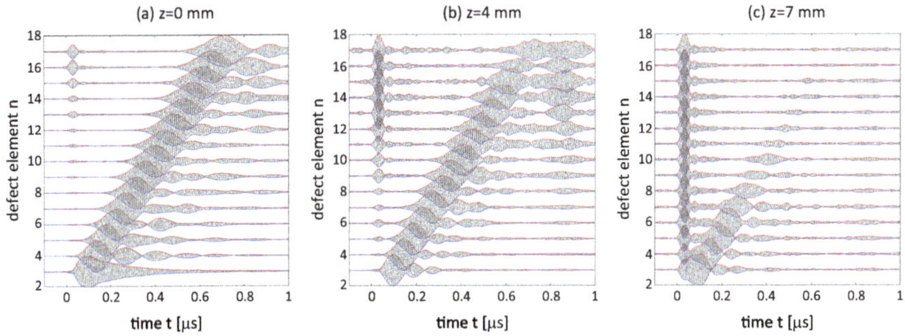

Fig. 24 Same as Fig. 22 but with both unperturbed and perturbed experimental signals filtered prior to the subtraction procedure

As a result, there is a small but observable time mismatch in the range of tens of picoseconds between any two measurement sets. As the unperturbed and the perturbed signals are slightly out of sync, subtracting one from another results in an artefact at $t = 0$ where the signals should coincide. The appearance of the artefact at $t = 0$ can impede localisation, in particular, when reflection signals are weak as in case of Fig. 24c for $z = 7\,\text{mm}$.

To synchronise both the perturbed signal and the unperturbed signal we perform the following steps: a gradual time shift to the perturbed signal is introduced, ranging from 0 to 50 ps; we subtract the unperturbed signal from the perturbed signal, evaluate the resulting reflection signal at $t = 0$ and repeat the procedure until the reflection signal at $t = 0$ is minimised. Figure 25 illustrates the result of both filtering and synchronising, showing a notable improvement upon Figs. 22 and 24.

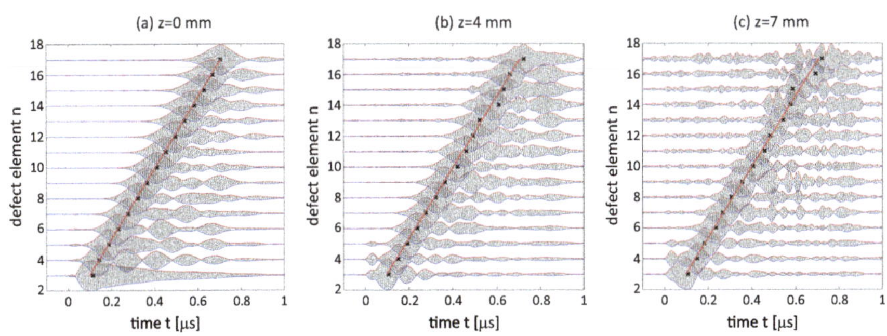

Fig. 25 Same as Figs. 22 and 24 but with both unperturbed and perturbed experimental signals filtered and synchronised prior to the subtraction procedure. Black crosses mark the experimental times of peak envelope and the red line is a reference showing the theoretical prediction for the times of peak envelope assuming signal travel speed of $v = 480\,\text{km/s}$

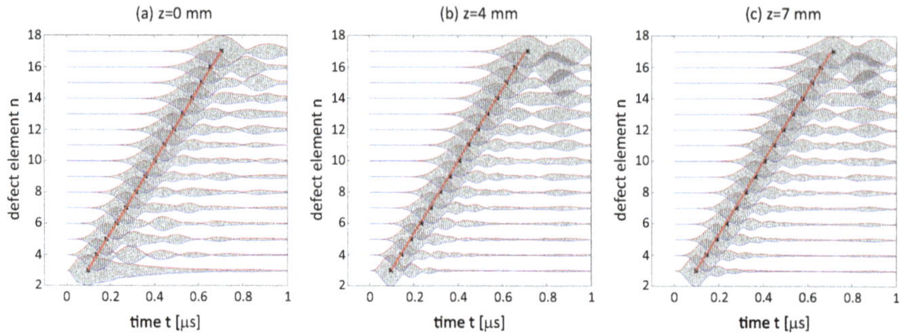

Fig. 26 Analytical model matching the post-processed experimental data of Fig. 25. Black crosses mark the theoretical times of peak envelope and the red line is a reference showing the theoretical prediction for the times of peak envelope assuming signal travel speed of $v = 480\,\text{km/s}$

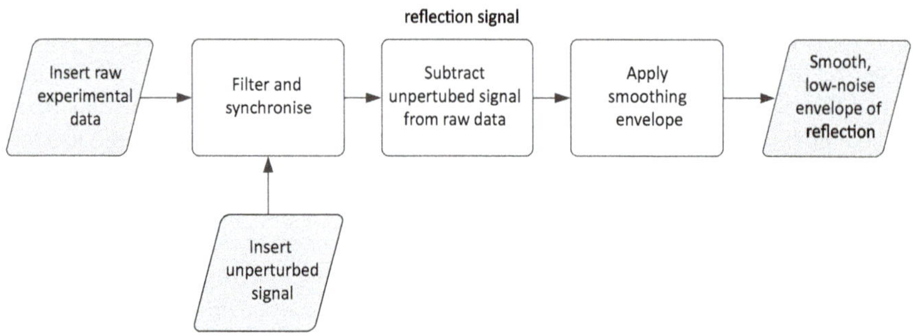

Fig. 27 Overall schema for the post-processing of experimental data

For comparison, the results of analytical modelling for the same bolt positions are shown in Fig. 26. There is a remarkable agreement between the post-processed experimental data of Fig. 25 and the analytical results of Fig. 26 confirming that the analytical model is suitable for prediction of the response of the MI waveguide to perturbations caused by the presence of metallic objects.

The outline of the overall post-processing for experimental data adopted in this work is summarised in Fig. 27. To determine which element corresponds to an identified reflection, the arrival time is compared to a look-up table of arrival times predicted by the analytical model, and the closest one is taken as outlined in the next subsection.

3.2.3 Automated Localisation Algorithm

To localise the position of the reflecting object, we must first correctly identify the corresponding peak in the signal envelope both in the experimental signal and in the analytical model—these are marked by black crosses in Figs. 25 and 26, respectively. Through comparison of the times of the experimental envelope peaks with the theoretical predictions, an automated algorithm can infer how far down the line the conductive object is located. The red line in Figs. 26 and 25 shows the theoretically calculated arrival times versus horizontal defect locations. It serves as a guide to the eye to visualise how far off the experimental peak envelopes (black crosses) deviate from the theoretical prediction. We note that for more remote bolt positions (both in the vertical and in the horizontal direction) there are noticeable deviations from the theoretical model. Hence at large vertical positions of the bolt as well as for bolt positions further away from the receiver in the horizontal direction, the MIW elements are not defected to a great enough degree to produce easily distinguishable reflections in the experimental data as they get masked by unfiltered noise and ultimately result in inaccurate localisation.

The accuracy of our localisation method is shown Fig. 28a, showing the difference between the actual and the found horizontal position of the defect, as a colour scale

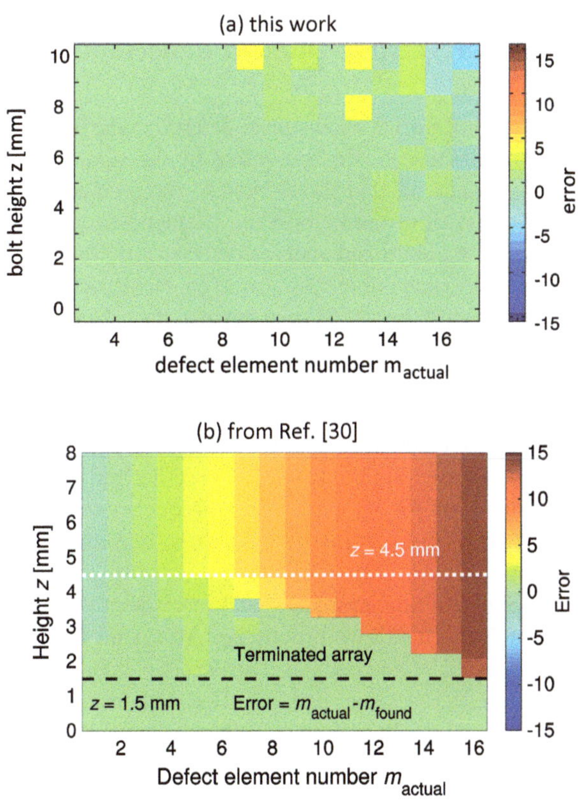

Fig. 28 **a** Time-domain localisation error in this work. The localisation error is shown as a colour scale plot showing the difference between the actual and the found horizontal position of the defect. **b** Frequency-domain localisation error using a terminated 16 element array, reprinted from [30]. Ref. [30] is licensed under a Creative Commons Attribution 4.0 License. © 2021 by J. Yan, C. J. Stevens, and E. Shamonina

plot for all actual horizontal defect locations from 3 to 17 and for all bolt heights above the MI waveguide from 0 to 10 mm. The overall green colour of the colour scale plot corresponds to zero error, i.e. to the accurate localisation achieved for heights up to 2 mm for every element, and up to 7 mm for elements 3–12. To evaluate the performance of the time-domain localisation, Fig. 28b shows, for comparison, the results from our previous work on frequency-domain localisation [30]. The geometry of the setup are comparable in both cases and hence the significant improvement of the localisation method achieved here by employing the time-domain localisation (a) instead of the frequency-domain localisation (b) is quite impressive. The reason is that, in the frequency-domain reflectometry, all reflections are detected within the same frequency range, with multiple resonance occurring within a narrow passband of MI waves [30], whereas in the time-domain reflectometry used here, reflections are separate events, taking place at their well defined times enabling high sensitivity and high accuracy of localisation.

4 Characterisation by Time-Domain Reflectometry

The localisation method of Sect. 3 allows us to determine the horizontal position of the bolt by measuring the time of flight required by the signal to reach the defect created by the bolt, bounce off and travel back to the receiver. The time-domain localisation has been shown to outperform the frequency-domain localisation—it is more sensitive, more accurate and enables detection of weaker reflections. Our goal in this section is to expand our method to the characterisation of the conductive object, by analysing not only the arrival time but also the magnitude of the reflected signal. As the conductive object has a known shape (bolt) and a known conductivity (stainless steel), the missing quantity we wish to determine is its vertical position—the height—above the MI waveguide. In Sect. 2.5 we quantified how the conductive object affects the nearby element of the MI waveguide by measuring $f_{0,\mathrm{def}}$ and Q_{def} versus bolt height z (see Fig. 8). In Sect. 2.6 we derived an expression for the reflection coefficient spectra for MI waves in the frequency domain. In this section we shall make the transition back to the time domain by introducing the time-domain reflection coefficient.

As discussed in Sect. 3 a time-domain signal in element 2 contains two contributions, each arriving within its own timeframe. At around $t = 0$ we observe the unperturbed signal injected in element 1 as it arrives in element 2, whereas at a later time we observe the reflected signal arriving from a defect created further down the line or from the end of the line. We denote by A the peak value of the injected signal and by B the peak value of the reflected signal and introduce the time-domain reflection coefficient as:

$$\rho = B/A\,. \tag{38}$$

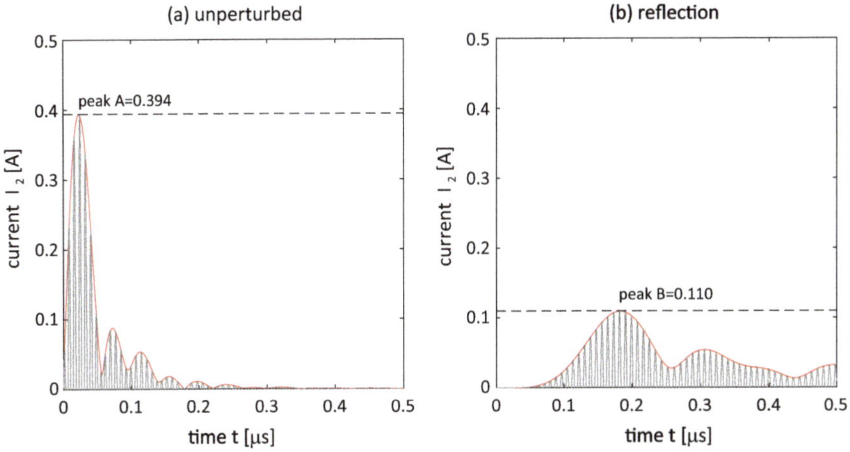

Fig. 29 Analytical model. Current in element 2 (grey) with upper envelope (red) for unperturbed signal (**a**) and reflection signal **b** for the bolt above element 5 at a height $z = 0$ mm. Horizontal dashed lines show the level of the peak value in each signal

An illustration of this is provided in Fig. 29 showing a theoretical example for the bolt above element 5 at the height $z = 0$, resulting in the peak values of $A = 0.394$ and $B = 0.110$, for the incident and the reflected signal, respectively, yielding the reflection of $\rho = 0.28$.

Figures 30 and 31 show the time-domain reflection coefficients extracted from the post-processed experimental data (a) and from the analytical model (b). In Fig. 30 the reflection coefficient ρ is plotted versus the horizontal defect position for varying bolt height and in Fig. 31 the same data are presented as the reflection coefficient ρ plotted versus the bolt height for varying horizontal defect position. It is gratifying to see

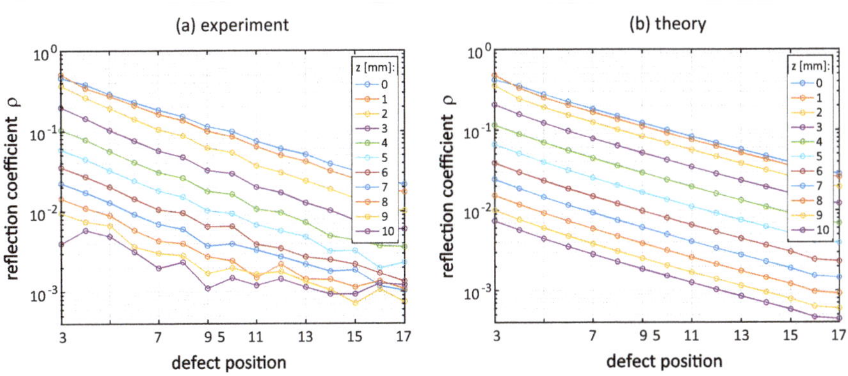

Fig. 30 Time-domain reflection coefficient ρ versus defect position for bolt height varying from 0 to 10 mm. Experiment (**a**) and theory (**b**)

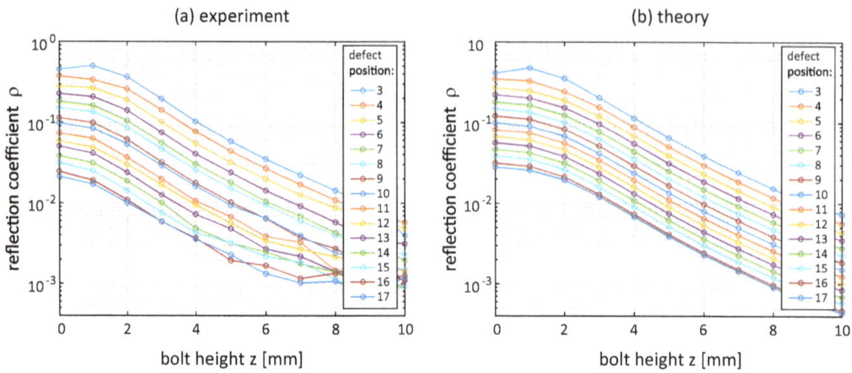

Fig. 31 Time-domain reflection coefficient ρ versus bolt height for horizontal defect position varying from 3 to 17. Experiment (**a**) and theory (**b**)

that for stronger reflection coefficients (i.e. for stronger defects) there is quantitative agreement between theory and experiment for heights up to approximately $z = 5$ mm. Agreement between theory and experiment becomes less accurate at greater heights and/or more remote horizontal defect positions. The reason is two-fold, first, when the reflection signal becomes too weak it is more affected by the noise still present after the post-processing, and, second, the localisation from the time of flight, being less accurate for these cases, may result in incorrect identification of the reflection peak, thus also leading to errors in the reflection coefficient.

To characterise a given conductive object the calculated reflection coefficient is compared to the set of analytical reflection coefficients for the element we have localised the defect to. The object is then assigned the height of the analytical reflection coefficient which is closest to it. The absolute error for the vertical position in mm when using this simple matching is plotted in Fig. 32. Using this method successful characterisation of conductive object is possible with ± 1 mm accuracy for all defect positions except a very few cases for the height of $z = 10$ mm and for the most distant defect location 17 in the horizontal direction, for which the accuracy is ± 2 mm. We should of course bear in mind that Fig. 32 must be considered in conjunction with Fig. 28 as successful characterisation requires accurate localisation first. This is a promising result, particularly given no attempt was made to adjust the fit of the reflection coefficient curves or using a look-up table.

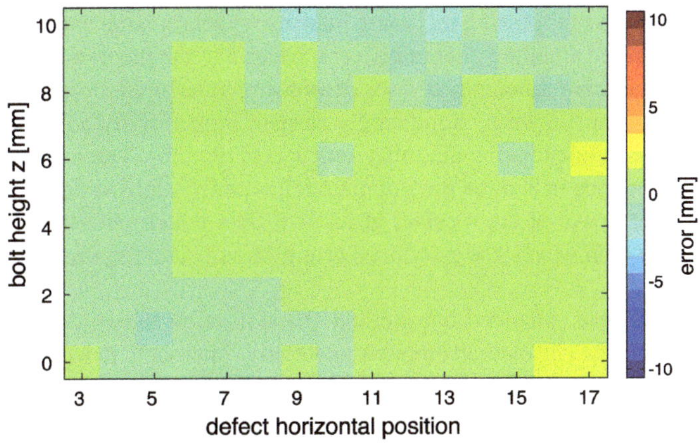

Fig. 32 Characterisation error (in mm) when determining the vertical position of the metal bolt from experimental data through direct comparison to the analytical model. Characterisation error is shown as a colour scale plot of the difference between the actual and the found vertical position of the bolt

5 Conclusions

This chapter lays foundations for development of sensing of metallic objects using time-domain reflectometry of magnetoinductive waves, developing an analytical model for localisation and characterisation of a metallic object and presenting experimental verification. The method relies on propagation of slow magnetoinductive (MI) waves along a chain of magnetically coupled resonant elements (MI waveguide). A conductive object placed in the vicinity of an element of the array creates a perturbation in the waveguide causing reflections of MI waves within the array. The localisation relies on detecting the time of flight for the reflection echoes whereas the characterisation requires quantifying the amplitude of the reflection signal.

The analytical model for the time-domain reflectometry uses generalised Kirchhoff's equations for arrays of coupled meta-atoms treated as resonant LCR circuits capable of propagating MI waves within a passband centred around the resonance frequency of individual meta-atoms. The presence of a metallic object is modelled by modifying the impedance of the element underneath it while assuming all the other elements of the array unaffected.

A periodic voltage excitation, applied to element 1 of the MI waveguide, is decomposed into its Fourier spectrum, and, for each frequency component of the excitation signal, a coupled system of Kirchhoff's equations is solved, enabling calculation of current spectra in the elements of the array. Inverse Fourier transform is then used to reconstruct the currents in the time domain.

In the experiment, element 1 of the array was connected to a signal generator and the magnetic field above element 2 was sensed by a small loop probe connected to an oscilloscope. A post-processing including filtering, synchronising and smoothing of the data was applied enabling quantitative comparison to the theoretical predictions.

Using time-domain reflectometry, we demonstrated accurate defect localisation and characterisation shown to significantly surpass in sensitivity and accuracy the previously used frequency-domain reflectometry (FDR) [30]. The reason of such a superior performance is that the time-domain reflectometry allows for each individual reflection to arrive at the receiver at its own time which obeys the dispersion characteristics of MI waves, the associated group velocity and the attenuation during propagation.

Recently we have expanded our method of localisation to two defects [76, 78] (not presented here). The method employs an iterative approach, in which defects are localised one by one and the corresponding individual reflections are subtracted iteratively from the overall signal. We expect that this approach can be further generalised to include also characterisation by quantifying the reflection magnitude and also to more than two defects or to generally inhomogeneous conductive environment.

Potential applications for real-time contactless monitoring of inhomogeneous conductive environments ranging from in-situ 3D printing quality monitoring to medical imaging are envisaged. This chapter may serve as a useful reference guide to those developing imaging modalities relying on the time-domain reflectometry using MI waves.

References

1. Solymar L, Shamonina E (2009) Waves in metamaterials. Oxford University Press, Oxford
2. Engheta N, Ziolkowski RW (2006) Metamaterials: physics and engineering explorations. Wiley, Hoboken, NJ
3. Marques R, Martin F, Sorolla M (2008) Metamaterials with negative parameters: theory, design and microwave applications. Wiley, New York
4. Cai W, Shalaev V (2010) Optical metamaterials: fundamentals and applications. Springer, New York
5. Craster RV, Guenneau S (2013) Acoustic metamaterials: negative refraction, imaging, lensing and cloaking. Springer, Dordrecht
6. Deymier PA (2013) Acoustic metamaterials and phononic crystals. Springer, Heidelberg
7. Simovski C, Tretyakov S (2020) An introduction to metamaterials and nanophotonics. Cambridge University Press, Cambridge
8. Smith DR, Vier DC, Kroll N, Schultz S (2000) Direct calculation of permeability and permittivity for a left-handed metamaterial. Appl Phys Lett 77:2246–2248
9. Smith DR, Padilla WJ, Vier DC, Nemat-Nasser SC, Schultz S (2000) Composite medium with simultaneously negative permeability and permittivity. Phys Rev Lett 84:4184–4187
10. Shelby RA, Smith DR, Schultz S (2001) Experimental verification of a negative index of refraction. Science 292:77–79
11. Lee SH, Park CM, Seo YM, Kim CK (2010) Reversed Doppler effect in double negative metamaterials. Phys Rev B 81:241102–1–4
12. Berman PR (2002) Goos-Hänchen shift in negatively refractive media. Phys Rev E 66:067603–1–3
13. Lakhtakia A (2003) On planewave remittances and Goos-Hänchen shifts of planar slabs with negative real permittivity and permeability. Electromagnetics 23:71–75

14. Ziolkowski RW (2003) Pulsed and CW Gaussian beam interactions with double negative metamaterial slabs. Opt Express 11:662–681
15. Pendry J (2000) Negative refraction makes a perfect lens. Phys Rev Lett 85:3966–3969
16. Shamonina E, Kalinin VA, Ringhofer KH, Solymar L (2001) Imaging, compression and Poynting vector streamlines for negative permittivity materials. Electron Lett 37:1243–1244
17. Chen H, Chan CT, Sheng P (2010) Transformation optics and metamaterials. Nat Mater 9:387–396
18. Ma HF, Cui TJ (2010) Three-dimensional broadband ground-plane cloak made of metamaterials. Nat Commun 1:21–1–6
19. Landy N, Smith DR (2013) A full-parameter unidirectional metamaterial cloak for microwaves. Nat Mater 12:25–28
20. Shamonina E, Kalinin VA, Ringhofer KH, Solymar L (2002) Magneto-inductive waveguide. Electron Lett 38:371–372
21. Shamonina E, Kalinin VA, Ringhofer KH, Solymar L (2002) Magnetoinductive waves in one, two, and three dimensions. J Appl Phys 92:6252–6261
22. Sydoruk O, Zhuromskyy O, Shamonina E, Solymar L (2005) Phonon-like dispersion curves for magnetoinductive waves. Appl Phys Lett 87:072501–1–3
23. Sydoruk O, Radkovskaya A, Zhuromskyy O, Shamonina E, Shamonin M, Stevens CJ, Edwards DJ, Faulkner G, Solymar L (2006) Tailoring the near-field guiding properties of magnetic metamaterials with two resonant elements per unit cell. Phys Rev B 73:224 406–1–12
24. Syms RRA, Shamonina E, Kalinin V, Solymar L (2005) A theory of metamaterials based on periodically loaded transmission lines: interaction between magnetoinductive and electromagnetic waves. J Appl Phys 97:064 909–1–6
25. Shamonina E (2012) Magnetoinductive polaritons: hybrid modes of metamaterials with interelement coupling. Phys Rev B 85:155146–1–10
26. Gorkunov M, Lapine M, Shamonina E, Ringhofer K (2002) Effective magnetic properties of a composite material with circular conductive elements. Eur Phys J B 28:263–269
27. Radkovskaya A, Petrov P, Kiriushechkina S, Satskiy A, Ivanyukovich M, Vakulenko A, Prudnikov V, Kotelnikova O, Korolev A, Zakharov P (2018) Magnetic metamaterials: coupling and permeability. J Magn Magn Mater 459:187–190
28. Sydoruk O, Zhuromskyy O, Radkovskaya A, Shamonina E, Solymar L (2009) "Magnetoinductive waves I: theory," in theory and phenomena of metamaterials. CRC Press, New York ch. 36
29. Sydoruk O, Radkovskaya A, Zhuromskyy O, Shamonina E, Solymar L (2009) "Magnetoinductive waves II: applications," in applications of metamaterials. CRC Press, New York ch. 14
30. Yan J, Stevens C, Shamonina E (2021) A metamaterial position sensor based on magnetoinductive waves. IEEE Open J Antennas Propag 2:259–268
31. Syms R, Solymar L, Young I, Floume T (2010) Thin-film magneto-inductive cables. J Phys D: Appl Phys 43:055 102–1–7
32. Syms R, Floume T, Young I, Solymar L, Rea M (2010) Flexible magnetoinductive ring MRI detector: design for invariant nearest-neighbour coupling. Metamaterials 4:1–14
33. Vallecchi A, Cadman D, Whittow WG, Vardaxoglou J, Shamonina E, Stevens CJ (2019) 3-D printed bandpass filters with coupled vertically extruded split ring resonators. IEEE Trans Microw Theory Tech 67:4341–4352
34. Tatartschuk E, Gneiding N, Hesmer F, Radkovskaya A, Shamonina E (2012) "Mapping inter-element coupling in metamaterials: scaling down to infrared. J Appl Phys 111:094 904–1–9
35. Yan J, Radkovskaya A, Solymar L, Stevens C, Shamonina E (2022) Switchable unidirectional waves on mono- and diatomic metamaterials. Sci Rep 12:16845-1-14
36. Sydoruk O, Shamonina E, Solymar L (2007) Parametric amplification in coupled magnetoinductive waveguides. J Phys D: Appl Phys 40:6879–6887
37. Freire MJ, Marques R (2005) A planar magnetoinductive lens for 3D subwavelength imaging. Appl Phys Lett 86:182 505–1–3

38. Zhuromskyy O, Shamonina E, Solymar L (2005) 2D metamaterials with hexagonal structure: spatial resonances and near field imaging. Opt Express 13:9299–9309
39. Sydoruk O, Shamonin M, Radkovskaya A, Zhuromskyy O, Shamonina E, Trautner R, Stevens CJ, Faulkner G, Edwards DJ, Solymar L (2007) "A mechanism of subwavelength imaging with bi-layered magnetic metamaterials: theory and experiment. J Appl Phys 101:073 903–1–8
40. Stevens CJ, Chan CW, Stamatis K, Edwards DJ (2010) Magnetic metamaterials as 1-D data transfer channels: an application for magneto-inductive waves. IEEE Trans Microw Theory Tech 58:1248–1256
41. Chan CW, Stevens CJ (2011) Two-dimensional magneto-inductive wave data structures. In: 2011 fifth European conference on antennas and propagation (EUCAP). IEEE, pp 1071–1075
42. Wang B, Teo KH, Nishino T, Yerazunis W, Barnwell J, Zhang J (2011) Experiments on wireless power transfer with metamaterials. Appl Phys Lett 98:254 101–1–3
43. Stevens CJ (2014) Magnetoinductive waves and wireless power transfer. IEEE Trans Power Electron 30:6182–6190
44. Freire M, Marques R, Medina F, Laso M, Martín F (2004) Planar magnetoinductive wave transducers: theory and applications. Appl Phys Lett 85:4439–4441
45. Nefedov I, Tretyakov S (2005) On potential applications of metamaterials for the design of broadband phase shifters. Microw Opt Technol Lett 45:98–103
46. Shamonina E, Solymar L (2013) Superdirectivity by virtue of coupling between meta-atoms. In: Seventh international congress on artificial materials for novel wave phenomena (Metamaterials), Bordeaux, France, pp 97–99
47. Radkovskaya A, Kiriushechkina S, Vakulenko A, Petrov P, Solymar L, Li L, Vallecchi A, Stevens C, Shamonina E (2018) Superdirectivity from arrays of strongly coupled meta-atoms. J Appl Phys 124:104 901–1–11
48. Yan J, Radkovskaya A, Kiriushechkina S, Khromova I, Stevens C, Solymar L, Shamonina E (2020) A method for optimising superdirectivity of coupled meta-atoms via planar directivity evaluation. IEEE Open J Antennas Propag 1:300–308
49. Sun Z, Akyildiz I (2010) Magnetic induction communications for wireless underground sensor networks. IEEE Trans Antennas Propag 58:2426–2435
50. Chu S, Luloff M, Yan J, Petrov P, Stevens C, Shamonina E (2021) Magnetoinductive waves in attenuating media. Sci Rep 11:7679–1–12
51. Dodd C, Deeds WE (1968) Analytical solutions to eddy-current probe-coil problems. J Appl Phys 39:2829–2838
52. Chu S, Vallecchi A, Stevens C, Shamonina E (2018) Fields and coupling between coils embedded in conductive environments. EPJ Appl Metamaterials 5:2–1–8
53. Vallecchi A, Chu S, Solymar L, Stevens C, Shamonina E (2019) Coupling between coils in the presence of conducting medium. IET Microw Antennas Propag 13:55–62
54. Vallecchi A, Shamonina E, Stevens C (2019) Wireless power transfer in the presence of a conducting interface: analytical solution. IET Microw Antennas Propag 13:725–731
55. Moulder JC, Uzal E, Rose JH (1992) Thickness and conductivity of metallic layers from eddy current measurements. Rev Sci Instrum 63:3455–3465
56. Bowler N, Huang Y (2005) Electrical conductivity measurement of metal plates using broadband eddy-current and four-point methods. Meas Sci Technol 16:2193–2200
57. Yin W, Peyton AJ, Zysko G, Denno R (2008) Simultaneous non-contact measurement of water level and conductivity. IEEE Trans Instrum Meas 57:2665–2669
58. Pan Y-L, Tai C-C (2012) Thickness and conductivity analysis of molybdenum thin film in CIGS solar cells using resonant electromagnetic testing method. IEEE Trans Magn 48:347–350
59. Wang C, Fan M, Cao B, Ye B, Li W (2018) Novel noncontact eddy current measurement of electrical conductivity. IEEE Sensors J 18:9352–9359
60. Chady T, Enokizono M, Sikora R (1999) Crack detection and recognition using an eddy current differential probe. IEEE Trans Magn 35:1849–1852
61. Tian GY, Sophian A, Taylor D, Rudlin J (2005) Multiple sensors on pulsed eddy-current detection for 3-D subsurface crack assessment. IEEE Sensors J 5:90–96

62. Zenzinger G, Bamberg J, Satzger W, Carl V (2007) Thermographic crack detection by eddy current excitation. Nondestruct Test Eval 22:101–111
63. Chana K, Cardwell D (2008) The use of eddy current sensor based blade tip timing for FOD detection. In: ASME turbo expo: power for land, sea, and air, Berlin, Germany, pp 169–178
64. Kuyvenhoven N, Dean C, Melton J, Schwannecke J, Umenei A (2011) Development of a foreign object detection and analysis method for wireless power systems. In: IEEE symposium on product compliance engineering proceedings (ISPCE). IEEE, San Diego, CA, USA, pp 1–6
65. Cheng L, Tian GY (2011) Surface crack detection for carbon fiber reinforced plastic (CFRP) materials using pulsed eddy current thermography. IEEE Sensors J 11:3261–3268
66. Ali KB, Abdalla AN, Rifai D, Faraj MA (2017) Review on system development in eddy current testing and technique for defect classification and characterization. IET Circuits Devices Syst 11:338–351
67. Cung LT, Dao TD, Nguyen PC, Bui TD (2018) A model-based approach for estimation of the crack depth on a massive metal structure. Meas Control 51:182–191
68. Yu Y, Gao K, Theodoulidis T, Yuan F (2019) Analytical solution for magnetic field of cylindrical defect in eddy current nondestructive testing. Phys Scr 95:015 501–1–12
69. Nakagawa N (2021) A model of eddy current nondestructive evaluation for a current leaking crack. NDT E Int 123:102 517–1–14
70. Sun H, Yi J, Wang T, Wang Y, Qing X (2021) A new interleaving eddy current array-based sensing film for fatigue crack quantification of bolted joints. J Intell Mater Syst Struct 32:1867–1877
71. Pasupathy P, Zhuzhou M, Neikirk D, Wood S (2008) "Unpowered resonant wireless sensor nets for structural health monitoring. In: IEEE Sensors Conference. Italy, Lecce, pp 697–700
72. Chen Y, Munukutla S, Pasupathy P, Neikirk D, Wood S (2010) Magneto-inductive waveguide as a passive wireless sensor net for structural health monitoring. In: Sensors and smart structures technologies for civil, mechanical, and aerospace systems conference, vol 7647. SPIE, San Diego, CA, USA, pp 1259–1267
73. Chen Y, Pasupathy P, Trivedi T, Neikirk D, Wood S (2012) Improved magneto-inductive waveguide as wireless sensor net for structural health monitoring. In: Nondestructive characterization for composite materials, aerospace engineering, civil infrastructure, and homeland security conference, CA, USA, San Diego, vol 83472, pp 607–614
74. Chen Y (2014) Design of magneto-inductive waveguide for sensing applications, PhD dissertation, University of Texas
75. Trivedi T, Chen Y, Williamson I, Pasupathy P, Neikirk D (2021) Group velocity estimation and defect localization in magneto-inductive waveguides. IEEE Trans Microw Theory Tech 69:2072–2077
76. Dhayaa D, Radkovskaya A, Yan J, Dima G, O'Hara E, Solymar L, Shamonina E (2022) Metamaterials for sensing conductive objects using time-domain reflectometry of magnetoinductive waves. In: Sixteenth international congress on artificial materials for novel wave phenomena (Metamaterials), Siena, Italy, pp 125–127
77. Long C, Radkovskaya A, Yan J, Dima G, O'Hara E, Solymar L, Shamonina E (2023) Metamaterials for characterisation of conductive objects using time-domain reflectometry of magnetoinductive waves. In: Seventeenth international congress on artificial materials for novel wave phenomena (Metamaterials), Chania, Greece, pp 205–207
78. Dhayaa D, Long C, Radkovskaya A, Solymar L, Shamonina E (2023) Magnetoinductive reflectometry, PCT Patent Application PCT/GB2023/052200, filed on 24 Aug 2023
79. Syms RR, Voronov A, Sydoruk O (2022) HF RFID tag location using magneto-inductive waves. IEEE J Radio Freq Identif. 6:347–354
80. Zhuromskyy O, Shamonina E, Solymar L (2005) Effect of radiation on dispersion of magnetoinductive waves in a metamaterial. In: 2005 congress on optics and optoelectronics, vol 5955, pp 48–55
81. Radkovskaya A, Shamonin M, Stevens C, Faulkner G, Edwards D, Shamonina E, Solymar L (2006) An experimental study of the properties of magnetoinductive waves in the presence of retardation. J Magn Magn Mater 300:29–32

82. Zhuromskyy O, Sydoruk O, Shamonina E, Solymar L (2009) Slow waves on magnetic metamaterials and on chains of plasmonic nanoparticles: driven solutions in the presence of retardation. J Appl Phys 106:104 908–1–7
83. Shamonina E, Solymar L, Kalinin V (2021) On wireless power transfer between coils in the presence of radiation. J Phys D: Appl Phys 54:405 502–1–13
84. Hesmer F, Tatartschuk E, Zhuromskyy O, Radkovskaya A, Shamonin M, Hao T, Stevens CJ, Faulkner G, Edwards DJ, Shamonina E (2007) Coupling mechanisms for split ring resonators: theory and experiment. Phys Status Solidi B 244:1170–1175
85. Dima G, Radkovskaya A, Stevens CJ, Solymar L, Shamonina E (2023) Design of a remote, multi-range conductivity sensor. Sensors 23:9711-1-18
86. Syms RRA, Shamonina E, Solymar L (2006) Magneto-inductive waveguide devices. IEE Proc Microw Ant Prop 153:111–121

Author Index

F. Martín and E. Bronchalo (eds.), *Coupled Structures for Microwave Sensing*,
Lecture Notes in Electrical Engineering 1150,
https://doi.org/10.1007/978-3-031-53861-2

GPSR Compliance
The European Union's (EU) General Product Safety Regulation (GPSR) is a set of rules that requires consumer products to be safe and our obligations to ensure this.

If you have any concerns about our products, you can contact us on

ProductSafety@springernature.com

In case Publisher is established outside the EU, the EU authorized representative is:

Springer Nature Customer Service Center GmbH
Europaplatz 3
69115 Heidelberg, Germany

www.ingramcontent.com/pod-product-compliance
Ingram Content Group UK Ltd.
Pitfield, Milton Keynes, MK11 3LW, UK
UKHW022216051026
15624UKWH00001B/195
9783031538605